W9-CSA-990

Techniques of
Model-based Control

PRENTICE HALL INTERNATIONAL SERIES
IN THE PHYSICAL AND CHEMICAL ENGINEERING SCIENCES

NEAL R. AMUNDSON, SERIES EDITOR, *University of Houston*

ADVISORY EDITORS

ANDREAS ACRIVOS, *Stanford University*
JOHN DAHLER, *University of Minnesota*
H. SCOTT FOGLER, *University of Michigan*
THOMAS J. HANRATTY, *University of Illinois*
JOHN M. PRAUSNITZ, *University of California*
L. E. SCRIVEN, *University of Minnesota*

Techniques of
Model-based Control

Coleman Brosilow
Case Western Reserve University and Ben Gurion University of the Negev
Babu Joseph
Washington University in St. Louis and University of South Florida

ISBN 0-13-028078-X

Prentice Hall PTR
Upper Saddle River, New Jersey 07458

Library of Congress Cataloging-in-Publication Data

Brosilow, Coleman.
  Techniques of model-based control / Coleman Brosilow, Babu Joseph.
    p. cm. -- (Prentice-Hall international series in the physical and chemical engineering sciences)
  Includes bibliographical references and index.
  ISBN 0-13-028078-X
  1. Process control. I. Joseph, Babu, 1950- II. Title. III. Series

TS156.8 .B755 2001
670.42'7--dc21

                                                             200105133

Production Supervisor: Wil Mara
Publisher: Bernard M. Goodwin
Editorial Assistant: Michelle Vincenti
Marketing Manager: Dan DePasquale
Manufacturing Manager: Alexis R. Heydt-Long
Cover Designer: Talar Boorujy

© 2002 Prentice Hall PTR
A Division of Pearson Education, Inc.
Upper Saddle River, NJ 07458

The publisher offers discounts on this book when ordered in bulk quantities. For more information contact: Corporate Sales Department, Prentice Hall PTR, One Lake Street, Upper Saddle River, NJ 07458. Phone: 800-382-3419; FAX: 201-236-7141; E-mail: corpsales@prenhall.com.

Printed in the United States of America

10 9 8 7 6 5 4 3 2 1

ISBN 0-13-028078-X

Pearson Education LTD.
Pearson Education Australia PTY, Limited
Pearson Education Singapore, Pte. Ltd
Pearson Education North Asia Ltd
Pearson Education Canada, Ltd.
Pearson Educación de Mexico, S.A. de C.V.
Pearson Education—Japan
Pearson Education Malaysia, Pte. Ltd
Pearson Education, Upper Saddle River, New Jersey

To

My wife Rosalie,
My children Rachelle and Benjamin, and
My grandchildren Adam, Maya, Tomer and Elah.
My parents Ethel and Samuel, may his memory be for a blessing.
Coleman Brosilow

My wife Philomina and
My children Mili, Neeraj, and Sonia
Babu Joseph

T A B L E   O F   C O N T E N T S

3 One-Degree of Freedom Internal Model Control 39

4 Two-Degree of Freedom Internal Model Control 65

5 Model State Feedback Implementations of IMC 85

6 PI and PID Parameters From IMC Designs 113

7 Tuning and Synthesis of 1DF IMC for Uncertain Processes 135

8 Tuning and Synthesis of 2DF IMC for Uncertain Processes 179

9 Feedforward Control 221

10 Cascade Control 241

11 Output Constraint Control (Override Control) 269

12 Single Variable Inferential Control (IC) 283

13 Inferential Estimation Using Multiple Measurements 311

14 Discrete-Time Models 343

15 Identification: Basic Concepts 363

16 Identification: Advanced Concepts 387

Appendices

Preface

The design and tuning of any control system is always based on a model of the process to be controlled. For example, when an engineer tunes a PID control system online the controller gain and the integral and derivative time constants obtained from the tuning depend on the local behavior of the process, and this local behavior could, if desired, be well approximated by a mathematical model. It turns out, however, that even in the prosaic task of tuning a PID controller, much better control system behavior can be obtained if the local mathematical model of the process is actually obtained, and the PID tuning is based on that model (see Chapter 6 on PID tuning). The foregoing not withstanding, in this text the term model-based controller is used primarily to mean control systems that explicitly embed a process model in the control algorithm. In particular, we consider control algorithms such as internal model control (IMC), inferential control (IC) and model-predictive control (such as dynamic matrix control or DMC), which have found applications in the process industry over the last few decades.

The book focuses on techniques. By this we mean how the algorithms are designed and applied. There is less emphasis on the underlying theory. We have also used simple examples to illustrate the concepts. More complex and realistic examples are provided in the text as case study projects.

We have written the text with two types of audience in mind. One is the typical industrial practitioner engaged in the practice of process control and interested in learning the basics behind various controller tuning methods as well as advanced control strategies beyond traditional PID feedback control. Our aim is to provide sufficient understanding of the methodologies of model-based control to enable the engineer to determine where and when such control strategies can offer substantial improvement in control as well as how to implement and maintain such strategies. The second audience that we have in mind is students in senior or graduate level advanced process control courses. For such students, we have tried to provide homework exercises and suggested projects to enhance the learning process. It is assumed throughout that the student has convenient access to modern computing systems along with the necessary software. Most of the problems and examples cannot be carried out without the use of such tools. We have organized the book as shown in the following figure.

BASIC CONCEPTS

1 Introduction
2 Continuous-Time Models
3 One-Degree of Freedom Internal Model Control

CONTINUOUS-TIME MODEL-BASED CONTROL

4 Two-Degree of Freedom Internal Model Control
5 Model State Feedback Implementations of IMC
6 PI and PID Parameters From IMC Designs
7 Tuning and Synthesis of 1DF IMC for Uncertain Processes
8 Tuning and Synthesis of 2DF IMC for Uncertain Processes
9 Feedforward Control
10 Cascade Control
11 Output Constraint Control (Override Control)
12 Single Variable Inferential Control
13 Inferential Estimation Using Multiple Measurements

DISCRETE TIME MODEL-BASED CONTROL

14 Discrete-Time Models
15 Identification: Basic Concepts
16 Identification: Advanced Concepts
17 Basic Model Predictive Control
18 Advanced Model Predictive Control
19 Inferential MPC

SUPPORTING APPENDIX MATERIAL

A Review of Basic Concepts
B Frequency Response Analysis
C Review of Linear Least-Squares Regression
D Random Variables and Random Processes
E MATLAB and Control Toolbox Tutorial
F SIMULINK Tutorial
G Tutorial on IMCTUNE Software
H Identification Software
I SIMULINK Models For Case Studies

SUPPORTING SOFTWARE

IMCTUNE, MODELBUILDER, POLYID, AND PIDTUNER

Overview of Material in Text

Chapter 1 gives an overview of the hierarchical approach to process control. Chapter 2 reviews the various types of models used in process control with emphasis on continuous time models. Chapter 3 gives the development of the basic model-based control structure (IMC). The latter two chapters form the basis of the further developments in both continuous and discrete time implementations.

From this point on the reader can take two possible paths: the first path focuses on the development of theory and structures for continuous-time implementation using the IMC structure. The second path focuses on discrete-time (computer-based) implementation of model-based control.

Chapters 4 through 13 cover the first path. Chapter 4 focuses on simultaneous setpoint and disturbance rejection using a two-degree of freedom control structure. Chapter 5 shows how to handle control effort constraints using model state feedback. Chapter 6 shows the relationship between classical PID controllers and the internal model control structure. Chapter 7 shows how to design one-degree of freedom controllers in the presence of model uncertainty. Chapter 8 shows how to tune two-degree of freedom controllers. Chapters 9 through 11 focus on multiloop control structures such as feedforward, cascade, and constraint control. Chapter 12 discusses control using secondary measurements (called inferential control). Chapter 13 extends the concepts of Chapter 12 to inferential control using multiple secondary measurements using disturbance estimation methods.

Chapters 14 through 19 cover the second path dealing with discrete-time computer implementation of model-based controllers. Chapter 14 introduces the models used in discrete-time representation and Chapters 15 and 16 discuss algorithms used to identify such models from plant test data. Chapters 17 and 18 discuss the computer implementation of model-based control using the model-predictive control framework. Chapter 19 extends this to inferential control using secondary measurements.

As an aid to the various possible readers, we have provided an extensive set of appendices that contain the background material necessary for the material in the main body of the text. We have indicated at the beginning of each chapter the prerequisite material, and we suggest that the appropriate appendices be reviewed prior to reading the chapter. The following material is reviewed:

- Laplace transforms and block diagrams.
- Frequency response methods.
- Linear least square regression.
- Probability theory and random variables.
- MATLAB® and SIMULINK® software[1].

We have used the MATLAB/SIMULINK software system as the platform upon which to develop software that provides added functionality and a convenient interface for solving otherwise complex problems. Among the reasons for this choice is that the MATLAB platform provides the tools required to implement and test various control concepts with relatively little effort required to learn how to use the software. However, there is

---

[1] MATLAB and SIMULINK are trademarks of MathWorks, Inc.

other software that provides similar functionality, and the reader is encouraged to use which-ever tools are most comfortable.

The website[2] associated with this text contains the following material:

- **IMCTUNE:** A user-friendly interface to a set of MATLAB m-files that enables the reader to design and tune both IMC and PID controllers for single loop, feedfor-ward and cascade control systems. In addition to MATLAB and SIMULINK, IMCTUNE requires the Control System and Optimization toolboxes in addition to MATLAB 5.3.1 or later versions.

- **MODELBUILDER:** A set of MATLAB m-files to generate both discrete time and continuous time process models from input/output data. Requires the Identification Toolbox.

- **PIDTUNER:** A set of MATLAB m-files that implement and test classical PID tun-ing. Requires the Control System Toolbox.

- **SIMULINK case study models**: MATLAB/SIMULINK models of a number of process systems including the Shell fractionation column, a Naphtha cracker simu-lator and the Tennessee Eastman problem. These models can be used in course pro-jects.

- **Microsoft PowerPoint slides** containing the text material.

- **MATLAB m-files, mat-files and data files, SIMULINK mdl-files** and Microsoft Word files related to the examples and exercises in the text. These files are identi-fied by chapter and example number. The *.mat files for the examples in Chapters 3 through 11 are associated with IMCTUNE.

- **Syllabi** used by the authors in their respective undergraduate and graduate courses.

The reader is strongly encouraged to download the software listed above and to use it to reproduce results of examples in the text, and to solve the problems at the end of the chapters. Our experience with teaching using this text material indicates that hands-on exer-cises using simulated processes is important to get a good understanding of the concepts. Hence the reader is strongly urged to experiment with the software. We have provided a number of exercises that require computer implementation and testing of the concepts.

While the normal reader will have had an introductory course in process control, it is possible to use at least parts of this text in an introductory course. For example, Coleman Brosilow has used Chapters 3, 5, 6, 7, and parts of 9 and 10 in a first undergraduate course

---

[2] The website address is *http://www.phptr.com/brosilow/*

in process control for more than ten years. Babu Joseph has been using the Appendix material in the laboratory sessions associated with the undergraduate course. The entire material in this book can be included in a graduate course on process control. However, the book should be supplemented with some additional material on multivariable control.

The SIMULINK case studies provide comprehensive test beds for implementing and testing the various concepts and algorithms presented in the text. The visual feedback provided by the simulation case studies is valuable in understanding the performance and limitations of the control algorithms. Experience with the simulated examples can smooth the transition to real-world applications.

Acknowledgements

We want to thank the pioneers in the field of model-based control, who are listed in the references, and from whom we have borrowed extensively. We also want to thank the numerous anonymous reviewers who read this manuscript and gave very valuable suggestions. Also not to be overlooked are our many students who also made suggestions for improving the readability and clarity of the text. In particular we would like to thank our former graduate students Drs. Jiawen Dong, Frieda Wang-Hanratty, Peter Hanratty, Shi-Shang Jang, Tannakorn Kumsaen, Mario Laiseca, Srinivas Palavajjhala, Sairam Potaraju, Deepak Srinivasagupta, Karel Stryczek, Matthew Thomas, Srikanth Voorakaranam, and Chao-Ming Ying, who helped us develop the material in the book and the software at the website.

The IMCTUNE software development was started by Mario Laiseca, and continued, with many substantial changes and improvements, by Karel Stryczek, Jiawen Dong, and Tannakorn Kumsaen. It is not an exaggeration to say that without their work Coleman Brosilow's contribution to this text would not have been possible.

Dr. A. Bemporad (working with Dr. Morari and Dr. Ricker) provided us with beta test versions of the new SIMULINK blocks for MPC Toolbox, and we thank them for their help.

We acknowledge Drs. Downs and Vogel for permission to include a SIMULINK model of the Tennessee Eastman (TE) process on our website. Dr. Palavajjhala graciously agreed to provide a copy of a chapter from his thesis dealing with this problem on the website. He also prepared the MATLAB m-files that simulate the TE process. Undergraduate students Andrew Tillinghast and Nick Graham developed the SIMULINK files for this process.

Thanks are also due to Dr. Ying, who developed the core programs used in the identification software provided with this text.

Over the years, we have received financial support from Case Western Reserve University and Washington University in the form of sabbaticals, released time, and much encouragement, the National Science Foundation, The Camille and Henry Dreyfus Foundation, MEMC Electronic Materials, Inc. and the Dow Chemical Company.

Finally, but most importantly, we thank our families for supporting this endeavor despite the enormous time commitment it required.

Introduction

Objectives of the Chapter

- Introduce the hierarchical approach to process automation and control.
- Introduce the role of model-based control in the overall process control hierarchy.
- Introduce the concepts of model-based control.
- Discuss the outline of the book.

The term model-based control (MBC) is used in this text to mean control systems that explicitly embed a process model in the control algorithm. In particular, we consider control algorithms such as Internal Model Control (IMC) and Model-Predictive Control (MPC) which have found applications in the process industry. Beginning in the early 1970s, the concept of using a predictive model of the process to determine the control actions to be taken has been accepted widely by industrial practitioners. The impetus for this came from the difficulties encountered with decentralized PID (proportional integral derivative) based control approaches that had been the norm in the past. The emergence of computer-based control allowed complex calculations to be performed online. At the same time, the increased cost of energy and raw materials provided greater economic incentives to push the plants to their optimal operating points.

The objective of this book is to provide an introduction to the concepts and practice of model-based process control. A prior beginning-level undergraduate course in process control is assumed.

Model-based control is not a replacement for traditional single-loop controllers. Rather, it complements the traditional methods. The theory of model-based control gives additional valuable insights into the way traditional single-loop PID controllers are tuned and operated.

We begin by considering the process control problem as a subset of the overall plant operations problem.

1.1 NATURE OF THE PROCESS CONTROL PROBLEM

The overall objective of plant operations is to achieve maximum profitability (optimum plant performance) subject to operational constraints related to safety, economics, government regulations, and equipment limitations. This complex problem is tackled by breaking it down into simpler subproblems that are tackled more easily. Figure 1.1 shows the resulting hierarchy of functions. This structure was proposed by Mesarovic et al. (1970). Popovic and Bhatkar (1990) and Williams (1983) discuss this concept in more detail. A brief discussion of each of the layers is given next.

At the lowest level we have the *instrumentation layer,* which consists of devices for acquiring data (sensors and transmitters), field display devices for displaying process variables, and hardware safety interlocks for ensuring emergency actions in case of safety violations. The devices at this level sense the process variables and transmit the data to the computer control system running the plant. Instrument engineers use the local display devices to troubleshoot the problems in the plant. With the introduction of smart sensors (computerized sensing devices) there is a trend towards creating a local area network (LAN) of the sensors and actuators. Data are collected at sampling rates on the order of one second or less. Some equipment might require smaller sampling times.

The instrumentation layer reports the data to the regulatory control layer which is implemented using control hardware known as DCS (Distributed Computer Control Systems, used primarily in large continuous production facilities) or PLCs (Programmable Logic Controllers, used primarily in discrete or batch process operations). See McMillan (1991) and Williams (1983) for a more detailed discussion of DCS and PLC control hardware.

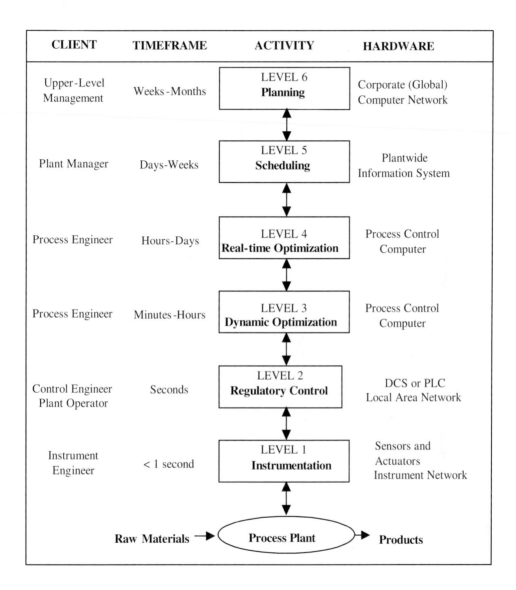

CLIENT	TIMEFRAME	ACTIVITY	HARDWARE
Upper-Level Management	Weeks-Months	**LEVEL 6** **Planning**	Corporate (Global) Computer Network
Plant Manager	Days-Weeks	**LEVEL 5** **Scheduling**	Plantwide Information System
Process Engineer	Hours-Days	**LEVEL 4** **Real-time Optimization**	Process Control Computer
Process Engineer	Minutes-Hours	**LEVEL 3** **Dynamic Optimization**	Process Control Computer
Control Engineer Plant Operator	Seconds	**LEVEL 2** **Regulatory Control**	DCS or PLC Local Area Network
Instrument Engineer	< 1 second	**LEVEL 1** **Instrumentation**	Sensors and Actuators Instrument Network

Raw Materials → **Process Plant** → **Products**

Figure 1.1 The six-layer hierarchical decomposition of control functions.

The function of the *regulatory control layer* (see Figure 1.2) is to maintain the process variables at their prescribed setpoints in spite of local disturbances that are occurring at a timescale of seconds to minutes. Such disturbances originate due to a variety of causes, such as changes in ambient conditions, changes in raw material properties, and startup and shutdown at other sections of the plant. In addition to this regulatory function, this layer allows the operator to take control of the plant in case the need arises.

The main objectives of this layer are:

1.   Provide operator interfaces.

2.   Provide regulatory control of process variables at their setpoints.

3.   Reject local, high-frequency disturbances (disturbances on a timescale of seconds to minutes).

4.   Follow the directives of the layers above.

The PID controller is widely used to implement the local single input single output (SISO) control loops. The DCS is also used to implement multiloop control strategies such as feedforward, cascade, ratio, and override control.

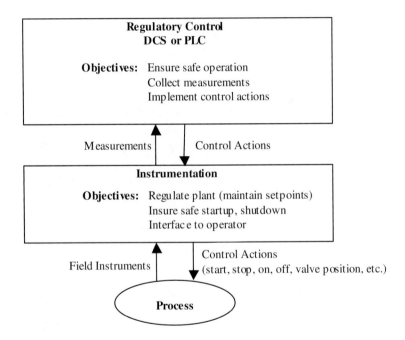

Figure 1.2   Lower levels of the control hierarchy.

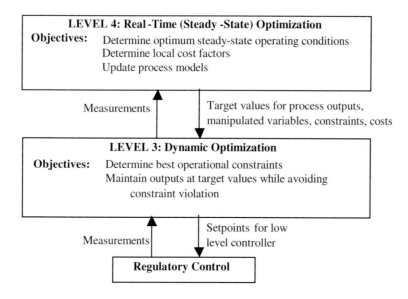

Figure 1.3 The middle layers of the control hierarchy.

The functions of the next two levels are shown in Figure 1.3. The function of the *dynamic optimization (DO)* layer is to keep the process operating near optimum efficiency by constantly adjusting the setpoints and responding to longer duration disturbances that could shift the optimum from one constraint to another. The problems here are typically multivariable and constrained in nature. Due to the possible interaction among the variables, all the control moves must be calculated simultaneously. The control actions are taken to accommodate longer duration disturbances entering the process (at a timescale on the order of minutes).

This layer uses the process model to guide its actions so that the plant operates within the constraints and moves to the most economic operating point given the current status of the disturbances. Process dynamics are important, and hence a dynamic process model is necessary. Some authors refer to this as the predictive control layer. The output of this layer is used by the DCS or PLC system as setpoints. An implicit assumption here is that the DCS or PLC is capable of responding to the changes requested by the layer before the next set of model-predictive control (MPC) calculations are initiated. The response time of the process variables is used to guide the allocation of control loops between the regulatory control layer and the MPC layer.

The *real-time optimization (RTO)* layer is charged with the responsibility of determining the optimum steady-state operating conditions for the plant given the current production requirements and factors such as material costs, utility costs, and product demand. This layer employs a detailed, first principles, steady-state model of the plant to determine the optimum operating conditions. This model is continuously updated online to reflect the changing process parameters and to account for errors in the process model. The timescale of the disturbances considered at this layer allows the plant to reach and maintain a steady-state between actions.

The cost factors, production capacity targets, raw material availability and production schedule are determined by the *scheduling layer*. For example, a refinery will optimize the operating schedule of the crude units, reforming units, cracking units, and blending units to meet the market demand for the coming weeks or months, based on the crude availability and prevailing prices. Large-scale linear programming (LP) is used at this level. This is run on a plantwide information management system, which collects and disseminates the information over all the control facilities distributed throughout the plant. Costs of utilities and other items exchanged between processing units are determined at this scheduling level (see Figure 1.4).

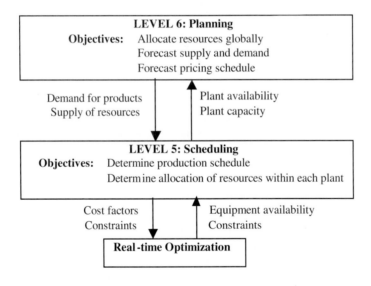

Figure 1.4   Top layers of the control hierarchy.

The *planning layer* is implemented at the corporate level. This layer would determine the allocation of resources at a global level. Because of the long timescales involved, changes in the planning layer are made over much longer periods of time, ranging from weeks to months. Planning requires forecasting supply and demand.

Another way of looking at the hierarchical decomposition is along the geographical domain of influence of each layer as illustrated in Figure 1.5. This figure shows the various layers of control hierarchy in a global oil company. The highest level of planning could cover all of the global facilities owned by the company. The scheduling level is executed locally at each plant location. The plant itself could consist of many distinct processes, each with its own set of process control computers. Dynamic optimization takes place at this level. Tightly coupled units (through mass or energy exchange) in a process are grouped together for multivariable control purposes. Individual DCSs operate for each processing unit set to enforce regulatory control. Multiloop and single-loop control strategies are applied to each unit operation individually at this level using local control units.

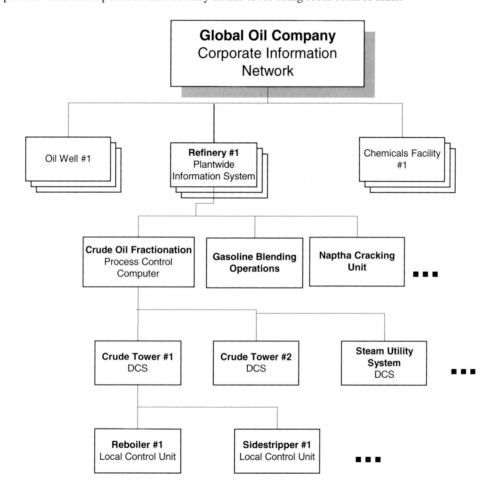

Figure 1.5 Geographical decomposition of the hierarchical control structure as applied to an oil company.

The concept of tying all the layers together to operate seamlessly (without manual intervention) is called *Computer Integrated Manufacturing* (CIM). A company that embraces CIM is able to respond to changing market conditions rapidly and hence operate more efficiently. Until recently, each of the layers tended to be supported by different vendors and computer systems, making it virtually impossible to maintain a seamless flow of information from top to bottom and vice versa. The need for CIM is now widely accepted and the trend is towards DCS and PLC manufacturers (who typically manage the bottom two layers) to provide integration with the planning and scheduling software used in higher levels of management. In an ideal CIM system, when a customer places an order, the information is immediately passed up and down the hierarchical chain so all affected parties may respond to it immediately.

The middle layers (RTO, DO) are still in a state of development. Lack of good process models is usually an impediment to the implementation of these layers. Most advanced control projects involve the development and implementation of these layers. According to industry experts, this layer can increase the profitability of a process by 7% to 10%. Payback times of six months to a year are often claimed on such projects.

Example 1.1 Application to a Microbrewery

Consider a microbrewery manufacturing different blends of beer at three different locations. Let us consider how the hierarchical control strategy applies to this firm.

Level 1. Instrumentation layer: This layer will include the sensors to measure process variables such as temperature, pressure flow, and pH, the control and solenoid valves, safety interlock devices that prevent safety violations, and local alarm devices.

Level 2. Regulatory control layer: This layer will include the PLC system that controls the brewing process. The PLC system is responsible for implementing the batch recipe for each product, regulating the process variables at their setpoints using feedback control (typically PID controllers), startup, shutdown, and emergency shutdowns. It will also create logbooks and provide operator interfaces. It will accept new recipes from the layer above.

Level 3. Dynamic optimization layer: The formation of this layer will be to monitor the process during a batch and to make midcourse corrections if necessary so that the product specification targets can be met. For example, it might decide to alter the recipe slightly to maintain desired pH level of the product at the end of the batch. This layer usually employs a dynamic process model to predict and control product quality.

Level 4. Real-time optimization layer: This layer is responsible for creating a batch recipe for a specified brand of beer that will maximize the yield or profit, given the product demand, product and raw material prices, and cost of operation. It may alter the fermentation conditions and raw material proportions so as to minimize cost while meeting specifications.

Level 5. Scheduling layer: This layer is responsible for deciding the schedule of manufacturing different brands of beer, given the capacity limits in the plant and market constraints (demand, availability, cost of products and raw materials). It must also take into account cost of inventory.

Level 6. Planning layer: This layer is responsible for making decisions such as capacity expansion, new plant construction, resource allocation, and long-term forecasting. Inputs to this layer are projected demand and supply, interest rates, past and projected government regulations, and projected growth of the company.

◆

1.2 OVERVIEW OF MODEL-BASED CONTROL

Process models are used in each of the layers of the hierarchical structure. In the regulatory control layer, models are employed in the tuning of PID and internal model controllers (IMCs), the design of feedforward, cascade, and override controllers, and the design of inferential controllers (control using secondary measurements). In the dynamic optimization layer, models are used to predict the behavior of the process into the future, to compute control actions, and to determine local economic optimum point given process constraints. Nonlinear steady-state models are used in the real-time optimization layer to compute steady-state optimum operating conditions. Linear models of the overall plant operating characteristics are used in the scheduling and planning layers.

In this book we will focus primarily on the use of models in regulatory and dynamic optimization layers.

Figure 1.6 shows the generic form of a model-based control strategy. There are three distinct blocks associated with a model-based control system:
1. *The process model*: This block computes predicted values of the process measurements.
2. *Disturbance estimation/model parameter adaptation*: This block makes adjustments to the disturbance estimate or model parameters so that the predicted values are brought closer to the actual measurements.
3. *Controller or optimizer*: This block computes the actions needed so that the selected outputs of the process will be driven to their desired or optimum setpoints, while avoiding constraint violations.

Implementations of the model-based control strategy vary greatly. In the simplest form, for a SISO system, the model is a linear transfer function, the adaptation is a linear correction, and the controller is an approximate inverse of the model transfer function. In such a case (we call it IMC) the strategy can be reduced to a conventional feedback control structure. An issue that arises here is the effect of model uncertainty on the controller stability and performance. We find that there is a tradeoff between performance and stability for real processes when model uncertainty is always present.

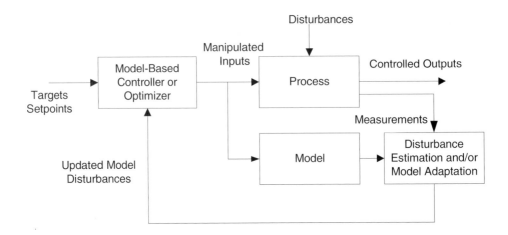

Figure 1.6 Generic form of the model-based control strategy.

In the dynamic optimization layer, a linear multiple-input multiple-output (MIMO) model is used. The controller is usually a linear least-squares calculation done at every sample time that minimizes future predicted variations of the output from their setpoints subject to process constraints. The adaptation consists of matching the model prediction and actual measurement through a linear correction to the predicted values. This layer also involves a computation of the optimum operating conditions given the estimate of the current and future values of disturbances at every sample time. Much of the economic incentive for implementing model-based control is derived from this secondary optimization of setpoints that allows the plant to follow the most economic process constraint (such as manipulated variable saturation, output limits, etc.) at any given time.

In the most general form, the model-based control consists of a nonlinear process model, a complex estimation scheme that computes process parameters to minimize plant/model mismatch, and a controller that computes an approximate inverse in an implicit form by exercising the model repeatedly (using a numerical search procedure) to achieve the economic goals of the process.

1.3 SUMMARY

In this chapter, we introduced the concept of model-based control in the context of the general process control problem. For ease of implementation, the control problem is tackled using a hierarchical decomposition of the control tasks resulting in six clearly identifiable layers:

1. Instrumentation layer
2. Regulatory control layer
3. Dynamic optimization layer
4. Real-time (steady-state) optimization layer
5. Scheduling layer
6. Planning layer

The type of models used is dependent on the application layer. The focus of this text is on the regulatory control layer and the dynamic optimization layer.

Problems

1.1 Consider the following control applications. Classify each application to one of the levels of the control hierarchy. Identify the type of control hardware where these applications will reside (DCS system, safety interlock system, process control computer, plantwide computer, etc.). Also identify whose job function it is to maintain these applications (instrument engineer, control engineer, process engineer, plant manager, etc.).

  a. Temperature control in a heat exchanger
  b. Safety release valve in a reactor
  c. Optimum operating temperature for a reactor
  d. Production rate in a crude oil distillation plant
  e. The product distribution from a blending unit in an oil refinery
  f. The scheduling of crude oil shipped to various refineries from a production well

1.2 Consider the operation of a continuous stirred tank reactor. Consider the following control objectives. Identify which layer of the control hierarchy will address each objective.

  a. Temperature and level control
  b. Feedforward compensation for variations in coolant inlet temperature
  c. Keep the level in the tank from running too low
  d. Keep the pressure in the tank from exceeding a certain limit
  e. Adjust the reactor temperature to compensate for variations in the production rate

1.3 Apply the hierarchical decomposition of control functions to the following integrated plants. Show specific objectives of each layer as it applies to that particular plant. Give an example of the application of each level in the control hierarchy to the operation of the plant. Where applicable, discuss the hardware platform used for implementing the objectives.

  a. A paper mill manufacturing paper from wood
  b. A pharmaceutical plant making medical drugs
  c. A steel processing plant making steel from iron ore
  d. A pet food manufacturing plant making pet foods from cornmeal and meat

References

Marlin, T. E. 1999. *Process Control: Designing Processes and Control Systems for Dynamic Performance.* McGraw-Hill, NJ.

McMillan, G. K. 1991. *Distributed Control Systems: Selection, Implementation and Maximization.* Instrument Society of America, Research Triangle Park, NC.

Mesarovic, M. D., D. Macko, and Y. Takahara. 1970. *Theory of hierarchical multilevel systems.* Academic Press, NY.

Popovic, D., and V. P. Bhatkar. 1990. *Distributed Computer Control Systems in Industrial Automation.* Marcel Dekker, NY.

Seborg, D. E., T. F. Edgar, and D. A. Mellichamp. 1989. *Process Dynamics and Control.* John Wiley & Sons, NY.

William, T. J. 1984. *The Use of Digital Computers in Process Control.* Instrument Society of America, Instrument Learning Modules Series, Research Triangle Park, NC.

Williams, T. J. 1983. *Analysis and Design of Hierarchical Control Systems.* Elsevier, Amsterdam.

Continuous-Time Models

Objectives of the Chapter

- Review the types of continuous-time model representations frequently employed in model-based control systems.
- Discuss common transfer function models.
- Identify first order plus dead time models from process data.

Prerequisite Reading

Appendix A, "Review of Laplace Transforms and Feedback Control Concepts"
Appendix E, "MATLAB and Control Toolbox Tutorial"
Appendix F, "SIMULINK Tutorial"

2.1 INTRODUCTION

The process model is the most important component of a model-based control system. Hence we begin by a study of the various types of process models employed in model-based control. The discussion in this chapter is restricted to those types of models that have found widespread applications in process control. The discussion focuses mainly on those aspects of modeling most relevant to process control.

Consider a process with input variables $u(t)$ and output variables $y(t)$ as shown in Figure 2.1. A process can be a unit operation (e.g., a reactor) or a part of a unit operation (e.g., a reboiler attached to a distillation column). The inputs are external variables that can affect the internal state of the system. Outputs are variables that have some external consequence. Inputs may be classified as manipulated variables (if they can be deliberately changed) or disturbance variables (if their variation is not under control). Outputs may be measured or unmeasured.

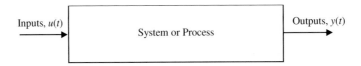

Figure 2.1   A process with inputs $u(t)$ and outputs $y(t)$.

A mathematical model is a relationship between y and u that can be expressed in terms of some mathematical equations. Models can be classified according to the relationship between y and u. A few common classifications are discussed next. More detailed descriptions and examples are given in the sections that follow.

2.2. PROCESS MODEL REPRESENTATIONS

Process models may be expressed in many different ways. *First principles models* (models derived by writing the fundamental relationships such as mass and energy balances combined with thermodynamic, transport, and kinetic equations) are expressed as differential equations in the time domain. These may be ordinary differential equations (ODE) models sometimes also called *lumped parameter* models. If we consider spatial variations of the variables, then we get *distributed parameter* models (partial differential equations in space and time). Such models may be classified into *linear* and *nonlinear* models. *Linear* models are characterized by the following:

1. If $y_1(t)$ is the response of the process to an input forcing function $u_1(t)$ then, $ky_1(t)$ is the response to an input forcing function $ku_1(t)$.

2. If $y_1(t)$ is the response to $u_1(t)$ and $y_2(t)$ is the response to $u_2(t)$, then, $y_1(t) + y_2(t)$ is the response to $u_1(t) + u_2(t)$.

Although most real processes are nonlinear, for small variations in $u(t)$ these processes can be approximated by linear constant coefficient models through a procedure of *linearization,* about a steady-state (nominal) operating point, as explained below. The equations are written using deviation variables, which represent the variation from this nominal operating point. Sometimes these linearized models are referred to as *small signal models*. Most model-based controllers employ linear constant coefficient models. Such models may also be arrived at by empirical model fitting (identification) using input-output response data (called data-driven or regression models).

Linear constant coefficient approximations are useful because in regulatory control we remain near or around a nominal operating point. In this case *u* and *y* are taken to be *deviations* or *perturbations* from these nominal steady-state values. Thus, if *u* and *y* remain small, then the linear approximation is valid.

Linear models in the form of linear constant coefficient differential equations with time as the independent variable are called *state-space representations*. By applying the Laplace transform to the linear model, we can express the model in the Laplace domain (also called the *s-domain*). The model is written in the form of an algebraic relationship between the input and output variables. Such a model is called a *transfer function model*.

For the study of computer control systems, it is common to express the model as a relationship between the discrete inputs coming from the computer and the sampled values of the output measurements on the process. Such a model is called the *discrete time model* of the process. Discrete time models are expressed as *difference equations* or in the form of a discrete time Laplace transform called the z-transform. The latter models are called *z-transfer functions*. Discrete models are discussed in more detail in Chapter 14.

Often, a fundamental understanding of the process is not available. In this instance, one must resort to *blackbox* modeling that attempts to capture the input-output behavioral characteristics. Such *data-driven* models are built empirically from input/output data and curve fitting. If a structure is assumed for the input/output model, then we have *parametric* models. In this instance, the model parameters are fitted to the data. If no structure is assumed, then it is called a *nonparametric* model. Impulse response and step response models fall under this category (see Ch. 14).

Figure 2.2 shows a hierarchy of process models used in process control.

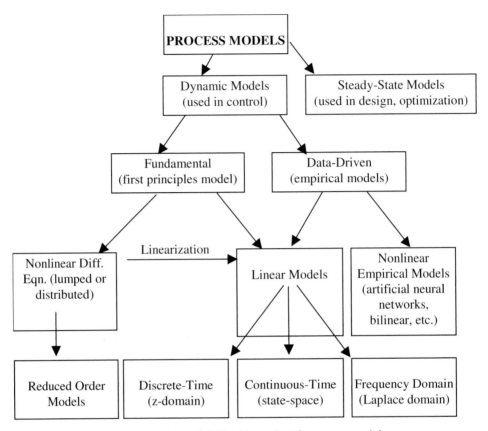

Figure 2.2 The hierarchy of process models.

2.3 TIME DOMAIN MODELS

An example of a time domain representation is the following process model:

$$\frac{dx}{dt} = f(x,u)$$

$$y = g(x,u),$$

(2.1)

where $u(t)$ is an input to the process, $x(t)$ represents a state of the system, and $y(t)$ is the output. The functions f and g can be nonlinear. This type of process model is frequently derived from a first principles analysis of a system. A small signal model can be obtained from the above equation by a linearization approximation as follows. Let

$$u(t) = \overline{u} + u_d(t).$$

The above perturbation in control effort causes

$$x(t) = \overline{x} + x_d(t)$$
$$y(t) = \overline{y} + y_d(t),$$

where

$\overline{u}$	=	the initial nominal steady-state value of the input variable,
$\overline{x}$	=	the initial nominal steady-state value of the state variables,
$\overline{y}$	=	the initial nominal steady-state value of the output variable,
$y_d(t)$	=	the change in y from the initial value, called the deviation of y,
$u_d(t)$	=	the change in u from the initial value , called the deviation of u.

Substituting these into the differential equation, we get (we have used a simplified notation by dropping the explicit dependency on time in some places)

$$\frac{d(\overline{x} + x_d(t))}{dt} = f(\overline{x} + x_d(t), \overline{u} + u_d(t)) .$$

Expanding in a Taylor series,

$$\frac{dx_d}{dt} = f(\overline{x}, \overline{u}) + \left.\frac{\partial f}{\partial x}\right|_s x_d(t) + \left.\frac{\partial f}{\partial u}\right|_s u_d(t) + higher\ order\ terms$$

$$\overline{y} + y_d(t) = g(\overline{x}, \overline{u}) + \left.\frac{\partial g}{\partial x}\right|_s x_d(t) + \left.\frac{\partial g}{\partial u}\right|_s u_d(t) + higher\ order\ terms.$$

(2.2)

The subscript s in the partial derivatives indicates that they are evaluated at the nominal steady-state conditions. If we start from a steady-state, then $f(\overline{x}, \overline{u}) = 0$ and $\overline{y} = g(\overline{x}, \overline{u})$. Neglecting higher order terms, we get the approximate linear model of the process as

$$\frac{dx_d}{dt} = Ax_d(t) + Bu_d(t)$$

$$y_d(t) = Cx_d(t) + Du_d(t),$$

(2.3)

where A, B, C, and D are constants as implied by Eq. (2.2). The procedure can be extended to multivariable systems as well. Essentially, the same procedure as above is used except that both x and u are vectors and the derivatives are replaced by the corresponding Jacobian matrices. This type of representation is called the **state-space** model of the system.

2.4 LAPLACE DOMAIN MODELS

Consider the state-space model:

$$dx(t)/dt = Ax(t) + Bu(t)$$
$$y(t) = Cx(t) + Du(t). \tag{2.4}$$

Here we have used $\dot{x}$ to denote dx/dt. Taking the Laplace transform of both sides and assuming $x(0) = 0$, we get (for convenience we have used the same symbol for the time domain and Laplace domain representation of a variable)

$$sx(s) = Asx(s) + Bu(s)$$
$$y(s) = Cx(s) + Du(s), \tag{2.5}$$

which can be solved to get

$$y(s) = \left[C(sI - A)^{-1} B + D \right] u(s)$$
$$= G(s)u(s). \tag{2.6}$$

The Laplace domain representation $G(s)$ is called the transfer function. If y and u are vectors then $G(s)$ is a matrix of transfer functions.

A typical SISO transfer function can be expressed in the form of a ratio of polynomials (the delay term has been approximated through a Padé approximation; see Appendix A):

$$G(s) = \frac{a_n s^n + a_{n-1} s^{n-1} + \ldots + a_1 s + a_0}{b_m s^m + b_{m-1} s^{m-1} + \ldots + b_1 s + b_0}$$
$$= \frac{N(s)}{D(s)}. \tag{2.7}$$

The steady-state gain of the process is given by the final value theorem of Laplace transforms as $G(0)$. For a physical process, the numerator polynomial degree must be less than or equal to the degree of the denominator polynomial. Otherwise, we say that the transfer function is nonrealizable. This relates to the fact that we cannot physically build a pure differentiator, and pure differentiators do not exist in nature.

Table 2.1 shows common transfer functions. The *gain* model is used whenever the dynamics of the system are negligible. The *integrator* represents an accumulation of mass or energy in the system. It appears in the models of tanks whose output flows are independent of the tank level. A *first-order lag* model appears whenever the system is dominated by one capacity term for accumulation of mass or energy. Such systems are characterized by a gain (K) and a time constant (τ). Second-order systems have a second-order polynomial in s in the denominator. Time delay (or dead time) is used to model transport lags caused by flow through pipes. A transfer function that is a combination of the first-order system with time delay term (called the *first-order plus dead time, or FOPDT, model*) is often used to characterize the response of open-loop systems (systems with no feedback control present). The next section discusses the characteristics of such models.

Table 2.1 Commonly Encountered SISO Transfer Functions.

Terminology	Laplace Domain Representation	Transfer Function
Gain	$y(t) = Ku(t)$	$y(s) = Ku(s)$
Integrator	$y(t) = \int u(t)dt$	$y(s) = u(s)/s$
First-Order Lag	$\tau\dfrac{dy}{dt} + y = Ku$	$y(s) = \dfrac{K}{\tau s + 1}u(s)$
Second-Order System	$\tau^2\dfrac{d^2 y}{dt^2} + 2\tau\xi\dfrac{dy}{dt} + y = Ku$	$y(s) = \dfrac{K}{\tau^2 s^2 + 2\tau\xi s + 1}u(s)$
Lead-Lag	$\tau_1\dfrac{dy}{dt} + y = \tau_2\dfrac{du}{dt} + u$	$y(s) = \dfrac{\tau_2 s + 1}{\tau_1 s + 1}u(s)$
Inverse Response (Whether we get an inverse response depends on the system parameters.)	$y = y_1 - y_2$ $\dfrac{dy_1}{dt} + y_1 = k_1 u_1,\ \dfrac{dy_2}{dt} + y_2 = k_2 u_2$	$y(s) = \left(\dfrac{k_1}{(\tau_1 s + 1)} - \dfrac{k_2}{(\tau_2 s + 1)}\right)u(s)$
Time-Delay	$y(t) = y(t-D)$	$y(s) = e^{-Ds}u(s)$

The *second-order* transfer function is important because the response characteristic of many closed-loop systems (systems with feedback control) can be approximated by a second-order transfer function. The response of the second-order system is characterized by the time constant τ (which determines the speed of response) and the damping coefficient ξ, which determines if the response is oscillatory (see Problem 2.2). The *lead-lag* model is often used to approximate differentiation in a controller by keeping the lag time constant smaller than the lead time constant. *Inverse response* is often obtained when there are two lags of opposite signs in parallel. Inverse response systems can be hard to control since the initial response is opposite to the final change in the variable. Such systems exhibit a zero in the right half of the *s* plane.

Figure 2.3 shows the step response characteristics of these common transfer functions. Each graph shows the response of the system to a step input of 1 occurring at $t = 0$.

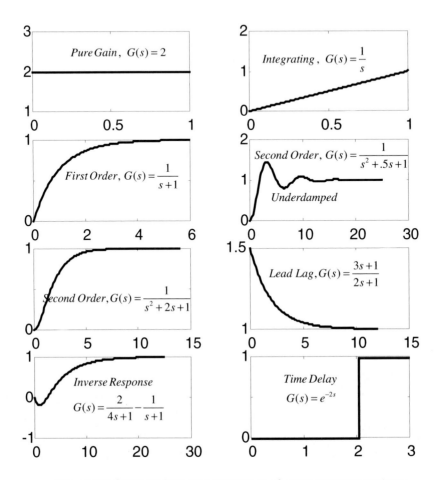

Figure 2.3 Characteristic step responses of some common systems.

Example 2.1 Converting Models

Consider the differential equation model (a_2, a_2, a_0, b are constants):

$$a_2 \frac{d^2 x}{dt^2} + a_1 \frac{dx}{dt} + a_0 x = bu(t), \quad x(0) = 0,\ dx/dt\,(at\,t = 0) = 0.$$

a. Convert this into a transfer function model.
b. Convert this into a state-space model.

Taking Laplace transform of both sides,

$$a_2 s^2 X(s) + a_1 s X(s) + a_0 X(s) = bU(s)$$

$$X(s) = \left(\frac{b}{a_2 \ s^2 \ + \ a_1 \ s \ + \ a_0} \right) U(s),$$

Define

$$x_1(t) = \frac{dx(t)}{dt}$$

$$x_2(t) = x(t),$$

then

$$\frac{dx_2}{dt} = x_1$$

$$a_2 \frac{dx_1}{dt} + a_1 x_1 + a_0 x_2 = bu(t)$$

or

$$\frac{dx_1}{dt} = - \frac{a_1}{a_2} \ x_1 - \frac{a_0}{a_2} x_0 + bu(t)$$

$$\frac{dx_2}{dt} = x_1,$$

which is in state-space form. Note that state-space forms are not unique. For example, we could have defined x_2 to be $x + dx/dt$ to get a different representation. However, for any given transfer function, there is a state-space realization that contains the smallest number of states. Such a representation is called the *minimal realization*. MATLAB has built-in functions to do the conversion between state-space and Laplace transform representations.

♦

Example 2.2 Building a Model from First Principles: A Radiant Heater

This example is taken from Seborg et al. (1989). Consider the temperature dynamics of a radiant heater, shown in Figure 2.4.

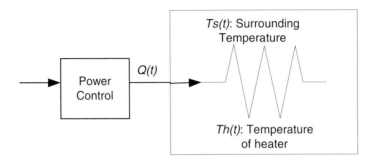

Figure 2.4 Radiant heater.

Assuming the surrounding temperature is constant, we can write the energy balance on the heating element as

Rate of Energy Accumulation in Heater = Rate of Energy Input into the Heater – Rate of Energy Output from the Heater

Accumulation of Energy in Heater = Input of Energy into Heater – Output of Energy from Heater

$$\frac{d}{dt}\left(MC_pT_h\left(t\right)\right) = Q(t) \; - \; k\left(T_h^4\left(t\right) \; - \; T_s^4\left(t\right)\right),$$

where

M = mass of heater,
$T_h(t)$ = temperature of heating element,
k = constant,
C_p = heat capacity of heating element,
$Q(t)$ = power supplied,
$T_s(t)$ = surrounding temperature.

To linearize this equation, define

$$Q(t) \;\; = \;\; \overline{Q} + Q_d\left(t\right), \, T_h = \overline{T}_h + T_{hd}\left(t\right) \text{ and } T_s = \overline{T}_s + T_{sd}\left(t\right),$$

where the bar denotes steady-state values and subscript d denotes the deviation variable. Substituting these, then expanding the terms in a Taylor series as outlined above and assuming system is at steady-state initially, we get

$$M\,C_p\,\frac{d\,T_{hd}\left(t\right)}{dt} = Q_d\left(t\right) - 4k\,\overline{T}_h^3\cdot T_{hd}\left(t\right) + 4k\overline{T}^{\,3}\cdot T_{sd}\left(t\right).$$

Since the system was initially at steady-state so that $T_{hd}(0) = Q_d(0) = 0$. Taking Laplace transforms of both sides, we get

$$M\,C_p s\,T_{hd}(s) = Q_d(s) - 4k\overline{T}_h^3\,T_{hd}(s) + 4k\overline{T}_s^3 T_{sd}(s)$$

or

$$T_{hd}(s) = \frac{Q_d(s)}{\mu\,C_p s + 4k\,\overline{T}_h^3} + \frac{4k\overline{T}_s^3\,T_{sd}(s)}{\mu\,C_p s + 4k\overline{T}_h^3}\,.$$

This relationship gives us the transfer functions relating Q_d and T_{sd} to T_{hd}. Note that the gain of the first transfer function has units of temperature divided by power, and is $1/(4k\overline{T}_h^3)$. The gain of the second transfer function is dimensionless, and has a numerical value of $(\overline{T}_s/\overline{T}_h)^3$. The time constant $\mu C_p / 4k\overline{T}_h^3$ is the same for both transfer functions.

♦

Example 2.3 Model of a Mixing Tank

Consider the mixing tank shown in Figure 2.5. Hot and cold fluids are mixed together in a stirred tank. To develop a mathematical model, we make the following assumptions: (a) well-mixed tank, (b) no heat losses to the surroundings, (c) constant density ρ, heat capacity C_p, (d) constant area of cross-section, A, (e) T_h and T_c are constants, and (f) accumulation of energy in the tank wall is negligible.

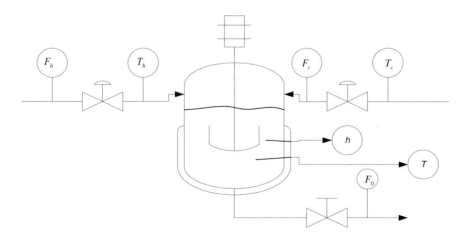

Figure 2.5 A mixing tank.

Total Mass Balance:
Rate of Mass Accumulation = Rate of Mass Input – Rate of Mass Output.

$$\frac{d}{dt}(\rho Ah(t)) = \rho F_h(t) + \rho F_c(t) - \rho F_o(t)$$

$$\frac{dh}{dt} = \frac{F_h}{A} + \frac{F_c}{A} - \frac{F_0}{A} = f_1(F_h, F_c, F_o)$$

Energy Balance:

$$\frac{d}{dt}(\rho Ah.C_p T) = \rho F_h C_p T_h + \rho F_c C_p T_c - \rho F_o C_p T$$

$$Ah\frac{dT}{dt} + AT\frac{dh}{dt} = F_h T_h + F_c T_c - F_o T$$

Substituting for *dh/dt* from the mass balance equation,

$$\frac{dT}{dt} = \frac{F_h}{Ah}(T_h - T) + \frac{F_c}{Ah}(T_c - T) = f_2\left(F_h, F_c, F_o, T_h, T_c\right).$$

The above constitutes a nonlinear, state-space model of the tank. For simulation purposes, we can set up a SIMULINK model of the nonlinear system, as shown in Figure 2.6. Such simulations are useful for testing control system designs, since SIMULINK provides a variety of tools to do control system studies.

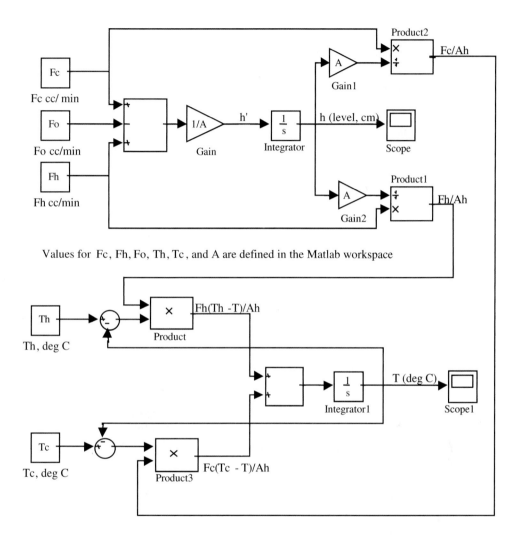

Figure 2.6 SIMULINK diagram of the model for the mixing tank.

The linear state-space model can be obtained by linearizing the above equations. We have the two nonlinear differential equations:

$$\frac{dh}{dt} = f_1(F_h, F_c, F_o)$$

$$\frac{dT}{dt} = f_2(F_h, F_c, F_o, T_h, T_c)$$

Applying the linearization procedure outlined above to this system around a nominal steady-state operating condition (denoted by the subscript s), we get

$$\frac{dh_d}{dt} = \frac{\partial f_1}{\partial F_h}\bigg|_s .F_{hd} + \frac{\partial f_1}{\partial F_c}\bigg|_s .F_{cd} + \frac{\partial f_1}{\partial F_o}\bigg|_s .F_{od}$$

$$\frac{dT_d}{dt} = \frac{\partial f_2}{\partial F_h}\bigg|_s .F_{hd} + \frac{\partial f_2}{\partial F_c}\bigg|_s .F_{cd} + \frac{\partial f_2}{\partial F_o}\bigg|_s .F_{od} + \frac{\partial f_2}{\partial h}\bigg|_s .h_d + \frac{\partial f_2}{\partial T}\bigg|_s .T_d,$$

where

$$\frac{\partial f_1}{\partial F_h}\bigg|_s = \frac{1}{A}, \; \frac{\partial f_1}{\partial F_c}\bigg|_s = \frac{1}{A} \; and \; \frac{\partial f_1}{\partial F_o}\bigg|_s = -\frac{1}{A}$$

$$\frac{\partial f_2}{\partial F_h}\bigg|_s = \frac{1}{Ah_s}\left(T_{hs} - T_s\right), \frac{\partial f_2}{\partial f_c}\bigg|_s = \frac{1}{Ah_s}\left(T_{cs} - T_s\right)$$

$$\frac{\partial f_2}{\partial h} = \frac{-F_{hs}(T_{hs} - T_s)}{Ah_s^2} - \frac{F_{cs}(T_{cs} - T_s)}{Ah_s^2}, \frac{\partial f_2}{\partial T}\bigg|_s = \frac{-F_{hs}}{Ah_s} - \frac{F_{cs}}{Ah_s}.$$

Define

$$\mathbf{x} = \begin{bmatrix} h_d \\ T_d \end{bmatrix}, \; \mathbf{u} = \begin{bmatrix} F_{hd} \\ F_{cd} \end{bmatrix}, \; d = F_{od}, \; \mathbf{f} = \begin{bmatrix} f_1 \\ f_2 \end{bmatrix}$$

$$\mathbf{A} = \frac{\partial \mathbf{f}}{\partial \mathbf{x}} = \begin{bmatrix} 0 & 0 \\ \dfrac{-F_{hs}(T_{hs} - T_s)}{Ah_s^2} - \dfrac{F_{cs}(T_{cs} - T_s)}{Ah_s^2} & \dfrac{-F_{hs} - F_{cs}}{Ah_s} \end{bmatrix}$$

$$\mathbf{B} = \frac{\partial \mathbf{f}}{\partial \mathbf{u}} = \begin{bmatrix} \dfrac{1}{A} & \dfrac{1}{A} \\ \dfrac{T_{hs} - T_s}{Ah_s} & \dfrac{T_{cs} - T_s}{Ah_s} \end{bmatrix}, \; \mathbf{E} = \frac{\partial \mathbf{f}}{\partial \mathbf{d}} = \begin{bmatrix} \dfrac{-1}{A} \\ 0 \end{bmatrix}.$$

Then the state-space model can be written as

$$d\mathbf{x}/dt = \mathbf{Ax} + \mathbf{Bu} + \mathbf{Ed}$$

                                                                  ◆

Example 2.4 Model of a Heated Mixing Tank

Consider the tank shown in Figure 2.7. A dynamic process model can be derived by writing mass and energy balances on the fluid in the tank. To describe changes in level, $h(t)$, we can write a total mass balance on the fluid.

Total Mass Balance

Rate of mass accumulation = Rate of mass input – Rate of mass output

$$\frac{d}{dt}\left(\rho\ Ah(t)\right) = \rho F_i(t) - \rho F_0(t); \quad \rho = \text{density, A} = \text{Area of cross section}$$

Energy Balance

Rate of energy accumulation = Rate of energy input – Rate of energy output

Assume that the energy of fluid is proportional to temperature (with constant heat capacity C_p). Neglecting heat losses to the surroundings, accumulation of energy in the wall, and that the fluid in the tank is well mixed (i.e., has a uniform temperature), we get

$$\frac{d}{dt}\left(Ah(t)\rho C_p T(t)\right) = \rho\ C_p F_i(t)T_i(t) - \rho C_p F_o(t)T(t) + Q(t).$$

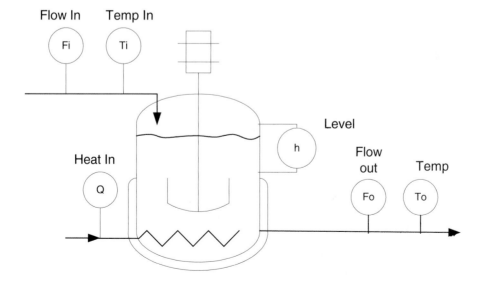

Figure 2.7 A heated mixing tank.

These equations can be simplified and linearized. Typically A, ρ, and C_p are treated as constants. $F_i(t), T_i(t), Q(t),$ and $F_0(t)$ are externally specified input variables. $T(t)$ and $h(t)$ are the state (and possibly measured output) variables. Reducing the above equations to state-space form gives:

$$\frac{d}{dt}\left(h(t)\right) = (F_i(t) - F_o(t))/A$$

$$\frac{d}{dt}\left(T(t)\right) = F_i(t)(T_i(t)-T(t))/(Ah(t)) + q(t)/(A\rho C_p h(t))$$

♦

2.5 THE FIRST-ORDER PLUS DEAD TIME PROCESS

In this section we will discuss the response characteristics of an FOPDT process. As mentioned above previously, type of model is often used to approximate many real systems (which may actually consist of many first-order systems in series). This model is shown in Figure 2.8.

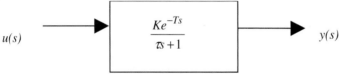

$$u(s) \qquad \frac{Ke^{-Ts}}{\tau s+1} \qquad y(s)$$

Figure 2.8 An FOPDT model.

If this process is perturbed using a step input of the form

$$\begin{aligned} u(t) &= 0 \quad 0 \le t \le t_o \\ &= A \quad t > t_o \end{aligned}$$
(2.8)

with transfer function given by $u(s) = \dfrac{A}{s}\, e^{-t_o s}$, then the output is given by

$$y(s) = \frac{AKe^{-(t_0+T)s}}{(\tau s+1)s},$$
(2.9)

with inverse given by

$$\begin{aligned} y(t) &= 0 & t \le t_o+T \\ &= AK\left[1-e^{-(t-T-t_0)/\tau}\right] & t > t_o+T. \end{aligned}$$
(2.10)

$y(t)$ is shown in Figure 2.9. This step response has the following properties:

(1) *Gain, K*.
Note that $\lim_{t \to \infty} y(t) = AK$, i.e., the steady-state change in $y(t)$ is AK. This allows us to calculate the gain from a given response.

$$\text{Gain, } K \approx \frac{\text{Steady-state change in } y(t)}{\text{Step change in input } u(t)} \tag{2.11}$$

(2) *Delay, T* defined as the time at which $y(t)$ begins to change minus the time at which the input step change was made.

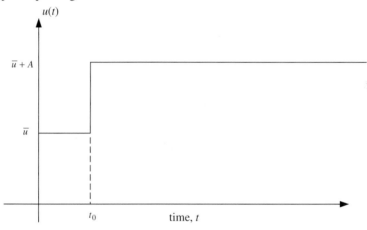

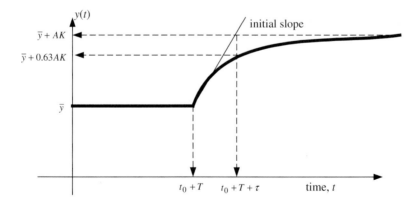

Figure 2.9   Step response of an FOPDT process.

(3) *Time Constant, τ.* The maximum slope of the curve occurs at $t = t_o + T$, and is given by $y' = AK / \tau$ at $t = t_0 + T$. (This can be verified by differentiating the equation for $y(t)$ above). This implies that the tangent to the response curve will cross the steady-state line at $t = t_o + T + \tau$ (see Figure 2.9). Another way to obtain the time constant is to observe that at $t = t_0 + T + \tau$, $y(t) = AK(1 - e^{-1}) = 0.63\ AK$.

Control engineers often use first-order plus dead time transfer functions to approximate the S-shaped time responses (see Example 2.5) of actual processes even if they are not true first-order plus dead time systems.

Example 2.5 Identifying an FOPDT Model

Often it is difficult to build a process model from fundamental principles. In that case one must resort to empirical tests to *identify* a process model. A simple test often used in industry is the step test. We will discuss this test method as it is applied to identify a FOPDT model. Consider the response of the system to a step change in input, as shown in Figure 2.10. This represents the response of a process to a step change of five units in the input occurring at $t_0 = 10$ sec.

We first draw a tangent of maximum slope of the response curve, as shown in Figure 2.10. The point at which it crosses the initial steady-state line is taken as the starting point of the response $(t_0 + T)$. The point at which it crosses the final steady-state value is taken as $t_0 + T + \tau$. For the graph shown we have $t_0 = 10$ sec, $t_0 + T \approx 21$ sec $\Rightarrow T \approx 11$ sec and $\tau \approx 21$ sec. The gain K is estimated as 4 using Eq. (2.11). This yields the approximate FODPT model of the process as

$$G(s) = \frac{4e^{-11s}}{21s + 1}.$$

Another estimate of the time constant can be obtained by looking at the 63% response time, when y reaches

$$y_{63} = \overline{y} + 0.63\ (y_\infty - \overline{y}) = 90 + 0.63\ (110 - 90) = 102.6$$

This occurs at approximately $t_{63} = 40$ sec, which yields another estimate of the time constant as $\tau = 19$. In Chapters 13 and 14 we will discuss more accurate ways of process identification from empirical test data.

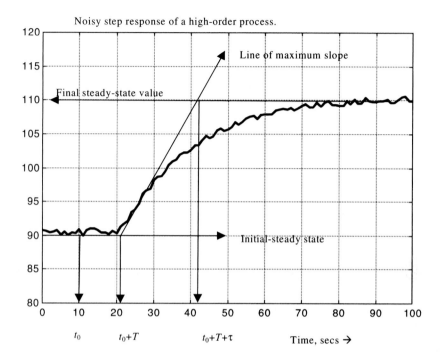

Figure 2.10 Step response of a real process. The step change in input is five units and occurs at $t = 10$ sec.

♦

2.6 SUMMARY

The objective of this chapter was to review continuous-time model representations used in control.

In particular, we discussed the following representations:

- nonlinear time domain models
- linear time domain (state-space) models
- Laplace domain models

Much of the control theory is developed around transfer function models. There is also a wealth of control theory built around linear constant coefficient state-space models, particularly in the electrical engineering domain. In this book, we shall use transfer function models to introduce model-based control. We will then discuss computer implementation of model-based control using state-space models.

First principles models are not yet used widely in model-based control. Industry appears reluctant to employ this modeling approach, primarily because of the high initial investment needed to build such models. In contrast, first principles models are widely employed in steady-state simulation and online optimization. Steady-state models can be obtained with mass and energy balances combined with relationships from thermodynamics, kinetics, and transport phenomena. To build dynamic models, we must add details about the sizes of equipment and how these are connected (length, diameter of pipes, etc.). Even if such a model could be built, often such models are too complex for use in an online model-based control strategy.

One frequently encountered problem in model-based control is the disparity (modeling error) that might exist between the model and the actual plant performance. Modeling errors can lead to less than perfect control or even instability. Quite often we find that a tradeoff exists between good control system performance and tolerance to modeling errors. This tradeoff between control system performance and stability is referred to as robust control. Control system design and tuning for robust control is discussed at some length in Chapters 7 and 8. Unless the control system is well behaved in the presence of model errors, operators are likely to turn it off completely and rely on their own skills to manually operate the process. This human acceptance factor is very important, especially when one designs large multivariable control systems that look like black boxes to the operator.

Problems

2.1. Consider the process modeled by

$$\frac{dy_1}{dt} + y_1 = u$$

$$\frac{dy_2}{dt} + y_2 = y_1$$

    a.   Derive the transfer function model between y2 and u. Assume $y_1(0) = y_2(0) = 0$.

    b.   Identify the matrices A, B, C, D in the state-space form of the model. Do this by hand and compare with the state-space model obtained using MATLAB to convert the transfer function model created in part (a) into state-space form. Note that space representation of a process is not unique. Hint: Use MATLAB function tf2ss.

2.2   Consider a blending process shown in Figure 2.11. Streams A and B are mixed in a tank of uniform cross section. The level in the tank and the volume fraction of A in the exit stream is to be controlled by manipulating the inlet flow of A and B. Write mass balance on A and the total mass balance to generate a dynamic model. Linearize the model. Express in deviation variables. Obtain transfer functions relating deviations in h and x_A to F_A and F_B respectively. Assume density of A and B are equal and constant.

F_A	= Flow rate of A into tank, cm^3/sec
F_B	= Flow rate of B into tank, cm^3/sec
h	= Height of liquid in the tank, cm
x_A	= Weight fraction of A in the tank
A	= Area of cross section, cm^2

Linearize the equations and derive a linear state-space model of the system.

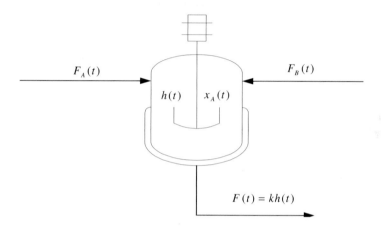

Figure 2.11 A mixing tank.

2.3 Obtain the step response of the following transfer function models using MATLAB, SIMU-LINK, or both.

  a. A first order system:

$$p(s) = \frac{1}{\tau s + 1} \; ; \; \tau = 0.1, \; 1, \; 10 .$$

  b. A second order system:

$$p(s) = \frac{1}{\tau^2 s^2 + 2\tau\xi s + 1} , \tau = 1 \quad and \quad 10, \quad \xi = 0.5, \; 1, \; 2 .$$

These represent the three types of second-order systems (underdamped, critically damped, and overdamped).

  c. Inverse response process:

$$p(s) = \frac{1}{s+1} - \frac{1}{\tau s + 1} \ , \quad \tau = 0.1, 10 \ .$$

Note the inverse response characteristics.

d.   A third order process:

$$p(s) = \frac{1}{(s+1)^3} \ .$$

Higher order systems response is similar to second order system response.

2.4   Consider the level in a tank modeled by

$$\frac{dh}{dt} = F_i - \sqrt{h}$$

where h is the level in the tank. F_i is the inlet flow. The outlet flow is proportional to the square root of the level in the tank.

a.   Derive a transfer function model relating F_{id} and h_d around the nominal steady-state $\overline{F}_i = 1.0, \overline{h} = 1.0$.

b.   Compare the step response of the linear approximation with the step response of the nonlinear equation obtained using a SIMULINK simulation for $F_{id} = 0.1$ and *10*. Discuss your results.

2.5   Hot and cold fluids are mixed in tank as shown in Figure 2.12. Derive a model to describe the variations in tank level and tank temperature. The flows of hot and cold fluid are manipulated input variables. The temperatures of the hot and cold fluid are disturbances. The outlet flow is also a disturbance variable (independent of level in the tank).

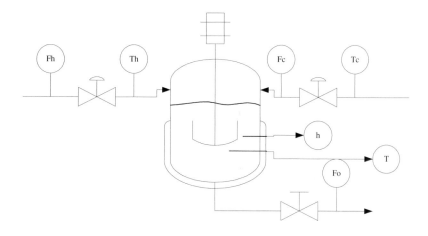

Figure 2.12 Hot and cold water mixing tank.

a.   Set up a SIMULINK model of the system for the following steady-state parameters:

$$\bar{F}_c = \bar{F}_h = 1 \; \text{lit}/\text{min}$$

$$\bar{T}_c = 20\degree C, \; \bar{T}_h = 50\degree C$$

$$\bar{h} = 50 \text{ cm}, \; \bar{T} = 35\degree C, \; \bar{F}_o = 2 \; \text{lit}/\text{min}$$

Area of cross section of tank = 300 cm^2

Heat capacity of fluid = 1 cal/gm $\degree$ C, Density of fluid =1 gm/cc.

b.   Obtain a linearized state-space model of the system above.

c.   Compare the step response of the nonlinear and linear models for a step change in temperature of the hot inlet stream from 50° to 60°C.

2.6   Consider the pressurized tank shown in Figure 2.13. Water is fed through a control valve and leaves the tank at a rate set by the hand-operated valve. Air leaves the tank through a control valve. The air pressure and water level in the tank are variable.

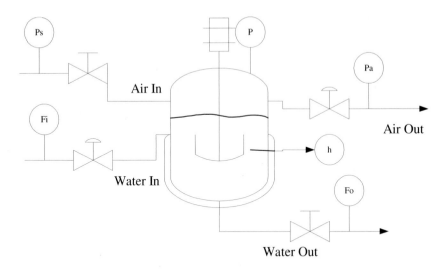

Figure 2.13 A pressurized tank with liquid flow.

Assume that the liquid flow out of the tank can be modeled as $F_o = k_o P$ cc/sec where $h =$ liquid level, $P =$ pressure in tank. The airflow out can be modeled using the equation $F_a = k_a (P - P_a)$ $gmoles/\text{sec}$, where P_a is outside pressure. Inlet flow is modeled using the equation $F_{ai} = k_{ai} (P_s - P)$ $gmole/\text{sec}$, where $P_s =$ air supply pressure.

    a.   Develop a model to describe the tank pressure and level variations. Cross-sectional area of the tank is constant. Hint: Air may be assumed to be an ideal gas: $PV = nRT$ with $T=25$ C constant.

    b.   Given that area of cross section of tank $= 100$ cm^2, height of tank $= 40$ cm,

$$\overline{F}_o = \overline{F}_i = 30 \ cc/\text{sec}, \ \overline{h} = 20 \ \text{cm}, \ \overline{F}_a = \overline{F}_{ai} = .001 \ gmole/\text{sec},$$

$$\overline{P} = 200 \ \text{KPa}, \ P_a = 100 \ \text{KPa}, \ \overline{P}_s = 300 \ \text{KPa}, \ \text{density} = 1 \ \text{gm/cc for water},$$

        develop a SIMULINK model of the system.

2.7   Consider the heated mixing tank model developed in Example 2.3. Set up a SIMULINK model of this tank using the following parameters:

    State: level $h(t)$, range: 0-100 cm,; $\overline{h} = 50$ cm

    State: temp $T(t)$, range: 0-100 °C; $\overline{T} = 50°$ C

    Disturbance: $T_i(t)$; range: 10-50 °C; $\overline{T}_i = 25°$ C

    Disturbance: $F_i(t)$; range: 0-30 cm$^3/\text{sec}$; $\overline{F}_i = 15$ cm$^3/\text{sec}$

    Input: $Q(t)$; range: 0-50 watts; $\overline{Q} = 25$ watts

Output flow: $F_0(t) = k\sqrt{h(t)}$, where $k = 0.3 \text{ cm}^3/\text{sec}/\text{cm}$

Area = 300 cm^2.

Use this model to obtain step response of the level $h(t)$ when $\Delta F_i = \pm 10 \text{ cm}^3/\text{sec}$. Fit an FOPDT model to this data and comment on the nonlinearity of the system by comparing the time constant for the step increase in flow with the time constant for the step decrease.

2.8 (Deriving transfer functions) A storage tank is 8 ft deep and open to the atmosphere. The uniform cross-section area is 100 ft^2. The normal flow to the tank is $100 \text{ ft}^3/\text{min}$ and the normal level is 4 ft. The inlet flow is given by $F = k_1\sqrt{p - h}$, where p = supply pressure. k_1 is a constant. Normal supply pressure is 20 ft of liquid. The output flow is proportional to the square root of the height of liquid in the tank, $F_o = k_2\sqrt{h}$.

   a.   Write a mass balance equation for the tank. Use the given steady-state conditions to solve for k_1 and k_2.

   b.   Linearize the resulting differential equation around the normal operating steady-state height and supply pressure.

   c.   Find a transfer function relating the changes in height of liquid in the tank to small changes in supply pressure p.

2.9 (Linearization) Air is leaking from a tire. The flow out can be modeled by the equation

$$q(t) = k\sqrt{P(t) - P_a} \text{ lbmoles/min}.$$

P_a is outside pressure (constant). Assume air is an ideal gas so that $PV = nRT$, where P = pressure in tire, V = Volume of tire assumed constant, n = number of lbmoles of air in tire, and T is the temperature which may be assumed to be constant.

   a.   Write a mass balance on the air in the tire and derive a differential equation that describes how $P(t)$ changes with time

   b.   Linearize the equation (P_d in the deviation of the pressure in the tank from the initial value).

   c.   If the initial pressure in the tire is 30 psig and 1 minute after the tire is punctured the pressure is 28 psig, how long will it take before the pressure drops to 20 psig? Use the linear approximation derived above to estimate your answer and compare with the numerical solution obtained by solving the nonlinear equation.

References

Bequette, W. 1998. *Process Dynamics, Analysis and Simulation*. Prentice Hall, NJ.

Chen, C. T. 1999. *Linear System Theory and Design*. 3rd Ed., Oxford University Press, NY.

Corripio, A. B., and C. A. Smith. 1997. *Principles and Practice of Automatic Process Control*, 2nd Ed., John Wiley & Sons, NY.

Franks, R. G. E. 1967. *Mathematical Modeling in Chemical Engineering*. John Wiley & Sons, NY.

Luyben, W. L. 1989. *Process Modeling, Simulation, and Control for Chemical Engineers*, 2nd Ed., McGraw-Hill, NJ.

Marlin, T. E. 1999. *Process Control: Designing Processes and Control Systems for Dynamic Performance*. McGraw-Hill, NJ.

Seborg, D. E., T. F. Edgar, and D. A. Mellichamp. 1989. *Process Dynamics and Control*. John Wiley & Sons, NY.

C H A P T E R **3**

One-Degree of Freedom Internal Model Control

Objectives of the Chapter

- Present the one-degree of freedom (1DF) Internal Model Control (IMC) structure and describe its properties.

- Provide design methods for 1DF IMC systems with perfect models that give the best possible performance consistent with noisy measurements for inherently stable processes.

Prerequisite Reading

Chapter 2, "Continuous-Time Models"
Appendix A, "Review of Basic Concepts"
Appendix B, "Frequency Response Analysis"

3.1   INTRODUCTION

This chapter introduces methods for designing feedback controllers to force the output of an inherently stable process to (1) respond in a desired manner to a setpoint change, and (2) counter the effects of disturbances that enter directly into the process output. To enable us to carry out a quantitative controller design, we assume that we have a mathematical model of the process that allows us to predict how the process output (sometimes also called the process variable) responds to the control effort (e.g., how a process flow responds to the opening or closing of a valve) and to disturbances. Further, to keep these initial discussions as simple as possible, we assume that (1) the mathematical model is a perfect representation of the process, (2) the process is linear, and (3) there are no constraints on the control effort so it can take on any value between plus and minus infinity. Chapter 7 extends the results of this chapter to the case of imperfect models and uncertain processes. Chapter 5 shows how the controller designs obtained in this chapter and Chapter 4 can be implemented so as to accommodate control effort saturation. Chapter 6 shows how the IMC designs of this and the next chapter can be converted into PID controllers. While the treatment of nonlinear process models is beyond the scope of this text, many of the controller design concepts for linear models carry over to fairly broad classes of nonlinear process models (see, for example, Kravaris, 1987).

The IMC structure (Garcia & Morari, 1982) given in Figure 3.1 is central to our discussions on the design of controllers. Its conceptual usefulness lies in the fact that it allows us to concentrate on the controller design without having to be concerned with control system stability *provided that the process model* $\tilde{p}(s)$ *is a perfect representation of a stable process p(s).*

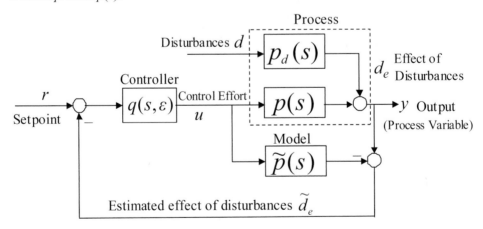

Figure 3.1 The IMC system.

As we shall see in Chapter 7, even if the model is imperfect, it is still possible to design the controller $q(s, \varepsilon)$ without concern for system stability, and then select the tuning parameter ε in $q(s, \varepsilon)$ to assure stability, provided only that the process $p(s)$ is inherently stable. In addition, if the controller gain is the inverse of the *model gain*, then the process output $y(s)$ will eventually reach and maintain the setpoint r (in the absence of new disturbances), provided only that the process and model gains have the same sign and that the controller is tuned so as to assure stability. For these reasons, the structure of Figure 3.1 is well suited for exploring ideal control system performance. Further, since the structure of Figure 3.1 can be rearranged into other structures, it can be used to obtain the controller for these other structures. We shall show how this is done for classical PID feedback control in Chapter 6.

While investigators have made use of concepts similar to those of IMC to design optimal feedback controllers since the late 1950s (Newton, Gould, & Kaiser, 1957), it was not until 1974 that the German investigator Frank first proposed utilizing the structure shown in Figure 3.1 to control processes. In 1979, Brosilow recognized that the IMC structure was at the core both his inferential control system (Brosilow, 1979; Joseph & Brosilow, 1978a, 1978b; Tong & Brosilow, 1978) and the Smith Predictor (Smith, 1957), and proposed methods for designing the controller $q(s, \varepsilon)$. Morari and his coworkers, in a series of papers (Garcia & Morari, 1982, 1985a, 1985b; Morari, 1983, 1985; Morari, Skogestad, & Rivera 1984; Morari & Zafiriou, 1989), greatly expanded on the design methods for $q(s, \varepsilon)$ and placed the methodology in a sound theoretical framework. During this period, it also became clear that the IMC structure underlies the industrially important model predictive controllers known as IDCOM (Richalet, 1978), DMC (Cutler & Ramaker, 1979), and QDMC (Garcia & Morshedi, 1986; Prett & Garcia, 1988).

3.2 PROPERTIES OF IMC

3.2.1 Transfer Functions

An easy way to develop the transfer functions between the inputs d and r and the process output y is to first redraw Figure 3.1 as a simple feedback system, as shown in Figure 3.2, and then apply the following rule:

The transfer function between any input and the output of a single-loop feedback system is the forward path transmission from the input to the output divided by one plus the loop transmission for negative feedback.

For the feedback controller $c(s)$ of Figure 3.2, the rule gives

$$c(s) \equiv \frac{u(s)}{e(s)} = \frac{q(s)}{1 - q(s)\, \tilde{p}(s)}. \tag{3.1}$$

The negative term in the denominator of Eq. (3.1) arises from the positive feedback around $q(s)$.

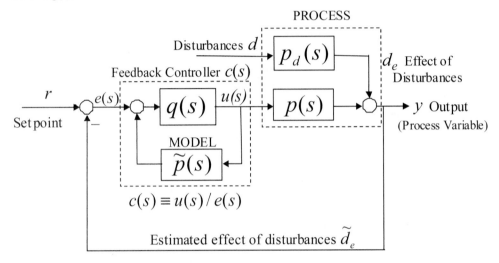

Figure 3.2 Alternate IMC Configuration.

The input-output relationships for Figure 3.2 are given by

$$\frac{y(s)}{r(s)} = \frac{p(s)c(s)}{1 + p(s)c(s)} \tag{3.2}$$

$$\frac{y(s)}{d(s)} = \frac{p_d(s)}{1 + p(s)c(s)} \tag{3.3}$$

$$\frac{u(s)}{r(s)} = \frac{c(s)}{1 + p(s)c(s)} = \left(\frac{y(s)}{r(s)} \right) p^{-1}(s) \tag{3.4}$$

$$\frac{u(s)}{d(s)} = \frac{-p_d(s)c(s)}{1 + p(s)c(s)} = \left(- \frac{y(s)}{d(s)} \right) c(s) \tag{3.5}$$

Substituting Eq. (3.1) into Equations (3.2) and (3.3) and clearing fractions gives

$$y(s) = \frac{p(s)q(s)r(s)}{1 + (p(s) - \tilde{p}(s))q(s)} \qquad (3.6a)$$

$$y(s) = \frac{(1 - \tilde{p}(s)q(s))p_d(s)d(s)}{1 + (p(s) - \tilde{p}(s))q(s)} \qquad (3.6b)$$

3.2.2 No Offset Property of IMC

The steady-state gain of any stable transfer function can be obtained by replacing the Laplace variable s with zero (see Chapter 2). If Equations (3.6a) and (3.6b) are stable, and if we choose the steady-state gain of the controller $q(0)$ to be the inverse of the model gain $(\tilde{p}(0)q(0) = 1)$, then the gain of the denominator of Equations (3.6a) and (3.6b) is $p(0)q(0)$. Thus the gain between the setpoint $r(s)$ and $y(s)$ is one; the gain between the disturbance $d(s)$ and $y(s)$ is zero, and there is no steady-state deviation of the process output from the setpoint.

An ideal control system would force the process output to track its setpoint instantaneously and perfectly suppress all disturbances so that they do not affect the output. That is, the ideal controller would accomplish

$$y(s) = r(s) \qquad (3.7a)$$

and

$$y(s)/d(s) = 0. \qquad (3.7b)$$

From Equations (3.6a) and (3.6b), the above requires that

$$p(s)q(s) = 1 \quad \text{and} \quad \tilde{p}(s) = p(s). \qquad (3.8)$$

Thus, for perfect control, we need a perfect model, and from Eq. (3.8), the controller must perfectly invert that perfect model. Unfortunately, one never has a perfect model, and if the model has any dynamics at all (i.e., it is not just a gain), no controller can perfectly invert the process model. A controller can, however, come very close to inverting the process model. Just how close it can come is the subject of our next section, where we discuss controller design assuming perfect models. The design methodology always incorporates a tuning parameter that can slow down the control system response sufficiently to accommodate most modeling errors for inherently stable processes. Calculation of the tuning parameter for various descriptions of anticipated modeling errors is treated in Chapter 7.

The next section is limited to controller design of stable processes. IMC controller design for unstable processes is discussed in Chapter 4.

3.3   IMC Designs for No Disturbance Lag

This section discusses the case where the disturbance lag $p_d(s)$ (c.f. Figure 3.2) is unity. For design methods when the disturbance lag is not one, see Chapter 4. The discussion begins by considering a process modeled by a first-order lag plus deadtime. Such models approximate the behavior of many chemical and petroleum processes. Since in this section we always assume that the model is perfect, we shall suppress the tilde notation.

Example 3.1 IMC Controller for an FOPDT Process

$$p(s) = \frac{Ke^{-Ts}}{\tau s + 1}; \quad p_d(s) = 1 \tag{3.9}$$

The inverse of the process $p(s)$ given in Eq. (3.9) is

$$p^{-1}(s) = \frac{\tau s + 1}{K} e^{Ts}. \tag{3.10}$$

The only term in Eq. (3.10) that can be realized (i.e., be physically constructed) directly in a controller is the inverse of the process gain K. The term e^{Ts} represents an unrealizable prediction of future outputs,[1] while the term $(\tau s + 1)$ requires an unrealizable pure (i.e., unfiltered) differentiation of the process output. A real-time unfiltered differentiation of a continuous time signal is not realizable,[2] and even if it were, it would not be implemented because of unacceptable amplification of the noise on the measured process output y. Thus the best[3] that can be done is to implement the controller $q(s)$ for the process given by Eq. (3.9) as

$$q(s) = \frac{\tau s + 1}{K(\varepsilon s + 1)}, \tag{3.11}$$

where ε = a filter time constant or tuning parameter chosen to avoid excessive noise amplification and to accommodate modeling errors.

For modest or small modeling errors, the filter time constant ε will be less than the process time constant τ and the controller $q(s)$, given by Eq. (3.11), will be a lead network. That is, the frequency response of the controller will show that its magnitude increases from

---

[1] The inverse transform of $f(s)e^{Ts}$ is $f(t+T)$. The IMC controller of Figure 3.1 operates on the setpoint minus the disturbance estimate. Therefore if the controller has a term of the form e^{Ts}, implementation of the controller would require a prediction of the disturbance T units of time in the future, and an exact prediction is impossible unless we have a priori knowledge of the disturbance.

[2] Cannot be implemented by any physical device, such as a digital or analog computer.

[3] Best in the sense that as ε approaches zero, the output response to a step setpoint change approaches a step output change after one dead time. It is not possible to improve on such a response without a priori knowledge.

$1/|K|$ at low frequencies to $\tau / \varepsilon |K|$ at high frequencies, while the phase angle goes from zero at low frequencies to $\tan^{-1} \tau/\varepsilon$ at high frequencies ($\tan^{-1} \tau / \varepsilon$ approaches $90°$ as ε approaches zero).

The perfect model loop response transfer function is given by $pq(s)$ (see Eq. (3.8) with $p(s) = \tilde{p}(s)$). Using Equations (3.9) and (3.11) for $p(s)$ and $q(s)$ gives

$$y(s) = \frac{e^{-Ts}}{(\varepsilon s + 1)} \, r(s) + \left(1 - \frac{e^{-Ts}}{\varepsilon s + 1}\right) d(s). \qquad (3.12)$$

The time response of y, given by Eq. (3.12), to a unit step change in setpoint r is shown in Figure 3.3 for $T = 1$ and $\varepsilon = .05$ and 1.0.

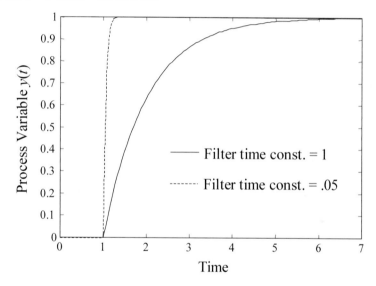

Figure 3.3   Perfect model IMC step response for first-order lag plus dead time.

The choice of the filter parameter ε in Eq. (3.12) depends on the allowable noise amplification by the controller and on modeling errors. Methods for choosing the filter time constant to accommodate modeling errors are discussed in Chapter 7. To avoid excessive noise amplification, we recommend that the filter parameter ε be chosen so that the high frequency gain of the controller is not more than 20 times its low frequency gain. For controllers that are ratios of polynomials, this criterion can be expressed as

$$|q(\infty)/q(0)| \le 20. \qquad (3.13)$$

The criterion given by Eq. (3.13) arises from the standard industrial practice of limiting the high frequency gain of a PID controller to no more than 20 times the low frequency controller gain, which is usually referred to simply as the controller gain (see Appendix A).

Factors of 5 and 10 are also frequently encountered in practice. While Eq. (3.13) limits only the infinite frequency gain of the controller, the complete design methodology presented later in this section limits the gain to less than 20 over all frequencies.

♦

The controller design method for the first order lag and dead time process generalizes easily to processes of the form

$$p(s) = \frac{N(s)}{D(s)} e^{-Ts}, \tag{3.14}$$

where $N(s)$ and $D(s)$ are polynomials in s.

The design of IMC controllers for Eq. (3.14) depends mainly on the characteristics of $N(s)$, as discussed in the following sections.

3.4 DESIGN FOR PROCESSES WITH NO ZEROS NEAR THE IMAGINARY AXIS OR IN THE RIGHT HALF OF THE S-PLANE

When $N(s)$ has no zeros in the right half of the s-plane or near the imaginary axis, the inverse of the model is stable and not overly oscillatory. In this case, the IMC controller for Eq. (3.14) can be chosen as

$$q(s) = \frac{D(s)}{N(s)\,(\varepsilon s + 1)^r}, \tag{3.15}$$

where r = the relative order[4] of $N(s)/D(s)$.

From Eq. (3.13), the filter time constant ε in Eq. (3.15) must satisfy

$$\varepsilon \geq \left(\lim_{s \to \infty} \frac{D(s)N(0)}{20 s^r N(s)D(0)} \right)^{1/r}. \tag{3.16}$$

As before, the limit given by Eq. (3.16) ensures that the high frequency gain of the controller is not more than 20 times its low frequency gain. The actual values of the filter time constant will more often be dictated by modeling errors and will be computed as shown in Chapter 7.

The form of the filter (i.e., $1/(\varepsilon s + 1)^r$) is somewhat arbitrary. It was chosen because it is the simplest form with a single adjustable parameter, ε, that provides an

---

[4] For transfer functions that are ratios of polynomials, relative order is defined as the order of the denominator minus the order of the numerator.

overdamped response[5] and makes $q(s)$ realizable. Such a filter has the great merit of simplicity at the possible price of being suboptimal. There is also no incentive to use a filter order, r, greater than the minimum required to make the IMC controller realizable, because when there are modeling errors, higher order filters lead to slower responses, as shown in Chapter 7. Choosing a filter whose order is the same as the relative order of the model leads to a controller, $q(s)$, whose relative order is zero.

3.5   DESIGN FOR PROCESSES WITH ZEROS NEAR THE IMAGINARY AXIS

If the term $N(s)$ in Eq. (3.14) contains complex roots with low damping ratios,[6] then such terms can cause an IMC controller formed like that given in Equations (3.15) and (3.16) to amplify noise excessively at intermediate frequencies. There are two relatively simple options to reduce such excessive noise amplification. The first option is to increase the filter time constant sufficiently to reduce the peak to an acceptable level. This option generally requires large filter time constants, excessively increasing the settling time of the control system, and is therefore *not recommended.* The second, and recommended, option is to not invert low damping ratio zeros. Rather, form a controller similar to that given by Eq. (3.15), but with the damping ratio in the original polynomials in $N(s)$ modified so as to be sufficiently large to avoid excessive noise amplification at intermediate frequencies. An example should help clarify the suggested design procedure.

Example 3.2 A Process with Low Damping Ratio Zeros

Consider the process given by

$$p(s) = \frac{s^2 + 0.001\,s + 1}{(s+1)^4}.$$ (3.17a)

By Equations (3.15) and (3.16), the controller and filter time constant would be

$$q(s) = \frac{(s+1)^4}{(s^2 + .001\,s + 1)\,(\varepsilon s + 1)^2}$$ (3.17b)

---

[5]Any transfer function of the form $1/\prod_i (\tau_i s + 1)$ is said to yield an overdamped response because the frequency response of such transfer functions is everywhere less than one. Therefore, a plot of the time response of the output of the transfer function is smoother than its input.

[6]The damping ratio ζ is defined for second-order polynomials of the form $D(s) = \tau^2 s^2 + 2\zeta\tau s + 1$. If $0 \leq \zeta < 1$, then the roots of this polynomial are complex conjugates. Damping ratios less than .4 (but greater than zero) are generally considered to be low because transfer functions of form $1/D(s)$ yield oscillatory responses to step inputs. A damping ratio of zero is characteristic of a system whose response to a step input oscillates continuously (i.e., without damping).

$$\varepsilon \geq 1/\sqrt{20} \cong .22. \tag{3.17c}$$

The frequency response of $q(s)$ given by Equations (3.17b) and (3.17c) is in shown Figure 3.4. The magnitude of the peak in Figure 3.4 is actually 3810. A filter time constant of 20 would be required to reduce the noise amplification at frequencies around 1.0 to a factor of 20. The settling time of the control system with such a controller exceeds 100 units. As we shall see, a controller with the same form as that given by Eq. (3.17b), but with a different damping ratio in the denominator, gives a much faster response.

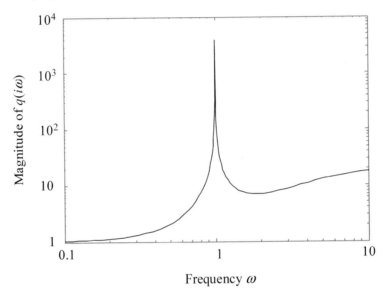

Figure 3.4 Frequency response of $q(s) = (s + 1)^4/((s^2 + 2\zeta s + 1)(.22s + 1)^2)$.

A better controller than that of Eq. (3.17b) is

$$q(s) = \frac{(s + 1)^4}{(s^2 + 2\zeta s + 1)(.22s + 1)^2}. \tag{3.18a}$$

To reduce the controller frequency response peak so that its magnitude is only 20 (with $\varepsilon = .22$, as before) requires a damping ratio, ζ, of 0.1. The resulting loop response $pq(s)$ is

$$p(s)\,q(s) = \frac{s^2 + .001\,s + 1}{(s^2 + 2\zeta s + 1)(.22s + 1)^2}. \tag{3.18b}$$

Figure 3.5 shows the loop response given by Eq. (3.18b) for a damping ratio of 0.1 as well as a damping ratio of 0.5. As is apparent from the figure, a controller damping ratio of .5 gives a less oscillatory response than that given by a controller damping ratio of .1. On the other hand, the response for $\zeta = .5$ is more sluggish than that for $\zeta = .1$. In such cases the engineer has to use process knowledge to select the most appropriate controller

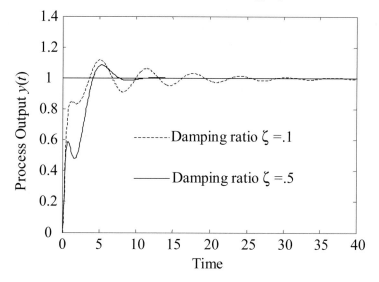

Figure 3.5 Perfect model loop response for $p(s) = (s^2 + .001s + 1)/(s + 1)^4$ with $q(s) = (s + 1)^4/((s^2 + 2\zeta s + 1)(.22s + 1)^2)$.

◆

3.6 DESIGN FOR PROCESSES WITH RIGHT HALF PLANE ZEROS

When $N(s)$ in Eq. (3.14) has factors of the form $(-\tau s + 1)$ or $(\tau^2 s^2 - 2\tau \zeta s + 1)$, with τ and ζ greater than zero, its inverse is unstable. In this case the IMC controller cannot be formed as given by Eq. (3.15). The integral square error (ISE)[7] optimal choice of controller for such cases is to invert that portion of the model which has zeros in the left half plane and add poles at the mirror image of the right half plane zeros (Morari & Zafiriou, 1989). That is, we assume that the model given by Eq. (3.14) can be rewritten as

$$p(s) = \frac{N_-(s)\, N_+(s)}{D(s)}\, e^{-Ts}, \qquad (3.19a)$$

---

[7] $ISE \equiv \int_0^\infty (y(t) - r(t))^2\, dt.$

where $N_-(s)$ contains only left half plane zeros, none of which have small damping ratios. $N_+(s)$ contains only right half plane zeros, and can be written as

$$N_+(s) = \prod_{i,j}(-\tau_i\, s + 1)\,(\tau_j^2\, s^2 - 2\tau_j\zeta_j\, s + 1) \tag{3.19b}$$

$$\tau_i, \tau_j > 0; \quad 0 < \zeta < 1.$$

Notice that the gain of $N_+(s)$ is one.

Before designing the IMC controller, we strongly recommend that the model be put in time constant form (i.e., the numerator and denominator are factored into products of the form $(\pm \tau s + 1)$, $(\tau^2 s^2 \pm 2\tau\zeta s + 1)$ so that it is easy to form $N_+(s)$ and $N_-(s)$. The MATLAB functions *tcf* and *tfn* provided with IMCTUNE were developed specifically to put transfer functions into time constant form, and to facilitate their manipulation in this form. There are also other software programs that can be used to accomplish the desired factorization as described in Section 3.9.

The ISE optimal IMC controller for Eq. (3.19a) is

$$q(s) = \frac{D(s)}{N_-(s)\, N_+(-s)\,(\varepsilon s + 1)^r}, \tag{3.20}$$

where the zeros of $N_+(-s)$ are all in the left half plane and are the mirror images of the zeros of $N_+(s)$. r = relative order of $N(s)/D(s)$ as before.

The choice of controller given by Eq. (3.20) results in a loop response given by

$$y(s) = pq(s) = \prod_{i,j}\left(\frac{-\tau_i s + 1}{\tau_i s + 1}\right)\left(\frac{\tau_j^2\, s^2 - \tau_j\zeta_j s + 1}{\tau_j^2\, s^2 + \tau_j\zeta_j s + 1}\right)\frac{e^{-Ts}}{(\varepsilon s + 1)^r} \tag{3.21}$$

$$\tau_i, \tau_j > 0; \quad 0 < \zeta_j < 1.$$

The loop response given in Eq. (3.21) is optimal in an ISE sense for a filter time constant ε of zero, and is suboptimal for finite ε. Also, when ε is zero, the loop transfer function given by Eq. (3.21) is called all-pass, since the magnitude of the frequency response is one over all frequencies.

Example 3.3 One Right Half Plane Zero

The process model is

$$p(s) = \frac{(s-1)}{27\,(s+1/3)^3}. \tag{3.22a}$$

Putting Eq. (3.22a) in time constant form yields

$$p(s) = \frac{-1(-s+1)}{(3s+1)^3}.$$ (3.22b)

The IMC controller is

$$q(s) = \frac{-1(3s+1)^3}{(s+1)(\varepsilon s+1)^2}.$$ (3.22c)

The resulting loop response is

$$pq(s) = \frac{(-s+1)}{(s+1)(\varepsilon s+1)^2}.$$ (3.22d)

Figure 3.6 compares the step response of the ISE optimal loop transmission given by Eq. (3.21) with $\varepsilon = 0$ to step responses of suboptimal responses obtained by increasing and decreasing the controller time constant. Notice that the faster response obtained with $pq(s) = (-s+1)/(.5s+1)$ comes at the expense of a more negative initial response. Thus, for this simple example, the ISE optimal response is also qualitatively the best compromise between a more sluggish response and a faster response with a more negative initial response.

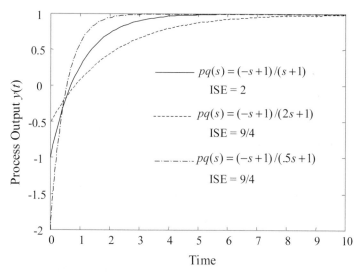

Figure 3.6 Response of processes with one right half plane zero.

The transfer function given by the controller of Eq. (3.22c) does not result in an optimal response to a step setpoint change unless ε is zero. We could get closer to an optimal transfer function by selecting the IMC controller as

$$q(s) = \frac{(3s+1)^3}{((1-2\varepsilon)s+1)(\varepsilon s+1)^2} \quad \text{for } \varepsilon \le 0.5. \qquad (3.22\text{e})$$

The controller given by Eq. (3.22e) was obtained by forcing the coefficient of the linear term of its expanded denominator to be one, which is the same as the linear term in the denominator of Eq. (3.22c) when ε is zero. The loop response then becomes that given by Eq. (3.22f).

$$p(s)q(s) = \frac{(-s+1)}{((1-2\varepsilon)s+1)(\varepsilon s+1)^2} = \frac{(-s+1)}{((1-2\varepsilon)\varepsilon^2 s^3 + (2-3\varepsilon)\varepsilon s^2 + s + 1)} . \qquad (3.22\text{f})$$

The cubic and quadratic terms in s in the denominator of Eq. (3.22f) are small relative to its linear term so that Eq. (3.22f) approaches the optimal transfer function given by Eq. (3.22d) with ε equal to zero. Notice, however, that Eq. (3.22e) is valid only for $\varepsilon \le .5$. For larger values of the filter time constant, Eq. (3.22e) is not stable.

The approach used to obtain the controller given by Eq. (3.22e) can be used to develop nearly optimal controllers for arbitrary nonminimum[8] phase processes. However, such controllers will generally be useful only for small filter time constants.

◆

Example 3.4 Two Right Half Plane Zeros

When the initial process response to a step is in a direction opposite to that of the final steady state, as in the previous example, the process is said to exhibit an inverse response. Processes with an odd number of right half plane zeros exhibit inverse responses. Processes with an even number of right half plane zeros do not have inverse responses, since the initial value at time zero plus is always in the direction of the steady-state[9] as shown in Figure 3.7 for the loop response given by

$$pq(s) = (-s+1)^2/(\tau s + 1)^2 \qquad (3.23)$$

---

[8] A minimum phase transfer function has only zeros in the left half of the s-plane (i.e., the zeros are all negative). A non-minimum phase transfer function has one or more zeros in the right half of the s-plane. The terminology arises from the fact that changing the time constants of the numerator from positive to negative always results in increasing the phase lag of the transfer function at all frequencies. For example, the phase lag of $(s + 1)/(2s +1)$ is always less than the phase lag of $(-s + 1)(2s + 1)$. A deadtime is often called a nonminimum phase element because it cannot be inverted and is in that way similar to a nonminimum phase transfer function, which also does not have a stable inverse.

[9] For a process with an odd number of right half plane zeros, the sign of the coefficient of the highest order s term is opposite to that of the zeroth order coefficient. For a process with an even number of right half plane zeros, the sign of the coefficient of the highest order s term is the same as that of the coefficient of the zeroth order coefficient. Therefore, by the initial value theorem, the sign of the first non-zero derivative is opposite to, or the same as, the sign of the steady-state gain in the case of an odd or even number of right half plane zeros, respectively.

with $\tau = .5$, 1, and 2.

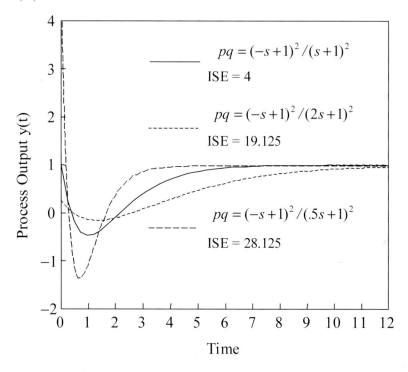

Figure 3.7 Response of processes with two right half plane zeros.

Notice that the ISE optimal response in Figure 3.7 is again that which also gives the qualitatively best compromise between a sluggish response and a response with too much initial overshoot.

♦

3.7 PROBLEMS WITH MATHEMATICALLY OPTIMAL CONTROLLERS

In both Figure 3.6 and Figure 3.7 the qualitatively and quantitatively optimal response is that given by the all-pass loop response. However, it can happen that the all-pass response, while ISE optimal, is not qualitatively the best response and may not even be an acceptable response, as shown by Example 3.5 following.

Example 3.5 A Process with an Infinite Number of RHP Zeros

The transfer function given by Eq. (3.24) below is one that might arise between a process variable and a control effort in a multivariable system, where one or more of the other process variables are under closed-loop control.

$$p(s) = \left(1 + 3e^{-s}/(s+1)\right)/(s+1) \tag{3.24}$$

$p(s)$ has two complex zeros in the right half plane at $0.214 \pm 2.10i$. These zeros were calculated by replacing the dead time in Eq. (3.24) by a five over five Padé[10] approximation and then factoring the resulting sixth order numerator of $p(s)$. The factored form of the approximated transfer function $\hat{p}(s)$ is

$$\hat{p}(s) = \frac{4(.225\,s^2 - .0964\,s + 1)\,(.0156\,s^2 + .0348\,s + 1)\,(.00235\,s^2 + .0616\,s + 1)}{(s+1)^2\,(.137\,s + 1)\,(.0175\,s^2 + 235\,s + 1)\,(.0138\,s^2 + .128\,s + 1)}. \tag{3.25}$$

The ISE optimal controller for (3.25) is

$$q(s) = \frac{.25(s+1)^2\,(.137s+1)\,(.0175s^2 + .235s + 1)\,(.0138s^2 + .128s + 1)}{(.225s^2 + .0964s + 1)\,(.0156s^2 + .0348s + 1)\,(.00235s^2 + .0616s + 1)\,(\varepsilon s + 1)}. \tag{3.26}$$

To achieve $|q(\infty)/q(0)| \le 20$ requires $\varepsilon \ge 0.2$ The resulting loop response is

$$\hat{p}(s)q(s) = \frac{(.225s^2 - .0964s + 1)}{(.225s^2 + .0964s + 1)\,(\varepsilon s + 1)}. \tag{3.27}$$

Figure 3.8a shows the response of Eq. (3.27) to a step with $\varepsilon = 0$ and $\varepsilon = 0.2$. The oscillatory response in Figure 3.8a is due to the low damping ratio in the denominator of Eq. (3.27), which is only 0.101. Increasing the damping ratio to 0.5 in the first term of the denominator of Eq. (3.26) by increasing the coefficient of 0.0964 to 0.48 gives a qualitatively better response, that is, one with less overshoot and no oscillations. However, this improved response has a *higher* ISE. The associated controller and loop response are given by

$$q(s) = \frac{.25(s+1)^2\,(.137s+1)\,(.0175s^2 + .235s + 1)\,(.0138s^2 + .128s + 1)}{(.225s^2 + .48s + 1)\,(.0156s^2 + .0348s + 1)\,(.00235s^2 + .0616s + 1)\,(.2s + 1)} \tag{3.28}$$

---

[10] A Padé approximation to the exponential e^{-Ts} is a ratio of polynomials of order m in the numerator, and n in the denominator, whose coefficients are chosen so that the ratio of polynomials approximates the exponential to within terms of order $n + m + 1$ in s. That is, the Maclaurin series expansion in s of the exponential and its Padé approximation agree through terms of order $m + n$. Padé approximations to the exponential e^{-Ts} of degree n over n (i.e., of order accuracy $2n$) for any $n > 0$ are available in the control toolbox of MATLAB and in Program CC.

$$\hat{p}(s)q(s) = \frac{(.225s^2 - .0964s + 1)}{(.225s^2 + .48s + 1)(.2\ s + 1)}.\tag{3.29}$$

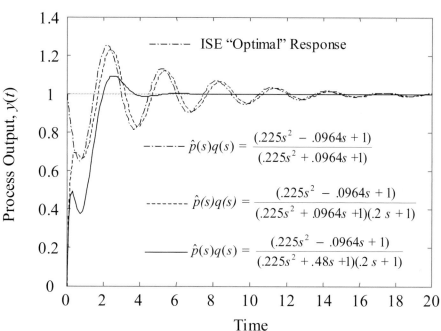

Figure 3.8a Optimal versus suboptimal responses to a step setpoint change for the process, $p(s) = (1 + 3e^{-s}/(s + 1))/(s + 1)$.

The responses in Figure 3.8a were obtained with a rather complex controller, as given by Eq. (3.26) and its modifications. It is reasonable to ask whether there is a much simpler suboptimal controller that might do nearly as well. Indeed, the control system that uses the simple controller given by

$$q(s) = \frac{.25(s + 1)}{(.05s + 1)}\tag{3.30}$$

produces a very good step setpoint response, as shown in Figure 3.8b. The above controller arises from the recognition that the numerical value of the process steady-state gain is four, that only the term $1/(s + 1)$ in $p(s)$ is easily invertible, and that the step response of $p(s)$ is monotonic. Figure 3.8b compares the loop responses using the controllers given by Eq. (3.26) with $\varepsilon = 0$, with Equations (3.28) and (3.30).

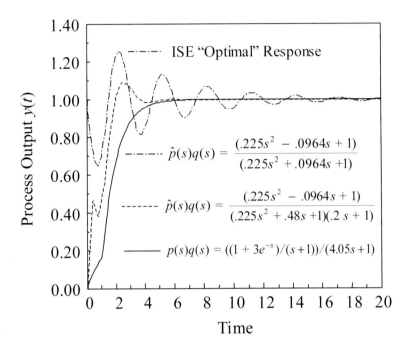

Figure 3.8b Optimal versus suboptimal responses to a step setpoint change for the process,
$p(s) = (1 + 3e^{-s}/(s + 1))/(s + 1)$

The controller given by Eq. (3.30) is not only simpler than that given by Eq. (3.26), but also yields a response with no overshoot and achieves the final steady state in about the same time. Many engineers, including the authors, will prefer the response obtained with the simple controller to that obtained with the more complex controller. Thus, this example demonstrates that optimality, like beauty, is in the eye of the beholder. As the foregoing examples attempt to demonstrate, controller designs that minimize the ISE can be quite useful, but they should not be applied by rote. The engineer should always look at the ideal loop response given by the loop transmission $pq(s)$ to judge whether the optimal controller gives desirable responses. If not, then a trial and error procedure is usually necessary to obtain a controller that does give an acceptable loop behavior. A systematic way to obtain such a design is to start with a controller $q(s)$ that is just the inverse of the model gain and then attempt to improve on that controller by including in it increasing portions of the model inverse.

Lest the reader get the misimpression that qualitative improvements in performance over that obtained with the ISE optimal controller are only possible when the optimal response is oscillatory, we offer the following example.

Example 3.6 A Process with an Infinite Number of RHP Zeros

$$p(s) = 1 - Ke^{-Ts} \; ; \; |K| > 1 \tag{3.31}$$

The above process has an infinite set of zeros in the right half plane at $(\ell nK - 2n\pi i)/T$ if $K > 1$, and at $(\ell n(-K) - (2n+1)\pi i)/T$ if $K < -1$. We can form a controller $q(s)$ with poles at the mirror image of the zeros of Eq. (3.31) as

$$q(s) = \frac{1}{(e^{-Ts} - K)}, \tag{3.32}$$

so that

$$p(s)q(s) = \frac{1 - Ke^{-Ts}}{e^{-Ts} - K}. \tag{3.33}$$

The response of Eq. (3.33) to a unit step is

$$y(t) = y(nT) \; ; \quad nT \leq t < (n+1)T$$
$$y(nT) = 1 - (1 + 1/K)(1/K)^n. \tag{3.34}$$

For the case $K = 2$, the first several values of $y(nT)$ are $y(0) = -1/2$, $y(T) = 1/4$, $y(2T) = 5/8$, and $y(3T) = 13/16$, as shown in Figure 3.9 for $T = 1$. Clearly, the optimal response approaches unity in a staircase fashion.

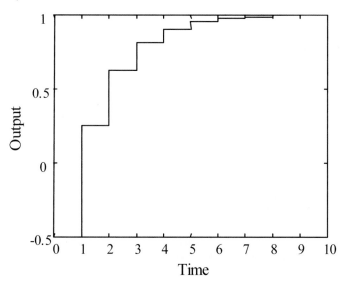

Figure 3.9 Loop response of Eq. (3.33) to a unit step input.

The ISE for the output $y(nT)$ of Eq. (3.33) is given by

$$ISE = T \sum_{n=0}^{\infty} (1 - y(nT))^2 .$$

Substituting Eq. (3.34) into the above gives

$$ISE = T \sum_{n=0}^{\infty} (1 + 1/K)^2 (1/K^2)^n .$$

Recognizing that $\sum_{n=0}^{\infty} (1/K^2)^n = \dfrac{1}{1 - 1/K^2}$ gives

$$ISE = T (1 + 1/K)^2 / (1 - 1/K^2). \tag{3.35}$$

For $K = 2$, the ISE from Eq. (3.35) is 3T. Now consider the controller that is simply the reciprocal of the model gain; that is

$$q(s) = \frac{1}{1 - K}. \tag{3.36}$$

The loop response for this controller is

$$y(t) = y(nT) \; ; \; nT \leq t \leq (n+1)T \tag{3.37}$$

$$y(0) = -1$$

$$y(nT) = 1, \; n = 1, 2, \ldots \infty.$$

The ISE for the above is $4T$ (vs. $3T$ before), but the output reaches and maintains its desired value after one dead time.

◆

3.8   MODIFYING THE PROCESS TO IMPROVE CONTROL SYSTEM PERFORMANCE

Another interesting feature of the process in Example 3.6 is that the response can be substantially improved by modifying the process. Let us rewrite Eq. (3.31) as

$$p(s) = e^{-T_0 s} - Ke^{-Ts} \; ; \; |K| > 1 \tag{3.38a}$$

If T_0 in Eq. (3.38a) is taken as zero, then Eq. (3.38a) is the same as Eq. (3.31). If, however, T_0 is taken as T, then Eq. (3.38a) becomes

$$p*(s) = e^{-Ts} - Ke^{-Ts} = (1 - K)e^{-Ts}, \qquad (3.38b)$$

where p* = the modified process.

Now, our simple controller, given by Eq. (3.36), gives

$$y(t) = 0; \quad 0 \le t < T$$
$$y(t) = 1; \quad T \le t. \qquad (3.38c)$$

The above has an ISE of T, which is substantially better than that given by Eq. (3.35) using the optimal controller given by Eq. (3.32). Recall that when $K = 2$, the optimal ISE was $3T$. This shows that small design changes to the process can lead to significant improvements in control system performance. Notice that the process given by Eq. (3.31) responds instantaneously to a step change in control effort. However, the process given by Eq. (3.38a), with $T_0 = T$, responds to a step change in control effort only after a dead time of T units. Thus, we have improved control system performance by, in effect, increasing the deadtime of the process in going from Eq. (3.31) to Eq. (3.38a). For more information on how to improve performance by process modifications that *increase* certain dead times, see Psarris & Floudas (1990).

3.9 SOFTWARE TOOLS FOR IMC DESIGN

An important aid in the design of IMC controllers is the ability to display the time response of the loop transmission pq as in Figures 3.5 to 3.8. This can be done quite conveniently using Program CC, MATLAB with SIMULINK or Vissim, or similar types of software. We provide a MATLAB version 5.3.1 (or later) suite of m-files IMCTUNE.m on the web site for this text[11] that allows the user to enter the process model $\widetilde{p}(s)$ and the portion of the model $p_m(s)$, inverted by the IMC controller in a convenient form. The program yields the output time response as one of its options under the Results and Simulations menu. IMCTUNE is also capable of producing PID controllers and of tuning the IMC controller to accommodate process uncertainty as described in Chapters 6 and 7. IMCTUNE requires the Optimization and Control System toolboxes in order to be used with MATLAB and SIMULINK.

To simplify the IMC design task, it is often convenient to convert a transfer function into time constant form wherein the numerator and denominator are factored into products of the form $(\pm \tau s + 1)$, $(\tau^2 s^2 \pm 2\tau\zeta s + 1)$. This can be done in Program CC simply by the command *tcf,g*, where *g* is the transfer function that is desired in time constant form. MATLAB m-files, *tcf.m* and *tfn.m*, provided with the IMCTUNE software, provide the same functionality. Information on how to use *tcf.m* and *tfn.m* can be obtained by typing *help*, followed by the function name (e.g., *help tcf*).

---

[11] http://www.phptr.com/brosilow/

3.10   SUMMARY

This chapter presents IMC design techniques for inherently stable linear processes where the disturbance enters directly into the process output (i.e., the disturbance lag is either unity or has very fast time constants relative to the process time constants). The controller design methods differ depending on the location of the zeros of the numerator of the process transfer function.

Controllers for processes with no right half plane zeros, or zeros near the imaginary axis, can be obtained simply by inverting the entire model except any multiplicative delay. That is, if $\tilde{p}(s) = g(s)e^{-Ts}$ then $q(s) = g^{-1}(s)f(s)$,

where $f(s) = 1/(\varepsilon s + 1)^r$,

r = the relative order of $g(s)$,

ε = an adjustable filter time constant.

We recommend choosing the filter time constant ε to avoid excessive high frequency noise amplification by using the criterion

$$|q(\infty)/q(0)| \leq N$$

where N is between 10 and 20.

If the term $g(s)$ has complex zeros with small damping ratios, then the controller, $q(s)$, might amplify noise more than is desirable at some midrange frequency (i.e., $|q(i\omega_c)/q(0)| > N$ for some ω_c). When this occurs, we recommend either (1) not inverting the zeros that cause the noise amplification, or (2) increasing the damping ratio of the denominator terms in the controller $q(s)$ that arise from inverting zeros with small damping ratios.

If the term $g(s)$ has right half plane zeros, then the ISE optimal IMC controller inverts all of $g(s)$ except the right half plane zeros, and poles are added to the controller $q(s)$ so that the loop transmission $\tilde{p}(s)q(s)$ is an all-pass system cascaded with the filter $f(s)$. That is, the poles of $\tilde{p}(s)q(s)$ are at the mirror image of the right half plane zeros of $\tilde{p}(s)q(s)$. As usual, the filter order is the relative order of the model, $g(s)$.

Finally, if the ISE optimal IMC controller described in the previous paragraph is overly complex (as might occur when the right half plane zeros of the model are due to transcendental terms such as $(p_1(s) + p_2(s)e^{-Ts})$, where $p_1(s)$ and $p_2(s)$ are polynomials in s, or if the loop response is undesirable because of excessive oscillations or overshoot, then we recommend rejecting the ISE optimal IMC controller in favor of a simpler controller. A systematic method of obtaining such a controller is to start with a controller that is just the inverse of the model gain, then attempt to improve the loop response $\tilde{p}(s)q(s)$ by including in the controller increasing portions of the model inverse.

Problems

3.1 Derive all the transfer functions from Eq. (3.1) to Eq. (3.6).

3.2 Use SIMULINK or other simulation software to obtain the step responses for each of the examples in the chapter. Note that only the transfer functions of the loop response need be simulated, since all of the examples are for perfect model responses to a step setpoint change.

3.3 Simulate a step disturbance response for Example 3.1 with $p_d(s) = 1$. How does this response relate to the step setpoint response?

3.4 Find an IMC controller for each of the following process models. The controller must satisfy

$$\max_{\omega} \left| \frac{q(i\omega)}{q(0)} \right| \le 20$$

with equality holding at some frequency (possibly infinity). State the rationale for your choice of the part of the model that the controller inverts. Recall the statement in Section 3.6 that "*Before designing the IMC controller, we strongly recommend that the model be put in time constant form.*"

a. $\dfrac{s^2 + 2s + .25}{s^4 + 6.5s^3 + 15s^2 + 14s + 4}$

b. $\dfrac{32s^2 + .8s + 2}{s^4 + 6.5s^3 + 15s^2 + 14s + 4}$

c. $\dfrac{(16s^2 + .4s + 1)}{16s^4 + 40.5s^3 + 18.25s^2 + 3s + 1}$

d. $\dfrac{(s-1)e^{-s}}{s^2 + s + 1}$

e. $3\dfrac{d^2 y}{dt} + 4\dfrac{dy}{dt} + 3y(t) = u(t) - \dfrac{d}{dt}u(t)$

f. $\dfrac{(s^2 + 2s + .25)e^{-s}}{s^4 + 6.5s^3 + 15s^2 + 14s + 4}$

g. $\dfrac{(s-1)(s^2 - s + 1)}{s^4 + 6.5s^3 + 15s^2 + 14s + 4}$

h. $\dfrac{s^3 - 1.9s^2 + .8s - 2}{(s^2 + .2s + 1)(3s + 1)^2 (s + 1)}$

3.5 Find an IMC controller for each of the following process models. The IMC controllers should satisfy the same noise amplification criterion as that for Problem 3.4. Note, however, that the inverse of the numerator in each of the following process models can be inverted using a simple negative feedback loop that has unity in the forward path and Ke^{-Ts} in the feedback path, with appropriate values of K and T. This is not to say, however, that such inverses will not be very oscillatory, or even unstable. Thus, in developing an IMC controller for each of the following, the student should decide whether or not to invert the numerator exactly; approximate it by replacing the exponential with a Padé approximation, and then invert all or part of the numerator; or perhaps simply invert the numerator gain. To complicate matters further, the filter time constant necessary to satisfy the same noise amplification criterion as that for Problem 3.4 depends on which option is selected, and if a Padé approximation is used, it even depends on whether highest order terms in the Padé polynomials are odd or even. We suggest that before designing a controller for the following problems, the student get a step response of the process model.

The student should note that, as in most engineering situations, there is no single correct answer. In such cases the simplest controller is often the best controller. To compare controllers, we recommend that the student simulate the output response to a step setpoint change.

a. $$\frac{1+0.5e^{-3s}}{(s+2)^2(s+.5)^2}$$

b. $$\frac{1+.95e^{-3s}}{(s+2)^2(s+.5)^2}$$

c. $$\frac{1-.95e^{-3s}}{(s+2)^2(s+.5)^2}$$

d. $$\frac{1+2e^{-s}}{s+1}$$

3.6 Why do the IMC controller filter time constants in the IMC controllers for all the processes of Problem 3.5 change depending on whether an even or odd order Padé approximation for the exponential term is used in forming the controller?

References

Brosilow, C. B. 1979. "The Structure and Design of Smith Predictors from the Viewpoint of Inferential Control." *Joint Automatic Control Conference*, June 17, Denver.

Cutler, C. R., and B. L. Ramaker. 1980. "Dynamic Matrix Control—A Computer Control Algorithm." AIChE National Meeting, Houston, TX; also *Proc. Joint Automatic Control Conf.*, San Francisco.

Garcia, C. E., and A. M. Morshedi. 1986. "Quadratic Programming Solution of Dynamic Matrix Control (QDMC)." *Chemical Engineering Communications* 46, 73–87.

Garcia, C. E., and M. Morari. 1982. "Internal Model Control –1. A Unifying Review and Some New Results." *Ind. Eng. Chem. Process Des. & Dev.* 21, 308–323.

Joseph, B., and C. B. Brosilow. 1978a. "Inferential Control of Processes, Part I, Steady-state Analysis and Design." *AIChE J.* 24(3), 485–492.

Joseph, B., and C. B. Brosilow. 1978b. "Inferential Control of Processes, Part III, Construction of Optimal and Suboptimal Dynamic Simulators," *AIChE J.24* (3), 500–509.

Kravaris, C., and C. B. Chung. 1987. "Nonlinear State Feedback Synthesis by Global Input-Output Linearization." *AIChE J.*, 33, 592–603.

MATLAB with SIMULINK, © The MathWorks, *www.mathworks.com.*

Morari, M. 1983. "Design of Resilient Processing Plants-III: A General Framework for the Assessment of Dynamic Resilience." *Chem. Eng. Sci.*, 38, 1881–1891.

Morari, M., S. Skogestad, and D E. Rivera. 1984. "Implications of Internal Model Control for PID Controllers." *Proc. of American Control Conference*, San Diego, CA, 661–666.

Morari, M. 1985. "Robust Stability of Systems with Integral Control." *IEEE Trans. Autom. Control*, AC-30, 574–577.

Morari, M., and F. Zafiriou. 1989. *Robust Process Control.* Prentice-Hall, NJ.

Newton, G. C., L. A. Gould, and J. F. Kaiser. 1957. *Analytic Design of Feedback Controls*, John Wiley & Sons, NY.

Psarris, A., and C. Floudas. 1990. "Improving Dynamic Operability in MIMO Systems with Time Delays." *Chem. Eng. Sci.* 45, 3505–3524.

Prett, D. M., and C. E. Garcia. 1988. *Fundamental Process Control.* Butterworths, MA.

Program CC, © Systems Technology, *www. systemstech.com/ progcc1.htm.*

Richalet, J. A. 1978. "Model Predictive Heuristic Control: Applications to an Industrial Process." *Automatica* 14, 413–428.

Smith, O. J. M. 1975. "Closer Control of Loops with Dead Time." *Chem. Eng. Progress* 53(5).

Tong, M., and C. B. Brosilow. 1978. "Inferential Control of Processes, Part II, The Structure and Dynamics of Inferential Control Systems." I. 24(3), 492–500.

VisSim, © Visual Solutions, *www.vissol.com*.

Two-Degree of Freedom Internal Model Control

Objectives of the Chapter

- Introduce the two-degree of freedom (2DF) IMC system and elucidate its advantages over 1DF IMC when step disturbances enter through process lags.

- Provide design methods for 2DF IMC systems with perfect models that give the best possible performance consistent with noisy measurements for inherently stable and inherently unstable processes.

Prerequisite Reading

Chapter 3, "One-Degree of Freedom Internal Model Control"

4.1   INTRODUCTION

The IMC design methods presented in Chapter 3 assume that step disturbances enter into the process output without passing through the process (i.e., $p_d(s) = 1$) in Figure 3.1, which is reproduced in Figure 4.1.

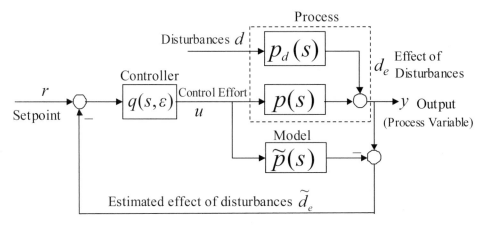

Figure 4.1 The 1DF IMC block diagram (from Chapter 3).

Assuming that the disturbance enters directly into the output allows the controller design to focus on achieving a good response to a step setpoint change. Such a controller equally well suppresses a step disturbance because the signal that enters the controller is the setpoint minus the disturbance estimate. Therefore, as far as the controller is concerned, a positive step disturbance is the same as a negative step setpoint change and vice versa. However, when the disturbance enters through the process, the disturbance transfer function is not one, and a controller designed to suppress a step disturbance will apply an inadequate control effort. The resulting response is often much more sluggish than is desirable. Consider Example 4.1, where the disturbance enters the process in the same way as the control effort so that the disturbance and the process transfer functions are the same.

Example 4.1 1DF Response to Process Disturbance

Process and models are

$$\widetilde{p}_d(s) = p_d(s) = \widetilde{p}(s) = p(s) = e^{-s}/(4s+1) \qquad (4.1a)$$

The single-degree of freedom IMC controller is (see Chapter 3).

$$q(s) = (4s+1)/(0.2s+1). \qquad (4.1b)$$

(0.2 is the filter-time constant for a noise amplification of 20.)

The resulting control effort, $m(s)$, and output, $y(s)$, for a step disturbance are:

$$m(s) = -\tilde{p}_d(s)q(s)/s \;\; = \;\; -e^{-s}/s(0.2s+1) \qquad (4.1c)$$

$$y(s) = (1-\tilde{p}(s)q(s))\tilde{p}_d(s)/s \;\; = \;\; \left(1-\frac{e^{-s}}{(0.2s+1)}\right)\frac{e^{-s}}{(4s+1)s}. \qquad (4.1d)$$

Figure 4.2 shows the time responses of the output and control effort of Equations (4.1c) and (4.1d). The long tail on the response of the output $y(t)$ can be ascribed to the fact that the control effort reaches steady state in about one time unit, while the output, delayed by 1 time unit, has not yet stopped rising. The output response from about 2.4 time units onward is similar to that of an unforced process, starting with an initial condition of about 0.24 and moving to a steady state of zero.

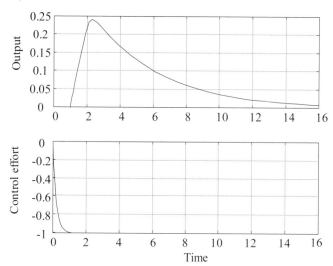

Figure 4.2 Response of 1DF IMC to a disturbance that enters through the process.

Another perspective on the output response can be obtained from the following partial fraction expansion of Eq. (4.1d).

$$y(s) = \left(\frac{1}{s}-\frac{4}{4s+1}\right)e^{-s}-\left(\frac{1}{s}+\frac{.010s}{.2s+1}-\frac{4.21}{4s+1}\right)e^{-2s} \qquad (4.1e)$$

Inverting the above and collecting terms for $t \geq 2$ gives

$$y(t) = 0.274e^{-.25(t-2)}-0.0526e^{-5(t-2)};t \geq 2 \qquad (4.1f)$$

The term $0.274e^{-.25(t-2)}$ gives rise to the long tail in the response shown in Figure 4.2. To remove this term, we can choose the IMC controller so that the term $(1-\tilde{p}(s)q(s))$ has a zero at $s = -.25$, thereby canceling the pole in the disturbance lag at $s = -.25$. However, choosing the IMC controller so that the term $(1-\tilde{p}(s)q(s))$ has a zero at the pole of the disturbance lag will generally seriously degrade the setpoint response. (Recall that the setpoint to output transfer function is $\tilde{p}(s)q(s)$.) To avoid degrading the setpoint response in favor of the disturbance response, we use the two-degree of freedom control structure of Figure 4.3.

◆

4.2   STRUCTURE OF 2DF IMC

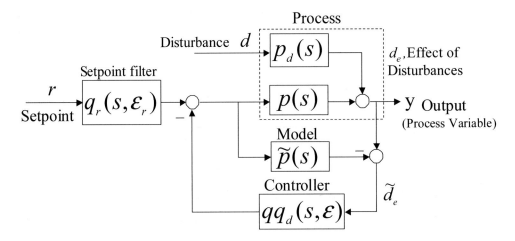

Figure 4.3 2DF IMC structure.

The controller $qq_d(s,\varepsilon)$ in Figure 4.3 is designed to reject disturbances while the setpoint controller $q(s,\varepsilon)$ is designed to shape the response to setpoint changes. Henceforth, we will refer to the setpoint controller as the setpoint filter in order to be consistent with industrial terminology.

The perfect model output and control effort responses for Figure 4.3 are

$$y(s) = \tilde{p}(s)q(s,\varepsilon_r)r(s) + (1 - \tilde{p}(s)qq_d(s,\varepsilon))p_d(s)d(s) \qquad (4.2a)$$

$$m(s) = q(s,\varepsilon_r)r(s) + qq_d(s,\varepsilon)p_d(s)d(s). \qquad (4.2b)$$

Returning to Example 4.1, Figure 4.4 shows the response of the control system of Figure 4.3 to a step disturbance with q_d chosen as

$$qq_d(s) = (4s+1)(1.19s+1)/(.2s+1)^2, \qquad (4.2c)$$

the filter-time constant in Eq. (4.2c) is 0.2 (i.e., $\varepsilon = 0.2$). There is a substantial improvement in the speed with which the disturbance on the output is eliminated. Also note, however, that the improved output response requires a substantially more aggressive control effort.

The next section shows how Eq. (4.2c) was obtained, and presents design procedures for both $qq_d(s,\varepsilon)$ and $q_r(s,\varepsilon_r)$ for a perfect model. Tuning both $qq_d(s,\varepsilon)$ and $q_r(s,\varepsilon_r)$ to accommodate model uncertainty is discussed in Chapter 7.

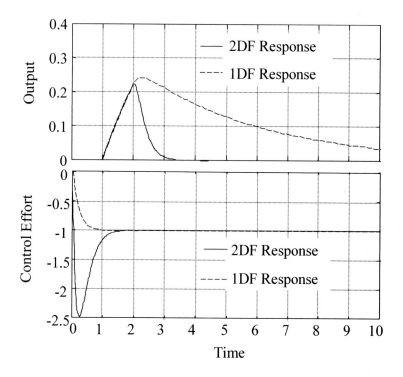

Figure 4.4 Comparison of 1DF and 2DF IMC responses to a step disturbance to the process of Example 4.1, $qq_d(s,.2) = (4s+1)(1.19s+1)/(.2s+1)^2$.

4.3 DESIGN FOR STABLE PROCESSES

4.3.1 Design of the Setpoint Filter, $q(s, \varepsilon_r)$

The setpoint filter $q_r(s, \varepsilon_r)$ in Figure 4.3 is designed like a single-degree of freedom controller using the methods in Chapter 3. However, since there is generally no noise on the setpoint, there is no noise amplification limit on ε_r. Nonetheless, very small values of ε_r are not recommended because of the likelihood of control effort saturation. Unless a model state feedback implementation is used (see Chapter 5), smaller filter-time constants can actually lead to slower output responses because of control effort saturation.

4.3.2 Design of the Feedback Controller, $qq_d(s, \varepsilon)$

The transfer function between output and disturbance for Figure 4.3 for a perfect model is

$$y(s) = (1 - \tilde{p}(s)qq_d(s, \varepsilon))\tilde{p}_d(s)d(s). \tag{4.3}$$

To design $qq_d(s, \varepsilon)$ for a perfect model, it is convenient to consider $qq_d(s, \varepsilon)$ to be composed of two terms, $q(s, \varepsilon)$ and $q_d(s, \varepsilon)$. The design then proceeds as follows:

1. Select $q(s, \varepsilon)$ as in Chapter 3. That is, $q(s, \varepsilon)$ inverts a portion of the process model $\tilde{p}(s)$. Select the controller filter as $1/(\varepsilon s + 1)^r$, where r is the relative order of the part of the process model that is inverted by $q(s, \varepsilon)$.

2. Select $q_d(s, \varepsilon)$ as

$$q_d(s, \varepsilon, \alpha) = \frac{\sum\limits_{i=0}^{n} \alpha_i s^i}{(\varepsilon s + 1)^n}; \quad \alpha_0 \equiv 1, \tag{4.4}$$

    where n is the number of poles in $\tilde{p}_d(s)$ to be cancelled by the zeros of $(1 - \tilde{p}(s)qq_d(s))$.

3. Select a trial value for the filter-time constant ε.

4. Find the values of α_i by solving Eq. (4.5) for each of the n distinct poles of $\tilde{p}_d(s)$ that are to be removed from the disturbance response.

$$(1 - \tilde{p}(s)qq_d(s, \varepsilon, \alpha))\big|_{s=-1/\tau_i} = 0; \quad i = 1, 2, \ldots n, \tag{4.5}$$

    where τ_i is the time constant associated with the ith pole of $\tilde{p}_d(s)$.

If any of the poles occur in complex conjugate pairs, then both the real and imaginary parts of equation are set to zero for one of the complex conjugate pairs. The set of equations given by Eq. (4.5) are linear in the parameters α_i, no matter whether any of the poles of $\widetilde{p}_d(s)$ are real or complex.

If $\widetilde{p}_d(s)$ contains repeated poles (e.g., $\widetilde{p}_d(s) = 1/(\tau_j s + 1)^r$), then the derivatives of Eq. (4.5) are set to zero, up to order one less than the number of repeated poles. For example, if $\widetilde{p}_d(s) = 1/(\tau_j s + 1)^r$ then we solve

$$(1 - \widetilde{p}(s)qq_d(s,\varepsilon,\alpha))\Big|_{s=-1/\tau_j} = 0 \tag{4.6a}$$

$$\frac{d^k}{ds^k}(\widetilde{p}(s)qq_d(s,\varepsilon,\alpha))\Big|_{s=-1/\tau_j} = 0; \quad k = 1, 2,....r - 1 \tag{4.6b}$$

5. Adjust the value for ε, and repeat step 4 until the desired noise amplification is achieved. A few trials are usually sufficient to achieve a noise amplification factor close enough to the desired value. Of course, one could solve simultaneously for the α_i that satisfy step 4 and the desired noise amplification. However, solving simultaneously for α_i and ε requires the solution of a set of nonlinear equations.

The software IMCTUNE automatically provides values for all α_i for up to fourth order disturbance lags in $\widetilde{p}_d(s)$ given a value for the filter-time constant ε.

To illustrate the procedure, we consider again the process of Example 4.1.

Example 4.2 Solving for the Feedback Controller Numerator Coefficient(s) α_i

In step 1, we select the setpoint filter $q(s)$ to invert the invertible portion of the process model $\widetilde{p}(s)$.

$$\widetilde{p}_d(s) = p_d(s) = \widetilde{p}(s) = p(s) = e^{-s}/(4s + 1) \tag{4.7a}$$

$$q(s) = (4s + 1)/(.2s + 1) \tag{4.7b}$$

In step 2 we design $q_d(s)$ so that the zeros of $(1 - \widetilde{p}(s)qq_d(s))$ cancel the poles of $p_d(s)$. Since $p_d(s)$ has a single pole at -1/4, we select $q_d(s)$ as

$$q_d(s) = \frac{(\alpha s + 1)}{(\varepsilon s + 1)}. \tag{4.7c}$$

The constant α is chosen so that $(1 - \widetilde{p}(s)qq_d(s))$ has a zero at $s = -\frac{1}{4}$. That is,

$$(1-\tilde{p}(s)qq_d(s,\varepsilon,\alpha))\big|_{s=-1/4} = \left(\frac{1-(-\alpha/4+1)e^{1/4}}{(\varepsilon s+1)}\right) = 0 \qquad (4.7d)$$

Selecting $\varepsilon = 0.2$ in Eq. (4.7d) gives $\alpha = 1.189$. Expanding Eq. (4.3) in partial fractions with $\alpha = 1.189$ for a unit step disturbance gives

$$y(s) = \left(\frac{1}{s} - \frac{1}{s+.25}\right)e^{-s} - \left(\frac{1}{s} - \frac{1.301}{(s+.5)^2} - \frac{.2213}{(s+5)} - \frac{.7787}{s+.25}\right)e^{-2s}. \qquad (4.7e)$$

Inverting Eq. (4.7e) and collecting terms for $t \geq 2$ gives

$$y(t) = (1.39(t-2)-.221)e^{-s(t-2)} - 2\times10^{-4}e^{-.25(t-2)} \qquad (4.7f)$$

The coefficient of the term $e^{-.25(t-2)}$ can be made as small as we like by carrying more significant figures in α. The graph of $y(t)$ given by Eq. (4.7f) is the solid line in Figure 4.4 labeled as "2DF Response."

In the foregoing, we chose the filter-time constant ε as 0.2 because this was the value that yields a noise amplification factor of 20 for the IMC controller $q(s)$. However, the noise amplification factor that is really of interest is that of $qq_d(s)$, and this is given by the maximum value of $|qq_d(j,\omega)/qq_d(0)|$ over all frequencies, ω. Usually this maximum occurs at $\omega = \infty$. For $\varepsilon = .2$ and $\alpha = 1.189$, the noise amplification factor is 119. Thus, this filter-time constant is much too small. Solving for a filter-time constant, ε, and α that satisfies the noise amplification factor and satisfies Eq. (4.7d), gives $\varepsilon = .59$ and $\alpha = 1.736$. The response of the process output to a step disturbance entering through $p_d(s)$ is shown in Figure 4.5. Clearly, insisting on not amplifying noise by more than a factor of 20 has lost some of the advantage of the two-degree of freedom control system. However, a settling time of 6 time units is still much better than a settling time of 20 time units.

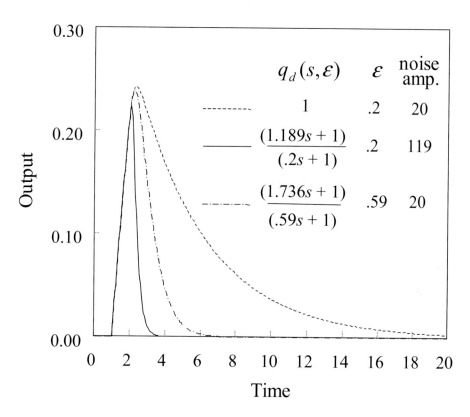

Figure 4.5 Comparison of 1DF and 2DF IMC responses to a step disturbance to the process of Example 4.2, $q(s) = (4s + 1)/(\varepsilon s + 1)$

♦

Example 4.3 A Lead Process

This example demonstrates that for a lead process (one whose frequency response initially increases before eventually decreasing), there is no advantage to using a 2DF over 1DF control system.

The process and disturbance lag are

$$p_d(s) = p(s) = \frac{(2s+1)e^{-s}}{(s+1)^2}. \tag{4.8a}$$

The one-degree of freedom IMC controller for the above process is

$$q(s) = \frac{(s+1)^2}{(2s+1)(.025s+1)}. \tag{4.8b}$$

A two-degree of freedom controller to cancel the linear term of the disturbance lag (i.e., $2s + 1$) is

$$qq_d(s) = \frac{(s+1)^2(0.969s+1)}{(2s+1)(0.156s+1)^2}.$$ (4.8c)

A two-degree of freedom controller to cancel both disturbance lags (i.e., $(s+1)^2$) is

$$qq_d(s) = \frac{(s+1)^2(0.519s^2+1.35s+1)}{(2s+1)(0.235s+1)^3}.$$ (4.8d)

Figure 4.6 gives the time responses to a step disturbance for each of the controllers.

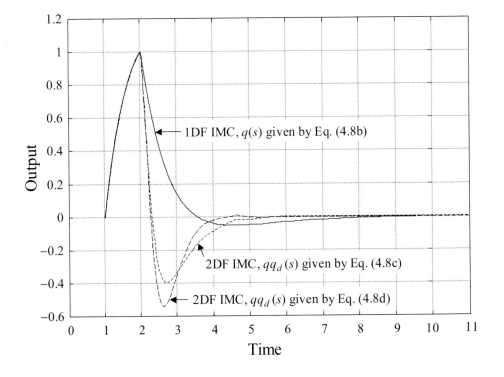

Figure 4.6 Comparison of 1DF and 2DF IMC responses to a step disturbance to the process of Example 4.3.

Clearly, for this example, the one-degree of freedom controller gives superior performance.

♦

Example 4.4 An Underdamped Process

The purpose of this example is to demonstrate the advantage of a two-degree of freedom control system over a single-degree of freedom control system when the disturbance passes through an underdamped process (i.e., a process with complex poles). We also demonstrate the computation of the q_d part of the feedback controller for such a process.

The process and disturbance lag are

$$p_d(s) = p(s) = e^{-s}/(s^2 + .2s + 1). \tag{4.9a}$$

The 1DF controller for the above process that has a noise amplification factor of 20 is

$$q(s) = \frac{(s^2 + .2s + 1)}{(.22s + 1)^2}. \tag{4.9b}$$

The two-degree of freedom controller for the above process that has a noise amplification factor of 20 is

$$qq_d(s) = \frac{(s^2 + .2s + 1)(2.4s^2 + .32s + 1)}{(.6s + 1)^4}. \tag{4.9c}$$

Figure 4.7 shows the responses of the single and two-degree of freedom control systems for Example 4 for a step disturbance. The response of the two-degree of freedom control system is clearly much superior to that of the single-degree of freedom system.

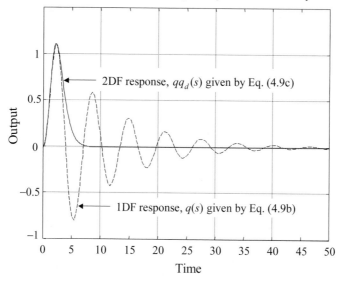

Figure 4.7 1DF and 2DF IMC responses to a step disturbance for Example 4.4.

4.4   DESIGN FOR UNSTABLE PROCESSES

4.4.1 Internal Stability

When the process is unstable, both 1DF and 2DF IMC systems are internally unstable. That is, applying only bounded inputs will cause one or more signals in the block diagram (see Figures 4.1 and 4.3) to grow without bound no matter how the controllers $q(s)$ and $qq_d(s)$ are chosen. To show that the IMC structures are internally unstable, we follow the approach in Morari and Zafiriou (1989).

By definition, a control system is internally stable if bounded inputs, injected at any point of the control system, generate bounded responses at any other point. A linear time invariant control system is internally stable if the transfer functions between any two points of the block diagram are stable. Figure 4.8 is the same as Figure 4.3, except for the addition of the two inputs u_1 and u_2. The inputs are the setpoint r, the disturbance d, and two error, or noise, inputs u_1 and u_2.

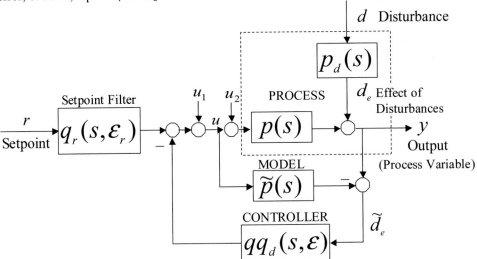

Figure 4.8 Block diagram of 2DF IMC system with additional inputs (u_1 and u_2) for deriving internal stability conditions.

For the analysis of internal stability, it is sufficient to take the control system outputs as the process outputs y, the model output $\tilde{y}$, the control effort u, and the estimate of the effect of disturbances on the process output $\tilde{d}_e$. The input and output transfer functions for a perfect model are

$$
\begin{bmatrix} y \\ \widetilde{y} \\ u \\ \widetilde{d}_e \end{bmatrix} = \begin{bmatrix} pq & (1-pqq_d)p_d & p & (1-pqq_d)p \\ pq & pp_d qq_d & p & p^2 qq_d \\ q & p_d qq_d & 1 & pqq_d \\ 0 & p_d & 0 & p \end{bmatrix} \begin{bmatrix} r \\ d \\ u_1 \\ u_2 \end{bmatrix}.
\tag{4.10}
$$

If p, p_d, q, and q_d are all stable, then all of the transfer functions in Eq. (4.10) are stable. However, if p, p_d, or q is unstable, then small changes in the inputs r, d, u_1, and u_2 will cause the outputs, y, $\widetilde{y}$, u, and $\widetilde{d}_e$ to grow without bound. Thus, for unstable processes, 1DF or 2DF control systems should not be implemented as shown in Figures 4.1 or 4.3, no matter how q and q_d are selected. All is not lost, however. An internally stable 2DF control system for unstable processes obtained from an IMC design can be implemented in the form of a single-loop feedback control system as discussed below.

4.4.2 Single-loop Implementation of IMC for Unstable Processes

Just as was done for 1DF control systems in Chapter 3 (see Figure 3.2), the two-degree of freedom control system of Figure 4.3 can be collapsed into a single-loop feedback control system. First, Figure 4.3 is reconfigured into Figure 4.9 by moving the controller $qq_d(s,\varepsilon)$ out of the feedback path.

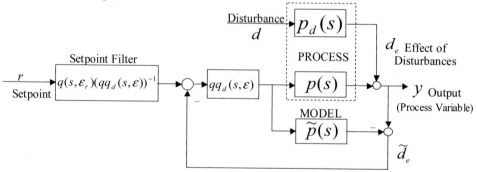

Figure 4.9 Single-loop configuration of a 2DF IMC system.

Next, the feedback loop around the model is collapsed to obtain Figure 4.10.

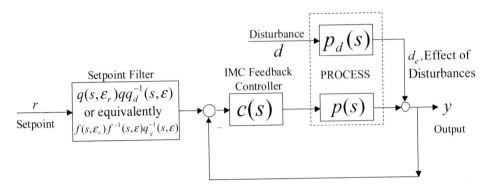

Figure 4.10 Feedback form of 2DF IMC system.

The feedback controller $c(s)$ in Figure 4.10 is

$$c(s) = \frac{qq_d(s,\varepsilon)}{(1 - \tilde{p}(s)qq_d(s,\varepsilon))}. \qquad (4.11a)$$

The setpoint filter of Figure 4.9 is converted into the setpoint filter of Figure 4.10 using the relationship

$$q(s,\varepsilon_r)q^{-1}(s,\varepsilon) = f(s,\varepsilon_r)f^{-1}(s,\varepsilon), \qquad (4.11b)$$

where $f(s,\varepsilon) \equiv 1/(\varepsilon s + 1)^r$.

Morari and Zafiriou (1989) show that the feedback system given in Figure 4.10 with $p_d = p$ and no setpoint filter (i.e., $q(s,\varepsilon_r)qq_d^{-1}(s,\varepsilon) = 1$) will be internally stable for a perfect model provided that the terms $\tilde{p}(s)c(s)/(1 + \tilde{p}(s)c(s))$, $\tilde{p}(s)/(1 + \tilde{p}(s)c(s))$, and $c(s)/(1 + \tilde{p}(s)c(s))$ are all stable. A necessary condition for the foregoing terms to be stable is that the term $(1 + \tilde{p}(s)c(s))$ has no right half plane zeros.

Substituting the controller $c(s)$, given by Eq. (4.11a), into the conditions given above places the following conditions on the design of the IMC controllers (Morari and Zafiriou, 1989).

1. $qq_d(s,\varepsilon)$ must be stable.

2. $\tilde{p}(s)qq_d(s,\varepsilon)$ must be stable. This requires that the zeros of $qq_d(s,\varepsilon)$ cancel the unstable poles of $\tilde{p}(s)$.

3. $(1 - \tilde{p}(s)qq_d(s,\varepsilon))\tilde{p}(s)$ must be stable. This requires that the zeros of $(1 - \tilde{p}(s)qq_d(s,\varepsilon))$ also cancel the unstable poles of $\tilde{p}(s)$.

In addition, in order for Figure 4.10, as drawn, to be internally stable, $q_d^{-1}(s,\varepsilon)$ must be stable, and the unstable poles of $p_d(s)$ must be *exactly* the same as the poles of $p(s)$ (i.e., any unstable poles in $p_d(s)$ must arise because the disturbance passes through an unstable part of the process in addition to any other stable lags that the disturbance may pass through).

It is important to note that satisfying conditions 1 through 3 does not necessarily lead to a stable controller, $c(s)$, as given by Eq. (4.11a). The problem is that the term $(1 - \tilde{p}(s)qq_d(s,\varepsilon))$ can have right half plane zeros in addition to those needed to cancel the right half plane poles of $\tilde{p}(s)$. While it is possible to have a stable closed-loop control system with an unstable controller and unstable process, it is generally not advisable to implement such a control system. The necessary conversion of the controller given by Eq. (4.11a) into a single transfer function usually requires a Padé approximation of a dead time. When the controller is unstable, this approximation can lead to closed-loop instability. As we shall see in the example that follows, for stable controllers $qq_d(s,\varepsilon)$ it is usually possible to avoid $(1 - \tilde{p}(s)qq_d(s,\varepsilon))$ having additional, undesired right half plane zeros by increasing the filter-time constant ε.

Since increasing the filter-time constant beyond that required to satisfy maximum noise amplification specifications or to accommodate process uncertainty degrades control system performance, Brosilow and Cheng (1987) and Berber and Brosilow (1999) developed an internally stable method of implementing 2DF IMC systems. A discussion of their method is, however, beyond the scope of this text.

The following example demonstrates the design of a 2DF controller for an unstable process and illustrates how the behavior of unstable processes under feedback control differs from that of stable processes.

Example 4.5 Control of a FOPDT Unstable Process

The process and model are

$$p(s) = \tilde{p}(s) = \frac{e^{-s}}{(-s+1)}. \tag{4.12}$$

Following the procedure outlined in Section 4.3.2 and selecting a filter-time constant for a maximum noise amplification of approximately 20 yields the following IMC controllers:

$$q(s) = \frac{(-s+1)}{(0.5s+1)}; \quad q_d(s) = \frac{(5.1162s+1)}{(0.5s+1)}. \tag{4.13}$$

The ideal disturbance response of the control system of Figure 4.9 is obtained from Eq. (4.3), which is repeated below.

$$y(s) = (1 - \tilde{p}(s)qq_d(s,\varepsilon))\tilde{p}(s)d(s), \tag{4.14}$$

where $y(s)$ is the process output and $d(s)$ is a step disturbance (i.e. $d(s) = 1/s$). Using a five over five Padé approximation for the dead time of one in $\tilde{p}(s)$, and being careful to cancel the common factors in Eq. (4.14) yields

$$y(s) = \frac{-3.12(0.00168s+1)(0.00497s^2 - 0.00993s + 1)(0.0318s^2 - 0.0684s + 1)}{(0.0138s^2 + 0.128s + 1)(0.0175s^2 + 0.235s + 1)(0.137s + 1)(0.5s + 1)^2}. \qquad (4.15)$$

The quadratic terms in Eq. (4.15) with complex roots arise from the Padé approximation of the dead time. Notice that $y(s)$ is stable. The time response of Eq. (4.15) is given in Figure 4.11 along with the control effort that is obtained from

$$u(s) = -\tilde{p}(s)qq_d(s,\varepsilon)d(s) \qquad (4.16)$$

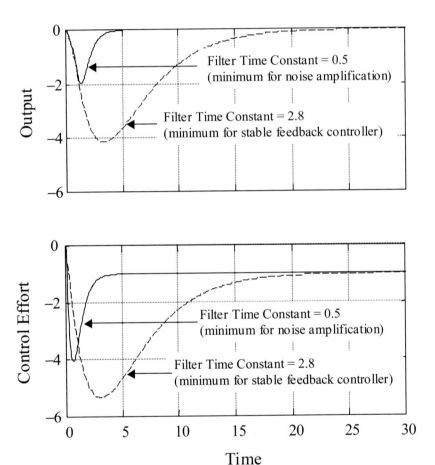

Figure 4.11 Comparison of responses of the unstable process of Example 4.5 for different IMC filter-time constants.

We must emphasize that the response given by Eq. (4.15), which has a filter time constant of 0.5, is the response that is obtained from an internally stable implementation of a 2DF IMC system such as that suggested by Berber and Brosilow (1999). If we attempt to implement the control system via Figure 4.10, then we find that the feedback controller, $c(s)$, associated with a filter-time constant of 0.5 is unstable, as shown in Eq. (4.17). The resulting closed-loop control system also turns out to be unstable.

$$c(s) = qq_d(s,.5)/(1 - \widetilde{p}(s)qq_d(s,.5))$$

$$= \frac{-.321(.0138s^2 + .128s + 1)(.0175s^2 + .235s + 1)(.137s + 1)(.512s + 1)}{(.00497s^2 - .00993s + 1)(.0318s^2 - .0684s + 1)(.0168s + 1)s}. \qquad (4.17)$$

To get a stable controller, $c(s)$, it is necessary to increase the filter-time constant from 0.5 to 2.8. The controller is given by Eq. (4.18).

$$c(s) = qq_d(s,2.8)/(1 - \widetilde{p}(s)qq_d(s,2.8))$$

$$= \frac{-.0316(.0138s^2 + .128s + 1)(.0175s^2 + .235s + 1)(.137s + 1)(38.25s + 1)}{(.00618s^2 + .0283s + 1)(.0424s^2 + 4.66 \times 10^{-5}s + 1)(.0312s + 1)s}. \qquad (4.18)$$

The closed-loop control system using the controller given by Eq. (4.18) is stable, and the disturbance response is that shown in Figure 4.11. Clearly, there is a significant performance incentive for implementing an internally stable 2DF IMC system for this example.

◆

Note that the responses in Figure 4.11 are different from the responses for stable processes. For a stable process, decreasing the IMC filter-time constant makes the control effort more aggressive. That is, both the magnitude and the speed of response of the control effort are increased. For an unstable process, decreasing the filter-time constant increases the speed of response of the control effort, but the magnitude of the control effort decreases. This difference, as well as other important differences between control systems for inherently stable and inherently unstable processes, changes the method of tuning the control system to account for process uncertainty. Therefore, in Chapter 8, we separate the discussion of tuning control systems for uncertain inherently stable and inherently unstable processes into different sections.

4.5 SOFTWARE TOOLS FOR 2DF IMC DESIGNS

2DF IMC controllers for both stable and unstable systems with perfect models can be designed with the aid of the IMCTUNE software included with this text. The software computes the q_d portion of the feedback controller given the form of q and a filter-time constant. It will also compute the filter-time constant that achieves any noise amplification factor specified in the Default Values window, as described in Appendix F. The software provides step setpoint and disturbance responses for inherently stable and unstable processes using the control system configurations of Figures 4.3 and 4.10 respectively. For example, the responses in figures 4.4 through 4.7 and the response for a filter-time constant of 2.8 in Figure 4.10 either were, or could have been, obtained from IMCTUNE.

4.6   SUMMARY

2DF control systems should be used for inherently unstable processes, or for stable processes where the disturbances pass through the process or through a lag whose time constants are on the order of, or larger than, the process lag time constants. There is usually no advantage of a 2DF controller over a 1DF controller when the disturbance lag time constant is small relative to the process lag time constants or when the disturbance passes through a stable process whose lead time constants are larger than its lag time constants (i.e., one or more process zeros lie to the right of the process poles). When in doubt, it is advisable to design a 2DF control system and compare its behavior to that of a 1DF control system. Only when there is a substantial improvement in the disturbance response should the 2DF controller be implemented.

The setpoint filter and $q(s)$ portion of the feedback controller in Figure 4.3 are designed using the 1DF methods in Chapter 3. The $q_d(s,\varepsilon)$ portion of the feedback controller is chosen as

$$q_d(s,\varepsilon,\alpha) = \frac{\sum_{i=0}^{n} \alpha_i s^i}{(\varepsilon s + 1)^n}, \tag{4.4}$$

where n is number of poles in $\tilde{p}_d(s)$ to be cancelled by the zeros of $(1 - \tilde{p}(s)qq_d(s))$. The constants α_i are chosen to cancel selected poles or other portions of the denominator of the disturbance transfer function $p_d(s)$.

The IMC configurations of Figures 4.1 and 4.3 are internally unstable for inherently unstable processes, and therefore cannot be used to control such processes. They can, however, be used for the purpose of analysis and controller design. Either the configuration given in Figure 4.10, or the algorithm suggested by Berber and Brosilow (1999) can be used to control unstable processes.

Problems

4.1 Compare the 1DF and 2DF disturbance responses for Problems 3.4 and 3.5 of Chapter 3, assuming that the disturbances pass through the process. Suggestion: For processes whose denominators are higher than second order, design the feedback controller to cancel either the one or two largest time constant poles or the linear or quadratic terms in the denominator.

4.2 Design and implement controllers for each of the following processes. Compare the ideal response using a filter-time constant that yields a noise amplification factor of 20, with the minimum filter-time constant for a stable control system using the configuration of Figure 4.10.

a.
$$p_d(s) = p(s) = e^{-s}/(s+1)(s^2 + .1s + 1)$$

b.
$$p_d(s) = p(s) = \frac{e^{-s}}{s(s+1)}$$

c.
$$p_d(s) = p(s) = \frac{2}{2s-1}e^{-0.5s}$$

d.
$$p_d(s) = p(s) = \frac{2(-.5s+1)}{(2s-1)(s+1)}$$

References

Berber, R., and C. Brosilow. 1999. "Algorithmic Internal Model Control," Proceedings of the 7th Mediterranean Conference on Control and Automation. Haifa, Israel.

Brosilow, C., and C.M. Cheng. 1987. "Model Predictive Control of Unstable Systems," presented at the Annual AIChE Meeting, NY.

Morari, M., and F. Zafiriou. 1989. *Robust Process Control*, Prentice-Hall, NJ.

Model State Feedback Implementations of IMC

Objectives of the Chapter

- To introduce MSF implementations of 1DF and 2DF IMC systems that yield improved output responses for constrained (i.e., real) control efforts.
- To provide examples of the dynamic simulation of systems (e.g., processes and their feedback control systems) described by linear differential equations with time delays and control effort limits.
- To introduce the diagrams used to input structure and data to dynamic simulators and digital control systems.

Prerequisite Reading

Chapter 2, "Continuous Time Models"
Chapter 3, "1DF Internal Model Control"
Chapter 4, "2DF Internal Model Control"
Appendix F, "SIMULINK Tutorial"

5.1   MOTIVATION

All physical control efforts, such as valves, pumps, steering wheels, and so on, have limits (often called constraints, or *hard* constraints) that they cannot exceed. Such control effort constraints can seriously degrade the speed of response, the amount of overshoot, and the shape of the output response of IMC systems. The degradation in output response is because the IMC controller assumes that (1) the calculated control effort has actually been applied, and (2) any needed future control can be applied. Both of these assumptions are false. The former assumption can lead to very sluggish behavior, while the latter can lead to both sluggish behavior and substantial overshoots of the setpoint, depending on the current operating point. To illustrate, let us consider the performance of a 1DF IMC system for the following process

Example 5.1 Comparison of IMC Implementations

Process and models are

$$p(s) = \tilde{p}(s) = 2e^{-s}/(s+1). \tag{5.1a}$$

The single-degree of freedom IMC controller is (see Chapter 3)

$$q(s) = (s+1)/2(0.05s+1). \tag{5.1b}$$

(0.05 is the filter time constant for a noise amplification factor of 20.)

The control effort $u(t)$, is constrained by

$$0 \le u(t) \le 1.5. \tag{5.1c}$$

Figure 5.1a shows the IMC diagram for the process.

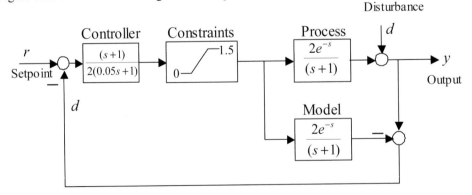

Figure 5.1a The IMC system with control effort constraints for Example 5.1.

Figure 5.1b is a SIMULINK[1] diagram for the IMC block diagram of Figure 5.1a. This figure is simplified in that the model and feedback loop are eliminated, since the disturbance estimate is perfect when the model is perfect, and in a simulation it is a simple matter to sum the disturbance input d into both the output and the setpoint. The diagram in Figure 5.1b is constructed by dragging and dropping icons onto a blank SIMULINK window. The labels without dotted boxes around them are default labels that SIMULINK provides automatically. The labels with dotted boxes around them are additional labels added to clarify the relationships between the blocks in Figure 5.1b and those in Figure 5.1a. The data contained in each SIMULINK block can be accessed and, if desired, modified by double-clicking on the block. In the case of transfer function blocks, the data is displayed in the block, as shown. The saturation block transmits its input signal unchanged if it is within constraints. Otherwise, the output signal is at the binding constraint. The transport delay is the dead time block, which is implemented as an array of memory elements of whatever size is necessary to realize the delay. The scopes display the time responses of the simulation and, if desired, send their input signals to the MATLAB workspace as a matrix, with time in the first column and the variable in the second column.

Refer to Appendix F for a more complete tutorial on SIMULINK. Future examples will assume a familiarity with SIMULINK beyond that provided by the current example.

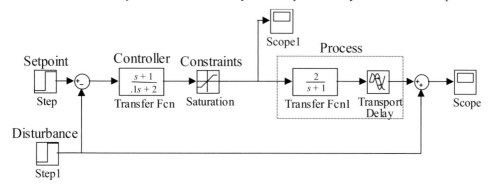

Figure 5.1b A SIMULINK diagram for the IMC system of Figure 5.1a.

Figure 5.1c shows an MSF implementation of the same IMC system as in Figures 5.1a and 5.1b. The advantage of the implementation in Figure 5.1c over that of Figures 5.1a and 5.1b is that it takes into account the control effort that was actually applied, rather than computing the control based only on the disturbance estimate and the setpoint as in Figures 5.1a and 5.1b. Figures 5.2a and 5.2b demonstrate the advantage of computing the control effort from the model output as in Figure 5.1c. Nonetheless, in the absence of constraints the transfer functions between output and setpoint or output and disturbance are exactly the same for Figures 5.1a, 5.1b, and 5.1c.

---

[1] © The MathWorks, Inc.

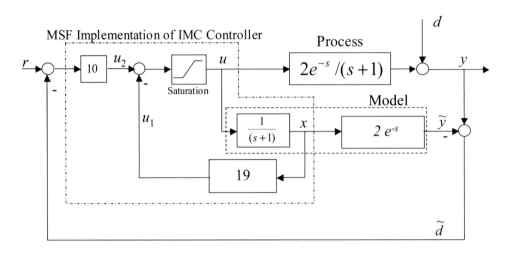

Figure 5.1c MSF implementation of IMC system for Example 5.1.

We show how to construct such an MSF diagram in Section 5.2. At this point, all that is desired is that the reader be convinced that in the absence of control effort constraints, the transfer functions between output and setpoint or output and disturbance are the same for all three figures.

Figure 5.2a shows the output responses of constrained and unconstrained lead-lag implementations of IMC systems, that is, $q(s)$ given by Eq. (5.1b) and an MSF implementation of IMC for a unit step change in setpoint. As can be seen from the figure, the effect of saturation is to make response of the constrained lead-lag control system much more sluggish than it would otherwise be.

Figure 5.2b shows the control efforts associated with Figure 5.2a. Notice that the MSF implementation of Figure 5.1c holds the control effort at its upper bound much longer than the lead-lag implementation. This is because the control effort in MSF is formed from the state of the lag portion of the model, which in this case is just the output of the transfer function $1/(s + 1)$. Since the output of the lag depends on the applied control effort, rather than on the calculated control effort as in Figures 5.1a and 5.1b, the control effort is held at its constraint longer, and the resulting output response is much more rapid.

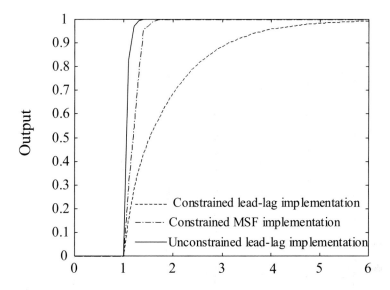

Figure 5.2a Comparison of lead-lag and MSF output responses to a step change in setpoint with and without control effort saturation for Example 5.1.

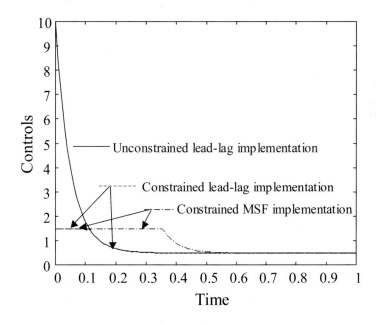

Figure 5.2b Control efforts associated with the responses in Figure 5.3a.

The next two examples illustrate how control effort saturation can result in either a significant overshoot for a setpoint change or a pseudo-inverse response for lead-lag IMC implementations, depending on the control system operating point.

Example 5.2 Constraints Can Lead to Overshoots

Process and models are:

$$p(s) = \tilde{p}(s) = \frac{1}{(4s+1)(3s+1)(2s+1)(s+1)}. \qquad (5.2a)$$

The single-degree of freedom IMC controller is:

$$q(s) = \frac{(4s+1)(3s+1)(2s+1)(s+1)}{(.8s+1)^4}. \qquad (5.2b)$$

The control effort, $u(t)$, is constrained by

$$0 \le u(t) \le 100. \qquad (5.2c)$$

The filter time constant of 0.8 in Eq. (5.2b) yields a noise amplification factor of almost 59, which is quite high. We are using such a small filter time constant so as to amplify the phenomena that we wish to demonstrate using a relatively simple example.

The SIMULINK block diagram structure for this example is the same as that of Figure 5.2a except, of course, the controller, process, and saturation blocks have different data. Figure 5.3a shows both the constrained and unconstrained IMC output responses for this example.

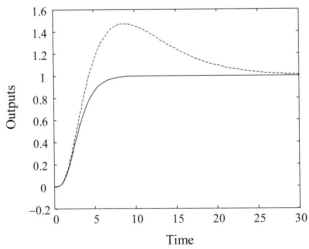

Figure 5.3a Output responses for IMC systems with and without control effort saturation for Example 5.2.

The overshoot in Figure 5.3a is caused by the fact that the operating point is such that there is unlimited control effort in the positive direction, but inadequate control effort in the negative direction. Figure 5.3b shows the control efforts that produced the output responses of Figure 5.3a.

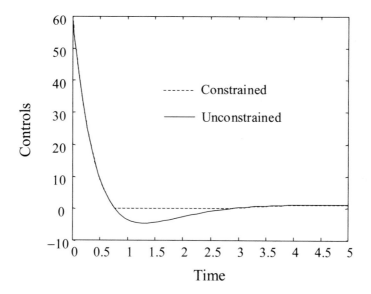

Figure 5.3b Control efforts associated with the output responses of Figure 5.3a.

◆

As a final example, we show that a simple change in operating point results in very much different behavior than that shown in Figures 5.3a and 5.3b.

Example 5.3 Constraints Can Lead to a Pseudo-Inverse Response

This example is the same as Example 5.2 except that the operating point has changed and the output and control effort have been renormalized so that at the current steady state they again have numerical values of zero. While the range of the control effort remains 100, the limits are now

$$-98.5 \le u(t) \le 1.5. \tag{5.3}$$

Figures 5.4a and 5.4b show the responses to a unit step setpoint change.

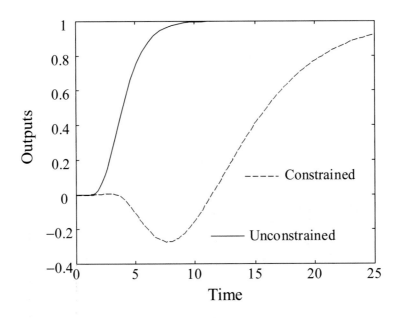

Figure 5.4a SIMULINK output responses for IMC system with and without control effort saturation for Example 5.3.

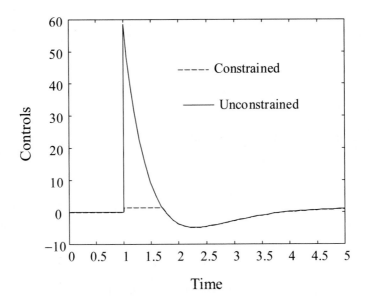

Figure 5.4b SIMULINK control effort responses for IMC system with and without control effort saturation for Example 5.3.

The pseudo-inverse output response in Figure 5.4a arises from the fact that the control effort is severely constrained in the positive direction, but, as shown by Figure 5.4b, does not encounter a constraint in the negative direction. Thus, the output response looks as though it came from a non-minimum phase process. The reality is that the control effort constraints have induced the pseudo-inverse response.

♦

The MSF implementations of IMC described in the next section can overcome the defects of lead-lag implementations of IMC. As an added bonus, they are also easier to implement in most digital control systems than lead-lag IMC controllers. However, the reader may find the algebra necessary to obtain the required MSF gains a bit messy. Happily, all the messy algebra can be automated, and once the IMC controller is designed using the methods in Chapters 3 and 4, IMCTUNE can be used to obtain the needed MSF gains.

5.2   MSF IMPLEMENTATIONS OF 1DF IMC

5.2.1 MSF for Models of the Form of Eq. (5.4)

The process model for this section is

$$y(s) = \frac{N(s)e^{-Ts}}{D(s)} u(s), \quad D(0) = 1, \tag{5.4}$$

where $y(s)$ and $u(s)$ are the Laplace transforms of the model output and the control effort, $D(s)$, is an nth order polynomial in s, and $N(s)$ is an mth order polynomial in s ($m \leq n$). For convenience, we take the gain of $D(s)$ in Eq. (5.4) to be one (i.e., $D(0) = 1$ and $\tilde{p}(0) = N(0)$).

In model state feedback, the IMC control effort, $u(t)$, for the above model is formed as a linear combination of the states of the model[2] plus a constant times the setpoint minus disturbance signal. That is, the control effort is formed as[3]

$$u(t) = -\sum_{i=0}^{n-1} k_i \frac{d^i}{dt^i} x(t) + k_{sp} \left[r(t) - \tilde{d}(t) \right], \tag{5.5}$$

where

$$x(s) \equiv \frac{1}{D(s)} u(s) . \tag{5.6}$$

---

[2] The states of an nth order model (i.e., one where $D(s)$ is an nth order polynomial) are normally taken to be any nonsingular transformation of the model output and its $n - 1$ derivatives.

[3] Eq. (5.5) is written in the time domain because all controller realizations are in the time domain. The Laplace transform notation is often convenient, but ultimately it has to be interpreted and implemented in the time domain.

From Equations (5.6) and (5.4), the model output $y(s)$ can be formed as

$$y(s) = N(s)e^{-Ts}x(s). \tag{5.7}$$

Notice that the definition of the states of the model allows the model output $y(t)$ to be formed as a linear combination of $x(t - T)$ and its m derivatives (e.g., if $N(s) \equiv \sum_{i=0}^{m} \gamma_i s^i$,

then $y(t) = \sum_{i=0}^{m} \gamma_i \dfrac{d^i}{dt^i} x(t - T)$).

The constants k_i, $i = 0, 1, 2 \ldots n - 1$, and K_{sp} in Eq. (5.5) are chosen so that *in the absence of constraints the control effort, u(t), is exactly the same as that formed by the IMC controller q(s)* in Figure 3.1. The advantage of such an implementation is that it makes use of the model's knowledge of past controls that is contained in its state. The MSF controller can thereby compensate for past control effort saturation, unlike the standard lead-lag IMC controller implementation which computes its output undeterred by whether or not the computed controls are actually applied.

One method of computing the constants, k_i and K_{sp}, in Eq. (5.5) is to equate the control effort given by Eq. (5.5) to the output of the IMC controller $q(s)$ in Figure 3.1. However, since the model output $x(t)$ in Eq. (5.5) depends on the control effort $u(t)$ by Eq. (5.6), the first task is to eliminate $x(t)$ in favor of $u(t)$.

Substituting Eq. (5.6) into the Laplace transform of Eq. (5.5) and solving for $u(s)$ gives

$$u(s) = \frac{K_{sp}}{(1 + K(s)/D(s))}(r(s) - \tilde{d}(s)). \tag{5.8}$$

where:

$$K(s) = \sum_{i=0}^{n-1} k_i s^i.$$

The control effort computed by the IMC controller in Figure 3.1 is

$$u(s) = q(s)(r(s) - \tilde{d}(s)). \tag{5.9}$$

Equating the control efforts in equations (5.8) and (5.9) and solving for $K(s)$ gives

$$K(s) = \frac{K_{sp}D(s)}{q(s)} - D(s). \tag{5.10}$$

The term on the left-hand side of Eq. (5.10) is an $(n - 1)^{th}$ order polynomial in s. Therefore the term on the right must also be an $(n - 1)^{th}$ polynomial. To see that the term on the right of Eq. (5.10) is indeed a polynomial, we note that from Chapter 3, for a model of the form of Eq. (3.14) (or, equivalently, Eq. (5.4)) the IMC controller is formed as

$$q(s) = \frac{D(s)}{N_m(s)(\varepsilon s + 1)^r},$$

(5.11)

where $N_m(s)$ has the same order as $N(s)$, but has only left half plane zeros, and $N_m(0) = \tilde{p}(0)$. Substituting Eq. (5.11) into Eq. (5.10) gives

$$K(s) = K_{sp} N_m(s)(\varepsilon s + 1)^r - D(s).$$

(5.12)

The polynomials $K_{sp} N_m(s)(\varepsilon s + 1)^r$ and $D(s)$ on the right-hand side of Eq. (5.12) are n^{th} order, while the polynomial $K(s)$ on the left-hand side of Eq. (5.12) is only $(n-1)^{th}$ order. Therefore, for consistency, the parameter K_{sp} must be chosen so that the coefficient of the nth order term of $K_{sp} N_m(s)(\varepsilon s + 1)^r - D(s)$ is zero. That is, we require that

$$K_{sp} = \lim_{s \to \infty} \frac{D(s)}{N_m(s)(\varepsilon s + 1)^r} = q(\infty).$$

(5.13)

The limit in Eq. (5.13) yields the ratio of the highest order term in $D(s)$ to that in $N_m(s)(\varepsilon s + 1)^r$. Since this ratio is the same as that for the IMC controller $q(s)$, we find that the constant K_{sp} is obtained simply by evaluating the IMC controller at infinity. The constants k_i, $i = 0, 1, \ldots, (n-1)$ are obtained from the coefficients of the polynomial on the right hand side of Eq. (5.12) after substituting for K_{sp} from Eq. (5.13).

Figure 5.5 is a schematic diagram of a MSF implementation for the process of Eq. (5.4) with the model given by Equations (5.6) and (5.7) and the control law given by Eq. (5.5).

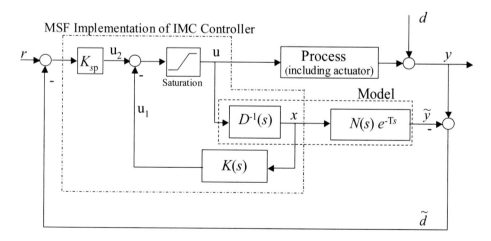

Figure 5.5 Schematic representation of MSF.

The arguments of the process variables in Figure 5.5 have been purposely omitted. In a simulation they would all be functions of time. However, it is more convenient to represent the operators such as $K(s)$ in the Laplace domain rather than in the time domain. Section 5.3 shows how such operators are actually described to a dynamic simulator or a digital control system.

Example 5.4 Computation of MSF Gains

Here we compute the MSF gains for the process of Eq. (5.14a).

$$\tilde{p}(s) = \frac{(s^2 + 2s + .25)e^{-s}}{s^4 + 6.5s^3 + 15s^2 + 14s + 4}. \tag{5.14a}$$

The IMC controller for the above process with a noise amplification of 20 is

$$q(s) = \frac{s^4 + 6.5s^3 + 15s^2 + 14s + 4}{(s^2 + 2s + .25)(.0559s + 1)^2}. \tag{5.14b}$$

From Eq. (5.13), we have

$$K_{sp} = q(\infty) = 1/(.0559)^2 = 320.02 . \tag{5.14c}$$

At this point we could substitute Equations (5.14a), (5.14b), and (5.14c) into Eq. (5.12) and compute the MSF gains. However, because the gain of $D(s)$ is 4 rather than 1, the feedback gains computed in this manner will be different from those computed by IMCTUNE because IMCTUNE normalizes the gain of $D(s)$ to 1 (i.e., $D(0) = 1$). Further, all the SIMULINK diagrams of the next sections also normalize the gain of $D(s)$ to 1. With the gain of $D(s)$ normalized to 1, substituting Equations (5.14a), (5.14b), and (5.14c) into Eq. (5.12) gives

$$K(s) = (.25s^4 + 9.449s^3 + 97.96s^2 + 162.2s + 20) - (.25s^4 + 1.625s^3 + 3.756s^2 + 3.5s + 1)$$

$$= 7.824s^3 + 94.20s^2 + 158.7s + 19 \equiv k_3s^3 + k_2s^2 + k_1s + k_0 \tag{5.14d}$$

so

$$[k_3 \ k_2 \ k_1 \ k_0] = [7.824 \quad 94.20 \quad 158.7 \quad 19]. \tag{5.14e}$$

♦

5.2.2 MSF for Models with Transcendental Numerators

Consider

$$y(s) = \frac{N(s, e^{-s})e^{-Ts}}{D(s)} u(s), \quad D(0) = 1 \tag{5.15}$$

where $N(s,e^{-s})$ is polynomial in s and e^{-s}, and time has been normalized so that all dead times can be represented as a power of the unit dead time. Further, the highest power of s in $N(s,e^{-s})$ is not multiplied by an exponential term.

Models of the form of Eq. (5.15) can occur whenever two or more processes interact to form the process output.

Following the methods of Chapter 3, the IMC controller for the above process can be written as

$$q(s) = \frac{D(s)}{N_m(s,e^{-s})(\varepsilon s + 1)^r} \qquad (5.16)$$

where all the zeros of $N_m(s,e^{-s})$ are in the left half of the s-plane.

For the above process and controller, MSF requires feeding back the current and delayed values of the model state. That is,

$$K(s,e^{-s}) = K_{sp} N_m(s,e^{-s})(\varepsilon s + 1)^r - D(s). \qquad (5.17a)$$

The constant K_{sp} is again chosen to make the coefficient of the nth order term of the polynomial in s vanish so that

$$K_{sp} = \lim_{s \to \infty} \frac{D(s)}{N_m(s,e^{-s})(\varepsilon s + 1)^r} = q(\infty). \qquad (5.17b)$$

Figure 5.6 gives a schematic diagram for MSF for the above process and controller of Equations (5.15) and (5.16).

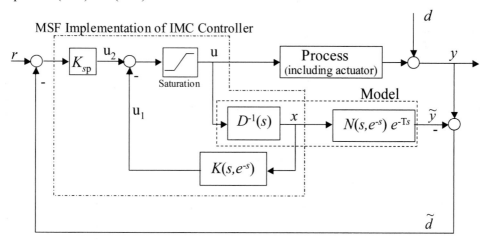

Figure 5.6 Schematic representation of MSF for the process of Eq. (5.15).

Example 5.5 Computation of MSF Gains for Processes with Transcendental Numerators

Compute the MSF gains for the process of Eq. (5.18a).

$$\tilde{p}(s) = \frac{1+0.5e^{-3s}}{(2s+1)^2(.5s+1)^2}.$$
(5.18a)

One possible choice for an IMC controller is

$$q(s) = \frac{(2s+1)^2(.5s+1)^2}{(1+0.5e^{-3s})(\varepsilon s+1)^4}.$$
(5.18b)

The maximum noise amplification in Eq. (5.18b) occurs when $s = (2n + 1)\pi i \omega$, $n = 1, 2, \ldots$ for large ω. (Here, $i \equiv \sqrt{-1}$.) At such points, $q(\infty)/q(0) = 1.5/.5\varepsilon^4$. So, for a maximum noise amplification of 20, $\varepsilon^4 = 3/20$ and $\varepsilon = 0.622$.

In the above model, $N_m(s) = (1+.5e^{-3s})$ and Eq. (5.17a) becomes

$$K(s) = K_{sp}(1+.5e^{-3s})(\varepsilon s+1)^4 - (2s+1)^2(.5s+1)^2$$
$$= \{K_{sp}(\varepsilon s+1)^4 - (2s+1)^2(.5s+1)^2\} + K_{sp}(.5e^{-3s})(\varepsilon s+1)^4$$
(5.19)

Only the fourth order term inside the braces in Eq. (5.19) can be made to disappear by appropriate selection of K_{sp}. The fourth order term multiplied by the exponential cannot be removed. However, the fourth derivative of the output delayed by the three units is available from past data and so, there is no algebraic loop in the MSF diagram resulting from this term. The value of K_{sp} required to cancel the fourth order term in the braces is $K_{sp} = 1/\varepsilon^4 = q(\infty) = 20/3$. Substituting the foregoing into Eq. (5.19) gives

$$K(s) = 1.427s^3 + 7.242s^2 + 11.60s + 5.667 + 3.333(s^4 + 5s^3 + 8.25s^2 + 5s + 1)e^{-3s}.$$
(5.20)

The current version of IMCTUNE (2001) does not allow entry of the IMC controller given by Eq. (5.18b). Therefore $K(s)$ given by Eq. (5.20) cannot be obtained directly from IMCTUNE. However, IMCTUNE can be "fooled" into providing the first four terms on the right of Eq. (5.20) by entering

$$p(s) = 1/(2s+1)^2(.5s+1)^2; \quad q(s) = (2s+1)^2(.5s+1)^2/(\varepsilon s+1)^4 \text{ with } \varepsilon = 0.6223.$$

♦

5.3   SIMULINK REALIZATIONS OF MSF IMPLEMENTATIONS OF IMC

5.3.1 Overview

Most simulation and digital control software use block diagrams rather than differential equations to describe the process models and the control system to the computer. The reasons for this approach are both historical and practical. Historically, dynamic simulations were originally performed on analog computers, and the blocks of the simulation diagrams actually represented individual analog components. The block diagram descriptions were carried over into digital simulations first because of their familiarity, and second because they facilitate the debugging and modularization of the simulation or the control system. In addition, the block diagrams provide an unambiguous description of the system to the computer that can be readily transformed into algorithms for computing the desired variables. The next sections describe how the schematic diagrams of Figures 5.5 and 5.6 are converted into SIMULINK realizations of MSF control systems.

5.3.2 Model Realization

In order to implement the MSF law given Eq. (5.5), it is necessary that the method used to simulate the model produce $x(t)$ as defined by Eq. (5.6) and its first $(n-1)$ derivatives. That is, we want to simulate

$$\beta_n x^{(n)}(t) + \beta_{n-1} x^{(n-1)}(t) + \cdots \beta_1 x'(t) + x(t) = u(t), \tag{5.21a}$$

where

$$D(s) \equiv \sum_{i=0}^{n-1} \beta_i s^i, \quad \beta_0 = 1. $$

A simple way of describing Eq. (5.21a) to SIMULINK (and indeed to most other simulators) is to solve Eq. (5.21a) for the highest order derivative and then use integrators to form all the lower order derivatives. That is,

$$x^{(n)}(t) = \{u(t) - (\beta_{n-1} x^{(n-1)}(t) + \cdots \beta_1 x'(t) + x(t))\} / \beta_n. \tag{5.21b}$$

Figure 5.7 shows the SIMULINK diagram for Eq. (5.21b) for n = 3.

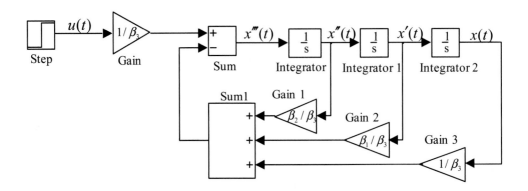

Figure 5.7 A SIMULINK diagram for a third order model of the form of Eq. (5.21a).

A somewhat simpler, and therefore often more convenient, form for the SIMULINK diagram results if we form the state vector using SIMULINK's multiplexer ("Mux") and "Matrix Gain," as shown in Figure 5.8.

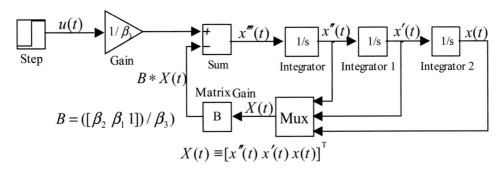

Figure 5.8 An alternate SIMULINK diagram for a third order model of the form of Eq. (5.21b).

Please note that in Figure 5.8, $X(t)$ is a column vector, the prime symbol ' denotes differentiation, and ^{T} denotes transpose. Also, since B is a row vector, the product $B*X(t)$ is a scalar.

In order to complete the simulation diagram for the model given by Eq. (5.4), we note that from the definition of $x(s)$ given by Eq. (5.6), the model output $\tilde{y}$ is given by Eq. (5.7) which is repeated below

$$\tilde{y}(s) = e^{-Ts} N(s) x(s).$$

If we write $N(s)$ as

$$N(s) \equiv \sum_{i=0}^{n-r} c_i s^i \qquad (5.22)$$

then $\widetilde{y}$ can be written as

$$\widetilde{y}(t+T) = \sum_{i=0}^{n-r} c_i x^{(i)}(t), \qquad (5.23)$$

where r is the relative order of $N(s)/D(s)$.

Figure 5.9 gives the SIMULINK diagram for the complete model.

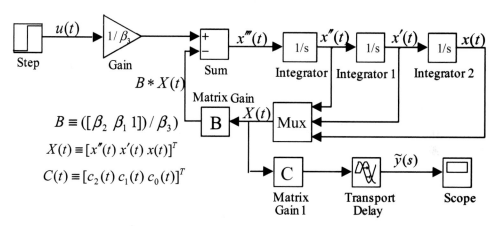

Figure 5.9 A SIMULINK diagram for a third-order plus dead time model of the form of Eq. (5.4).

5.3.3 Computing MSF Gains

The software system IMCTUNE will compute MSF gains once the model, controller, filter order, and filter time constant have been entered. The model numerator and denominator polynomials can be entered in any form (e.g., as single polynomials, in time constant form, or in mixed forms). The MSF computations automatically normalize the denominator polynomial so that it has a gain of one for the MSF gain calculation. Thus, to use the MSF gains provided by IMCTUNE, it is necessary that β_0 be one as in Figures 5.8 and 5.9.

5.3.4 State Feedback Realizations

To get the MSF realization of an IMC controller from the diagram given by Figure 5.9 it is only necessary to multiply the state vector $X(t)$ by the feedback gains calculated from Eq. (5.12) and subtract this quantity from $K_{sp}(r - \widetilde{d})$. The completed diagram is given in Figure 5.10.

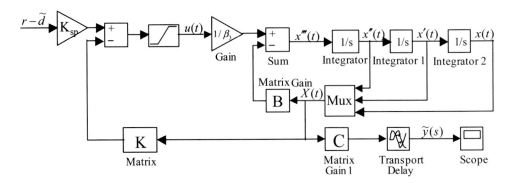

Figure 5.10 A SIMULINK MSF realization of an IMC system.

Example 5.6 A MSF Implementation for Example 5.4

Recall that the process and controller for Example 5.4 are given by Equations (5.14a) and (5.14b)

$$\widetilde{p}(s) = \frac{(s^2 + 2s + .25)e^{-s}}{(s^4 + 6.5s^3 + 15s^2 + 14s + 4)} \quad ;$$

$$q(s) = \frac{(s^4 + 6.5s^3 + 15s^2 + 14s + 4)}{(s^2 + 2s + .25)(.0559s + 1)^2} .$$

The MSF gains computed in Example 5.4a are given by Eq. (5.14e).

$$[k_3 \ k_2 \ k_1 \ k_0] = [7.824 \quad 94.20 \quad 158.7 \quad 19].$$

Figure 5.11 gives the entire MSF diagram for this example. Notice that to form the model output from the model state, we have used the fact that the gain of the numerator and denominator of the process in Eq. (5.20) have to be normalized so that the gain of the denominator is one.

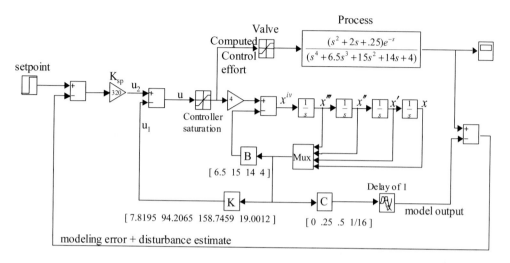

Figure 5.11 An MSF implementation for Example 5.4 and 5.6.

♦

Example 5.7 An MSF implementation for Example 5.5

The process model and controller for Example 5.5 are given by Equations (5.18a) and (5.18b).

$$\tilde{p}(s) = \frac{1 + 0.5e^{-3s}}{(2s+1)^2(.5s+1)^2} \ ; \ q(s) = \frac{(2s+1)^2(.5s+1)^2}{(1+0.5e^{-3s})(\varepsilon s+1)^4}.$$

The MSF gains computed in Example 5.4a are given by Eq. (5.20), which is repeated below.

$$K(s) = 1.427s^3 + 7.242s^2 + 11.60s + 5.667 + 3.333(s^4 + 5s^3 + 8.25s^2 + 5s + 1)e^{-3s}.$$

Figure 5.12 gives the complete MSF implementation for the model and controller of Equations (5.18a) and (5.18b).

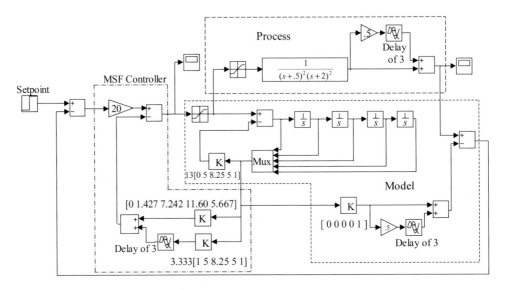

Figure 5.12 An MSF implementation for Examples 5.5 and 5.7.

♦

5.4   FINDING A SAFE LOWER BOUND ON THE MSF FILTER TIME CONSTANT

In Section 5.1 (Motivation), we pointed out that lead-lag implementations of IMC systems suffer from two problems: (1) The control system does not compensate for the fact that the calculated control effort may not have been applied because of constraints, and (2) the control system implicitly assumes that any needed future controls can be applied. An MSF implementation of IMC solves the first problem, but not necessarily the second. The MSF control system presented in the previous sections does not recognize that needed future controls may not be available due to constraints, and this defect can lead to performance degradation when the controller filter time constant is too "small." Fortunately, a filter time constant small enough to cause problems in MSF performance is usually smaller than the minimum filter time constant for a noise amplification factor of 20. Nonetheless, to be safe, we recommend using the algorithm of Mhatre and Brosilow (2000) to find a safe lower bound on the controller filter time constant. This algorithm is quite easy to apply, and is not overly conservative, as we demonstrate by example. The algorithm is summarized in the two steps below.

1. Transform the lower bound on the control effort to zero, set the upper bound to infinity, and set the filter time constant to its smallest allowable value.

2. Starting at steady state with all controls at zero, make an arbitrary step setpoint change in a direction so that the final value of the control effort is positive. If necessary, increase the filter time constant so that the computed control effort is never negative over its entire trajectory.

Transforming all control effort lower bounds to zero is convenient, but is not actually necessary. What is important is that the tuning starts at a steady state where all of the controls are at their lower (upper) bounds. Setting the upper (lower) bounds to infinity (minus infinity) is also not necessary provided that the setpoint changes are small enough that the control effort upper (lower) bounds are not encountered. The idea is that we wish to drive the process as hard as possible in the positive (negative) direction so as to find the smallest filter time constant such that none of the computed controls exceed their lower (upper) bounds.

The philosophy behind the above tuning procedure is to assure that the control system can apply enough control in the direction opposite to the initial control effort response, no matter how large that initial response. If the initial control effort hits the upper bound, and the MSF algorithm keeps it there for some period of time, the momentum built up by the system will still be less than that which would have been achieved in the absence of the upper bound constraint. Thus, if the subsequent control effort does not drop below zero (the new lower bound constraint) there is enough control effort in the "opposite" direction to recover from whatever momentum is developed.

Based on the following examples, and many others not shown here, we are confident that the above simple algorithm will find a non-conservative lower bound on the controller filter time constant that avoids performance degradation in MSF systems due to control effort constraints.

Example 5.8 Computation of Filter Time Constant
for Example 5.2

Recall that the process and controller for Example 5.2 are taken from Eq. (5.2a), Eq. (5.2b), with a variable filter time constant ε replacing the previous value of 0.8, and Eq. (5.2c).

$$\tilde{p}(s) = \frac{1}{(4s+1)(3s+1)(2s+1)(s+1)},$$

$$q(s) = \frac{(4s+1)(3s+1)(2s+1)(s+1)}{(\varepsilon s+1)^4},$$

$$0 \le u(t) \le 100.$$

A convenient way of applying the above algorithm to find the smallest "safe" filter time constant for Example 5.2 is to use LTIVIEW to obtain the step response of the controller given by Eq. (5.2b) for several trial values of ε. Notice that there is no need to impose constraints since the upper bound is infinite, and we seek the minimum filter time constant that causes the control effort to just touch the zero axis. Starting with a filter time constant 1.05 for a noise amplification of 20 gives a controller whose step response dips below zero by about 0.1 units. Increasing the filter time constant to 1.06 provides the desired response as shown by Figure 5.13. The MSF gains for a filter time constant of 1.06 are:

$$K = [40.5660 \quad 93.1595 \quad 70.6035 \quad 18.0102] \text{ and } K_{sp} = 19.010 \qquad (5.24)$$

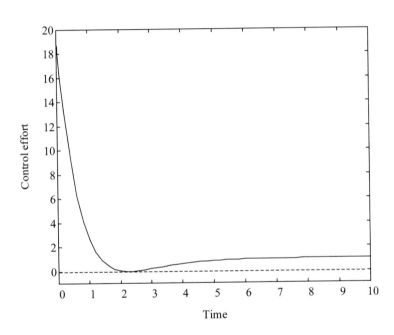

Figure 5.13 Response of control effort for the controller of Eq. (5.2b) with a filter time constant of 1.06.

The responses of the process output of the MSF control system for the foregoing IMC controller are shown in Figure 5.14a for constraints on the control effort of zero to 10 and zero to 2. The output response for control effort constraints between 0 and 10 are almost imperceptibly different from the output response for constraints between 0 and 100 for a filter time constant of 1.06. Notice that from Figure 5.13, constraints on the control effort from zero to 100 are never exceeded, so the response with such constraints is the unconstrained response. Control effort responses for constraints of zero to ten and zero to two are given in Figure 5.14b.

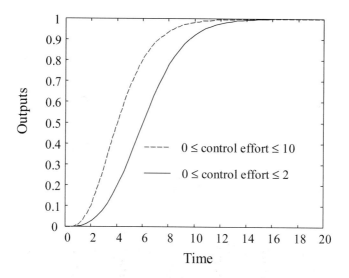

Figure 5.14a Output responses of a MSF control system with the model and controller given by Eq. (5.2a) and Eq. (5.2b) with a filter time constant of 1.06.

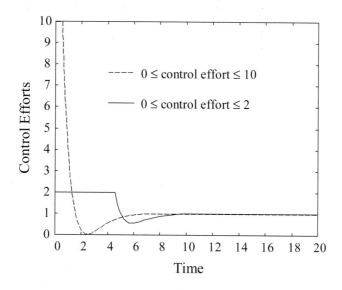

Figure 5.14b Control effort responses of a MSF control system with the model and controller given by Equations (5.2a) and (5.2b) with a filter time constant of 1.06.

♦

We now consider a process that has stable zeros very near the imaginary axis.

Example 5.9 Computation of Filter Time Constant for an Oscillatory Controller

Consider the process discussed in Chapter 3 and repeated below.

$$\tilde{p}(s) = \frac{s^2 + 0.001s + 1}{(s+1)^4} \tag{5.25a}$$

As discussed in Chapter 3, choosing a controller that inverts the process will result in oscillating control effort that amplifies noise more than is acceptable for inputs that have any noise around a frequency of one. However, we shall nonetheless use such a controller to provide an extreme test of the tuning procedure. The controller is then:

$$q(s) = \frac{(s+1)^4}{(s^2 + 0.001s + 1)(\varepsilon s + 1)^2} \tag{5.25b}$$

Let the lower bound on the control effort be 0 with no upper bound. Applying the tuning procedure gives $\varepsilon = 1.75$. Figure 5.15 shows the output and the control effort obtained for a perfect model.

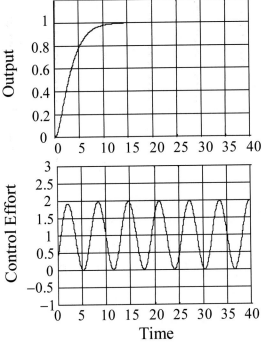

Figure 5.15 Response of the process and controller given by Equations (5.25a) and (5.25b) with $\varepsilon = 1.75$.

While the control effort is quite oscillatory, as might be expected for a controller with a damping ratio of .0005, the system is stable, and the control effort does settle in about 10,000 time units.

Figure 5.16 shows the response of the system with the same filter constant and an upper bound of 1.2 on the control effort. Notice that the control effort never hits the lower bound and the process remains stable. The control effort still takes about 10,000 time units to settle to one.

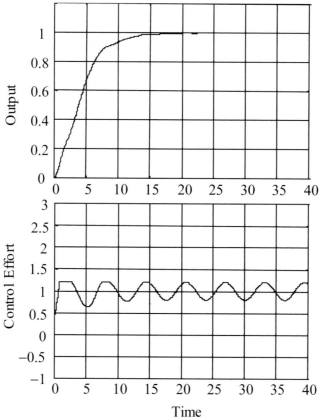

Figure 5.16 Response of the process and controller given by Equations (5.25a) and (5.25b) with ε = 1.75 and upper bound of 1.2 on the control effort.

To avoid an oscillating control effort, zeros near imaginary axis should not be inverted. Instead, we suggest using a controller of the form

$$q(s) = \frac{(s+1)^4}{(\varepsilon s+1)^4} \qquad (5.25c)$$

A maximum noise amplification factor of 20 for the above controller requires a filter time constant of 0.473 or greater. Tuning the controller for saturation yields a filter time constant of 0.5. As can be seen from Figure 5.17, the control effort from Eq. (5.25c) does not oscillate, and the output response is actually faster than that in Figure 5.16.

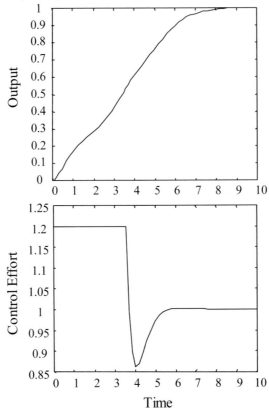

Figure 5 .17 Response of the process and controller given by Equations (5.25a) and (5.25c) with ε = 0.5 and upper bound of 1.2 on the control effort.

♦

5.5   MSF IMPLEMENTATIONS OF 2DF IMC

In order to implement a 2DF MSF system it is convenient to redraw Figure 4.2 as shown in Figure 5.18.

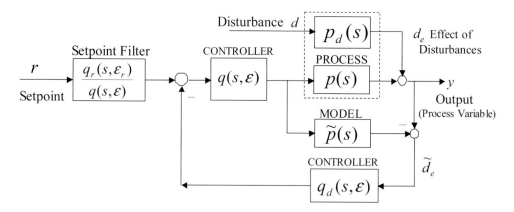

Figure 5.18 A 2DF control system in a form suitable for MSF.

The setpoint filter $q_r(s, \varepsilon_r) q^{-1}(s, \varepsilon)$ in Figure 5.18 is the same as $f(s, \varepsilon_r) f^{-1}(s, \varepsilon)$, where $f(s, \varepsilon)$ is the IMC controller filter. Replacing the IMC controller $q(s, \varepsilon)$ in Figure 5.18 with its MSF realization yields Figure 5.19

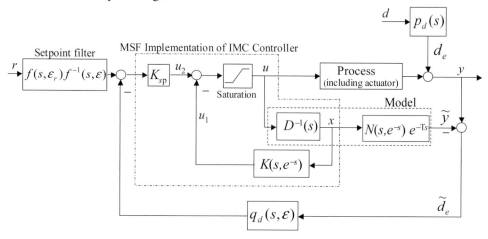

Figure 5.19 MSF implementation of a 2DF control system.

The selection of $K(s, e^{-s})$ in Figure 5.19 proceeds as described previously, as does the calculation of the minimum allowable controller filter time constant.

5.6 SUMMARY

MSF is the preferred method of implementing IMC systems. It is simpler to implement than is a lead-lag controller and provides substantially improved performance in the presence of control effort constraints. The method suggested for tuning the controller filter to accommodate control effort constraints is straightforward, and does not seem to increase the filter time constant much beyond that required for a maximum noise amplification factor of 20.

Problems

5.1 Compute the minimum MSF controller filter time constant for each of the processes of problems 3.4 and 3.5 in Chapter 3, and compare these time constants with those required for a noise amplification of 20.

5.2 Implement an MSF control system for each of the following processes, and compare its performance with a lead-lag implementation of an IMC system that uses the same filter time constant as the MSF system for a setpoint change from 0 to 1.

a.
$$\frac{s^3 - 1.9s^2 + .8s - 2}{(s^2 + .2s + 1)(3s + 1)^2(s + 1)} \qquad -1 \le u(t) \le 1$$

b.
$$\frac{1 + .95e^{-3s}}{(s + 2)^2(s + .5)^2} \qquad 0 \le u(t) \le 1$$

c.
$$\frac{1 - .95e^{-3s}}{(s + 2)^2(s + .5)^2} \qquad 0 \le u(t) \le 25$$

d.
$$\frac{1 + 2e^{-s}}{s + 1} \qquad 0 \le u(t) \le 0.5$$

e.
$$\frac{e^{-10s}}{(2s + 1)^5} \qquad 0 \le u(t) \le 1.5$$

References

Coulibaly, E., S. Maiti, and C. Brosilow. 1995 (Oct.). "Internal Model Predictive Control (IMPC)." *Automatica* 31.

Mhatre, S., and C. Brosilow. 2000 (Aug.). "Multivariable MSF: A Computationally Simple, Easy to Tune Alternative to MPC." *AIChE Journal*, Vol. 46, No. 8, 1566–1580.

PI and PID Parameters from IMC Designs

Objectives of the Chapter

- Provide methods for approximating most IMC controllers with I, PI, or PID controllers.
- Provide methods for approximating all other IMC controllers with PI or PID controllers cascaded with first-order or second-order lags.

Prerequisite Reading

Chapter 3, "One-Degree of Freedom Internal Model Control"
Chapter 4, "Two-Degree of Freedom Internal Model Control"
Appendix A.3, "P, PI, and PID Controller Transfer Functions"

6.1   INTRODUCTION

While IMC controller implementations are becoming more popular (see Chapter 5), the standard industrial controllers remain the proportional (P), proportional plus integral (PI), and the proportional plus integral and derivative (PID) controllers. Morari and Zafiriou (1989) and Rivera et al. (1986) show how to approximate the IMC controller for a limited class of processes with PI and PID controllers. To obtain a PID controller for the industrially important first-order lag and dead time process model, they approximate the dead time with a low-order Padé approximation. Because their approximations are surprisingly good, they conclude that there is relatively little to be gained in dynamic response by implementing the IMC controller rather than the PI or PID approximation.

While motivated by Morari's work (Morari, 1989), this chapter presents a more general, yet conceptually simple, approach to approximating IMC controllers with I, PI or PID controllers using Maclaurin series approximations. The PID parameters so obtained all depend on the IMC filter time constant. As the filter time constant gets larger, the PID controller goes smoothly into a PI and a floating integral controller, just as intuition says it should. Since our IMC design methods always result in an offset-free control system, our simple feedback approximations always have an integrating action, and one never obtains a simple proportional controller (which always leads to offset for non-integrating processes).

This chapter assumes that the IMC filter time constant is known a priori either from considerations of process uncertainty as discussed in Chapter 7, or from consideration of noise amplification in the case of a perfect model as discussed in Chapter 3. However, the filter time constant that provides the desired maximum noise amplification for the IMC system often yields a PID controller that results in excessive overshoot to a step setpoint change. We have found that increasing the filter time constant always reduces the overshoot to an acceptable level. With the aid of IMCTUNE two or three trials usually provide a filter time constant that yields the desired overshoot. We generally take a 10% overshoot to a step setpoint change for a perfect process model to be a good approximation of the IMC closed loop response.

The approximation methods of this chapter provide analytical expressions for the dependence of controller gain, integral and derivative time constants model parameters, and the filter time constant for simple, but industrially important, process models such as a first-order lag plus dead time. In general, however, the simplest method for obtaining the PID parameters associated with any IMC design is via a computer program such as IMCTUNE that carries out the necessary calculations. The algorithms used in this program are described in sections 3 and 4 of this chapter. The inputs required by IMCTUNE to compute PID parameters are the process model, the portion of the model to be inverted by the controller, and the IMC controller filter time constant.

In the following section we review the time domain and Laplace transform representations of a common form of PID controller that is implemented in IMCTUNE. There is a more extensive discussion of the various types of controllers in Appendix A.3, and readers not familiar with the different types of controllers are urged to review this appendix.

6.2   THE PID CONTROLLER

The form of the PID controller that we will use in this chapter is given by

$$u(s) = K_c(1 + \frac{1}{\tau_I s} + \frac{\tau_D s}{\alpha \tau_D s + 1})e(s) \qquad (6.1)$$

where α is a constant that is usually fixed between .05 and .2, depending on the manufacturer, to limit its high frequency gain to between 5 and 20 times its proportional gain K. Recall that in Chapter 3 on IMC controller design for perfect models, we adopted the same high frequency gain limitation for the IMC controller as is used for the PID controllers.

A time domain implementation of the PID controller given by Eq. (6.1) can be expressed in terms of the error between the setpoint and measured process output, $e(t)$, as follows

$$u(t) = K_c\left(e(t) + \tfrac{1}{\tau_I}\int e(t)dt + \tfrac{1}{\alpha}(e(t) - e_L(t))\right), \qquad (6.2)$$

Where $e_L(t)$ is a filtered version of $e(t)$, given by

$$\alpha \tau_D \frac{d}{dt} e_L(t) + e_L(t) = e(t). \qquad (6.3)$$

The PID controller implemented in IMCTUNE is that given by Eq. (6.1) and its time domain representation given by Eq. (6.2).

The next section shows how to use IMC controller designs to find the PID gain and integral and derivative time constants that give IMC-like closed-loop responses.

6.3   1DF PID PARAMETERS FROM 1DF IMC

6.3.1 1DF PID Parameters for General Process Models

As shown in Section 3.2, the IMC structure reduces to the classical feedback structure of Figure 6.1, with the feedback form of the IMC controller, $c(s)$, given by Eq. (3.1), which is repeated below.

$$c(s) = \frac{q(s)}{1 - \widetilde{p}(s)q(s)},$$

where

$$q(s) = \text{1DF IMC controller in Figure 6.1,}$$

$\tilde{p}(s) =$ process model in Figure 6.1.

The controller given by $c(s)$ is called the feedback form of the IMC controller (see Figure 6.1) because it gives the same response as the IMC system shown in Figure 3.1.

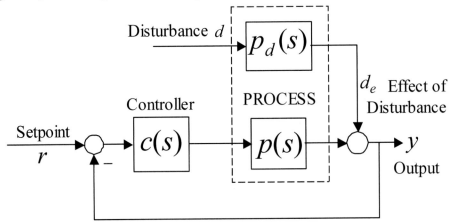

Figure 6.1 Simple 1DF feedback control system.

The design techniques developed in Sections 3.3, 3.4, and 3.5 for the 1DF IMC controller $q(s)$ all lead to controllers of the form

$$q(s) = \frac{p_m^{-1}(s)}{(\varepsilon s + 1)^r} \tag{6.4}$$

where

$p_m(s) =$ the portion of the process model inverted by the IMC controller, $q(s)$

$r =$ relative order of $p_m(s)$.

Substituting Eq. (6.4) into Eq. (3.1) gives

$$c(s) = \frac{p_m^{-1}(s)}{(\varepsilon s + 1)^r - p_A(s)} \tag{6.5}$$

where

$p_A(s) =$ portion of the process model that is not inverted.

Often, $p_A(s)$ will be an "all pass" ratio of polynomials cascaded with a dead time term, e^{-Ts}. However, in general, $p_A(s)$ can be any portion of the process model that the designer does not wish to invert, provided that the gain of $p_A(s)$ is one so that there is no offset[1]. That is,

---

[1] No offset means that the output tracks the setpoint exactly in the steady state and requires that $p_A(0) = 1$.

$$\widetilde{p}(s) \equiv p_m(s)p_A(s); \quad p_A(0) = 1. \tag{6.6}$$

Because $p_A(0)$ is 1, the denominator of Eq. (6.5) has zero at the origin, and $c(s)$ can be re-written as

$$c(s) \equiv \frac{1}{s}f(s), \tag{6.7}$$

where

$$f(s) \equiv \frac{sp_m^{-1}(s)}{((\varepsilon s+1)^r - p_A(s))}.$$

Expanding $f(s)$ in a Maclaurin series in s gives

$$c(s) = \frac{1}{s}(f(0)+f'(0)s+\frac{f''(0)}{2}s^2+\ldots), \tag{6.8}$$

$$= K_c(1+\frac{1}{\tau_I s}+\tau_D s+\ldots) \tag{6.9}$$

where

$$K_c = f'(0); \quad \tau_I = f'(0)/f(0); \quad \tau_D = f''(0)/2f'(0). \tag{6.10}$$

In order to evaluate the PID controller given by Eq. (6.10), we let

$$D(s) \equiv ((\varepsilon s+1)^r - p_A(s))/s. \tag{6.11a}$$

Then, by Maclaurin series expansion, we get

$$D(0) = r\varepsilon - p_A'(0), \tag{6.11b}$$

$$D'(0) = (r(r-1)\varepsilon^2 - p_A''(0))/2, \tag{6.11c}$$

$$D''(0) = (r(r-1)(r-2)\varepsilon^3 - p_A'''(0))/3. \tag{6.11d}$$

Using Eq. (6.11), the function $f(s)$ and its first and second derivatives, all evaluated at the origin, are given by

$$f(0) = \frac{1}{K_p D(0)}, \tag{6.12a}$$

$$f'(0) = \frac{-(p_m'(0)D(0)+K_p D'(0))}{(K_p D(0))^2}, \tag{6.12b}$$

$$f''(0) = f'(0)\left[\frac{p_m''(0)D(0) + 2p_m'(0)D'(0) + K_p D''(0)}{p_m'(0)D(0) + K_p D'(0)} + \frac{2f'(0)}{f(0)}\right], \quad (6.12c)$$

where

$$K_p \equiv p_m(0).$$

Equations (6.10) and (6.12) together can be used to obtain the controller gain and integral and derivative time constants as analytical functions of the process model and IMC controller parameters, including the IMC filter time constant. The PID parameters for a first-order lag dead time process are given below, and derived in the following subsection.

6.3.2   1DF PID Parameters for a First-Order Lag Plus Dead time Model

The process model is

$$p(s) = \frac{K_p e^{-Ts}}{(\tau s + 1)} \tag{6.13}$$

The 1DF PID parameters for the model in Eq. (6.13) are given in Eq. (6.14). The derivation of these parameters from Equations (6.10) and (6.12) follows in Section 6.3.3.

$$\tau_I = \tau + \frac{T^2}{2(\varepsilon + T)} \; ; \; K_c = \frac{\tau_I}{K_p(\varepsilon + T)} \; ; \; \tau_D = \frac{T^2(1 - \frac{T}{3\tau_I})}{2(\varepsilon + T)} \tag{6.14}$$

Notice that, as one might expect, all the above PID parameters depend on the filter time constant. Further, the gain and derivative time constants go to zero as the filter time constant gets large. The integral time constant, on the other hand, approaches the model time constant.

6.3.3   Derivation of the 1DF PID Parameters of Eq. (6.14) for a First-Order Lag Plus Dead time Model

Following the IMC controller design procedures in Chapter 3, we select

$$p_m(s) \equiv \frac{K_p}{(\tau s + 1)} \qquad p_A(s) \equiv e^{-Ts} \tag{6.15a}$$

Therefore,

$$p_m(0) = K_p; \quad p'_m(0) = -K_p\tau; \quad p''_m(0) = 2K_p\tau^2 \tag{6.15b}$$

$$p_A(0) = 1; \quad p'_A(0) = -T; \quad p''_A(0) = T^2; \quad p'''_A(0) = -T^3 \tag{6.15c}$$

Evaluating Equations (6.11b), (6.11c), and (6.11d) using Eq. (6.15c) and the fact that the relative order of the transfer function in Eq. (6.13) is one (i.e. $r = 1$) gives

$$D(0) = \varepsilon + T; \quad D'(0) = -T^2/2; \quad D''(0) = T^{3/3}. \tag{6.15d}$$

Substituting the above into Eq. (6.12) gives

$$f(0) = \frac{1}{K_p(\varepsilon + T)}, \tag{6.15e}$$

$$f'(0) = \frac{\tau(\varepsilon + T) + \dfrac{T^2}{2}}{K_p(\varepsilon + T)^2}, \tag{6.15f}$$

$$f''(0) = f'(0)\left[2\tau_I + \frac{2K_p\tau^2(\varepsilon + T) + K_p\tau T^2 + K_p T^3/3}{K_p\tau(\varepsilon + T) + K_p T^2/2}\right]. \tag{6.15g}$$

Finally, after some simplification, we get Eq. (6.14), which is repeated below.

$$\tau_I = \tau + \frac{T^2}{2(\varepsilon + T)}; \quad K_c = \frac{\tau_I}{K_p(\varepsilon + T)}; \quad \tau_D = \frac{T^2(1 - \dfrac{T}{3\tau_I})}{2(\varepsilon + T)}.$$

   To illustrate how well the PID parameters given by Eq. (6.14) perform in a control system we consider two particular FOPDT processes given by Eq. (6.13). We assume a perfect model in each case so that only noise amplification considerations limit how small we can make the filter time constant.

Example 6.1   PID Parameters for FOPDT with $K_p = \tau = T = 1$

In this case an allowable noise amplification factor of twenty for the IMC system dictates a filter time constant, ε, of .05. This gives

   $K = 1.406$, $\tau_I = 1.476$ and $\tau_D = .3687 : \alpha = .05$

   Using the controller form given by Eq. 6.1 with $\alpha = .05$ gives the response shown in Figure 6.2a. This response has an excessive overshoot. However, as stated previously, the overshoot of an IMC-induced PID control system can generally be reduced by increasing the IMC filter time constant. For the current model parameters increasing the filter time constant to 0.27 (found by trial and error using IMCTUNE) yields the closed-loop response of the PID control system that has only a 10% overshoot as shown in Figure 6.2b. The controller parameters are:

$$K = 1.097, \tau_I = 1.394, \tau_D = 0.2995, \alpha = .05$$

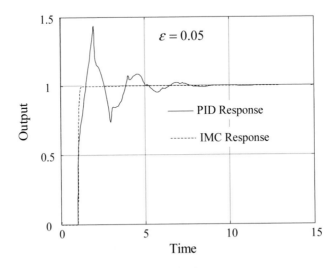

Figure 6.2a Perfect model closed-loop FOPDT response to step setpoint change.

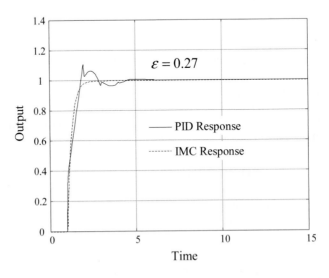

Figure 6.2b Perfect model closed-loop FOPDT response to step setpoint change.

Continuing to increase the filter time constant to .5 gives the control system response shown in Figure 6.2c.

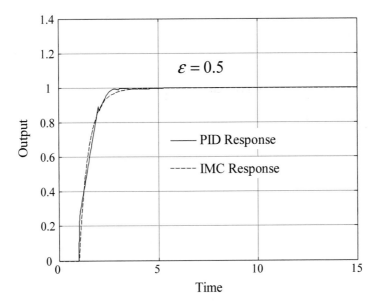

Figure 6.2c Perfect model closed-loop FOPDT response to step setpoint change.

The controller parameters for this case are

$$K = .889; \ \tau_I = 1.33; \ \tau_D = .25, \ \alpha = .05.$$

As seen from Figure 6.2c, there is little difference between the IMC and PID controllers for an IMC filter time constant of 0.5. Thus, we conclude that for a FOPDT process model *there is no essential difference between the performance of an IMC controller and the IMC induced PID controller for large enough values of the filter time constant,.*

Figure 6.2d compares the performance of PID control systems for this case when the PID parameters are obtained by the following methods:

1. The Ziegler-Nichols Reaction Curve Method (Appendix A and Seborg et al., 1989, page 303)

$$K_c = \frac{1.2\tau}{K_p T} = 1.2 \quad \tau_I = 2T = 2 \quad \tau_D = .5T = 0.5.$$

2. The Cohen and Coon Tuning Method (Appendix A and Seborg et al., 1989, page 282)

$$K_c = \frac{\tau}{K_p T}(1.33 + \frac{T}{4\tau}) = 1.58 ; \quad \tau_I = \frac{T(32\tau + 6T)}{(13\tau + 8T)} = 1.81 ; \quad \tau_D = \frac{4\tau T}{(11\tau + 2T)} = 0.308$$

3. The PID tuning obtained by Rivera et al. (1986) by approximating the IMC controller using a second-order Padé approximation for the dead tune.

$$K_c = (2\tau + T)/2(\varepsilon + T) = 1.2 \text{ for } \varepsilon = .25; \quad \tau_I = \tau + T/2 = 1.5 ; \quad \tau_D = \frac{\tau T}{(2\tau + T)} = .333$$

for a PID controller of the form:

$$K_c(1 + \frac{1}{\tau_I s} + \tau_D s)/(\alpha \tau_D s + 1),$$

where

$$\alpha = \frac{\varepsilon(\tau + .5T)}{\varepsilon + T} = 0.3 \text{ , and } \varepsilon = .25.$$

The Ziegler-Nichols and Cohen and Coon responses shown in Figure 6.2d were obtained using a controller of the form given by Eq. (6.1) with an α of 0.1. The PID tuning developed by Rivera et al. (1986) uses the controller shown under his tuning parameters. The minimum value of the filter time constant ε recommended by Rivera is 0.25, and so this value was used to obtain the curve shown in Figure 6.2d.

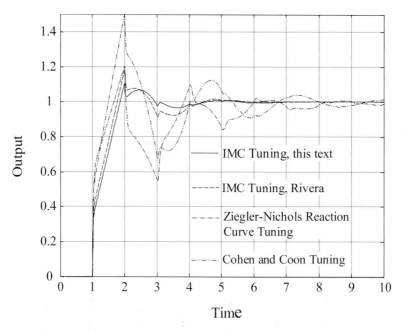

Figure 6.2d Comparison of tuning methods for perfect FOPDT model.

The response of a control system tuned by the Zeigler-Nichols continuous cycling method (Seborg et al., 1989) lies roughly between the responses of the Zeigler-Nichols reaction curve method and the Cohen and Coon method. It is evident from Figure 6.2d that the IMC tuning methods provide better closed-loop behavior than the traditional tuning methods for the example process.

♦

Example 6.2 PID Parameters for FOPDT with K_p = τ = T = 10

In this case the process model is almost a pure delay. A few trial responses with different filter time constants shows that to limit the overshoot of the response of a PID control system to a step setpoint change to 10% or less requires a filter time constant of 2.5 or greater.

Figure 6.3 shows the PID response for a filter time constant of 2.5, which, from Eq. (6.14) gives

$$K_c = .40, \ \tau_I = 5.0 \ \text{and} \ \tau_D = 1.33$$

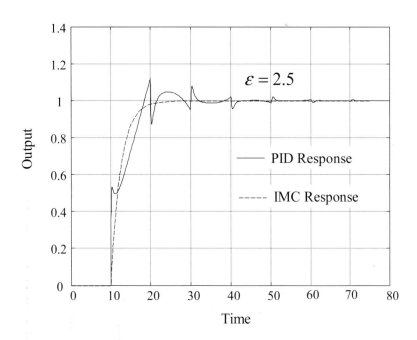

Figure 6.3 Response to step setpoint change for Example 6.2.

As can be seen from Figures 6.2a through 6.2d, the IMC-induced PID controllers perform quite well using filter time constants only slightly larger than the minimum filter time constants required to limit noise amplification to a factor of 20. Thus, for these examples and indeed for most single input, single output systems, there is little or no advantage in dynamic response in implementing an IMC system rather than an IMC-induced PID control system. However, when the dead time is very large relative to the sum of the process time constants, as in Example 6.2, then there is a significant advantage to applying an IMC system. Figure 6.3 shows that the response time of even a well tuned PID control system is at best twice that of an IMC system which uses the minimum filter time constant for the allowable noise amplification (i.e., 0.05 for Example 6.2).

♦

6.4 ALGORITHMS AND SOFTWARE FOR COMPUTING PID PARAMETERS

The analytical expressions for the PID parameters derived in Section 6.3.3 require evaluating several derivatives related to the process model and the feedback form of the IMC controller. Any software package that is capable of evaluating derivatives analytically, such as Mathematica or Maple, can be used to automate this procedure for general models. However, analytical expressions for the PID parameters in terms of the model parameters for general models and controllers are of limited usefulness, and one is usually quite satisfied with numerical values for the PID parameters for a specific model, IMC controller, and filter time constant. In this case there is no need to carry out analytical differentiations, and there are many software tools that can be used to get the desired PID parameters. Recall that the feedback form of the IMC controller, $c(s)$, is given by Eq. (6.5) and is

$$c(s) = \frac{p_m^{-1}(s)}{(\varepsilon s + 1)^r - p_A(s)}$$

Substituting Padé approximations for the exponential terms in $p_m(s)$ and $p_A(s)$ allows us to express these terms as

$$p_m(s) = \frac{K_p N_m^*(s)}{D_m(s)}; \quad p_A(s) = \frac{N_A^*(s)}{D_A^*(s)}, \qquad (6.16)$$

where

$$K_p \equiv p(0), \quad N_m^*(0) = N_A^*(0) = D_m(0) = D_A^*(0) = 1.$$

Substituting Eq. (6.16) into Eq. (6.5) and clearing fractions gives

$$c(s) = \frac{D_A^*(s)D_m(s)}{K_p N_m^*(s)D_A^*(s)(\varepsilon s + 1)^r - N_A^*(s)}. \qquad (6.17a)$$

Since the process model $\tilde{p}(s)$ is assumed stable, the use of stable Padé approximates for e^{-Ts} assures that the numerator terms in Eq. (6.17a) (i.e., $D_A^*(s)$ and $D_m(s)$) are polynomials with roots in the left half of the s-plane.

The denominator term in Eq. (6.17a) is a polynomial with a zero at the origin, whose coefficients all depend on the filter time constant ε. This polynomial has only stable (i.e., left half plane) zeros once the zero at the origin is factored out, provided that $p_A(s)$ is selected so that the magnitude of its frequency response is non-increasing. In this case application of the Nyquist criterion shows that the denominators of Equations (6.5) and (6.17a) have no right half plane zeros. This will always be the case using the design methods suggested in Chapter 3. Therefore, the controller $c(s)$ given by Eq. (6.17a) can be rewritten as

$$c(s) = \frac{k(\varepsilon)\sum_{i=0}^{n}\alpha_i s^i}{s\sum_{i=0}^{n-1}\beta_i(\varepsilon)s^i}, \quad \alpha_0 = \beta_0 = 1 \qquad (6.17b)$$

where

$$\alpha_i, \ \beta_i \geq 0 \text{ and } \beta_i \ (\varepsilon), \ k(\varepsilon) \text{ are functions of } \varepsilon.$$

Dropping terms higher than second-order in the numerator and higher than third-order in the denominator gives

$$c(s) \cong \frac{k(\varepsilon)(\alpha_2 s^2 + \alpha_1 s + 1)}{s(\beta_2 s^2 + \beta_1 s + 1)}. \qquad (6.18a)$$

The controller given by Eq. (6.18a) can be viewed as an ideal PID controller cascaded with a second-order lag. The ideal PID controller parameters are

$$\tau_i = \alpha_1; \quad K_c = k(\varepsilon)\tau_i; \quad \tau_D = \alpha_2/\alpha_1. \qquad (6.18b)$$

The second-order lag is given by:

$$1/(\beta_2 s^2 + \beta_1 s + 1). \qquad (6.18c)$$

Expanding (6.18a) in a Maclaurin series, as was done in Section 6.3.1, gives

$$c(s) \cong \frac{k[(\alpha_2 + \beta_1^2 - \beta_2 - \alpha_1\beta_1)s^2 + (\alpha_1 - \beta_1)s + 1]}{s}. \qquad (6.19)$$

From Eq. (6.19), the parameters of an ideal PID controller are

$$\tau_i = \alpha_1 - \beta_1; \quad K_c = k\tau_i; \quad \tau_D = (\alpha_2 + \beta_1^2 - \beta_2 - \alpha_1\beta_1)/\tau_i. \qquad (6.20)$$

If a sufficiently high-order Padé approximation (a five over five Padé approximation is more than adequate) is used in obtaining Eq. (6.19), then the PID parameters given by

Eq. (6.20) are the same as those given by the combination of Eq. (6.10) as evaluated from Eq. (6.12).

IMCTUNE carries out the manipulations necessary to obtain Eq. (6.17b) from Eq. (6.5) and from there gets the PID parameters given by Eq. (6.20). The input required by IMCTUNE is the process model $\widetilde{p}(s)$, the portion of the model to be inverted by the controller $p_m(s)$, and the IMC controller filter time constant ε.

6.5 ACCOMMODATING NEGATIVE INTEGRAL AND DERIVATIVE TIME CONSTANTS

The derivative and/or integral time constants computed from Eq. (6.10) or from Eq. (6.20) can be negative for some process models, independent of the choice of filter time constant. This often occurs when the process model has one or more dominant lead time constants, as in Example 6.3.

Example 6.3 A Process that Yields Negative PID Constants

The process is given by

$$\widetilde{p}(s) = \frac{s^2 + 2s + 1}{s^4 + 6.5s^3 + 15s^2 + 14s + 1} = \frac{.0625(7.46s + 1)(.536s + 1)}{(2s + 1)(.5s + 1)^3} \qquad (6.21)$$

The open-loop response of the process in Eq. (6.21) to a unit step change in control effort is given in Figure 6.4. Notice the very large overshoot of the final steady state caused by the strong lead action of the term $(7.46s + 1)$ in the numerator of Eq. (6.21).

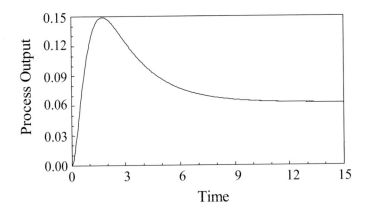

Figure 6.4 Step response for $\widetilde{p}(s) = (s^2 + 2s + .25)/(s^4 + 6.5s^3 + 15s^2 + 14s + 1)$.

Using Eq. (6.20) to compute PID parameters with an IMC filter time constant of .2 yields a PID controller with $\tau_I = -4.60$ and $\tau_D = -7.87$.

$\blacklozenge$

To avoid computing negative controller time constants, as in Example 6.3, we recommend replacing the simple PID controller with a PID controller cascaded with a first-order or second-order lag of the form $1/(\gamma s + 1)$ or $1/(\beta_2 s^2 + \beta_1 s + 1)$ respectively. The *PID-lag* controller given by Eq. (6.18a) is already in the form of a PID controller cascaded with a second-order lag. Furthermore, the time constants α_i and β_i are positive, as discussed previously. Unfortunately the controller given by Eq. (6.18) usually requires relatively large filter time constants to yield responses with less than a 10% overshoot for setpoint changes. A generally better approximation to the feedback form of the IMC controller, $c(s)$, given by Eq. (6.5) can be obtained by slightly modifying the Maclaurin series expansion presented in Section 6.3.1. Starting from Eq. (6.5) we let

$$c(s) \equiv \frac{f(s)}{s} = \frac{f(s)h(s)}{sh(s)}; \quad h(s) \equiv \gamma s + 1. \tag{6.22}$$

Expanding the quantity $f(s)\,h(s)$ in a Maclaurin series gives

$$c(s) = \frac{f(0) + (f'(0) + \gamma f(0))s + (f''(0) + 2\gamma f'(0))s^2/2 + (f'''(0) + 3\gamma f''(0))s^3/6 + O(s^4)}{s(\gamma s + 1)}.$$

$$\tag{6.23}$$

Choosing γ to make the third-order term zero gives

$$\gamma = -f'''(0)/3f''(0). \tag{6.24}$$

Dropping the terms in s^4 and higher in Eq. (6.23) gives the desired *PID-lag* controller.

Example 6.4 A PID-lag Controller for a Process Whose Transfer Function Is a Lead

PID-lag controller for $\tilde{p}(s) = \dfrac{s^2 + 2s + 1}{s^4 + 6.5s^3 + 15s^2 + 14s + 1}$ from Eq. (6.23) becomes:

$$PID\text{-}lag = \frac{40(1.91s^2 + 2.86s + 1)}{s(7.47s + 1)}. \tag{6.25}$$

The response of the control system to a step setpoint change is given in Figure 6.5.

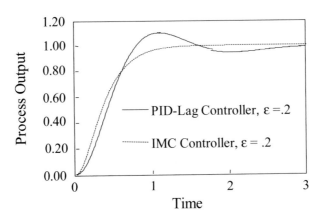

Figure 6.5 Feedback control of $\tilde{p}(s) = \dfrac{(s^2 + 2s + .25)}{(s^4 + 6.5s^3 + 15s^2 + 14s + 4)}$.

Notice that the lag time constant of 7.47 in Eq. (6.25) is nearly the same as the lead time constant in the process model, which is 7.46. Indeed, very nearly the same controller would have been obtained by finding the PID controller for the process given by Eq. (6.21), but without the lead. However, to obtain this controller it is necessary to use a second-order filter with a time constant of 0.2 (i.e., the IMC controller is improper). For Example 6.4, the *PID-lag* controller given by Eq. (6.25), is substantially better than using the *PID-lag* controller given by Eq. (6.18) since the latter controller requires a filter time constant of 0.55 to achieve a 10% overshoot. The settling time of the control system using the controller of Eq. (6.18) is about 2½ times as long as that for the control system using Eq. (6.25).

♦

A PID controller with a second-order lag is most useful when the process model has a strong second-order lead with complex zeros, as in Example 6.5.

Example 6.5 A Process with Complex Zeros

Consider the process given by

$$\tilde{p}(s) = \frac{.5(16s^2 + .4s + 1)}{(2s+1)(.5s+1)^3}.$$ (6.26)

Using a filter time constant of 0.5 with the controller given by Equations (6.10) and (6.12) yields an integral time constant of 2.85 and a derivative time constant of −4.98. The lag time constant computed from Eq. (6.24) is −2.75, and so the controller given by Eq. (6.23) also cannot be used. Finally, the controller given by Eq. (6.18) (for a filter time constant of 0.5) is

$$PID\text{-}lag = \frac{2(3.75s^2+3.5s+1)}{s(16.1s^2+.65s+1)}. \tag{6.27}$$

Notice that the denominator lag of Eq. (6.27) is very close to the numerator lead in the process model given by Eq. (6.26). Here again, if one eliminates the numerator term $(16s^2 + 4s + 1)$, the result is

$$PID\text{-}lag = \frac{2(2.94s^2+3.25s+1)}{s(16s^2+4s+1)}. \tag{6.28}$$

The control system response using Eq. (6.28) is very similar to that using Eq. (6.27).

♦

6.6   2DF PID PARAMETERS FROM 2DF IMC

To convert from a two-degree of freedom IMC controller to a PID controller, it is convenient to redraw the IMC system, as shown in Figure 6.6.

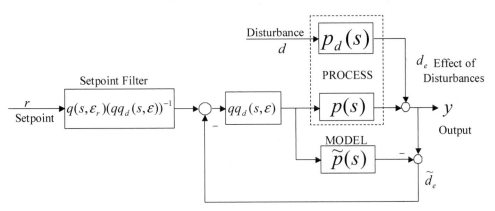

Figure 6.6 Alternate configuration of a two-degree of freedom IMC system.

The diagram in Figure 6.6 can be further rearranged into a form suitable for approximation with a PID control system, as given by Figure 6.7. The rearrangement is the same as that used to obtain the feedback form of a one-degree of freedom controller. The setpoint filter in Figure 6.7 is obtained by noting that

$$q(s,\varepsilon_r)q^{-1}(s,\varepsilon) = f(s,\varepsilon_r)f^{-1}(s,\varepsilon), \tag{6.29}$$

where

$$f(s,\varepsilon) \equiv 1/(\varepsilon s+1)^r.$$

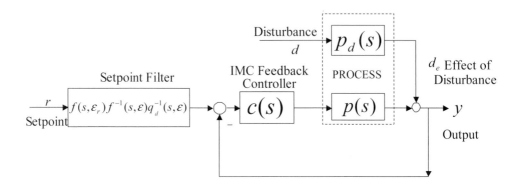

Figure 6.7 Two-degree of freedom feedback form of an IMC system.

The IMC feedback controller $c(s)$ in Figure 6.7 is given by

$$c(s) = qq_d(s,\varepsilon)/(1 - qq_d(s,\varepsilon)\widetilde{p}(s)). \qquad (6.30)$$

The feedback controller $c(s)$ in Eq. (6.30) can be approximated by a PID controller by expansion in a Maclaurin series just as was done for a single-degree of freedom controller in Section 6.3.1. There are two relatively minor changes from the procedure in Section 6.3.1. First, the controller $qq_d(s,\varepsilon)$ replaces the controller $q(s,\varepsilon)$. Second, care must be taken to cancel common factors in the numerator and denominator *before* expansion in the Maclaurin series. Common factors will arise whenever the controller $qq_d(s,\varepsilon)$ is chosen so that the zeros of both $q(s)$ and $(1 - qq_d(s,\varepsilon)\widetilde{p}(s))$ are the same as the common poles in $\widetilde{p}(s)$ and $\widetilde{p}_d(s)$. This type of controller design is generally used when the disturbance passes through the process.

IMCTUNE software will automatically compute the PID controller and the setpoint filter for a two-degree of freedom control system, given the process model and the disturbance lags to be cancelled by the zeros of $(1 - qq_d(s,\varepsilon)\widetilde{p}(s))$, the controller q(s), and the filter time constant.

6.7   SATURATION COMPENSATION

No PID controller should be implemented without compensation for the fact that all real control efforts have limits. For example, when a valve opens fully, the flow through the valve usually stops increasing fairly rapidly. When it shuts, the flow through it goes to zero. Most control valves in the U.S. are air operated diaphragm types that operate over an air pressure range of 3 to 15 pounds air pressure (psig) (Smith and Corripio, 1985). The decision whether the valve should be closed (i.e., "air to open") or open (i.e., "air to close") at 3 psig is based on maintaining safe operation should air pressure be lost. In any case, the limits of 3 and 15 psig air pressure constitute what are commonly called hard limits, since even if the air pressure can be increased or decreased, the valve ceases to either open or close further. A fully open or closed valve, or indeed any actuator at its limit, is often said to be "saturated."

Actuator saturation can cause serious problems in the performance of even well-tuned PID control systems that lack special saturation compensation. To illustrate the potential problem, we consider the operation of a PID control system with no saturation compensation. We assume that a disturbance has occurred that cannot be compensated with the available control effort. In this case the error between the setpoint and the measured process variable will be either positive or negative, and the control effort will be at its limit. The uncompensated PID controller will continue to integrate this nonzero error as long as the disturbance persists, and the controller output will continue to grow towards plus or minus infinity. Let us assume that the disturbance decreases to the point where the setpoint can be achieved by the available control effort. Now, however, the controller output signal is substantially above or below the saturation limits of the actuator. Since the disturbance that originally caused actuator saturation is now gone, the error between the setpoint and the process variable has changed sign and the controller has started integrating in the opposite direction. Therefore after some, possibly long, period of time, the controller output will drop below the actuator saturation limit, and the actuator will move away from its limit. The period between the time when the error changed sign and when the actuator moved away from saturation is a controller-induced dead time during which the output was needlessly away from its setpoint. Worse still, if the process variable was sufficiently far from its setpoint during the controller-induced dead time, it is possible that the control effort will saturate at the opposite limit, and the control system will enter a "limit cycle" wherein the control effort cycles continuously between limits. The problem just described is often called *reset windup*, as the integral action of a PID controller was, and sometimes still is, referred to as *reset action*.[2] A PID controller that has compensation to prevent reset windup is often said to have anti-reset windup compensation.

---

[2] This terminology probably dates back to the earliest proportional-only controllers (gain-only controllers). The operator would adjust (reset) the control effort bias (the value of the controller output when the controller input error is zero) to remove the steady-state error (offset) between the setpoint and process variable. Such offsets are characteristic of proportional-only control systems when there are changes in disturbances or the setpoint.

There are two types of anti-reset windup compensation. The most common type of compensation is a digital implementation of an older analog implementation of the integral action, wherein the integral action is obtained by positive feedback of a first-order lag around a limiter that has a gain of one in its linear region. This type of anti-reset windup compensation will be described in some detail in this section. A less common form of anti-reset windup is one where the integration stops as soon as the controller output saturates. The integration is restarted when either the error between the setpoint and process variable changes sign, or the control effort comes off saturation, possibly due to a change in control effort bias or in some other external signal, such as that from a feedforward controller.

Figure 6.8 shows an analog type of anti-reset windup PI controller. The auto-manual switch is shown in the manual position. In this position the process is under manual control by the operator, who supplies the control effort bias. Before switching to automatic control, the operator normally adjusts the control effort manually to bring the process close to setpoint. Then, when the switch is closed, the control effort will remain at the value supplied by the operator because the output of the feedback lag is the same as the control value supplied by the operator. This type of controller is commonly said to have "bumpless transfer."

When the auto-manual switch is closed, the transfer function of the closed-loop between the output of the controller gain $K_c e$ and the controller output m is $(\tau_1 s + 1)/\tau_1 s$. If the error e between the setpoint and the measured process output is zero, then the state of the integrator is the value of the current control effort m.

The saturation limits are usually set to the saturation limits of the actuator. However, they can be set to any values within the actual limits should safe plant operation so require. When the limits in the saturation block are reached, the output of the feedback lag in Figure 6.8 soon becomes that of the saturated control effort. As soon as the error signal changes sign, the signal into the saturation block is no longer at a limit, and the controller again acts as a PI controller.

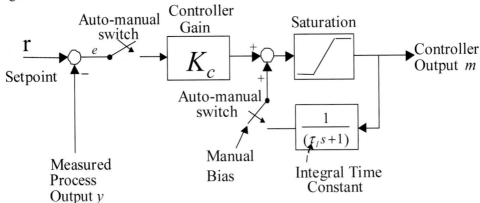

Figure 6.8 Anti-reset windup PI controller with bumpless transfer.

Figure 6.9 shows one method of incorporating the anti-reset windup configuration of Figure 6.8 into a PID controller of the form $K_c(1 + 1/\tau_I s + \tau_D s/(\alpha\tau_D s + 1))$.

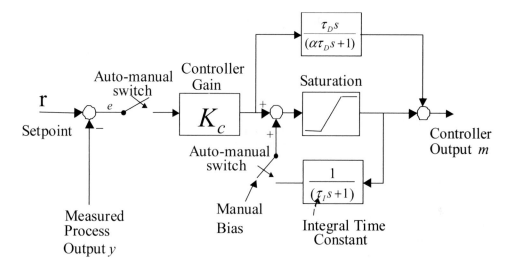

Figure 6.9 Anti-reset windup PID controller with bumpless transfer.

6.8 SUMMARY

PID parameters are obtained from a Maclaurin series expansion of the feedback form of the IMC controller given by Eq. (6.7).

$$c(s) \equiv \frac{1}{s} f(s),$$

where

$$f(s) \equiv \frac{sp_m^{-1}(s)}{((\varepsilon s + 1)^r - p_A(s))}.$$

The parameters are

$$K_c = f'(0); \quad \tau_I = K_c/f(0); \quad \tau_D = f''(0)/2K_c.$$

The derivatives of the function $f(s)$ evaluated at $s = 0$ are given by Equations (6.11) and (6.12). An alternate representation of the feedback form of IMC controller obtained by replacing the dead time terms (e^{-Ts}) by Padé approximates is given by Eq. (6.18a).

$$c(s) \cong \frac{k(\varepsilon)(\alpha_2 s^2 + \alpha_1 s + 1)}{s(\beta_2 s^2 + \beta_1 s + 1)}.$$

In terms of Eq. (6.18a), the PID parameters are given by Eq. (6.20).

$$\tau_i = \alpha_1 - \beta_1; \quad K_c = k\tau_i; \quad \tau_D = (\alpha_2 + \beta_1^2 - \beta_2 - \alpha_1\beta_1)/\tau_i.$$

If either, or both, τ_I or τ_D are negative, then a *PID-lag* controller should be used. The controllers, in order of preference, are given by Equations (6.23) and (6.18a). If any of the time constants in Eq. (6.23) are negative, then use the controller given by Eq. (6.18a).

Problems

6.1 Show that the combination of Equations (6.2) and (6.3) yields Eq. (6.1).

6.2 Derive an analytical expression for the PID parameters in terms of the process parameters for

$$p(s) = \frac{K\, e^{-TS}}{\left(\tau^2 s^2 + 2\zeta\, \tau s + 1\right)}.$$

6.3 Derive the PI parameters for

    a) a pure time delay process

    b) the process given in Problem 6.2.

6.4 Select PID or *PID-lag* controllers for each of the problems in Chapter 3. Use the smallest filter time constant that limits the closed-loop overshoot to 10% for a step setpoint change. Note that such a filter time constant can actually be smaller than the IMC filter time constant required by the IMC controller, since a PID controller never amplifies high frequency noise by more than 20 by the manner in which the derivative action is implemented.

6.5 Select two-degree of freedom PID controllers where appropriate for each of the problems in Chapter 3, assuming that the disturbance enters through the process. If a two-degree of freedom PID controller is not appropriate, explain why.

6.6 Derive the closed-loop transfer function for Figure 6.9.

References

Morari, M., and F. Zafiriou. 1989. *Robust Process Control*. Chapter, Prentice-Hall, NJ

Rivera, D. E., S. Skogestad, and M. Morari. 1986. "Internal Model Control 4. PID Controller Design," *I&EC Chem. Proc. Des. & Dev.* 25, 252–265.

Seborg, D. E., T. F. Edgar, and D. A. Mellichamp. 1989. *Process Dynamics and Control*, John Wiley & Sons, NY.

Smith, C. A., and A. B. Corripio. 1985. *Principles and Practice of Automatic Process Control*. John Wiley & Sons, NY.

Tuning and Synthesis of 1DF IMC for Uncertain Processes

Objectives of the Chapter

- Introduce the concept of process uncertainty and explore its effect on IMC system stability and performance.
- Present a tuning method for adjusting the IMC filter time constant that accomplishes a desired relative stability for all processes in a predefined uncertainty set.
- Explore the effect of uncertainty on controller design and model selection.

Prerequisite Reading

Chapter 3, "One-Degree of Freedom Internal Model Control"
Appendix A, "Review of Basic Concepts"
Appendix B, "Frequency Response Analysis"

7.1   INTRODUCTION

Chapter 3 discusses the design and tuning of a linear IMC controller when the linear model used in the IMC system is assumed to be a perfect representation of the process. This chapter treats the realistic situation that the process model is not the same as the process. Generally, the greatest contribution to the mismatch between the model and the process in a linear IMC system is the fact that the model and controller are linear while the process is nonlinear and time varying. While it is possible to design nonlinear IMC systems (Kravaris and Kantor, 1990), such control systems are not yet widely used in industry because the improved process performance over *a well-designed and well-tuned* linear control system does not usually justify the time and expense necessary to design and maintain a nonlinear control system. Of course, the key question is, how does one achieve a well-designed and well-tuned linear IMC system when the actual process is nonlinear? In order to accomplish this objective, we will approximate the nonlinear process as a set of linear processes with constant coefficients. Because the process is nonlinear, the parameters of the local, linear descriptions of the process change over time due to changes in operating point. The process operating point changes due to both external disturbances, such as changes in feed composition and ambient conditions, and internal changes, such as heat exchanger fouling and catalyst aging. Approximating the process as a set of linear processes with constant coefficients ignores the behavior of the process during parameter changes and focuses instead on the behavior of the process about all its steady-state operating points. This approximation is useful in that it allows us to use the powerful tools of linear mathematics to carry out control system analysis and design. Such an approximation is reasonable provided that the disturbances are such that the process spends most of the time operating about steady states rather than moving from one steady state to another.

Control system responses to setpoint changes and disturbances change as the local description of the process changes. Therefore, in tuning a control system, one must consider the entire range of possible responses rather than focusing on a single response. Figure 7.1 shows a typical range of responses of a well-tuned control system to step setpoint changes for a linear process at different operating points (i.e., with different values for the local process parameters). Based on such a range of responses, we can qualitatively define IMC controller tuning and synthesis objectives. Our IMC tuning objective is to select the smallest IMC filter time constant for which no setpoint response overshoots the setpoint by more than a specified amount and no response becomes too oscillatory. Our IMC controller synthesis objective is to choose both the IMC controller and the process model so as to speed up the slowest closed loop responses as much as possible without violating our overshoot and relative stability (i.e., not too oscillatory) objectives.

It is quite difficult to make the above qualitative time domain tuning and synthesis objectives sufficiently precise so as to be useful in obtaining numerical values for the IMC controller filter time constant, and model parameters. To illustrate one of the difficulties, notice that it is hard to select the slowest response in Figure 7.1. Curve 3 is the slowest response up to 20 time units; curve 4 yields the slowest response between 20 time units and 50 time units. After 50 time units, both curves approach steady state at the same rate.

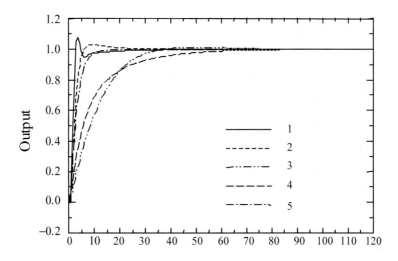

Figure 7.1 Responses to a step setpoint change at different operating points.

It turns out to be much easier to develop quantitative controller tuning and synthesis objectives and procedures for achieving such objectives in the frequency domain (i.e., in the domain of open-loop and closed-loop frequency responses). There is substantial literature on H_∞ frequency domain methods for the analysis and synthesis of control systems for processes described by sets of linear, constant coefficient systems. Kwakernaak (1993) gives a good, relatively brief overview of these methods for single-input single-output (SISO) systems. The texts by Morari and Zafiriou (1989), Doyle et al. (1992), and Dorato et al. (1992) provide more complete expositions. Unfortunately, all of the aforementioned texts require substantial expertise from the reader in order to understand and apply the H_∞ methods presented. In addition, the methods of these authors usually require various approximations before they can be applied to typical chemical process descriptions. For example, dead times must be replaced with finite dimensional approximations (e.g., Padé approximations). We have elected to present a related but simpler approach to tuning and synthesis, which we call Mp tuning and synthesis. Mp tuning aims to find the smallest IMC filter time constant that assures that (1) the magnitude of all closed-loop frequency responses between output and setpoint have magnitudes less than a specified value (usually taken as 1.05) at any frequency, and (2) any oscillations in the curve of the maximum magnitude of all closed-loop frequency responses do not have peaks higher than specified (usually about 0.1) from the highest adjoining valley. As we will discuss later, a control system tuned in this manner will usually satisfy the qualitative time domain tuning objectives discussed previously. Also, a by-product of the Mp tuning procedure is an estimate of the speed of response of the fastest and slowest responses of the control system for all processes in the set of possible processes (i.e., in the uncertainty set).

The next section discusses various process uncertainty descriptions. Section 7.3, which follows, presents the Mp tuning algorithm, describes a method for mitigating the fact that our uncertainty descriptions are themselves not known very precisely, and describes an inverse tuning algorithm that finds an uncertainty region over which a given controller will perform as specified. The two sections following Mp tuning provide justification for the tuning algorithm. Section 7.4 gives the conditions under which any Mp specification greater than one is achievable. Section 7.5 discusses the theoretically important property of *robust stability*. Robust stability means that the control system is stable for all processes in the uncertainty set. Any practical control system must be robustly stable. Since the conditions imposed on the process, model, and controller in order to safely apply the Mp tuning algorithm will automatically be met in most practical situations, those readers interested mainly in applications can skim Sections 7.4 and 7.5, paying attention only to Table 7.3, which limits selection of the model gain in order to be able to achieve Mp specifications arbitrarily close to one.

Mp synthesis, in Section 7.6, addresses the question of what controller and model to choose for the IMC system when the process can be any in the uncertainty set. The criterion that we select for choosing the IMC controller and model is that they speed up the slowest closed-loop responses as much as possible. It turns out that process uncertainty has a profound influence on both controller design and model selection. An important observation in this regard is that the traditional engineering approach of fitting a first-order plus dead time model to high-order overdamped processes, and designing the controller based on that model, often yields a control system that *performs better* than a system *based on a process model of the correct order* when other process parameters such as gain and dead time are sufficiently uncertain.

An important application of Mp tuning and synthesis applied to IMC systems is to convert the resulting IMC controller into an equivalent PID controller, using the methods described in Chapter 6. PID controllers are by far the most widely used industrial control systems and are likely to remain so for the foreseeable future. The IMCTUNE software provided with this text permits the user to automatically obtain PID parameters from the tuned IMC controller.

Another potentially important application of Mp tuning and synthesis is to determine the limits of linear, fixed parameter, control system performance. Such limits determine what incentive, if any, exists for the implementation of more complex nonlinear and adaptive control systems (Åström, 1995; Kravaris and Kantor, 1990; Seborg et al., 1989; Ljung 1987). The aim of such control systems is to improve the speed and quality of the control system response by substantially reducing process/model mismatch and by basing controller design and parameters on a nonlinear or an updated linear model. Such approaches have yet to be widely applied in the process industries, probably because the perceived benefits do not yet justify the added complexity. Further, even nonlinear and adaptive models are approximations to the actual process, and uncertainty in the process parameters will still need to be accounted for in the control system design and tuning. However, a discussion of the tuning of nonlinear and adaptive controllers to accommodate model uncertainty is beyond the scope of this text.

7.2   PROCESS UNCERTAINTY DESCRIPTIONS

The tuning and synthesis techniques in the following sections require frequency domain descriptions of the process uncertainty set. There are two convenient descriptions of the uncertainty set: (1) bounds on transfer function parameters and (2) bounds on the gain and phase of transfer functions over all frequencies of interest. Unfortunately, neither of these uncertainty descriptions is readily available from process data or from first principles. However, process engineers can often estimate ranges for the parameters of simple process models based on a combination of process observations and an understanding of how the process operates. Further, the uncertainty inherent in such estimates can be at least partially accounted for using multiple uncertainty regions, as discussed in Section 7.2.2. Gain and phase bounds as functions of frequency are usually more difficult for plant operating personnel to estimate. However, with some effort, such bounds can be obtained from input-output tests on the plant at different operating points. One advantage of gain and phase bounds is that they do not require a priori postulation of a model. While uncertainty bounds cannot be obtained with precision, they are nonetheless useful for obtaining safe controller tunings and in improving controller performance.

7.2.1 Parametric Uncertainty

The set of transfer functions with uncertain parameters, Π, is defined as

Π = the set of all transfer functions, $p(s, \beta(\alpha))$, with the vector of parameters α lying in the set S_α.

S_α = set of all vectors α with $\underline{\alpha}_i \, \alpha_i \le \overline{\alpha}_i$

where $\overline{\alpha}_i$ = upper-bound on the parameter α_i

$\underline{\alpha}_i$ = lower-bound on the parameter α_i

$\beta(\alpha)$ = a vector of parameters which are continuous functions of the vector α.

In addition, we will also require that

$$p(0, \beta(\alpha)) > 0 \quad \text{or} \quad < 0 \text{ for all } \alpha \in S_\alpha. \tag{7.1}$$

The restriction given by Eq. (7.1) means that all process gains in the uncertainty set Π have the same sign. Such processes are called integral controllable because the restriction given by Eq. (7.1) is a necessary condition for a no offset IMC controller[1] to be stable for any process in Π. Integral controllability is discussed more completely in Section 7.4.1. Typical examples of parametric uncertainty descriptions follow.

---

[1] Recall that IMC controller has no offset if $q(0) = \tilde{p}^{-1}(0)$.

Example 7.1 A FOPDT Process with Uncorrelated Uncertainty

$$p(s) = \frac{K\,e^{-Ts}}{\tau s + 1}\,, \quad \underline{K} \leq K \leq \overline{K}, \quad \underline{\tau} \leq \tau \leq \overline{\tau}, \quad \underline{T} \leq T \leq \overline{T}, \tag{7.2}$$

where $\underline{\sigma} \equiv$ the lower-bound on any parameter, σ,

$\quad\quad\quad\overline{\sigma} \equiv$ the upper-bound on any parameter, σ.

◆

In Example 7.1 all three parameters vary independently. However, it quite often happens that one parameter depends on another, in Example 7.2.

Example 7.2 A FOPDT Process with Correlated Uncertainty

$$p(s) = \frac{K\,e^{-Ts}}{\tau s + 1}\,, \quad \underline{K} \leq K \leq \overline{K}, \quad T(K) = \underline{T} + 2(K - \underline{K}) \tag{7.3}$$

◆

In Example 7.2 there is only one uncertain parameter, K. The function $\beta(\alpha)$ in the definition of the uncertainty set Π was included to accommodate correlated uncertainty such as that in the example. A common source of correlation between the parameters of a transfer function is the use of first principles modeling to obtain the transfer function. Consider Example 7.3.

Example 7.3 The Two-Tank Process

For Figure 7.2, we wish to obtain the transfer function between changes in inflow (Δq_i) and changes in the level of fluid in tank 1 (Δh_1). The differential equations that relate the flows and levels are

$$A_1 \frac{dh_1(t)}{dt} = q_i - q_1\,; \quad q_1 = C_V \sqrt{h_1 - h_2}\,,$$
$$A_2 \frac{dh_2(t)}{dt} = q_1 - q_2\,; \quad q_2 = C_V \sqrt{h_2}\,. \tag{7.4}$$

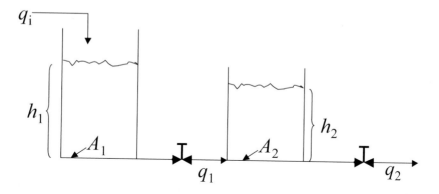

Figure 7.2 A two-tank process.

Linearizing the equations about the steady-state defined by $\bar{q}_i$, and solving for the desired transfer function gives

$$\frac{\Delta h_1(s)}{\Delta q_i(s)} = \frac{K(\tau_d s + 1)}{(\tau^2 s^2 + 2\varsigma\tau s + 1)},\qquad(7.5)$$

where

$$K = 4\bar{q}_i / C_V^2 ; \qquad \tau_d = A_2\bar{q}_i / C_V^2$$
$$\tau = 2(\sqrt{A_1 A_2})\bar{q}_i / C_V^2 ; \qquad \varsigma = (2A_1 + A_2)/(2\sqrt{A_1 A_2}).$$

Notice that of the four parameters in Eq. (7.5), three are uncertain and all three depend on only one uncertain parameter, $\bar{q}_i$. Once we have an estimate of the range of variation of $\bar{q}_i$, the uncertain process is completely specified.

◆

7.2.2 Frequency Domain Uncertainty Bounds

Most methods for treating process uncertainty accommodate frequency domain uncertainty bounds much more readily than parametric uncertainty bounds (Morari and Zafiriou, 1989; Doyle et al., 1992; Dorato et al., 1992). The general form of frequency domain uncertainty bounds for an uncertain process, $p(s)$, is given by

$$\underline{M}(\omega) \leq | p(i\omega)| \leq \overline{M}(\omega),\qquad(7.6a)$$

$$\underline{\phi}(\omega) \leq \text{Angle } p(i\omega) \leq \overline{\phi}(\omega),\qquad(7.6b)$$

where $\overline{M}\,(\omega)$ and $\underline{M}\,(\omega)$ = upper and lower magnitude bounds on $p(i\omega)$,

                         $\overline{\phi}(\omega)$ and $\underline{\phi}(\omega)$ = upper and lower phase bounds the angle of $p(i\omega)$.

Values for the upper and lower magnitude and phase bounds as functions of frequency can be obtained by multiple identifications of the process frequency response at different operating points. More commonly, the bounds are either estimated based on experience or obtained from parametric uncertainty bounds. However, frequency response bounds obtained from parametric bounds are always more conservative uncertainty descriptions than the original parametric bounds, and therefore lead to more sluggish control system designs. To see why this is so, consider the following first-order process with an uncertain time constant:

$$p(s) = 1/(\tau s + 1); \quad 1 \leq \tau \leq 5. \tag{7.7}$$

The frequency response of Eq. (7.7) is

$$|\,p(i\omega)\,| = (\tau^2\omega^2 + 1)^{-1/2}, \tag{7.8a}$$

$$Angle\; p(i\omega) = -\tan^{-1}\tau\omega. \tag{7.8b}$$

The set of all possible gain and phases of $p(i\omega)$ as the time constant τ ranges between one and five are given by the shaded areas of Figure 7.3, which are obtained from the maximum and minimum of Equations (7.8a) and (7.8b) and are given by Eq. (7.9).

$$\overline{M}(\omega) = (\omega^2 + 1)^{-1/2}, \tag{7.9a}$$

$$\underline{M}(\omega) = (25\omega^2 + 1)^{-1/2}, \tag{7.9b}$$

$$\overline{\phi}(\omega) = -\tan^{-1}\omega, \tag{7.9c}$$

$$\underline{\phi}(\omega) = -\tan^{-1}5\omega. \tag{7.9d}$$

The uncertainty bounds given by Eq. (7.9) are not the same as those given by Eq. (7.8) because the bounds given by Eq. (7.9) allow the magnitude and phase of $p(i\omega)$ to vary independently, whereas for the process given by Eq. (7.7), the magnitude and phase of $p(i\omega)$ are related through Eq. (7.8). For example, at a frequency of one radian/unit time, Eq. (7.9) shows that the magnitude can vary between .196 and .707, while the phase can take on any value between $-45.0°$ and $-78.7°$ (see Fig. 7.3). According to Eq. (7.8), when the magnitude is $1/(\tau^2 + 1)^{1/2}$, the phase is $-\tan \tau$ (e.g., if $\tau = 1$ the magnitude is .707 and the phase is $-45.0°$). Therefore, the magnitude and phase bounds given by Eq. (7.9) describe more processes than those given by Eq. (7.8). That is, the uncertainty set given by Eq. (7.9) is larger than that given by Eq. (7.8). As we shall see in Section 7.3.2, the larger the uncertainty set, the more sluggish the controller must be in order to meet closed-loop specifications. Therefore, a controller designed using the gain and phase bounds given by

Eq. (7.9) generally will be more sluggish than a controller designed using the original parametric bounds.

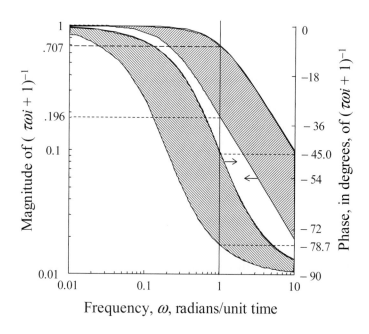

Figure 7.3 Frequency response of $1/(\tau s + 1)$ for $1 \le \tau \le 5$.

Many of the uncertainty descriptions in the literature make use of magnitude only uncertainty bounds as given by Eq. (7.6a). In such cases the implicit assumption is that the phase of the uncertain process can be anywhere within $\pm\,360°$. The advantage of such descriptions is that they sometimes lead to convex optimization problems for tuning or designing the controller. The disadvantage is that the resulting uncertainty set is even larger than that using gain and phase bounds, and therefore the final control system is likely to be significantly more sluggish than needed for uncertainty due only to parametric variations.

Because parametric uncertainty bounds generally lead to the least conservative controller tunings, and are usually the easiest to obtain, the next section deals only with parametric uncertainty. However, the methodology (but not the IMCTUNE software) also treats frequency domain bounds, should these be useful in particular situations.

7.3   Mᴘ Tᴜɴɪɴɢ

7.3.1 The Problem Statement

The aim in tuning any control system is to achieve desirable time domain closed-loop performance, such performance being measured by the speed of response, how oscillatory it is, and how much the response overshoots the setpoint. One can estimate such time domain performance measures most easily from the closed-loop frequency responses between output and setpoint. This frequency response is often called the complementary sensitivity function. The maximum magnitude of the complementary sensitivity function, which we will refer to as the Mp, generally gives a good indication of the overshoot to setpoint changes and/or the magnitude of oscillations in the time response. The frequency at which the maximum occurs is generally a good indication of frequency of the time domain oscillations. Finally, the inverse of the "break frequency" is a good estimate of the time constant of the fastest time domain response. The "break frequency" is usually taken as the intersection of the asymptote to the high-frequency portion of the frequency response with a horizontal line of magnitude one. This definition assumes a closed-loop gain of one (i.e., an integral control system). The justification for the foregoing statements is that the magnitude of the *closed-loop* frequency responses between output and setpoint for most control systems can be reasonably approximated by the magnitude of the frequency response of a second-order system of the form $1/(\tau^2 s^2 + 2\zeta\tau s + 1)$. For such a second-order system, the fractional overshoot to step inputs can be directly related to the maximum peak of the magnitude of its frequency response through the relationship:

$$OS = e^{-\pi \text{Mp}(1-(1-1/\text{Mp}^2)^{.5})}, \quad \text{Mp} > 1, \tag{7.10}$$

where

$\quad OS \equiv$ fractional overshoot

$\quad\quad \equiv$ (maximum change in output – change in setpoint)/change in setpoint.

$\quad$ Mp $\equiv$ maximum magnitude of the frequency response.

The damping ratio ζ of the second-order response is related to the Mp by

$$\zeta = ((1-(1-1/M_p^2)^{.5})/2)^{.5}. \tag{7.11}$$

Figures 7.4a and 7.4b show the time and frequency responses for $1/(\tau^2 s^2 + 2\zeta\tau s + 1)$ with $\tau = 1$ and $\zeta = .5$. Note that the unity gain arises because we are generally dealing with closed-loop systems with no offset.

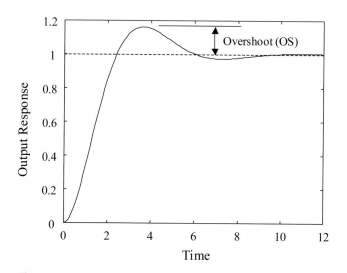

Figure 7.4a Response of $1/(s^2 + s + 1)$ to a unit step setpoint change.

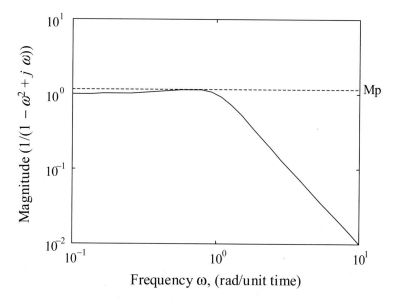

Figure 7.4b Frequency response of $1/(s^2 + s + 1)$.

The overshoot (OS) and damping ratio ζ are plotted versus Mp in Figure 7.5. As can be seen from Figure 7.5, an Mp of 1.05 corresponds to a 10% overshoot and a damping ratio of about 0.6. Higher values of Mp yield greater overshoots and lower damping ratios.

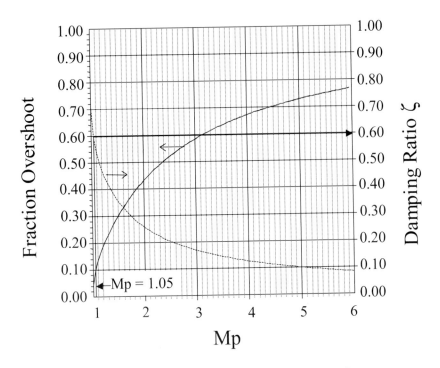

Figure 7.5 Overshoot and damping ratio versus Mp for $1/(s^2 + 2\zeta s + 1)$.

The Mp tuning problem is to find the smallest IMC filter time constant that assures that no closed-loop frequency response, from setpoint to output, will have more than the specified Mp (i.e., the specified maximum peak). The numerical value specified will approximately limit the maximum overshoot, as given by Figure 7.5. A typical specification is an Mp of 1.05, which leads to worst case overshoots of about 10%.

A formal statement of the Mp tuning problem follows:

Select the filter time constant $\varepsilon*$ for the IMC controller $q(s,\varepsilon)$ so that the magnitude of the complementary sensitivity function $CS(i\omega)$ is equal to or less than a specified Mp for all processes, $p(s)$, in a predefined set Π. For at least one process in Π, the magnitude of $CS(i\omega)$ must equal the specified Mp at one or more frequencies. That is,

$$\left| CS(i\omega, \varepsilon^*) \right| \le Mp \quad \forall p(s) \in \Pi, \text{ and } \forall \omega \qquad (7.12a)$$

where $\forall \equiv$ for all; $\in \equiv$ contained in, and

$$\left| CS(i\omega^*, \varepsilon^*) \right| = Mp \qquad (7.12b)$$

for at least one $p \in \Pi$ and some frequencies, $\omega_1^*, \omega_2^*, \dots$.

The motivation for the condition in Eq. (7.12b) is to avoid introducing any conservatism into the tuning beyond that of the specification.

For the IMC configuration of Figure 7.6 (reproduced from Chapter 3, Figure 3.1), the closed-loop transfer function between output and setpoint (i.e., the complementary sensitivity function) is given by

$$CS(s,\varepsilon) = \frac{y(s)}{r(s)} = \frac{p(s)q(s)}{(1+(p(s)-\widetilde{p}(s))q(s,\varepsilon))}, \qquad (7.13)$$

where $p(s)$ = any plant in the set of allowable processes Π

$\widetilde{p}(s)$ = a nominal model

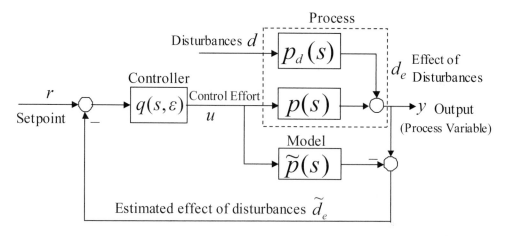

Figure 7.6 1DF IMC system.

The problem statement of Eq. (7.12) and above can be reformulated as the following optimization problem:

Find ε^* such that

$$\max_{\alpha,\omega}| CS(i\omega,\varepsilon^*,\alpha)| = \mathrm{Mp}, \qquad (7.14)$$

where α = vector of uncertain process parameters $\in \Pi$,

ω = frequency.

The solution of Eq. (7.14) yields values for the filter time constant ε^*, which satisfies Eq. (7.12) as well as values of the parameters α^* and ω^*, which solve the maximization problem given by Eq. (7.14).

The Mp tuning problem can be solved for very general processes, with uncertain parameters, using the IMCTUNE software associated with this text, and described in

Appendix G. The next section contains a brief description of the tuning algorithm used by the IMCTUNE software. Section 7.3.2 contains an example of Mp tuning, solved using IMCTUNE. Section 7.4 provides the theoretical justification for the Mp tuning algorithm presented in this section.

Occasionally, the Mp tuning procedure results in an upper-bound frequency response with one or more peaks where the valley to peak height is high enough to cause undesirable time domain oscillations. To avoid such time domain oscillations, we recommend restricting the magnitude of the upper-bound frequency domain peaks to not more than 0.1 above the highest adjoining valley. The IMCTUNE software checks the magnitude of any frequency domain peaks and adjusts the filter time constant to keep such peaks below a user specified value (the default value is 0.1).

7.3.2 The IMCTUNE Algorithm for Solving the Mp Tuning Problem

The following algorithm is implemented in MATLAB 5.3.1 for parametric uncertainty in the IMCTUNE software associated with this text.

Input data

1. A process model.
2. An uncertainty description in terms of upper- and lower-bounds on process parameters.
3. An initial value for the filter time constant. The default is the value of filter time constant that satisfies the maximum noise amplification specification (see item 5).
4. An M_p specification and tolerance. The defaults are

   $M_p = 1.05$ and Tolerance $= \pm 0.005$.

5. The maximum allowable high frequency controller noise amplification (i.e., $|q(\infty,\varepsilon)/q(0,\varepsilon)|$). The default is 20 (see Chapter 3).
6. Upper- and lower-bounds of the frequency range for the optimization. The defaults are
   Low frequency: reciprocal of 10 times the largest time constant or dead time
   High frequency: 1,000 times the low frequency
   Number of points and scale for plotting: 30, logarithmic
7. Upper- and lower-bounds of the frequency range for plotting. The defaults are
   Low Frequency: one-tenth the break frequency (see Figure 7.6)
   High Frequency: 100 times the break frequency
   Number of points and scale for plotting: 30, logarithmic

Output Data

1. The filter time constant that achieves the specified Mp ± Tolerance or the specified Maximum Peak Height ± Tolerance.
2. Curves of the upper- and lower-bounds of the magnitude of the complementary sensitivity function for the computed value of the filter time constant.
3. Tables listing which processes give the upper- and lower-bound at each frequency.
4. Output and control effort time responses for the constrained IMC system with the specified process parameters in the uncertainty sets.
5. PID parameters induced by the IMC controller and process model.
6. Output and control effort time responses for the constrained PID control system with the specified process parameters in the uncertainty sets.
7. Output and control effort time responses for the constrained model state feedback IMC system with the specified process parameters in the uncertainty sets.

Stryczek et al. (2000) give a detailed description of the Mp tuning algorithm.

Example 7.4 Tuning an Uncertain FOPDT Process

Find the filter time constant ε for the process given by Eq. (7.15), and model and controller given by Equations (7.16a) and (7.16b).

$$p(s) = \frac{K\,e^{-Ts}}{\tau s + 1}, \tag{7.15}$$

where

$$0.2 \le K \le 1; \ \ 8 \le \tau \le 14; \ \ .4 \le T \le 1.$$

Choosing a mid-range model and following the IMC controller design methods of Chapter 3 gives a model and IMC controller of

$$\widetilde{p}(s) = \frac{.6e^{-.7s}}{(11s + 1)}, \tag{7.16a}$$

$$q(s) = \frac{11s + 1}{.6(\varepsilon s + 1)}. \tag{7.16b}$$

To achieve an Mp specification of 1.05, the IMCTUNE software computes a value of 2.81 for the filter time constant ε^*. Using this filter time constant and IMCTUNE to compute the lower-bound of the complementary sensitivity function over all uncertain parameters, and requesting a plot of both upper- and lower-bounds, produces the curves in Figure 7.7.

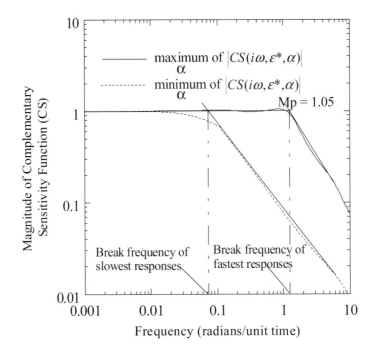

Figure 7.7 Upper- and lower-bounds of the complementary sensitivity function.

◆

7.3.3   Interpretation of the Results of Mp Tuning

Figure 7.7 can be used to estimate the range of closed-loop time responses. The reciprocal of the break frequency of the upper-bound curve is an estimate of the time constant for the fastest responses, while the reciprocal of the lower-bound curve is an estimate of the time constant of the slowest responses. The break frequencies[2] of the upper-bound and lower-bound frequency responses are approximately 1.2 and .07. The reciprocal of these break frequencies (1/1.2 and 1/.07, respectively) provides estimates of the fastest and slowest time constants of the closed-loop, which are .83 and 14. Settling times[3] should therefore be about 2.5 time units for the fastest responses and about 43 time units for the slowest responses. Comparison of these estimates with Figure 7.1, which gives the closed-loop time responses for several processes within the set given by Eq. (7.15), shows that the foregoing

---

[2] The break frequency is the intersection of the asymptote to the high frequency portion of the frequency response with the magnitude = 1 line.
[3] The settling time is usually taken as three time constants.

estimates are quite accurate. The various process responses shown in Figure 7.1 were obtained from IMCTUNE and correspond to the parameters given in Table 7.1.

Table 7 .1 Process Parameters for Figure 7.1 and the Process Given by Eq. (7.15).

curve no.	process gain, K	time constant, τ	dead time, T
1	1	8	1
2	1	14	1
3	.2	14	.4
4	.2	8	.4
5	.6	11	.7

It must be emphasized that the upper- and lower-bound frequency response curves in Figure 7.7 are not the frequency responses of any single process. At any frequency, the curves in Figure 7.7 represent the maximum and minimum of the complementary sensitivity function over the process parameters K, τ, and T. Therefore, each point on the curve could represent a different process. Usually, however, the upper-bound and lower-bound frequency responses come from only a small, finite subset of the possible process parameters. For example, the process parameters for the lower-bound curve in Figure 7.7 are given in Table 7.2.

Table 7.2 Process Parameters Associated with the Lower-Bound Curve in Figure 7.7.

frequency	process gain, K	time constant, τ	dead time, T
0.001 to 0.0033	0.2	14	1.00
0.0044 to 0.0862	0.2	8	.400
0.116 to 4.10	0.2	14	.400
5.52	0.2	14	.870
7.43	0.2	14	.636
10	0.2	14	.472

The parameters associated with the maximum peak of 1.05 are $[K, \tau, T] = [1\ 8\ 1]$, which yields the response given by curve 1 in Figure 7.1. There is a local maximum with an Mp of about 1.02 in the upper-bound curve at a frequency of about 0.1. The process parameters associated with this local maximum are $[K\ \tau\ T] = [1\ 14\ 1]$, which yields the response given by curve 2 in Figure 7.1.

As we shall see in Section 7.5, the shape of the upper-bound and lower-bound curves for the complementary sensitivity function and their associated break frequencies depend on the choice of the model. The Mp synthesis problem is to find the model and controller that give a lower-bound break frequency as high as possible for the specified Mp. Such a model and controller will provide the *fastest possible slowest responses* for the postulated uncertainty.

The above discussion assumes that the upper-bound curve of the complementary sensitivity function is similar to the response of a second-order system of the form $1/(\tau^2 s^2 + 2\zeta\tau s + 1)$. However, not all complementary sensitivity functions behave in this ideal manner. Indeed, even the upper-bound curve in Figure 7.7 is not quite that of a second-order system. Close inspection of the upper-bound in Figure 7.7 shows that it has a relative maximum of 1.0224, a relative minimum of 0.9771, and a global maximum of 1.0498. Second-order systems do not exhibit such relative maxima and minima. For systems with such oscillatory frequency responses, we must add the condition that the magnitude of the frequency response not have any peaks higher than specified (usually about 0.1) from the highest adjoining valley. For example, in Figure 7.7, the upper-bound curve has a peak at a frequency of .871, which is .0727 units higher than its highest adjoining valley, which is .9771 (.0727 = 1.0498 – .9771). In this case the peak is less than 0.1, so it does not influence the computed filter time constant of 2.81. If the peak had been higher than 0.1, then we would have increased the filter time constant beyond 2.81 to bring it down to 0.1. Section 7.5 contains some examples in which it is necessary to increase the filter time constant because of peaking, even though the Mp is less than 1.05.

7.3.4 Use of Multiple Uncertainty Regions to Account for Uncertain Uncertainty

Generally, the range of variation of the uncertain parameters α in an uncertain process, $p(s, \beta(\alpha))$ (see Section 7.2.1), is itself uncertain.   This is particularly true when the variations of the parameters α arise due to the fact that the true process is nonlinear, as in Eq. (7.4), and the set Π contains the local linearizations of the process about all possible operating points, as in Eq. (7.5).   In such situations, it is sometimes advantageous to view the set of uncertain parameters S_α as being composed of several overlapping sets, $S_{\alpha j}, j = 1...$ where the set $S_{\alpha j}$ is defined as

$$S_{\alpha j} \equiv \text{the set of all vectors } \alpha_j \text{ with } \underline{\alpha}_{j,i} \leq \alpha_{j,i} \leq \overline{\alpha}_{j,i}.$$

The $\underline{\alpha}_{j,i}$ and $\overline{\alpha}_{j,i}$ are ordered so that $\overline{\alpha}_{j,i} < \overline{\alpha}_{j+1,i}$ and $\underline{\alpha}_{j,i} > \underline{\alpha}_{j+1,i}$. A pictorial representation of the boundaries of the sets $S_{\alpha j}$ is given in Figure 7.8. Notice that $S_{\alpha 1}$ is contained in $S_{\alpha 2}$, and $S_{\alpha 2}$ is contained in $S_{\alpha 3}$. The shaded areas in Figure 7.7 represent the difference between $S_{\alpha 3}$ and $S_{\alpha 2}$ (i.e., $S_{\alpha 3} - S_{\alpha 2}$) and between $S_{\alpha 2}$ and $S_{\alpha 1}$ (i.e., $S_{\alpha 2} - S_{\alpha 1}$). As we shall see in Example 7.5, by assigning different tuning objectives to the various shaded areas of Figure 7.8 it is possible to account somewhat for uncertainty regarding process parameter ranges.

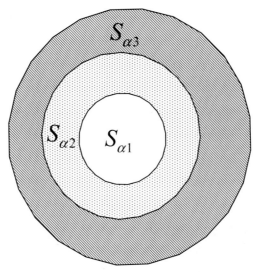

Figure 7.8 Multiple uncertainty sets.

Example 7.5 Use of Multiple Uncertainty Regions

Let us return to Example 7.4, where the uncertain process was given by Eq. (7.15):

$$p(s) = \frac{Ke^{-Ts}}{\tau s + 1},$$

where

$S_{\alpha 1}$ is $0.2 \leq K \leq 1$; $8 \leq \tau \leq 14$; $.4 \leq T \leq 1$; and $\underline{\alpha}_1 \equiv [\, .2\ 8\ .4\,]$, $\overline{\alpha}_1 \equiv [1\ 14\ 1]$.

Let us assume that the specified upper-bound of the parameters, $\overline{\alpha}_1$, are either not known very precisely, or may be violated for relatively brief periods of time. It is known, however, that the parameters never exceed 1.5, 20, and 1.5 for the gain, time constant, and dead time. That is,

$$S_{\alpha 2} \text{ is: } \underline{\alpha}_2 = [\, .2\ 8\ .4\,]), \quad \overline{\alpha}_2 = [\, 1.5\ 20\ 1.5\,]. \tag{7.17a}$$

The mid-range model for the above is

$$[\widetilde{K}, \widetilde{\tau}, \widetilde{T}] = [.85, 14, .95]. \tag{7.17b}$$

If we insist that no process in the enlarged uncertainty set $S_{\alpha 2}$ given by Eq. (7.17a) have an Mp greater than 1.05, and the maximum peak to valley height of 0.1, then IMCTUNE yields a filter time constant of 6.08 using a mid-range model of Eq. (7.17b) and its associated controller ($q(s) = (14s+1)/.85(6.08s+1)$). The slowest closed-loop time constant is now about 37, compared to slowest closed-loop time constant of about 14 for the original uncertainty set model and controller (see Example 7.4). For the larger uncertainty set, the maximum peak height of 0.1 is the binding constraint and the Mp is only 1.02.

Now let us assume that for the region given by ($S_{\alpha 2}$ – $S_{\alpha 1}$) (i.e., $1 \leq K \leq 1.5$; $14 \leq \tau \leq 20$; $1 \leq T \leq 1.5$), we are willing to accept overshoots of 50% and a fairly oscillatory response (e.g., ζ =.2). The Mp specification then becomes 2.4 in that region, and the maximum peak height specification must be 1.5 or higher, because the maximum peak height must be more than the Mp specification minus one. The filter time constant obtained using the same model and controller is 3.2 (i.e., $[\tilde{K}, \tilde{\tau}, \tilde{T}] = [.85, 14, .95]$; $q(s) = (14s+1)/.85(3.2s+1)$). The slowest closed-loop time constant for the foregoing tuning is about 25 units. Since a slowest time constant of 25 is slower than the slowest loop time constant of 14 for the previous design (see Example 7.4), we should expect that using the foregoing model and controller will yield an Mp smaller than 1.05 for the original uncertainty bounds, and this is indeed the case. Keeping the model and controller the same yields an Mp of 1.0 for the original uncertainty set. Thus, by separating the uncertainty region into two regions and allowing greater overshoots in the outer region, we have sped up the slowest responses from a closed-loop time constant of 37 to one of 25, or in terms of settling times, from about 110 units to 75 units.

The fact that the Mp for the current controller is only 1 means that if we were willing to tolerate even more overshoot in the outer uncertainty region (i.e., in ($S_{\alpha 2} - S_{\alpha 1}$)), then the filter time constant could be further reduced, thereby speeding up the slowest responses.

♦

7.3.4 The Inverse Tuning Problem

An alternate method of dealing with uncertainty about process parameter ranges is to specify what nominal closed-loop response time is acceptable, and then calculate the range of parameter variation that can be accommodated by a controller tuned to give that nominal response. This is equivalent to specifying the IMC controller filter time constant and calculating an uncertainty range that will yield that filter time constant for the specified Mp. Of course, it is necessary to recognize that actual response times will vary, possibly substantially, from the nominal (i.e., perfect model) response time. Further, the uncertainty range for a given filter time constant is not unique, because when more than one parameter varies, there is a multiple infinity of parameter ranges that yield the same filter time constant to achieve a specified Mp.

To associate an uncertainty range with an IMC controller filter time constant, it is necessary to specify a scalar function that tells how the model parameters vary within the uncertainty region. One convenient way of relating the parameters is to insist that all uncertain parameters vary by the same percentage from specified mid-range parameter values. Usually, the specified mid-range for parameter values will be the same as the model parameters. However, the IMCTUNE software permits the mid-range parameters supplied by the user to differ from the model parameters.

Example 7.6 Inverse Tuning

Again, we return to the process given by Eq. (7.15). The model parameters are as before (i.e., $[\,\tilde{K}\ \tilde{\tau}\ \tilde{T}\,] = [.6\ 11\ .7\,]$). Now, however, rather than calculating the IMC filter time constant to achieve an Mp of 1.05 for the uncertain parameter ranges given in Eq. (7.15), we shall assume that no uncertainty ranges are given, and our problem is to compute the uncertainty range for K, τ, and T that will yield a perfect model response time of about 8.4 time units to a unit step setpoint change. The foregoing problem specification is equivalent to specifying an IMC filter time constant, ε, of 2.8 (i.e., 2.8 = 8.4/3).

The IMCTUNE software calculates parameter ranges of $.35 \leq K \leq .85$; $6.5 \leq \tau \leq 15.5$; $.41 \leq T \leq .99$ for an Mp of 1.05 and an ε of 2.8. Notice that, except for the dead time, the calculated ranges are not the same as those for the process given by Eq. (7.15). This is because the calculated ranges have the same percent variation from the mid-range parameters ($\pm 41\%$), while in the original problem the gain varies by 66% and the time constant and dead time vary by $\pm 27\%$ from their mid-range values of .6 and 11.

◆

The next two sections provide the theoretical justification for the Mp tuning algorithm. Those readers willing to accept the algorithm on faith may skip or skim all but the restrictions given by Table 7.3. However, those readers who would like some insight into why the algorithm works should at least skim Sections 7.4 and 7.5.

7.4   CONDITIONS FOR THE EXISTENCE OF SOLUTIONS TO THE MP TUNING PROBLEM

The Mp tuning methodology described in the previous section implicitly assumes that (1) if all closed-loop complementary sensitivity functions have finite magnitudes, then the control system is stable for all processes in the uncertainty set, and (2) there always exists a filter time constant, which will cause the Mp for the complementary sensitivity function to lie below any preset limit greater than 1 for all processes in the uncertainty set. The following subsections give the conditions under which the foregoing statements are indeed true. As a prerequisite, however, we first present an abbreviated version of the Nyquist stability criterion discussed in Appendix B, Section B.4.

7.4.1 Statement of the Nyquist Stability Criterion

The number of zeros, Z, of the term $1 + kg(s)$ inside a closed contour, D, in the complex plane is equal to the number of poles of $1 + kg(s)$ inside of the closed contour D plus the number of clockwise (counterclockwise) encirclements of the point $(-1/k, 0)$ by the phasor $g(s)$ as s moves in a clockwise (counterclockwise) direction around the closed contour D. That is,

$$Z = N + P$$

where   Z = # of zeros of $1 + kg(s)$ inside D,

N = # of encirclements of $(-1/k, 0)$ point by $g(s)$ as s moves once around D,

P = # of poles of $1 + kg(s)$ inside D.

The control system of Figure 7.9 is stable if, and only if, the contour D encloses the entire right half of the s-plane and the number of zeros, Z, of the characteristic equation $1 + kg(s)$, as calculated above, is zero.[4]

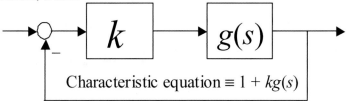

Figure 7.9 Feedback diagram for the Nyquist stability criterion.

If $g(s)$ has a pole at $s = 0$, then the contour D is usually taken as shown in Figure 7.10, and the radius δ is made to approach zero so that the contour encloses the entire right half of the s-plane.

---

[4] An excellent, intuitive proof of the Nyquist stability theorem can be found in Van de Vegte (1986).

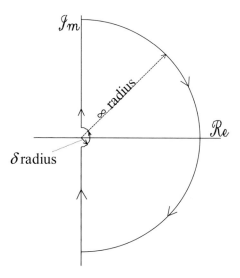

Figure 7.10 Nyquist D contour when there is a pole at the origin.

7.4.2 Integral Controllability

Any control system that forces the process output to exactly track the setpoint in the steady state (i.e., one that has no off-set) must have an integrator in the control loop. In IMC the required integration comes from the positive unity feedback loop formed by the controller and the model when the controller gain is the inverse of the model gain (see Chapter 6). For an uncertain process, the question arises as to what limits, if any, exist on the range of uncertain process parameters so that an integrating control system is stable over all possible process parameters. To address this question, Morari (1985) introduced the concept of integral controllability for uncertain processes. While the concept applies to general multivariable uncertain processes, here we shall use it only for SISO systems. Morari's definition of integral controllability follows

Integral Controllability: The open-loop stable uncertain system $h(s)$ in Figure 7.11 is called integral controllable if there exists a $k^* > 0$ such that the control system is stable for all k in the range $0 < k \leq k^*$.

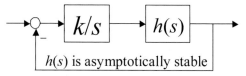

Figure 7.11 Feedback control diagram for integral controllability.

The uncertain system $h(s)$ in Figure 7.11 includes the process and actuator dynamics as well as a portion of the controller dynamics. Also, all the processes in $h(s)$ are assumed to be asymptotically stable.[5] (Such a system can have no poles in the right half of the s-plane or along the imaginary axis.)

An uncertain process that is integral controllable is one for which it is always possible to design a stable offset-free control system for all processes in the uncertainty set.

Integral Controllability Theorem: An uncertain system $h(s)$ is integral controllable if, and only if, the gains of $h(s)$, (i.e., $h(0)$) are positive for all plants in the uncertainty set.

The condition that all gains of $h(s)$ be positive effectively requires that all process gains $p(0)$ in the uncertainty set have the same sign and, of course, are never zero. If all process gains are negative, then the controller gain must also be negative. In Figure 7.11, the negative controller gain is included in $h(s)$ so that $h(0)$ is positive for all processes in the uncertainty set.

Morari uses the Nyquist Stability Criterion to prove the above theorem. To illustrate the nature of the proof, we consider the system $h(s) = h(0)/(s + 1)^3$ with $h(0)$ uncertain. The Nyquist diagram for $g(s) = h(s)/s$ for $h(0)$ equal one, and the D contour taken as shown in Figure 7.10 is given in Figure 7.12.

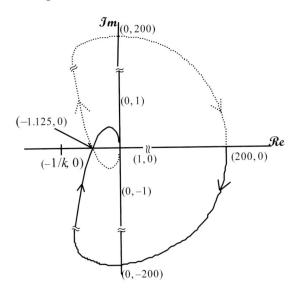

Figure 7.12 Nyquist diagram for $1/s(s + 1)^3$ as s travels about the contour, shown in Figure 7.10, with $\delta = .00492$.

The Nyquist diagram in Figure 7.12 is drawn with the central portion of the diagram (about the origin) much enlarged relative to the outer portion of the diagram. Also, the

---

[5] An unforced dynamic system is said to be asymptotically stable if it tends to the origin from any initial conditions.

diagram is distorted at the break points where the scale changes so that the diagram appears to be continuous, as it actually is, rather than discontinuous, as it would appear if two scales were rigorously adhered to. The point ($\mathcal{Re}$, $\mathcal{Im}$) = (200, 0) on the Nyquist diagram corresponds to the point $s = (\delta, 0)$ on the D contour of Figure 7.11, with $\delta = .00492$. As s moves along the curve $s = \delta e^{j\phi}$, with ϕ increasing from 0° to + 90°, the Nyquist diagram moves along the solid line. After ϕ reaches 90°, s moves along the curve $s = i\omega$ and the corresponding portion of the Nyquist diagram moves through the break in scale to the origin, as shown by the solid line in Figure 7.12. Along the infinite semicircle $1/s(s + 1)^3$ is zero. The dotted line in Figure 7.12 corresponds to the value of $1/s(s + 1)^3$ as s moves up the imaginary axis from $- \infty$, and then along the quarter circle back to $s = (\delta, 0)$.

Figure 7.12, shows that there will be no encirclements of the point $(-1/k, 0)$ provided that $-1/k$ is less than -1.125 which means that k must be greater than zero, but less than 1/1.125 (i.e., .889). It does not matter how small δ is taken to be, since smaller values of δ simply increase the radius of the right half plane semicircle in Figure 7.12. Changes in the gain $h(0)$ do not affect the general shape of the Nyquist diagram, provided all gains are positive. As long as $h(0)$ is greater than zero, changes in gain and time constants only change the point at which the Nyquist curve intersects the negative real axis. The intersection point is always finite, and so there always exists a k^* such that the control system is stable for $0 < k \leq k^*$. Should any of the gains of $h(0)$ be negative, however, then the Nyquist diagram is rotated about the imaginary axis, as shown in Figure 7.13.

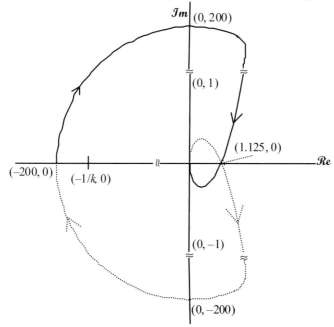

Figure 7.13 Nyquist diagram for $-1/s(s + 1)^3$ as s travels about the D contours shown in Figure 7.10 with $\delta = 0.00492$ and $h(0)$ negative.

From Figure 7.13, it is clear that if k is positive, the $-1/k$ point will always be encircled in a clockwise direction at least once if δ is allowed to approach zero, as it must if the D contour is to enclose the entire right half plane.

The above analysis can be repeated for any stable process and control system where the open-loop transmission has a pole at the origin. Morari uses an argument very similar to the foregoing in proving that the necessary and sufficient conditions for integral controllability are that the gain of $h(s)$ does not change sign.

For an IMC system, the equivalent definition of integral controllability is that an open-loop stable system is integral controllable if, and only if, there exists a filter time constant $\varepsilon^* > 0$ such that the closed-loop IMC system is stable for all finite ε greater than or equal to ε^*. Again, integral controllability requires that the process gain not change sign. To prove this statement, it is adequate to recognize the correspondence of an IMC system to an integral control system when the controller gain is the inverse of the model gain (see Chapter 6). However, we shall determine the necessary and sufficient conditions for integral controllability directly by application of the Nyquist criterion to the IMC system.

The characteristic equation of an IMC system such as that given by Figure 3.1 in Chapter 3 is given by (also see Eq. 3.6)

$$\text{Characteristic Equation} = (1 + (p(s) - \tilde{p}(s))q(s)). \tag{7.18}$$

This characteristic equation will have no right half-plane zeros only if the Nyquist diagram of $(p(s) - \tilde{p}(s))q(s)$ does not encircle the -1 point as s traverses a D contour which travels from $-i\infty$ to $+i\infty$ and then clockwise around the infinite semicircle back to $-i\infty$. We will investigate the behavior of Eq. (7.18) for filter time constants larger than ε^*. To start, we note that for low enough frequencies, the frequency response of all processes in the uncertainty set Π can be approximated by their steady-state gains. That is

$$p(i\omega) \cong p(0) \text{ and } \tilde{p}(i\omega) \cong \tilde{p}(0) \text{ for } \omega < \delta. \tag{7.19}$$

Therefore, for the controller filter time constant large enough so that

$$\varepsilon\delta \gg 1, \tag{7.20}$$

the terms pq and $\tilde{p}q$ can be approximated as

$$pq(s,\varepsilon) \cong \frac{p(0)/\tilde{p}(0)}{(\varepsilon s + 1)^r} \text{ and } \tilde{p}q(s,\varepsilon) \cong \frac{1}{(\varepsilon s + 1)^r} \quad s = i\omega, \ \omega < \delta. \tag{7.21}$$

Substituting Eq. (7.21) into Eq. (7.18) gives

$$\text{Characteristic Equation} = 1 + [p(0)/\tilde{p}(0) - 1]/(i\varepsilon\omega + 1)^r; \quad 0 \le \omega < \delta. \tag{7.22}$$

The stability of the control system is therefore completely determined by whether or not the function $[p(0)/\tilde{p}(0) - 1]/(i\varepsilon\omega + 1)^r$ encircles the -1 point on the Nyquist diagram as $\varepsilon\omega$ ranges from zero to large values. This function cannot encircle the -1 point so long as the ratio $p(0)/\widetilde{p}(0)$ is greater than zero. However, if any process gains are such that $p(0)/\widetilde{p}(0)$ is less than zero, then the Nyquist diagram will lie to the left of the -1 point at a frequency of zero and will spiral in a clockwise manner towards the origin as frequency increases. Thus, the Nyquist diagram will encircle the $(-1, 0)$ point, and the control system will be unstable. The necessary and sufficient conditions for integral controllability of an IMC system are therefore that all process gains must have the same sign and *the model gain must have the same sign as the process gains.*

7.4.3 Necessary and Sufficient Conditions for the Existence of a Solution to the Mp Tuning Problem for Any Mp Specification Greater than One

In this section we give the conditions under which any Mp specification can be achieved by choosing the IMC filter time constant large enough. As in Section 7.4.2, we consider all filter time constants, ε, greater than some ε^*, where ε^* is chosen so that $\varepsilon^*\delta$ is much greater than one and δ is small enough so that Eq. (7.19) is satisfied. Then the frequency response of the complementary sensitivity function $CS(i\omega,\varepsilon)$ (see Eq. 7.13) can be well approximated as follows:

$$CS(i\omega,\varepsilon) \cong \frac{(p(0)/\widetilde{p}(0)/(i\varepsilon\omega+1)^r}{1+[(p(0)/\widetilde{p}(0)-1]/(i\varepsilon\omega+1)^r}, \quad 0 \le \omega < \delta \tag{7.23}$$

$$\equiv [((i\Omega+1)^r - 1)\Phi + 1]^{-1}, \quad 0 \le \Omega \le \varepsilon\delta \gg 1, \tag{7.24}$$

where
$$\Omega \equiv \varepsilon\omega, \quad \Phi \equiv p(0)/\widetilde{p}(0)$$

The term $\varepsilon\delta$ in Eq. (7.24) is much greater than one because ε is chosen to be greater than ε^* and $\varepsilon^*\delta$ is much greater than one by choice of ε^*.

The behavior of $CS(i\Omega)$ given by Eq. (7.24) depends only on the ratio of process gain to model gain (i.e., $p(0)/\widetilde{p}(0)$). The upper-bound on the magnitude of $CS(i\Omega)$ (i.e., its M_p) will be less than one over all Ω greater than zero, only if the ratio of $p(0)$ to $\widetilde{p}(0)$ (i.e., Φ) is less than that shown in Table 7.3. Notice that for processes higher than second-order, it may not be possible to achieve an Mp arbitrarily close to one using the mid-range gain for the model if the range of possible process gains is large. For example, if the relative order r is 3 and the range of process gains is from 1 to 5, then the mid-range gain is three, and the ratio of $p(0)$ to $\widetilde{p}(0)$ can be as high as 5/3. For this case, the minimum achievable Mp is 1.015. Choosing a model gain of 10/3 allows one to achieve any Mp greater than 1 by choosing the filter time constant ε large enough. Of course, if the actual Mp specification is

1.05, then selecting a model gain of 3 will not cause a problem in meeting the Mp specification.

Table 7.3 Upper Limits on the Ratio of Process to Model Gain for Which It Is Possible to Achieve Mp Specifications as Close to One as Desired.

filter order, n	upper limit of the gain ratio $p(0)/\tilde{p}(0)$
1	∞
2	2
3	1.5
4	1.34
5	1.25

7.4.4 Justification for the Choice of Complementary Sensitivity Function in Mp Tuning

For the single-degree of freedom control system studied in this chapter, Mp tuning is based on the complementary sensitivity function rather than on the sensitivity function, which is commonly used in the literature to address issues of controller design with uncertain processes (Doyle et al., 1992). The sensitivity function $S(s)$ is the transfer function between the process output and the disturbance (see Figure 3.1). If the disturbance does not pass through a lag (i.e., if $P_d = 1$), then the sensitivity function is given by

$$S(s) \equiv \frac{(1 - \tilde{p}(s)q(s))}{1 + (p(s) - \tilde{p}(s))q(s)}. \tag{7.25}$$

From the definition of the complementary sensitivity function given by Eq. (7.13), the sum of the complementary sensitivity function and the sensitivity function given by Eq. (7.25) is one. That is,

$$CS(s) + S(s) = 1. \tag{7.26}$$

Therefore, from Equations (7.23) and (7.26), the expression for the sensitivity function for very large filter time constants is:

$$S(i\omega, \varepsilon) \cong \frac{1 - 1/(i\varepsilon\omega + 1)^r}{1 + [p(0)/\tilde{p}(0) - 1]/(i\varepsilon\omega + 1)^r}, \quad 0 \leq \omega < \delta. \tag{7.27}$$

The upper-bound of the magnitude of the sensitivity function given by Eq. (7.27) over all frequencies increases as the maximum of the ratio of $p(0)/\tilde{p}(0)$ increases. Therefore the maximum value of $|S(i\omega, \varepsilon)|$ is a minimum when the model gain $\tilde{p}(0)$ is chosen as the maximum process gain in the uncertainty set. In this case, $\max p(0)/\tilde{p}(0) = 1$, and the maximum of the sensitivity function is 1.15, 1.28, and 1.38 for r = 2, 3, and 4, respectively. Thus, unlike the situation for the complementary sensitivity function, there is no choice of model gain for which the maximum magnitude of the sensitivity function approaches one for large filter time constants. The corollary of this fact is that there is no single value for the maximum of the sensitivity function that corresponds to a specified overshoot for the time domain response to a step setpoint change. For this reason, it is more convenient to use the complementary sensitivity function to tune IMC controllers for good setpoint responses.

7.5   ROBUST STABILITY

This section introduces a theorem that states the conditions under which the Mp tuning algorithm in Section 7.3 yields stable control systems for all processes in the uncertainty set. A precise statement of the theorem and its proof can be found in Brosilow and Leitman (2001). The reason that such a theorem is required is that, by itself, a finite Mp does not guarantee stability. For example, the system $1/(-s + 1)$ has an Mp of one (i.e., upper-bound of $|1/(-i\omega + 1)|$ is 1), but is unstable. Similarly, a closed-loop control system for a stable process can have a finite Mp and yet the control system can be unstable. Consider the process and model given in Example 7.7.

Example 7.7 A Finite Mp, but an Unstable Closed-loop

$$p(s) = Ke^{-s} \quad 3 \leq K \leq 4 \tag{7.28a}$$

$$\tilde{p}(s) = e^{-s} \tag{7.28b}$$

$$q(s) = 1/(0.5s + 1) \tag{7.28c}$$

The Mp of the closed-loop system is 6.35, but *a Nyquist analysis of the control system for any process in the uncertainty set shows that the control system is unstable*. There are two right half plane poles for all processes with gains between 3 and 4. What's going on? We notice that the model gain is not in the uncertainty set, which is somewhat strange, but why should this matter? To answer such questions, we need the robust stability results from Brosilow and Leitman (2001), which we paraphrase below.

♦

7.5.1 A Frequency Domain Robust Stability Theorem for Infinite Dimensional Systems

A closed-loop transfer function, $H(s, \alpha)$, given by

$$H(s,\alpha) \equiv \frac{g(s,\alpha)}{1+h(s,\alpha)}, \qquad (7.29)$$

is stable for all parameter vectors, α, contained in the parameter set Π if $H(s, \alpha)$ *is stable for at least one parameter in the set* Π and if

$$\max_{\alpha \in \Pi} |H(i\omega, \alpha)| \text{ is finite } \forall \omega \geq 0 \qquad (7.30)$$

and the parameter set Π and the transfer functions $g(s, \alpha)$ and $h(s, \alpha)$ must satisfy the following conditions:

1. The parameter set Π is an *open, connected* subset of R^n. [6]
    Since Π is *connected*, for any two points $\tilde{\alpha}$ and $\tilde{\beta}$ in Π, there is a path σ in Π from $\tilde{\alpha}$ to $\tilde{\beta}$; that is, there is a continuous function $\sigma : [0,1] \to \Pi$ such that $\sigma(0) = \tilde{\alpha}$ and $\sigma(1) = \tilde{\beta}$. Since Π is *open*, a path σ from $\tilde{\alpha}$ to $\tilde{\beta}$ can be taken arbitrarily smooth.
2. There is a *connected, open* set $\hat{\Pi}$ in C^n such that $\hat{\Pi} \cap R^n = \Pi$ and the functions $g(s, \alpha)$, and $h(s, \alpha)$ have meromorphic extensions to $C^+ \times \hat{\Pi}$ [7] also denoted by g and h, where R, and C^n are real and complex n space and C^+ is the open complex half plane in C.
3. For each parameter $\alpha \in \Pi$, the function $h(s, \alpha)$ is not constant on C^+. [8]
4. For each parameter $\alpha \in \Pi$, the function $h(s, \alpha)$ is real-valued or ∞ on the positive real axis $\{s : Re(s) > 0, Im(s) = 0\}$.
5. The functions $g(s, \alpha)$, *and* $h(s, \alpha)$ are (jointly) continuous (in both s and α) functions from $\overline{P}^+ \times \Pi$ into P, where P is the complex projective sphere (Riemann sphere) and $\overline{P}^+$ represents the closure in P of the open hemisphere P^+.

---

[6] C^n denotes complex Euclidean n-space and R^n denotes its real Euclidean n-subspace.

[7] If $h(s, \alpha)$ is *meromorphic* for real parameters α in the sense of convergent power series expansions, this extension is always possible.

[8] If $h(s, \alpha)$ is the Laplace Transform of some real-valued *integrable* function and it is constant at α, it must be identically zero at α

Conditions (2), (3), and (4) are regularity or consistency conditions on g and h. However, conditions (1) and (5) on the parameter set and the continuity of g and h, especially on the imaginary axis, are structurally necessary for robust stability to hold.

A connected set is one in which there exists a path $\mathcal{P}$ between any two elements in the set, say, $\hat{\alpha}$ *and* $\hat{\beta}$. That is, there is a continuous function $\mathcal{P}$ such that $\mathcal{P}(0) = \hat{\alpha}$ and $\mathcal{P}(1) = \hat{\beta}$. The parameter set given by Eq. (7.15) is a particularly simple example of a connected set. It is not only connected, but also simply connected in that any closed curve in the set can be continuously contracted into a point. Figure 7.14 shows a connected set (shaded) that is not a simply connected set because of the hole in its center.

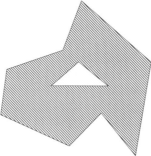

Figure 7.14 A connected set (but not simply connected).

We illustrate the concept of joint continuity by means of a transfer function that is not jointly continuous at the point $a = 0, s = 0$. Consider

$$h(s, a) = a(a - s)/(a(1 - a) + s(1 + a)) \qquad -1 \le a \le 1. \qquad (7.31)$$

At the point $a = 0$, $s \ne 0$, the value of the function is zero, even for very small s, since $h(s, 0) = 0/s$. For the point $s = 0$, $a \ne 0$, the value of the function is $a/(1 - a)$, which approaches one as a approaches zero. Therefore, $h(s, a)$ is not continuous at $s = a = 0$. Further, the point $a = 0$ cannot be removed from the uncertainty set because, if it were, the uncertainty set would not be connected. (The uncertainty set is just the section of the real line from -1 to 1.) As a consequence of the foregoing, the robust stability theorem cannot be applied to Eq. (7.31). We will have more to say about Eq. (7.31) in Chapter 8 in the section on the design and tuning of controllers for unstable, uncertain systems.

Conclusion of Example 7.7

Example 7.7 satisfies all of the conditions of the robust stability theorem. Application of the Nyquist criterion for a process that has a gain of 3 shows that the control system is unstable for this process. Since the Mp is finite (Mp = 6.36) we can conclude that the control system is unstable for all processes in the uncertain set, because if there were a process for which the control system is stable, the robust stability theorem tells us that the Mp would be infinite. Indeed, if the uncertainty set is expanded so that it includes the model gain (e.g., $1 \le K \le 4$), then all processes with gains sufficiently near 1 are stable (because there is effectively no modeling error), and the Mp for such an uncertain system is indeed infinite.

♦

For inherently stable processes in which the model parameters lie inside the connected uncertainty set, there will always be a set of processes for which the control system is stable. Therefore, if the Mp is finite, we can conclude that the control system is stable for all parameters in the uncertainty set. Of course, strictly speaking, we should check on all the conditions required by the robust stability theorem. However, it will be the very rare engineering model of a stable system that does not satisfy these conditions. We emphasize that the foregoing applies only to inherently stable processes. For inherently unstable processes, the control system can be unstable for all processes in the neighborhood of the model, while the control system is stable when the process equals the model. In such cases, the closed-loop transfer function is usually not jointly continuous at the point where the process parameters equal the model parameters. For inherently unstable processes, it is usually best not to consider the perfect model point to be in the uncertain parameter set, provided its (conceptual) removal does not create a disconnected set, as happened with Eq. (7.31).

7.5.2 A Heuristic Proof of the Robust Stability Theorem for Inherently Stable Processes

Let us assume that the robust stability theorem for inherently stable processes is false. Then, the Mp is finite for a system where there is a parameter vector β for which the system is stable and another parameter vector α for which the system is unstable. Now, for the parameter α, the Nyquist diagram of $(p(s) - \tilde{p}(s))q(s)$ must encircle the -1 point, since the control system is unstable. Similarly, for the parameter β the Nyquist diagram must not encircle the -1 point, since the control system is stable. Since the parameter set is open and connected, and the closed-loop transfer function is jointly continuous, there is a smooth path from α to β lying entirely inside the parameter set such that there is at least one parameter, say, γ, such that the Nyquist diagram of $(p(s) - \tilde{p}(s))q(s)$ passes right through the $(-1,0)$ point. So, there must also be a frequency $\hat{\omega}$ such that $1 + (p(i\hat{\omega}, \gamma) - \tilde{p}(i\hat{\omega}, \gamma))q(i\hat{\omega}) = 0$. Therefore, the maximum magnitude of the closed-loop frequency response over all parameters is not bounded. This contradicts our hypothesis that the robust stability theorem for inherently stable processes is false.

7.6   MP SYNTHESIS

Mp synthesis addresses the effect of uncertainty on controller design and model selection. The best model and controller combination is that which shifts the lower-bound break frequency furthest to the right without increasing the Mp, thereby speeding up the slowest closed-loop responses without increasing the overshoot to a step setpoint change. Example 7.8 explores the effect of changing just the controller and then changing both the controller and the model. We shall see that the effects are quite substantial when there is even a modest uncertainty in the model parameters.

Example 7.8 An Overdamped Second-Order Process Plus Dead Time

Process:
$$p(s) = \frac{K}{(8s+1)(6s+1)} e^{-Ts}, \quad 5 \le K, T \le 15 \qquad (7.32a)$$

Model:
$$\tilde{p}(s) = \frac{10}{(8s+1)(6s+1)} e^{-10s} \qquad (7.32b)$$

Controller 1:
$$q(s) = \frac{(8s+1)(6s+1)}{10(28.6s+1)^2} \qquad (7.32c)$$

Controller 2:
$$q(s) = \frac{(14s+1)}{10(31.6s+1)} \qquad (7.32d)$$

Controller 1, given by Eq. (7.32c), is the normal inverse of the invertible part of the model multiplied by a second-order filter to make the controller realizable. Controller 2, given by Eq. (7.32d) is the inverse of an approximate model with the original second-order lag replaced by a first-order lag whose time constant is the sum of the time constants of the second-order lag. Both controllers have been tuned to give an Mp of 1.05.

Figure 7.15a gives the upper-bound and lower-bound curves of the complementary sensitivity function for control systems that use the controllers 1 and 2. The model is that of Eq. (7.32b) for both control systems. Notice that both the upper-bound and lower-bound responses for Controller 2, Eq. (7.32d), are shifted to the right of those for Controller 1, Eq. (7.32c), implying that the fastest and slowest time domain responses for Controller 2 will be faster than those for Controller 1. Figure 7.15b bears this out.

Changing the model given by Eq. (7.32b) to $\tilde{p} = 10e^{-10s}/(14s+1)$, and using the controller given by Eq. (7.32d), does not change the control system responses significantly from those given in Figures 7.15a and 7.15b.

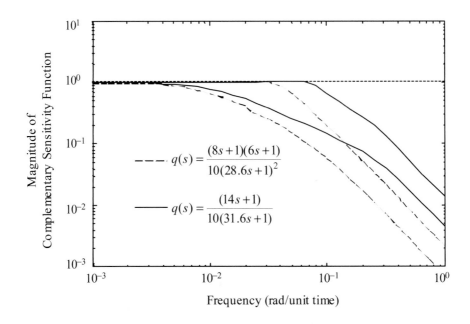

Figure 7.15a Comparison of the upper- and lower-bounds of the closed-loop frequency responses for the system of equation series (7.32).

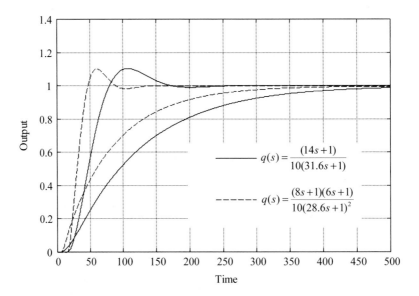

Figure 7.15b Fastest and slowest responses (i.e., K =15 and 5, respectively) to a step setpoint change for the control systems of equation series (7.32).

Standard engineering practice for the control of overdamped systems such as that given by Eq. (7.31a) is to fit a first-order model of the form of Eq. (7.33a) to the output response to a step change in the control effort, and then to tune the controller based on that model (Seborg et al., 1989).

$$\tilde{p}(s) = \frac{K}{(\tau s + 1)} e^{-(T+\varDelta)s},$$
(7.33a)

where T = the estimated process dead time,

   $\varDelta$ = an additional dead time arising from the fitting procedure,

   τ = an estimate of a first-order time constant which allows the approximate model step response to match that of the higher order process.

Using the reaction curve method (Seborg et al., 1989) yields $(T + \varDelta) = 12$, $\tau = 20$, and of course $K = 10$. Tuning the control system using this model and its model inverse controller yields a filter time constant of 38.6. The resulting controller is

$$q = \frac{(20s + 1)}{10(38.6s + 1)}.$$
(7.33b)

The model and a controller given by Equations (7.33a) and (7.33b) does not perform quite as well as the controller and model given by Equations (7.32b) and (7.32d). It turns out that the above model parameters do not give a very good fit to a step change in the control effort. A much better fit is obtained with $\Delta = 4$ and $\tau = 11$. The responses with this model and its associated controller are insignificantly different from those of Equations (7.32b) and (7.32d) (i.e., the solid line responses in Figure 7.15b).

The above models and controllers are all mid-range in the sense that the model gain is the mid-range gain, and the dead time is either the mid-range dead time or the dead time obtained by fitting the mid-range model. However, one can obtain performance that is slightly better than that obtained with Equations (7.32b) and (7.32d) with either of the following models and controllers.

$$\tilde{p}(s) = \frac{15}{(8s + 1)(6s + 1)} e^{-15s} \qquad q(s) = \frac{(8s + 1)(6s + 1)}{15(6.07s + 1)^2},$$
(7.34)

$$\tilde{p}(s) = \frac{15}{(8s + 1)(6s + 1)} e^{-15s} \qquad q(s) = \frac{(14s + 1)}{15(11s + 1)}.$$
(7.35)

The control systems of Equations (7.34) and (7.35) yield effectively the same frequency and time responses. Therefore Figures 7.16a and 7.16b show only the responses of the control system of Eq.(7.35).

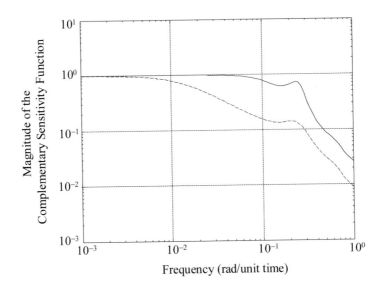

Figure 7.16a Upper-bound and lower-bound frequency responses for the process of Eq. (7.32a) and the control system of Eq. (7.35).

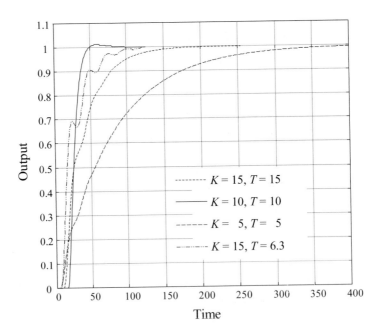

Figure 7.16b Time responses for the process of Eq. (7.32a) and the control system of Eq. (7.35).

Notice that in Figure 7.16a the Mp is 1, and the criterion that establishes the controller filter time constant is that no peak to valley height be greater than 0.1. The process at the local peak in Figure 7.16b has K = 15 and T = 6.3. This process yields the somewhat oscillatory response shown in Figure 7.16b. Also, comparison of the responses of mid-range process (i.e., $K = 10$, $T = 10$) using the control system of Eq. (7.35) with the control system of Equations (7.32b) and (7.32d), where the model is perfect, shows no significant difference. *Thus, for this example, there is no advantage to using the mid-range model, even if the process operates around the mid-range (i.e., K = 10, T = 10) far more often than at K = 15, T = 15.*

Changing the model in Eq. (7.35) to $\tilde{p}(s) = 15e^{-15s}/(14s+1)$ and retuning gives $q(s) = (14s+1)/15(10.6s+1)$. This controller model pair has an Mp of 1.05, and the lower-bound curve of the complementary sensitivity function actually lies slightly to the right of that in Figure 7.16a. Once again, the simpler model/controller is preferred.

As a check on the sensitivity of the control systems given by the equation series (7.32) to the assumed uncertainty bounds, we increased the upper- and lower-bounds by ± 20% to $\overline{K}$, $\overline{T} = 18$ and $\underline{K}$, $\underline{T} = 4$. This change causes worst-case overshoots of 30% for Eq. (7.32c) and a 35 % overshoot for Eq. (7.32d). The settling time of the slowest responses increases to 500 and 450 units respectively. Thus, the tuned control systems are not overly sensitive to modest errors in estimating the parameter uncertainty ranges.

The PID controllers that have the same performance as Equations (7.34) and (7.35) are respectively: $K_c = 0.0412$, $\tau_I = 16.8$, $\tau_D = 4.41$, and $K_c = 0.0422$, $\tau_I = 16.5$, $\tau_D = 4.42$.

PID controllers with the same performance as Equations (7.32c) and (7.32d) are respectively: $K_c = 0.00383$, $\tau_I = 2.57$, $\tau_D = 6.27$, and $K_c = 0.0338$, $\tau_I = 14.0$, $\tau_D = 1.73$.

If the above results extend to other similar processes, and we believe that they do, then they provide a strong justification for standard engineering practice of fitting a first-order lag plus dead time model to high order overdamped processes when there are significant process parameter variations. A reasonable question, however, is how much parameter variation is significant? For the above process, it turns out that the first-order model and controller performs as well as or better than a second-order model and controller for gain and dead time variations greater than ± 25%. We conjecture that the break points for going from a higher order to a lower order controller occur at uncertainty levels where the tuned higher order controller becomes a lag (i.e., $|q(i\omega)| \le 1$ for $\omega \ge 0$).

◆

Example 7.9 A Second-Order Process that Varies from Underdamped to Overdamped

This example consists of a second-order system with uncertainty in the time constant and damping ratio. The uncertain process has the following representation:

$$S = \left\{ p(s) \mid p(s) = \frac{2}{(\tau^2 s^2 + 2\zeta\tau s + 1)} \right\}, \quad 3 \leq \tau \leq 9, 0.5 \leq \zeta \leq 1.1. \quad (7.36)$$

The large uncertainty on the damping ratio makes the system underdamped for some plants ($\zeta < 1$) and overdamped for others ($\zeta > 1$).

Rather than varying the IMC controller separately from the model, as in the previous example, here we vary both simultaneously, with the controller taken as the inverse of the model so as to reduce the search space. Therefore, the controller model pair that is obtained is not necessarily the best possible. We start by designing the IMC controller using the mid-range plant so that we can use it as a reference to compare with other designs. The IMC model and controller are

$$\tilde{p}(s) = \frac{2}{(36s^2 + 9.6s + 1)}, \quad q(s) = \frac{36s^2 + 9.6s + 1}{2(\varepsilon s + 1)^2} \quad (7.37)$$

The filter time constant of the IMC controller $q(s)$ is chosen so that the robust performance condition is met:

$$|CS(i\omega)| \leq 1.05 \qquad \forall p \in S, \forall \omega, \quad (7.38)$$

$$|CS(i\omega_c)| = 1.05 \quad \text{for some } p \in S \text{ and some frequency, } \omega_c. \quad (7.39)$$

Figure 7.17 illustrates that the conditions given by Equations (7.38) and (7.39) are satisfied when epsilon is 10.96.

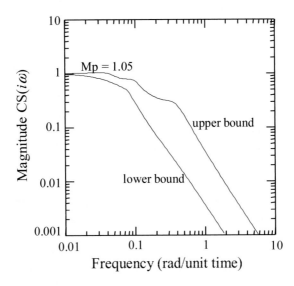

Figure 7.17 Maximum and minimum of $|CS(i\omega)|$ $\forall p \in S$ when $\varepsilon = 10.96$ $\tilde{\tau} = 6$ and $\tilde{\zeta} = 0.8$.

From Figure 7.17 we observe that the lower-bound curve break frequency is about 0.052 rad/sec and the maximum peak of $|CS(i\omega)| \ \forall p \in S$ is given by the plant with $\tau = 9$ and $\zeta = 1.1$. A lower-bound break frequency of 0.057 implies a settling time of 57 seconds for the slowest plant. This is in agreement with the closed-loop step responses shown in Figure 7.18. This figure also shows that the maximum overshoot to a unit step in setpoint change is 10.48%, produced by the upper extreme plant (i.e., $\tau = 9$ and $\zeta = 1.1$).

To improve the speed of response of the control system, we choose another model in the set S to attempt to shift the lower-bound break frequency to the right, thereby speeding up the slowest closed-loop responses while maintaining $|CS(i\omega)| \leq 1.05$. The fastest design is obtained with a model that has a time constant of 9 and a natural damping ratio of 0.8. Figure 7.19 shows the upper- and lower-bounds on the closed-loop frequency responses Figure 7.20 shows the closed-loop time responses to a unit-step setpoint change for several different plants. The maximum overshoot is 5.21%, which is produced by the upper extreme plant, and the slowest settling time is about 32 seconds.

Using the optimal model rather than the mid-range model reduces the slowest settling time from 52 seconds to 32 seconds. Moreover, since the maximum overshoot is 5.21%, the filter time constant can be further reduced, thereby further speeding the process response.

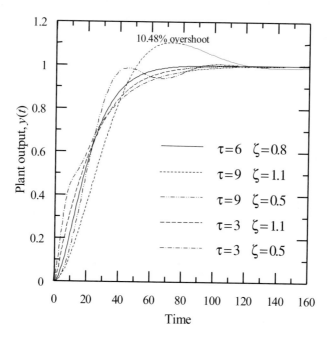

Figure 7.18 Closed-loop response to a unit-step change in setpoint for some plants in S when $\varepsilon = 10.96$, $\tilde{\tau} = 6$ and $\tilde{\zeta} = 0.8$.

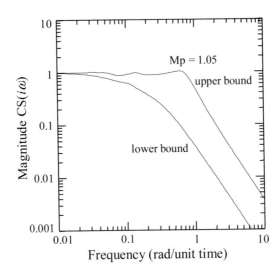

Figure 7.19 Maximum and minimum of $|CS(i\omega)| \; \forall p \in S$ when $\varepsilon = 4.93, \tilde{\tau} = 9, \tilde{\zeta} = 0.8$.

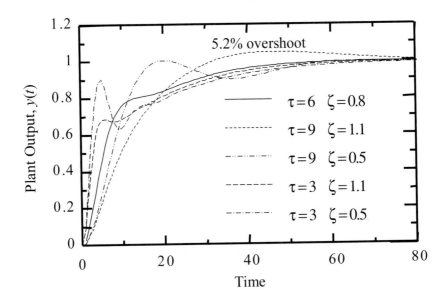

Figure 7.20 Closed-loop response to a unit-step change in setpoint for some plants in S when $\varepsilon = 4.927$, $\tilde{\tau} = 9$ and $\tilde{\zeta} = 0.8$.

♦

7.7   SOFTWARE FOR MP TUNING AND SYNTHESIS

The IMCTUNE software supplied with this text was used to generate all the figures displayed in this section. Appendix G provides a description of this software.

7.8   SUMMARY

Process uncertainty is described by allowing the process to have any parameter values in a predefined set. Tuning an IMC system for an uncertain process (Mp tuning) is carried out by finding the IMC controller filter time constant that yields a specified upper-bound on the maximum magnitude of the closed-loop transfer function between the output and the setpoint (i.e., on the complementary sensitivity function). A common specification, or Mp, for the magnitude of the complementary sensitivity function is 1.05 because this specification usually limits the maximum overshoot to step setpoint changes to about 10%. When, as is usually the case, parametric uncertainty bounds cannot be obtained precisely, then we recommend the use of multiple uncertainty regions, with different Mp specifications in each region, as described and illustrated in Section 7.3.4 and Example 7.5.

The robust stability theorem of Section 7.5 provides the theoretical foundation for Mp tuning. This theorem states that if the Mp is finite, then the control system for an inherently stable process is stable for any process parameter vector in the uncertain set. For the theorem to be applicable, the uncertain parameter set must be connected. (That is, there has to be a path between any two parameters that lies completely within the uncertainty set.) In addition, an uncertain process gain must not change sign within the uncertain set if a no offset control system is to be stable. The process model gain must obey the restrictions in Table 7.3 in order to be able to achieve an arbitrarily small overshoot to a step setpoint change for a large enough filter time constant.

Mp synthesis seeks to find an optimum controller and model for the IMC system. The model and controller are optimal in the sense that they maximize the speed of the slowest closed-loop responses subject to maintaining the specified Mp (i.e., the specified relative stability). An example of a second-order overdamped process with dead time shows that:

1. For gain and dead time uncertainty greater than $\pm$ 25%, a first-order controller performs as well as or better than a model inverse controller. The degree of improvement increases as the uncertainty increases.

2. The traditional engineering approach of fitting a first-order plus dead time model to high-order overdamped processes, and designing the controller based on that model, yields as good or better control system than that based on a process model of the correct order when the process parameters are sufficiently uncertain.

3. Limited numerical experiments, along with the analysis leading to Table 7.3, indicates that the optimum model gain is the upper-bound gain for inherently stable overdamped processes. This seems to be true even if the process operating point is most likely to be about its mid-range gain.

Problems

7.1 In each of the Problems in Chapter 3, assume that the gain and dead time (if any) are nominal values and that actual values of gain and dead time vary independently about the nominal value by $\pm$ 20%. What filter time constant is required to achieve an Mp of 1.05?

7.2 If in problem 7.1, the gain and dead time vary together, rather than independently what filter time constant is required to achieve an Mp of 1.05? That is, the gain K varies as $\overline{K}(1-x) \le K \le \overline{K}(1+x)$, and the dead time T varies as $\overline{T}(1-x) \le T \le \overline{T}(1+x)$, where x takes values between $\pm$ .2.

7.3 Find an optimal model for the following processes and compare the performance of each against a control system using a mid-range model.

$$\text{a. } p(s) = \frac{e^{-s}}{\tau^2 s^2 + 2\zeta\tau s + 1} \qquad \begin{array}{l} 3 \le \tau \le 9 \\ .5 \le \zeta \le 1.1 \end{array}$$

$$\text{b. } p(s) = \frac{e^{-s}}{\tau^2 s^2 + 2\zeta\tau s + 1} \qquad \begin{array}{l} 3 \le \tau \le 9 \\ .1 \le \zeta \le 1.1 \end{array}$$

$$\text{c. } p(s) = \frac{-\tau s + 1}{s + 1} e^{-s} \qquad .5 \le \tau \le 1.5$$

7.4 How many zeros does the following process have in the right half plane as K ranges from 1 to 10?

$$p(s) = \frac{3(s-1)}{(s+1)} - \frac{K(s-1)e^{-5s}}{(s+1)^2}$$

Hint: Convert p into a form that allows application of the Nyquist theorem

7.5 Sketch the slowest expected time response to a step setpoint change for a control system designed so that the fastest response to a step setpoint change does not overshoot by more than 20% for the following process:

$$p(s) = \frac{K(s^2 - s - 1)e^{-Ts}}{(3s+1)(2s+1)(s+1)}, \quad 3 \le K \le 7, \quad 2 \le T \le 6.$$

7.6 Derive Equations (7.13) and (7.25).

7.7 Show that for a second-order process, any Mp specification greater than 1 can be achieved, provided that $p(0)/\tilde{p}(0) < 2$.

The following questions assume familiarity with the material in Chapter 6.

7.8 For each of the problems in Chapter 3, assume that the gain and dead time (if any) are nominal values and that actual values of gain and dead time vary independently about their nominal values. How much variation is necessary before there is no advantage in using an IMC system over a PID control system?

7.9 Is it feasible to tune a PID controller so that the slowest closed-loop response for the process described below has a time constant of 100 minutes or less?

$$p_1(s) = \frac{K(s-1)e^{-Ts}}{(s+1)(3s+1)}, \quad 2 \leq K \leq 6, \quad 2 \leq T \leq 6$$

If so, what is one such tuning? If not, what is the best response that one can achieve, and what are the PID parameters for such a response? Show how you arrived at your result.

7.10 Over what range of process gains, K, is the closed-loop control system stable for the following process and controller?

$$process: \quad \frac{K(-2s+1)}{(2s+1)(5s-1)}$$

$$PID\ controller: \quad 2.37\left(1+\frac{1}{13.7s}+\frac{1.65s}{.082s+1}\right)$$

7.11 Your boss would like you to explore the possibility of speeding up a critical control loop by a factor of 2. The current PID controller tunings give a closed-loop response that can be well approximated by a first-order lag plus dead-time where the lag time constant is 20 minutes and the dead time is 15 minutes. That is

$$\frac{\Delta y(s)}{\Delta r} \cong \frac{e^{-15s}}{20s+1}, \quad \text{where } \Delta y = \text{change in output}, \quad \Delta r = \text{change in setpoint}$$

The process model associated with the above closed-loop response is believed to be of the form

$$\frac{\Delta y(s)}{\Delta m(s)} \cong \frac{10e^{-15s}}{15s+1} \qquad \text{where } \Delta m = \text{change in control effort}$$

The process operators are opposed to retuning the controller because they believe that the process gain can vary from 6 to 14. What recommendation would you give your boss and why? What PID controller tunings (i.e., gain, and integral and derivative time constants) would you recommend?

7.12 Design a PID controller for the following process using a nominal gain $\hat{K}$ of 10 for the model, and a specification that no process response to a setpoint change should overshoot the setpoint by more than 10%.

$$\frac{y(s)}{m(s)} = \frac{K(2s^3 - 3s^2 - 1)e^{-s}}{(s^2 + s + 1)(s + 3)^2}, \quad 5 \le K \le 15,$$

Over what range of process gains, K, is the system stable?

7.13 Select a PID controller for the process given by

$$p(s) = \frac{K(1 + 5e^{-3s})}{(2s + 1)^2} \qquad \text{for } 1 \le K \le 5.$$

References

Åström, K. J., G. C. Goodwin, and P. R. Kumar. 1995. *Adaptive Control, Filtering, and Signal Processing*. Springer-Verlag, NY.

Brosilow, C., and M. Leitman. 2001. "A Frequency Domain Robust Stability Theorem for Infinite Dimensional Systems with Parametric Uncertainty." *SIAM Journal on Control and Optimization*.

Dorato, P., L. Fortuna, and G. Muscato. 1992. *Robust Control for Unstructured Perturbations: An Introduction*. Springer-Verlag, NY.

Doyle, J. C., B. Francis, and A. Tannenbaum. 1992. *Feedback Control Theory*. Macmillan Publishing Company, NY.

Kravaris, C. and J. C. Kantor. 1990. "Geometric Methods for Nonlinear Process Control. 1. Background." Ind. Eng. Chem. Research, 29, 2295–2310.

Kravaris, C. and J. C. Kantor. 1990. "Geometric Methods for Nonlinear Process Control. 2. Controller Synthesis." Ind. Eng. Chem. Research, 29, 2311–2323.

Kwakernaak, H. 1993. "Robust Control and H∞ Optimization." Tutorial paper. Automatica. 29, 255–273.

Ljung, L. 1987. *System Identification—Theory for the User*. Prentice Hall, NJ.

Morari, M. 1985. "Robust Stability of Systems with Integral Control." IEEE Transactions on Automatic Control AC-30, 574–577.

Morari, M., and E. Zafiriou, E. 1989. *Robust Process Control*. Prentice Hall, NJ.

Seborg, D. E., T. F. Edgar, and D. A. Mellichamp. 1989. *Process Dynamics and Control*, John Wiley & Sons, NY.

Stryczek, K, M. Laiseca, M. Leitman, and C. Brosilow. 2000. "Tuning and Design of Single Input Single Output Control Systems for Parametric Uncertainty," *AIChE Journal* 46, 1616–1631

Van de Vegte, J. 1986. *Feedback Control Systems*. Prentice Hall, NJ.

Tuning and Synthesis of 2DF IMC for Uncertain Processes

Objectives of the Chapter

- Present and motivate 2DF IMC tuning algorithms for stable uncertain processes with large or oscillatory disturbance lags.
- Present and motivate 2DF IMC tuning algorithms for unstable uncertain processes.
- Explore the effect of uncertainty on 2DF IMC controller design and model selection.

Prerequisite Reading

Chapter 4, "Two-Degree of Freedom Internal Model Control"
Chapter 7, "Tuning and Synthesis of 1DF IMC Controllers for Uncertain Processes"
Appendix A, "Review of Basic Concepts"
Appendix B, "Frequency Response Analysis"

8.1   INTRODUCTION

Tuning a 2DF IMC system proceeds in two steps. First, the feedback loop in Figure 8.1 is tuned for good disturbance suppression by adjusting the filter time constant ε of the controller $qq_d(s,\varepsilon)$. Figure 8.1 is a reproduction of Figure 4.2 and is repeated here for convenience.

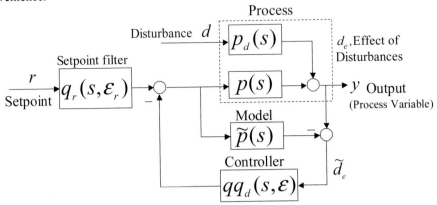

Figure 8.1 The 2DF IMC block diagram.

Second, the setpoint filter $q(s,\varepsilon_r)$ is tuned by adjusting ε_r to get the desired setpoint behavior using the Mp tuning method of Chapter 7. The tuning criterion for the feedback loop is based on the sensitivity function rather than on the complementary sensitivity function because it is the response to disturbances that is of interest, and the sensitivity function is the name given to the transfer function between the disturbance and the output. To get around the problem that the maximum peak of the sensitivity function is difficult to correlate with time domain behavior (c.f. Section 7.4.4), the next section, Section 8.2, introduces a related function that we call the partial sensitivity function.

     After a discussion of tuning using the partial sensitivity function, Section 8.3 discusses optimal model and controller selection for overdamped processes.

     Section 8.4 deals with the tuning and design of controllers for underdamped and unstable processes. For such processes we no longer attempt to meet an Mp specification, but rather suggest that the appropriate objective is to minimize the maximum magnitude of the sensitivity function. The last example in this section illustrates how crucially important it is to select the best model for an uncertain *unstable* process.

     An important property of 2DF control systems is that they reduce to 1DF control systems (see Figure 8.2, reproduced from Chapter 3, Figure 3.1), when $qq_d(s,\varepsilon)=q(s,\varepsilon)$ and $\varepsilon_r=\varepsilon$. That is, all of the transfer functions in Figures 8.1 and 8.2 between the inputs (i.e., the disturbance and the setpoint) and the outputs (i.e., the control effort and the process variable) are the same under the foregoing conditions. This means that the 2DF control system is more general than the 1DF control system and can therefore always be designed and/or tuned to perform at least as well as a 1DF control system. When the disturbance transfer function $p_d(s)$ is a lag (i.e., $\left|p_d(i\omega)\right| \leq 1$ for $\omega \geq 0$), then better performance can always be obtained from a 2DF control system with $qq_d(s,\varepsilon) = q(s,\varepsilon)$ by decreasing the filter time constant of the feedback controller below that required for good performance for a 1DF control system. The setpoint controller filter time constant ε_r is also adjusted for good setpoint response so that, in general $\varepsilon_r \geq \varepsilon$. How aggressive the tuning can be without resulting in undesirable oscillations in the output depends on how much the disturbance lag reduces the magnitude of disturbance frequencies that would otherwise cause oscillations in the output.

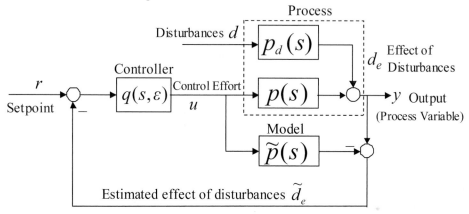

Figure 8.2 The 1DF IMC block diagram.

8.2   MP TUNING FOR STABLE, OVERDAMPED UNCERTAIN PROCESSES

8.2.1 Tuning the Feedback Loop via the Partial Sensitivity Function

As stated previously, the appropriate criterion for tuning the feedback loop of Figure 8.1 for good disturbance suppression is the sensitivity function, which is given by

$$\frac{y(s)}{d(s)} = \frac{(1-\widetilde{p}(s)qq_d(\varepsilon,s))p_d(s)}{(1+(p(s)-\widetilde{p}(s))qq_d(\varepsilon,s))} \equiv \text{Sensitivity Function} \tag{8.1}$$

The controller design and tuning most commonly found in the literature (Morari and Zafiriou, 1989; Dorato et al., 1992; Doyle et al., 1992) seeks to achieve a specified, frequency-dependent, upper bound on the magnitude of the sensitivity function. Unfortunately, for overdamped processes, it takes a good deal of experience to specify such an upper bound so as to achieve the desired closed-loop robustness and time domain disturbance rejection. As discussed previously, the reason is that, unlike the complementary sensitivity function, there is no fixed upper bound on the peak of the sensitivity function that assures achieving the desired closed-loop robustness and time domain behavior. Unlike the complementary sensitivity function, the maximum magnitude of the sensitivity function given by Eq. (8.1) is often greater than 1, even when there is a perfect model. Further, maximum magnitude of the sensitivity function is model dependent. Because it is not a simple matter to set bounds on the magnitude of the sensitivity function so as to achieve desirable time domain behavior, we introduce a modification of the sensitivity function that we call the partial sensitivity function that is easier to use. The partial sensitivity function is defined as

$$\text{Partial Sensitivity Function} \equiv \frac{p(s)qq_d(\varepsilon,s)p_d(s)/p_d(0)}{(1+(p(s)-\widetilde{p}(s))qq_d(\varepsilon,s))}. \tag{8.2}$$

The above definition is motivated by the following observations. The first term of the sensitivity function in Eq. (8.1), $p_d(s)/(1+(p(s)-\widetilde{p}(s))qq_d(\varepsilon,s))$, is the output response to the disturbance, as modified by the closed-loop. The second term, $p(s)qq_d(\varepsilon,s)p_d(s)/(1+(p(s)-\widetilde{p}(s))qq_d(\varepsilon,s))$, represents the response of the output to the control effort. This term is the negative of the model output response to a setpoint that is filtered by (i.e., cascaded with) the disturbance lag pd (s). If we replace the model in the numerator with p(s) and if the IMC controller is taken as qqd (s,ε), then this term is the same as a complementary sensitivity function filtered by a lag. This observation suggests the possibility of applying the Mp tuning algorithm to Eq. (8.2). Since a descriptive name such as "filtered complementary sensitivity function" is a bit of a mouthful, we shorten it to simply the partial sensitivity function.

The disturbance lag in Eq. (8.2) is divided by its gain so that the steady-state gain of the partial sensitivity function will be 1, and it can therefore be used for tuning just like the complementary sensitivity function. Since the magnitude of the disturbance is generally unknown, there is no loss in generality by this normalization.

We restrict the use of the partial sensitivity function given by Eq. (8.2) to overdamped, stable systems because for such systems the partial sensitivity function behaves very much like the complementary sensitivity function. Indeed, if $q_d = p_d = 1$, then the partial sensitivity function reduces to the complementary sensitivity function. However, in general, q_d is a lead and the perfect model *maximum* magnitude of the partial sensitivity function is generally unity only if the process is an overdamped lag. Said another way, the perfect model maximum magnitude of the partial sensitivity function is generally greater than 1 if the process is underdamped or unstable.

The Mp tuning algorithm of Chapter 7 includes a restriction on the allowable size of oscillations in the curve of the maximum magnitude of the closed-loop frequency response. The suggested limit is that no peak be higher than about 0.1 from the highest adjoining valley. While this restriction is not often encountered in tuning 1DF control systems, it will be frequently encountered in tuning 2DF control systems using the partial sensitivity function. The reason is that oscillations in the upper bound of the partial sensitivity function arise due to the fact that the disturbance lag allows the use of a smaller controller filter time constant than would be used in tuning a 1DF control system using the complementary sensitivity function. The smaller filter time constant makes the feedback loop less robust in the sense that the complementary sensitivity function (without the disturbance lag) can now have a maximum magnitude significantly higher than 1.05. By restricting the maximum peak size of oscillations in the magnitude of the closed-loop frequency response of the partial sensitivity function we are attempting to limit the maximum magnitude of the complementary sensitivity function. That is, we are trying to maintain a reasonable relative stability in the feedback loop of Figure 8.1 The value of 0.1 suggested as a limit on peak heights is based on experience with numerous examples, but it should not be taken as a hard and fast rule. After tuning, as described below, it is always advisable to check the maximum peak of the complementary sensitivity function as well as the time response of processes whose parameters are those that gave rise to the maximum peaks in the frequency domain. Example 8.1 below illustrates the use of Mp tuning using the partial sensitivity function.

Example 8.1 Tuning a FOPDT Process via the Partial Sensitivity Function

$$p(s) = \frac{Ke^{-Ts}}{(s+1)} \; ; \; p_d(s) = 1/(s+1) \quad .8 \le K \le 1.2; \quad .8 \le T \le 1.2, \quad\quad (8.3)$$

$$\tilde{p}(s) = \frac{e^{-s}}{(s+1)}, \qquad qq_d(s) = \frac{(s+1)(\alpha s+1)}{(\varepsilon s+1)^2}.$$

Applying Mp tuning using the partial sensitivity function produces the results shown in Figure 8.3. Figure 8.4 presents the disturbance responses of the control system for the tuning shown in Figure 8.3.

After tuning the feedback loop, the next step is to tune the setpoint controller $q(s, \varepsilon_r)$. Specifying an Mp of 1.05 for the complementary sensitivity function (defined as the transfer function between output and setpoint) results in a filter time constant, ε_r, of 1.03. The responses to a unit step setpoint change are given in Figure 8.5. Figure 8.6 shows the upper bound for the partial sensitivity function when the disturbance lag transfer function set to one (i.e., the complementary sensitivity function). This figure shows that the feedback loop has good relative stability as the Mp is only 1.8.

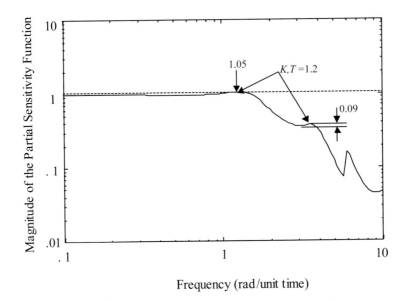

Figure 8.3 Upper bound of the partial sensitivity function for Example 8.1 with
$qq_d(s) = (s + 1)(0.942s + 1)/(0.602s + 1)^2$

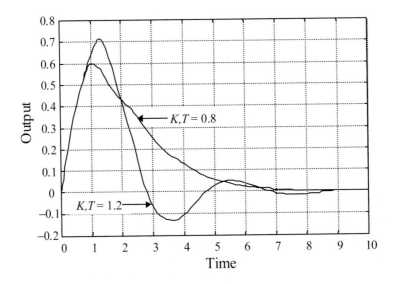

Figure 8.4 Control system responses to a unit step disturbance for Example 8.1,
$qq_d(s) = (s + 1)(0.942s + 1)/(0.602s + 1)^2$

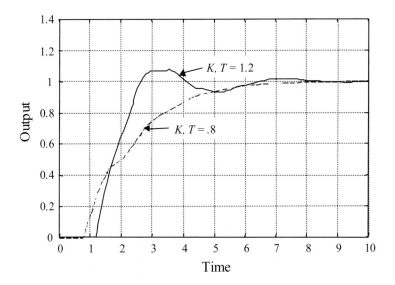

Figure 8.5 Setpoint responses for Example 8.1, with ε_r = 1.03.

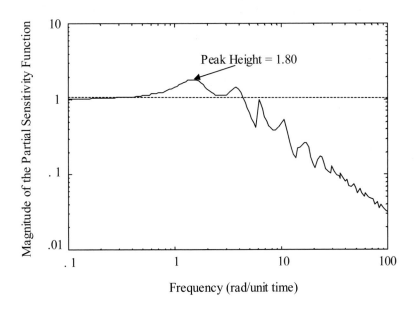

Figure 8.6 Magnitude of the partial sensitivity function for Example 8.1, with $p_d(s)$ = 1.

Example 8.2 Tuning a FOPDT Process with a Large Disturbance Lag

This example is the same as Example 8.1, except that the disturbance lag time constant is 4.

$$p(s) = \frac{Ke^{-Ts}}{(s+1)} \; ; \; p_d(s) = 1/(4s+1) \quad .8 \le K \le 1.2; \quad .8 \le T \le 1.2, \qquad (8.4a)$$

$$\tilde{p}(s) = \frac{e^{-s}}{(s+1)}, \qquad qq_d(s) = \frac{(s+1)(\alpha s+1)}{(\varepsilon s+1)^2}. \qquad (8.4b)$$

Tuning proceeds as before, and results are presented in Figures 8.7 through 8.10. The responses are quite similar to those of Example 8.1 except

(1) The peak of the output response in Figure 8.8 is significantly lower than that for Example 8.1 due to the stronger filtering action of the larger disturbance lag.

(2) The peak of the partial sensitivity function with the disturbance lag set to 1.0 shown in Figure 8.10 is 3.55 versus 1.80 for Figure 8.6 of Example 8.1. This is also due to the stronger filtering action of the larger disturbance lag.

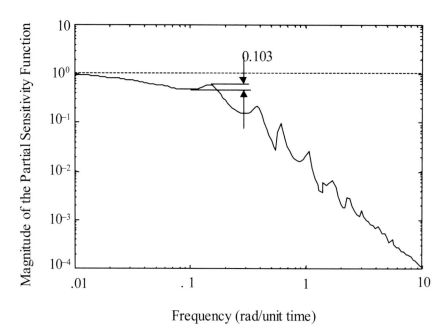

Figure 8.7 Upper bound of the partial sensitivity function for Example 8.2, with $qq_d(s) = (s + 1)(1.92s + 1)/(0.734s + 1)^2$

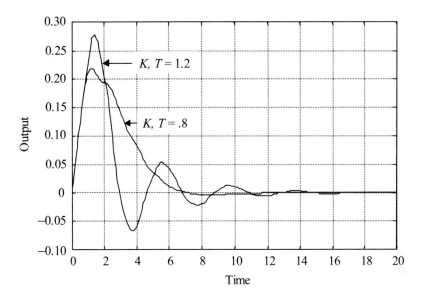

Figure 8.8 Control system responses to a unit step disturbance for Example 8.2, $qq_d(s) = (s + 1)(1.92s + 1)/(0.734s + 1)^2$

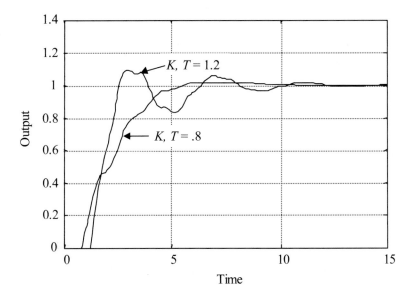

Figure 8.9 Setpoint responses for Example 8.2, with $\varepsilon_r = 1.03$.

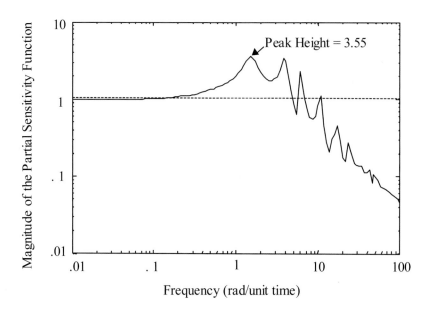

Figure 8.10 Magnitude of the partial sensitivity function for Example 8.2, with $p_d(s) = 1$.

From Example 8.2, we see that in 2DF control systems the relative stability, or robustness, of the feedback loops of two systems whose time responses are quite similar can be quite different. It is important therefore to check the relative stability of the feedback loop by computing the upper bound of the partial sensitivity function with the disturbance lag set to 1, before accepting a controller design. We recommend not accepting any design where the maximum of the partial sensitivity function is greater than 5.[1] Also, when comparing controllers, as in the next section, we recommend that they be compared with the same relative stability.

8.3 MP SYNTHESIS FOR STABLE, OVERDAMPED PROCESSES

The following discussion parallels the discussion in Section 7.5 on the effect of uncertainty on controller design and model selection for 1DF control systems for stable processes. In that section we established, by examples, that the performance of a first-order controller can be comparable to, or even significantly better than, that of a model inverse controller when there is significant process uncertainty. In this section we will demonstrate a similar effect

---

[1] A maximum magnitude of 5 for the complementary sensitivity function corresponds to a damping ratio of 0.1 for a second-order system.

of uncertainty on controller design for 2DF systems. However, the design is somewhat more complicated because there are more options for designing the controller. We begin the discussion by returning to Example 7.8, but this time, there is a significant disturbance lag.

Example 8.3 Effect of Model Selection on Performance, I

$$p(s) = Ke^{-Ts}/(8s+1)(6s+1); \quad p_d(s) = 1/(16s+1)(12s+1); \quad 5 \le K, T \le 15 \qquad (8.5)$$

Table 8.1 shows several control system designs, the first of which is the best 1DF controller found in Chapter 7 for the process given by Eq. (8.5). All of the 2DF feedback controllers in Table 8.1 were tuned to give a maximum peak of the partial sensitivity function of 0.1. Such tuning results in a closed-loop Mp for the complementary sensitivity function of 3.3 for the control systems in rows two and three, and 3.8 in row four.

Table 8.1 Controllers and Models for Example 8.3 Ordered by Performance.

Remarks	$qq_d(s,\varepsilon)$	$q(s,\varepsilon_r)$	$\tilde{p}(s)$
1DF $q_d = 1$, $\varepsilon_r = \varepsilon$	$\dfrac{(14s+1)}{15(10.6s+1)}$	$\dfrac{(14s+1)}{15(10.6s+1)}$	$\dfrac{15e^{-15s}}{(14s+1)}$
2DF 2nd order Model	$\dfrac{(8s+1)(6s+1)(1744s^2+26.5s+1)}{15(7.88s+1)^4}$	$\dfrac{(8s+1)(6s+1)}{15(9.07s+1)^2}$	$\dfrac{15e^{-15s}}{(8s+1)(6s+1)}$
2DF $q_d = 1$, $\varepsilon_r \ne \varepsilon$	$\dfrac{(14s+1)}{15(5.02s+1)}$	$\dfrac{(14s+1)}{15(16.4s+1)}$	$\dfrac{15e^{-15s}}{(14s+1)}$
2DF Best Response	$\dfrac{(14s+1)(22.0s+1)}{15(11.0s+1)^2}$	$\dfrac{(14s+1)}{15(16.5s+1)}$	$\dfrac{15e^{-15s}}{(14s+1)}$

Figure 8.11 compares the step disturbance responses for the slowest loop responses of the control systems in Table 8.1. Notice that the performance of all of the 2DF control systems is superior to that of the 1DF control system. Also notice that the best response is from the control system that uses a first-order model rather than the control system based on the actual second-order lag of the process. Here again we see that when there is a significant amount of uncertainty, simpler is better. Figure 8.12 shows the lower bound of the partial sensitivity function for the first and fourth controllers in Table 8.1. The frequency responses for the second and third controllers lie between those for controllers one and four.

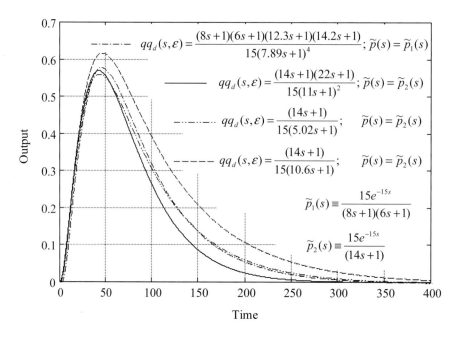

Figure 8.11 Slowest responses to a step disturbance for the process of Example 8.3 and the control systems in Table 8.1 (slowest responses occur when $K, T = 5$).

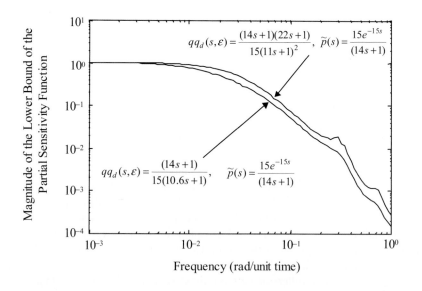

Figure 8.12 Lower bound of the partial sensitivity function for selected controllers.

The setpoint controllers $q(s, \varepsilon_r)$ in Table 8.1 were all tuned to achieve an Mp of 1.05. Figure 8.13 shows the step setpoint responses for the slowest processes (K,T = 5) for control systems one and four. Again, the responses for the other control systems lie between the two responses shown in the figure.

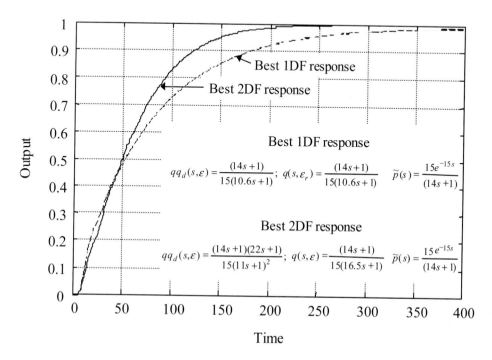

Figure 8.13 Slowest responses to a step setpoint change for Example 8.2 (*K,T* = 5).

Example 8.4 has the same process as Example 8.3, except that the disturbance lag is the same as the process lag.

Example 8.4 Effect of Model Selection on Performance, II

$$p(s) = Ke^{-Ts}/(8s+1)(6s+1); \quad p_d(s) = 1/(8s+1)(6s+1); \quad 5 \le K,T \le 15. \quad (8.6)$$

Figure 8.14 compares the step disturbance responses for the slowest loop responses of the best single-degree and 2DF control systems. As might be expected, the improvement in response time for the 2DF control system over that of the 1DF control system is less than in the previous example. Further, there is only a very slight degradation in response if 2DF controller in Figure 8.14 is replaced by

$$qq_d(s,\varepsilon) = \frac{(14s+1)}{15(5.9s+1)} \ (q_d(s,\varepsilon) = 1); \ q(s,\varepsilon_r) = \frac{(14s+1)}{15(11.5s+1)}; \ \tilde{p}(s) = \frac{15e^{-15s}}{(14s+1)}. \quad (8.7)$$

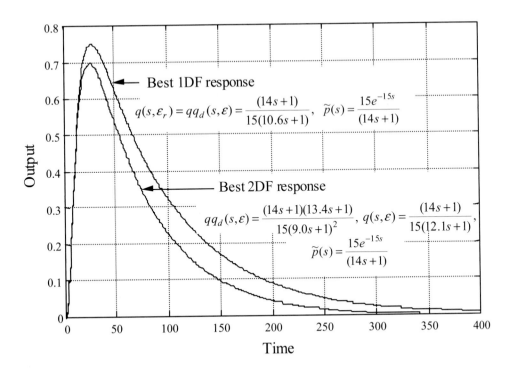

Figure 8.14 Slowest responses to a step disturbance for the process of Example 8.4 (slowest responses occur when K,T = 5).

♦

The degree of improvement in performance of the 2DF control systems over the best 1DF control system for uncertain processes demonstrated by the previous examples, while significant, is not all that great. Further, the improved responses come at the expense of a decrease in the relative stability of the feedback loop (i.e., an Mp of 1.05 for the 1DF control system versus an Mp of 3.8 and 2.04, respectively, for the 2DF control systems for Examples 8.3 and 8.4). Therefore, in many engineering situations where there is substantial process uncertainty, the benefits of the improved performance of a 2DF over a 1DF control system may not justify the more complicated implementation and reduced relative stability. This is perhaps one reason that 2DF control systems are not encountered very often in industrial applications of control for inherently stable overdamped processes. Another common, but less justifiable, reason is that implementation of a 2DF control system was never even considered.

In the previous examples, the disturbance lags were assumed to be known exactly. However, this is almost never the case in practice. At best, the variations of disturbance lag time constants are known. More often, the best that can be done is to approximate the disturbance lag as a first-order lag, and then estimate the smallest value for its time constant.

Actually, such information is usually quite adequate, since for any significant uncertainty in the process model parameters, such as in the two foregoing examples, the best 2DF controller design is that based on canceling only a single disturbance lag. Further, the appropriate lag time constant to use for design purposes is the smallest lag time constant. The reason for this is that use of a larger disturbance lag time constant will result in a more aggressive controller that will perform poorly for step disturbances that pass through disturbance lags whose time constants are significantly smaller than that used to design the controller.

8.4   TUNING FOR UNDERDAMPED AND UNSTABLE PROCESSES

8.4.1 Introduction

A key difference between overdamped stable processes and underdamped or unstable uncertain processes is that for the latter, it is generally not possible to achieve arbitrary relative stability specifications. That is, unlike with overdamped stable processes, one cannot necessarily achieve a desired closed-loop damping ratio simply by increasing the IMC controller filter time constant. The problem then becomes finding an IMC controller filter time constant that yields the best trade-off between speed of response and worst-case oscillations for a given uncertain process and model. Different situations may call for different trade-offs. However, as a starting point, we propose adjusting the filter time constant to minimize the maximum magnitude of the sensitivity function. The proposed procedure seems to produce very reasonable tunings for all the problems that we have tried.

8.4.2 Underdamped Stable Processes

Example 8.5 Behavior of the Sensitivity Function for an Underdamped, Stable Process

Consider Example 4.4 of Chapter 4, which is repeated here:

$$\widetilde{p}_d(s) = p_d(s) = \widetilde{p}(s) = p(s) = e^{-s}/(s^2 + .2s + 1) \, .$$

Table 8.2 shows the variation of the maximum magnitude of the sensitivity function (i.e., the Mp) and the numerator of $q_d(s, \varepsilon)$ with filter time constant ε. Figure 8.15 plots the data of Table 8.2.

Table 8.2 Variation of the Maximum Magnitude of the Sensitivity Function and $q_d(s,\varepsilon)$ with IMC Filter Time Constant ε for Example 8.5.

ε	Mp	$q_d(s)*(\varepsilon s+1)^2$
0	.473	$.431s^2+.849s+1$
.1	.696	$.758s^2+1.03s+1$
.3	1.03	$1.48s^2+1.08s+1$
.6	1.57	$2.40s^2+.317s+1$
.9	2.13	$2.40s^2+.317s+1$
1	2.31	$2.36s^2-2.27s+1$
2	3.69	$-14.5s^2-12.7s+1$
4	6.11	$-227.3s^2+46.7s+1$

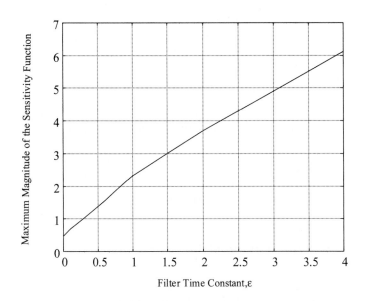

Figure 8.15 Variation of the maximum magnitude of the sensitivity function with IMC filter time constant, ε, for Example 8.5.

The graph in Figure 8.15 is composed of two nearly straight lines that intersect somewhere between 0.9 and 1, probably at the point where $q_d(s,\varepsilon)$ first becomes non-minimum phase.

♦

We emphasize that Table 8.2 and Figure 8.15 are both for a perfect model. When process uncertainty is introduced, we would expect the control system to become unstable for small filter time constants. Thus we expect that the curve of the maximum magnitude of the sensitivity function versus IMC filter time constant to have a minimum, provided that the function is continuous, which it usually will be. This observation and the results of numerous simulations are what led us to propose adjusting the filter time constant to minimize the maximum magnitude of the sensitivity function. As an example, we show the results of tuning the same process as in Example 8.5, except now including gain uncertainty.

Example 8.6 Effect of Gain Uncertainty on Performance

$$p(s) = Ke^{-s}/(s^2 + .2s + 1) \ ; \ .4 \le K \le 1.6 \ ; \ p_d(s) = 1/(s^2 + .2s + 1) \tag{8.8a}$$

$$\tilde{p}(s) = e^{-s}/(s^2 + .2s + 1) \ ; \ \tilde{p}_d(s) = p_d(s) = 1/(s^2 + .2s + 1) \tag{8.8b}$$

Tuning Eq. (8.8) yields a filter time constant of 0.601 and the graph of Figure 8.16.

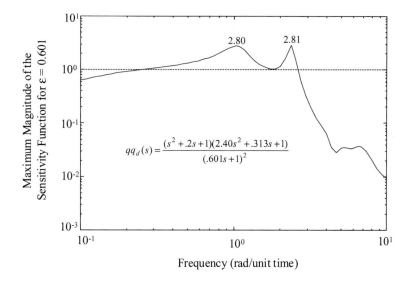

Figure 8.16 Maximum magnitude of the sensitivity function for Example 8.6.

The leftmost peak in Figure 8.16 is due mainly to the perfect model frequency response, while the rightmost peak is due to the introduction of uncertainty in the process gain. The time responses associated with Figure 8.16 are given in Figure 8.17.

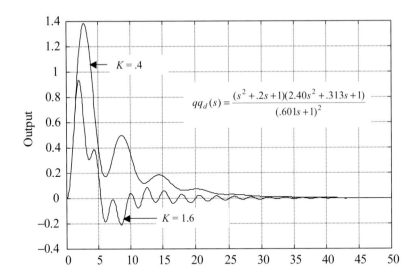

Figure 8.17 Response of the process output for Example 8.6 to step disturbances.

Figures 8.18 through 8.21 show the effect of changing the filter time constant to .7 and .55. (The control system is unstable for $\varepsilon = 0.5$.)

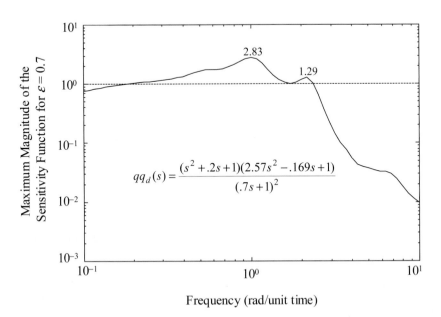

Figure 8.18 Maximum magnitude of the sensitivity function for Example 8.6.

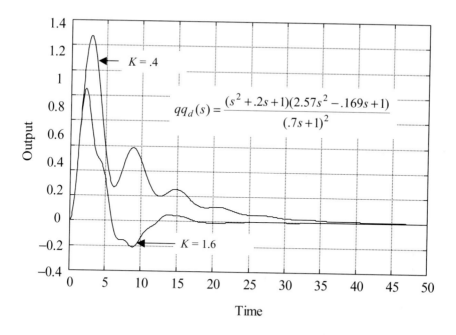

Figure 8.19 Response of the process output for Example 8.6 to step disturbances.

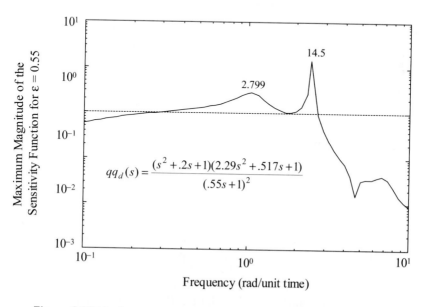

Figure 8.20 Maximum magnitude of the sensitivity function for Example 8.6.

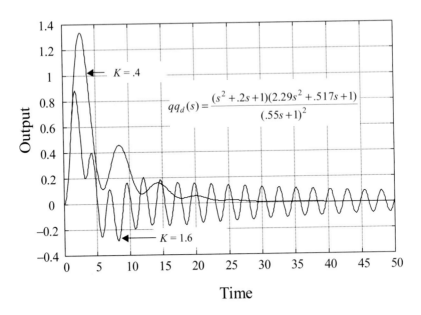

Figure 8.21 Response of the process output for Example 8.6 to step disturbances.

In most situations the responses given by Figure 8.19 would be preferred over those of Figures 8.17 and 8.21. Thus, the filter time constant that minimized the maximum of the sensitivity function had to be increased slightly to obtain the preferred control system. Also notice that the responses given in Figure 8.19 are still better than the responses of the 1DF control system for a perfect model given in Chapter 4, Figure 4.6, in spite of the significant process gain uncertainty.

♦

8.4.3 Unstable Processes

As pointed out in Chapter 4, for an unstable process, we must now ask how the control system is implemented before we can conclude anything about the control system stability. Since the IMC control system of Figure 8.1 is internally unstable, it cannot be used to get a stable control system in spite of the fact that *the IMC sensitivity function is itself stable even for imperfect models*, as we demonstrate below. Two alternative candidates for implementing stable control systems for unstable processes are the feedback form of the IMC control system shown in Chapter 4, Figure 4.9, and the internally stable model based control system proposed by Berber and Brosilow (1997, 1999).

The IMC sensitivity function is given by Eq. (8.1), which is repeated below, along with a slight rearrangement of terms in the denominator.

$$\frac{y(s)}{d(s)} = \frac{(1-\widetilde{p}(s)qq_d(\varepsilon,s))p_d(s)}{(1+(p(s)-\widetilde{p}(s))qq_d(\varepsilon,s))} \equiv \frac{(1-\widetilde{p}(s)qq_d(\varepsilon,s))p_d(s)}{((1-\widetilde{p}(s)qq_d(\varepsilon,s))+p(s)qq_d(\varepsilon,s))}$$

To show that the above IMC sensitivity function is stable, we start by noting that the unstable poles of $p(s)$ must be the same as the unstable poles of $p_d(s)$ in order to be able to design a controller that stabilizes the control system. Therefore, the unstable poles of $p(s)$ in the denominator of the sensitivity function exactly cancel the unstable poles of $p_d(s)$ in its numerator. We still need to show, however, that any unstable zeros in the numerator of $(1-\widetilde{p}(s)qq_d(\varepsilon,s))+p(s)qq_d(\varepsilon,s))$ are cancelled by zeros in the numerator of the term $(1-\widetilde{p}(s)qq_d(\varepsilon,s))$. Further, the term $(1-\widetilde{p}(s)qq_d(\varepsilon,s))$ must have no unstable poles. By the design of $q(s)$ the term $\widetilde{p}(s)qq_d(\varepsilon,s)$ has no unstable poles. Therefore, the term $(1-\widetilde{p}(s)qq_d(\varepsilon,s))$ has no unstable poles. The numerator of the term $(1-\widetilde{p}(s)qq_d(\varepsilon,s)+p(s)qq_d(\varepsilon,s))$ has zeros at the zeros of $(1-\widetilde{p}(s)qq_d(\varepsilon,s))$ because $q(s)$ has zeros at the unstable poles of $\widetilde{p}_d(s)$ and, by the choice of $q_d(s,\varepsilon)$, these are also the zeros of $(1-\widetilde{p}(s)qq_d(\varepsilon,s))$. Thus, the IMC sensitivity function is stable *provided that the term* $(1-\widetilde{p}(s)qq_d(\varepsilon,s))+p(s)qq_d(\varepsilon,s))$ *has no additional right half plane zeros beyond its zeros at the unstable poles of* $\widetilde{p}(s)$.

As we shall see in Example 8.8, it is possible for sensitivity function (i.e., Eq. (8.1)) to be stable (i.e., the term $(1-\widetilde{p}(s)qq_d(\varepsilon,s))+p(s)qq_d(\varepsilon,s))$ has no unstable zeros) even when the term $(1-\widetilde{p}(s)qq_d(\varepsilon,s))$ has right half plane zeros in addition to the zeros at the unstable poles of $\widetilde{p}(s)$. In this case, however, the controller in the feedback form of the IMC system of Figure 4.9 (i.e., $c(s)=qq_d(\varepsilon,s)/(1-\widetilde{p}(s)qq_d(\varepsilon,s))$ is unstable because of the additional right half plane zeros of $(1-\widetilde{p}(s)qq_d(\varepsilon,s))$. In such situations we recommend that the reader consider implementing the control algorithm of Berber and Brosilow (1997, 1999) rather than attempting to stabilize the controller $c(s)$ of Figure 4.9 by increasing the filter time constant.

There is a significant difference between tuning inherently unstable processes versus inherently stable processes. Recall from Chapter 7, equations (7.29) and (7.30), that the robust stability theorem states that a closed-loop transfer function, $H(s,\alpha)$, given by

$$H(s,\alpha) \equiv \frac{g(s,\alpha)}{1+h(s,\alpha)},$$

is stable for all parameter vectors α contained in the parameter set Π if $H(s,\alpha)$ is stable for *at least one parameter vector in the set Π, different from the model parameters,* and if

$$\max_{\alpha\in\Pi}|H(i\omega,\alpha)| \text{ is finite } \forall \omega \geq 0.$$

Also, the parameter set Π and the transfer functions g(s,α) and h(s,α) must satisfy the relatively mild conditions given in Section 7.4.

For inherently stable processes, we paid no attention to the statement in italics requiring that $H(s,\alpha)$ be stable for at least one parameter vector in the set Π, different from the model parameters. This is because for a stable process, we can always choose a process whose parameters are so close to the model parameters that the closed-loop control system will be stable for that process no matter how small the controller filter time constant. Thus, if the above conditions are satisfied we can conclude that the closed-loop is stable for all uncertain parameters, assuming that the other conditions are satisfied, which they usually will be. For an inherently unstable process, however, the closed-loop control system can be unstable no matter how closely the process parameters approach the model parameters. The point where the process parameters are equal to the model parameters can be a singular point where the control system is stable by virtue of exact cancellation of the unstable process poles. Everywhere else, such exact cancellation cannot occur, and the control system can be unstable for all other parameters in the uncertainty set. Such closed-loop systems will have $\max_{\alpha\in\Pi}|H(i\omega,\alpha)|$ finite everywhere. That is, a finite $\max_{\alpha\in\Pi}|H(i\omega,\alpha)|$ can now indicate that the closed-loop system is either stable for all uncertain parameters or unstable for all uncertain parameters. To find out which, we must check the stability of the control system for any vector of process parameters in the uncertainty set. A convenient method for determining stability is via the Nyquist criterion. If the control system is stable for the chosen parameter vector, then we conclude that the closed-loop control system is stable for all uncertain parameters. If the control system is unstable for the chosen parameter vector, then we conclude that the closed-loop control system is unstable for all uncertain parameters. In Example 8.8, we illustrate the foregoing. In Example 8.7 we show the similarity in behavior of the sensitivity function for the perfect model case of an unstable process with that of the underdamped stable process of Example 8.5.

There is yet another important difference between stable and unstable processes. For inherently stable processes, we could always stabilize the control system by increasing the filter time constant sufficiently. For inherently unstable processes the control system will be unstable both for very small and very large filter time constants. For large filter time constants the control action may be simply too slow to stabilize the process. Indeed, there may be no filter time constant that stabilizes all processes in the uncertainty set. The following examples illustrate this fact.

Example 8.7 Sensitivity Function for a FOPDT Unstable Process

$$p(s) = \widetilde{p}(s) = p_d(s) = \widetilde{p}_d(s) = e^{-s}/(-s+1) \tag{8.9}$$

Table 8.3 and Figure 8.22 show the variation of the maximum magnitude of the sensitivity function (Mp) and the numerator of $q_d(s,\varepsilon)$ with filter time constant ε.

Table 8.3 Variation of the Maximum Magnitude of the Sensitivity Function and $q_d(s,\varepsilon)$ with IMC Filter Time Constant ε for Example 8.7.

ε	Mp	$q_d(s)*(\varepsilon s+1)$
0	1.91	$1.72s+1$
.01	1.96	$1.77s+1$
.05	2.12	$2.00s+1$
.1	2.23	$2.29s+1$
.5	2.68	$5.12s+1$
1	3.27	$9.87s+1$
2.5	5.21	$32.3s+1$
4	7.22	$67.0s+1$
5	8.50	$96.9s+1$

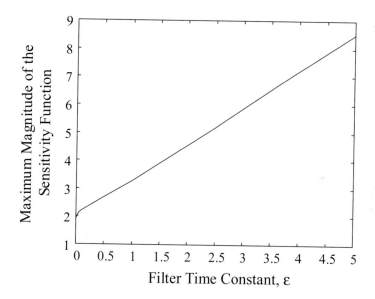

Figure 8.22 Variation of the maximum magnitude of the sensitivity function with IMC filter time constant, ε for Example 8.7.

The graph of Figure 8.22 is, as before, effectively a straight line for filter time constants greater than .05. Once again, the introduction of process uncertainty will cause the maximum magnitude of the sensitivity function to have a minimum for one or more values of the filter time constant.

Example 8.8 Tuning an Uncertain FOPDT Unstable Process

The process is that of Eq. (8.10) below, and it is one that is not easy to control, in spite of the relatively small range of uncertain gains, because of the relatively large dead time to time constant ratio.

$$p(s) = Ke^{-s}/(-s+1) \; ; \; p_d(s) = 1/(-s+1) \; ; \; .9 \le K \le 1.1 \tag{8.10}$$

Selecting a mid-range gain for the model and tuning to get the smallest Mp of the sensitivity function yields a filter time constant of 0.6. Figure 8.23 shows the closed-loop frequency response for this filter time constant as well as for filter time constants of 0.5 and 0.7. The process model and feedback controllers are

$$p_d(s) = \widetilde{p}_d(s) = 1/(-s+1) \; ; \; \widetilde{p}(s) = e^{-s}/(-s+1) \tag{8.11a}$$

$$qq_d(s,.5) = (5.116s+1)(-s+1)/(.5s+1)^2, \tag{8.11b}$$

$$qq_d(s,.6) = (5.959s+1)(-s+1)/(.6s+1)^2, \tag{8.11c}$$

$$qq_d(s,.7) = (6.856s+1)(-s+1)/(.7s+1)^2. \tag{8.11d}$$

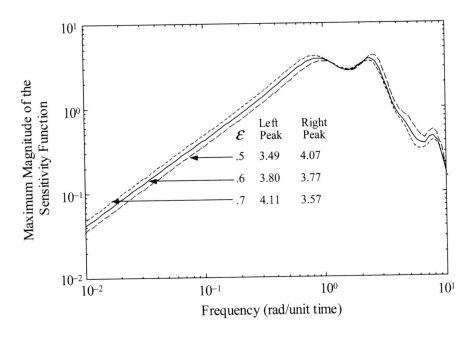

Figure 8.23 Upper bound of the magnitude of the sensitivity function for Example 8.8.

Computations of the maximum magnitude of the sensitivity function several frequency decades higher and lower than those shown in Figure 8.23 show that the maximum magnitude continues to drop off in both directions. We therefore feel safe in claiming that the closed-loop maxima are as shown in Figure 8.23. We now need to check whether the sensitivity function (i.e., Eq. (8.1)) is stable for at least one process in the set of uncertain processes. Choosing a process with a dead time and a process gain of 1.1, the controller filter time constant as .6, and using a 5/5 Padé approximation for the dead times in Eq. (8.1) gives

$$\frac{y(s)}{d(s)} \cong \frac{-3.42(.0186s+1)(.00512s^2-.00446s+1)(.0332s^2-.0622s+1)s}{(.00285s^2+.0752s+1)(.0148s^2.0530s+1)(.147s^2+.172s+1)(1.74s+1)}. \qquad (8.12)$$

Since the denominator of Eq. (8.12) has no right half plane zeros, we can conclude from Figure 8.23 that the control system is stable for all process gains between 0.9 and 1.1. A similar analysis for controller filter time constants of .5 and .7 also shows that the closed-loop systems are stable. Figures 8.24 and 8.25 show the time responses for process gains of .9 and 1.1, which are associated with the left and right peaks respectively in Figure 8.23.

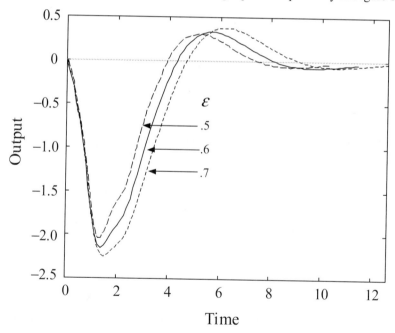

Figure 8.24 Step disturbance time responses for Example 8.8 for $K = 0.9$.

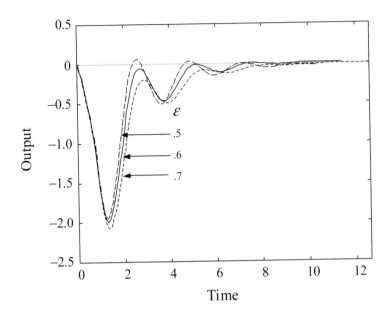

Figure 8.25 Step disturbance time responses for Example 8.8 for K = 1.1.

Comparing the responses in Figures 8.24 and 8.25 shows perhaps a slight preference for using a filter time constant of 0.5.

We now turn our attention to the feedback form of the IMC system given in Chapter 4 by Figure 4.9. The feedback controller is given by

$$c(s) = qq_d(s,\varepsilon)/(1 - \tilde{p}(s)qq_d(s,\varepsilon)). \tag{8.13}$$

By the design of $qq_d(s)$, both the numerator and denominator of $c(s)$ have a zero at the pole of $\tilde{p}(s)$. Unfortunately, however, for a filter time constant of 0.6, and a 5/5 Padé approximation for the dead time, the denominator of $c(s)$ has four additional complex right half plane poles, as shown by Eq. (8.14).

$$c(s) = \frac{-0.266(.0138s^2 + .128s + 1)(.0175s^2 + .235s + 1)(.137s + 1)(5.96s + 1)}{s(.0186s + 1)(.00512s^2 - .00446s + 1)(.0332s^2 - .0622s + 1)}. \tag{8.14}$$

Figures 8.26a and 8.26b show the Nyquist diagram for the open-loop transfer function $c(s)p(s)$. The process $p(s)$ is given by Eq. (8.10), with K = 1.1. This diagram shows three counter-clockwise encirclements of the (–1, 0) point. Thus, the closed-loop control system is unstable, with two poles in the right half of the s-plane. For the closed loop system to be stable there must be 5 counter-clockwise encirclements of the (–1, 0) point because the open loop system has 5 right half plane poles. Four right half plane poles come from the controller, and one comes from the process.

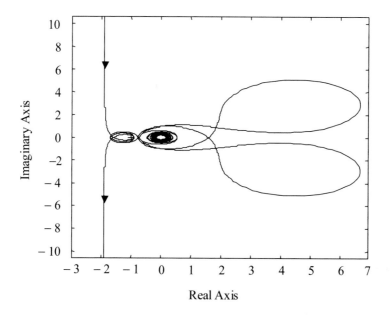

Figure 8.26a Nyquist diagram of $c(s)p(s)$ with $p(s)$ given by Eq. (8.10) with $K = 1.1$, and $c(s)$ given by Eq. (8.14).

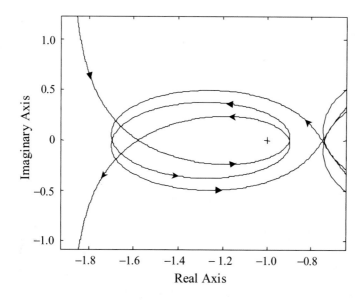

Figure 8.26b Enlargement of Figure 8.26a around the –1, 0 point.

To achieve a stable closed-loop control system, it is necessary to increase the IMC filter time constant to 1.6. The feedback controller then becomes

$$c(s) = \frac{-0.0759(.0138s^2 + .128s + 1)(.0175s^2 + .235s + 1)(.137s + 1)(17.4s + 1)}{s(.0276s + 1)(.00586s^2 + .0193s + 1)(.0398s^2 - .0220s + 1)}. \tag{8.15}$$

Figure 8.27 shows the Nyquist diagram for the open-loop transfer function $c(s)p(s)$ with $c(s)$ as given by Eq. (8.15) and with $p(s)$ given by Eq. (8.10) with $K = 1.1$. There are three counter-clockwise encirclements of the $(-1,0)$ point, and no clockwise encirclements. Thus, the control system is stable for all process gains, even though both the controller and the process are unstable.

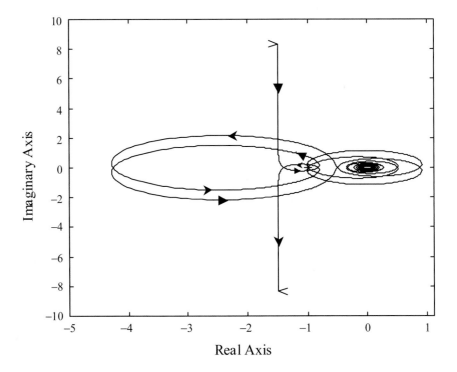

Figure 8.27 Nyquist diagram of $c(s)p(s)$ with $p(s)$ given by Eq. (8.10) with $K = 1.1$ and $c(s)$ given by Eq. (8.15).

Figure 8.28 compares the step disturbance responses for the control systems with filter time constants of 0.6 and 1.6. Clearly, there is a significant performance penalty associated with implementing the feedback form of the IMC controller rather than an internally stable form of the IMC controller proposed by Berber and Brosilow (1999).

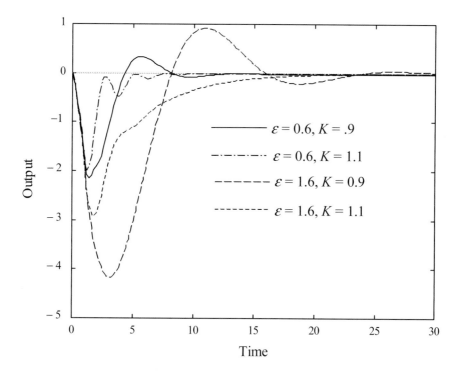

Figure 8.28 Comparison of the step disturbance responses for the control systems with filter time constants of 0.6 and 1.6.

It is also possible to approximate the feedback controller given by Eq. (8.15) with the 2DF PID controller of Equations (8.16a) and (8.16b) (see Chapter 6). The pre-filter time constant of 3.63 was obtained by tuning the setpoint filter of Figure 8.1 to achieve an Mp of 1.05 for the complementary sensitivity function.

$$PID = -1.35(1+1/17.85s+0.466s/(.0233s+1)),$$ (8.16a)

$$\text{Pre-filter} = (1.6s+1)^2/(3.63s+1)(17.4s+1).$$ (8.16b)

The PID control system responses for process gains of 0.9 and 1.0 are reasonably close to those of the IMC controller given by Eq. (8.15). However, for process gain of 1.1 the PID responses as shown in Figure 8.29 are not very good. To get a good PID approximation to the IMC responses, it is necessary to increase the filter time constant to 2.8, at which point the feedback IMC controller is stable.

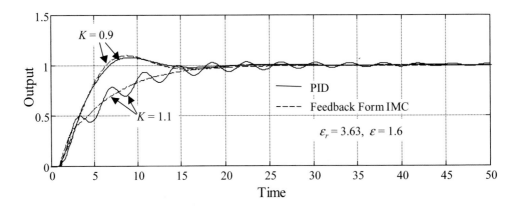

Figure 8.29a Comparison of closed-loop setpoint responses of the IMC controller of Eq. (8.15) and the PID controller of Equations (8.16a) and (8.16b).

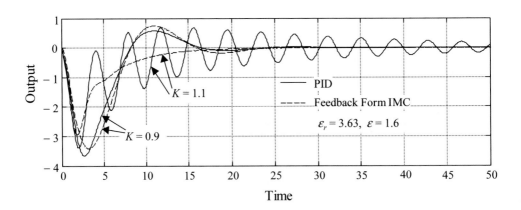

Figure 8.29b Comparison of closed-loop disturbance responses of the IMC controller of Eq. (8.15) and the PID controller of Equations (8.16a) and (8.16b).

The setpoint and disturbance responses at a filter time constant of 2.8 are displayed in Figures 8.30a and 8.30b for the upper-bound and lower-bound process gains. The IMC feedback controller and setpoint filter are

$$c(s) = \frac{-0.0316(.0138s^2 + .128s + 1)(.0175s^2 + .235s + 1)(.137s + 1)(38.3s + 1)}{s(.0312s + 1)(.00618s^2 + .0283s + 1)(.0424s^2 + .0000466s + 1)}, \quad (8.17a)$$

$$\text{Pre-filter, } q(s, \varepsilon_r)qq_d^{-1}(s, \varepsilon) = \frac{(2.8s + 1)^2}{(8.85s + 1)(38.3s + 1)}; \varepsilon_r = 8.85, \ \varepsilon = 2.8, \quad (8.17b)$$

The associated PID controller is

$$c(s) \approx PID(s) = -1.22(1 + 1/(38.7s) + .436s/(.0218s + 1)). \tag{8.18}$$

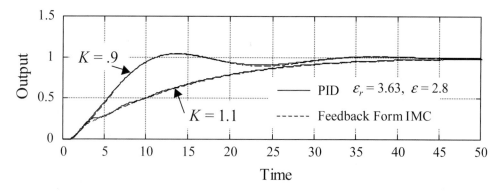

Figure 8.30a IMC and PID closed loop setpoint responses associated with controllers given by Equations (8.17a), (8.17b), and (8.18).

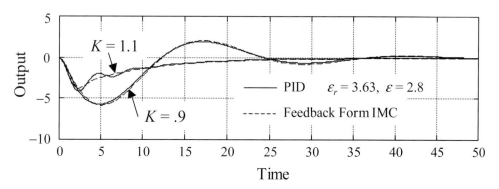

Figure 8.30b IMC and PID closed loop disturbance responses associated with controllers given by Equations (8.17a), (8.17b), and (8.18).

Notice that while the above PID control system response for a filter time constant of 2.8 is less oscillatory than the PID responses for a filter time constant of 1.6, it is also slower and drops lower than the response for a filter time constant of 1.6.

♦

The four examples in this section (i.e., examples 8.5 through 8.8) have taken the model gain to be the mid-range gain. The next section addresses the question as to whether the choice of mid-range gain for the model is an optimal choice when the performance criterion is the best time responses.

8.5 Mp Synthesis for Underdamped and Unstable Processes

8.5.1 Introduction

The behavior of underdamped and inherently unstable, uncertain processes requires a somewhat different approach from that of Mp synthesis for overdamped, inherently stable processes. In the latter case, we sought to find the model and controller order and parameters that move the lower bound of the magnitude of the complementary sensitivity function as far to the right in the frequency domain as possible while maintaining a specified upper bound Mp. The aim of the foregoing was to speed up the slowest time responses as much as possible while limiting the overshoot of any process to a step setpoint change. However, for underdamped or inherently unstable, uncertain processes, the time responses found in the previous section show little difference in settling time for the fastest and slowest processes (i.e., those processes with the maximum and minimum gains). Therefore, it seems reasonable to focus on selecting a model and controller that improve the speed of response of the fastest process, in anticipation that this will also speed the response of all other processes. Another criterion is to increase the relative stability of the closed-loop system as measured by the maximum magnitude of the sensitivity function. Happily, based on the following examples, it seems that the foregoing criteria are consistent with each other.

Example 8.9 Optimum Model for an Underdamped Process

This example is the same as Example 8.6, except that we now seek to find the optimum model and controller gain. Recall the process description given by Equations (8.8a) and (8.8b):

$$p(s) = Ke^{-s}/(s^2 + .2s + 1) \; ; \; .4 \leq K \leq 1.6 \; ; \; p_d(s) = 1/(s^2 + .2s + 1),$$

$$\widetilde{p}_d(s) = p_d(s) = 1/(s^2 + .2s + 1).$$

An exhaustive search for the model gain that moves the sensitivity function to the right without increasing the maximum peaks produced a model gain of 1.3 and a filter time constant of 0.45. Figure 8.31 compares the frequency response for this tuning with that found previously using the mid-range gain of 1.0. Note the appearance of a third peak in the sensitivity function for the model gain of 1.3. It is the growth of this third peak that limits increasing the model gain beyond 1.3.

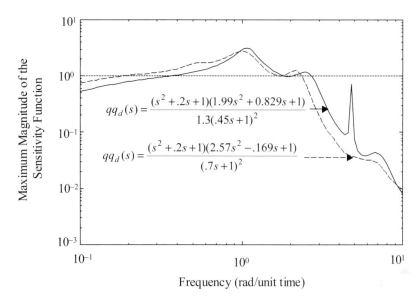

Figure 8.31 Comparison of closed-loop frequency responses for mid-range and optimal models.

Figures 8.32 and 8.33 compare the time responses to a step disturbance for the mid-range and optimal models.

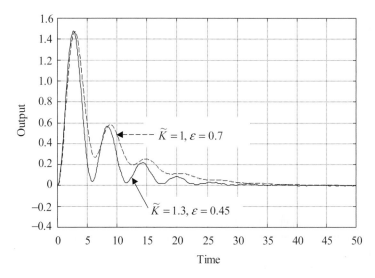

Figure 8.32 Comparison of the closed-loop time responses to a step disturbance for a process gain of 0.4 and the mid-range and optimal models.

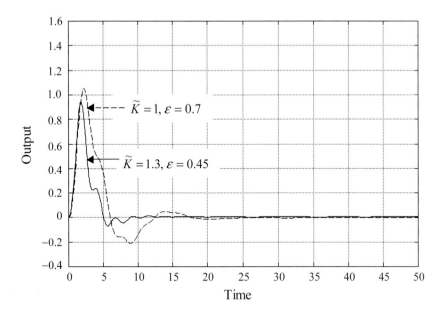

Figure 8.33 Comparison of the closed-loop time responses to a step disturbance for a process gain of 1.6 and the mid-range and optimal models.

♦

Example 8.10 Model Selection for an Unstable FOPDT Process

This example is the same as Example 8.8, except that the gain uncertainty is increased to vary between 0.8 and 1.2. This small change is almost enough to make the process uncontrollable using a mid-range model. As we shall see, very small changes in filter time constant away from its optimal value cause the control system to become unstable. The process is

$$p(s) = Ke^{-s}/(-s+1) \; ; \; .8 \le K \le 1.2 \qquad p_d(s) = \widetilde{p}_d(s) = 1/(-s+1)$$

$$\widetilde{p}(s) = e^{-s}/(-s+1) \tag{8.19}$$

The closed-loop upper bound curves for model gains of 1.0 and 1.05 and filter time constants that minimize the maximum magnitude of the sensitivity function are given in Figure 8.34.

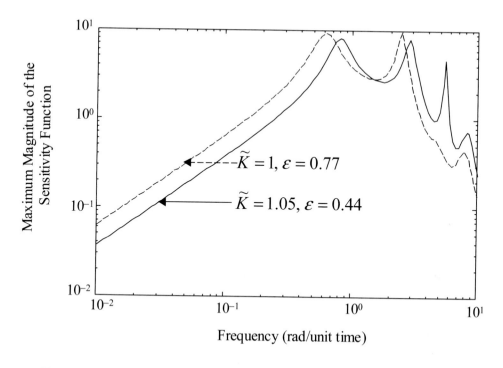

Figure 8.34 Comparison of upper bounds of the sensitivity function for different models.

For the system with a model gain of 1.0, increasing the filter time constant from 0.77 to 0.8 or decreasing it to 0.75 produces effectively unstable control systems with the peak of the maximum sensitivity function greater than 10^4 in each case. For the system with a model gain of 1.05, the filter time constant can be increased to 1.1 or decreased to .39 before the maximum peak reaches 100. In this sense, the control system with a model gain of 1.05 is much more stable than that with a model gain of 1.0. As in the previous example, the growth of the third peak in Figure 8.34 makes it inadvisable to increase the model gain much above 1.05. The responses to step disturbances for the systems of Figure 8.34 are given in Figures 8.35 and 8.36.

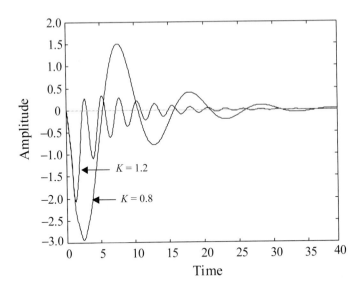

Figure 8.35 Step disturbance responses for a control system with a model gain of 1.0 and a filter time constant of 0.77.

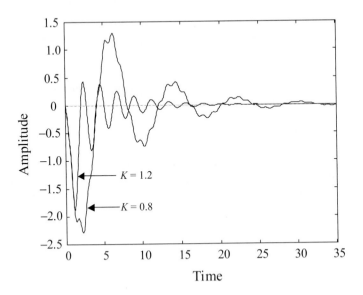

Figure 8.36 Step disturbance responses for a control system with a model gain of 1.05 and a filter time constant of 0.45.

Example 8.11 Controller Design for a "Benchmark" Problem

This example presents the solution for an unstable "benchmark" control problem proposed by Wie and Bernstein (1990, 1992). It consists of an undamped two-mass spring system modeled by

$$y(s) = p(s)u(s) + p_d(s)d(s), \qquad (8.20a)$$

$$p(s) = \frac{\frac{1}{2}}{s^2\left(\frac{1}{2k}s^2 + 1\right)}; \quad p_d(s) = \frac{\frac{1}{2}\left(\frac{1}{k}s^2 + 1\right)}{s^2\left(\frac{1}{2k}s^2 + 1\right)} \quad .5 \le k \le 2 \qquad (8.20b)$$

where k = process spring constant, which varies between 0.5 and 2,
  u = control effort,
  d = disturbance,
  y = position of the second mass.

The control objective is to design a robust controller to suppress impulse disturbances for all the plants in the uncertainty set. Further, the "nominal" process is to have a settling time of 15 seconds.

As before, we take the controller $qq_d(s, \varepsilon)$ as being composed of two terms, $q(s, \varepsilon)$ and $q_d(s, \varepsilon)$. The term $q(s, \varepsilon)$ is taken the inverse of the model of the process as given in Eq. (8.21):

$$q(s, \varepsilon) = \frac{s^2\left(\frac{1}{2\tilde{k}}s^2 + 1\right)}{\frac{1}{2}(\varepsilon s + 1)^4}, \qquad (8.21)$$

where $\tilde{k}$ = model spring constant,
  ε = controller filter time constant.

The above choice for $q(s, \varepsilon)$ reduces the problem to that of choosing $q_d(s, \varepsilon)$ so that the zeros of $(1 - q_d(s, \varepsilon)/(\varepsilon s + 1)^4)$ cancel the poles of $\tilde{p}_d(s)$. This requires that $q_d(s, \varepsilon)$ be at least third order, as given by Eq. (8.22):

$$q_d(s) = \frac{\tau_3 s^3 + \tau_2 s^2 + \tau_1 s + 1}{(\varepsilon s + 1)^3}. \qquad (8.22)$$

The constants τ_1, τ_2, and τ_3 in Eq. (8.22) are chosen so that the numerator of $(1 - q_d(s, \varepsilon)/(\varepsilon s + 1)^4)$ has zeros at $s = 0$ and at $s = \pm i\sqrt{2k}$. There is automatically an additional zero at $s = 0$ by virtue of the fact that $q_d(0, \varepsilon)$ is 1. Equating coefficients of the requisite polynomials gives

$$\tau_1 = 7\varepsilon, \tag{8.23a}$$

$$\tau_2 = 21\varepsilon^2 - 70\widetilde{k}\varepsilon^4 + 28\widetilde{k}^2\varepsilon^6, \tag{8.23b}$$

$$\tau_3 = 35\varepsilon^3 - 42\widetilde{k}\varepsilon^5 + 4\widetilde{k}^2\varepsilon^7. \tag{8.23c}$$

The constants in the equation series (8.23) depend on the value of the filter time constant and the model parameter $\widetilde{k}$. The filter time constant ε is selected to be 1 so that the nominal plant with $\widetilde{k} = 1$ has the desired settling time of about 15 seconds, as required in the benchmark problem specifications (Wie and Bernstein, 1990, 1992). Having thus chosen ε, our only remaining task is to find an optimal value for $\widetilde{k}$.

We will choose that value of $\widetilde{k}$ that minimizes the magnitude of the maximum peak sensitivity function of Eq. (8.1). By exploring values of the spring constant $\widetilde{k}$ between 0.5 and 2, we determined that the minimum maximum peak is obtained using a model spring constant of 0.7. The feedback controller for this spring constant from Eq. (8.13) is given by

$$c_1(s) = \frac{10.8(s^3 - 1.88s^2 + .93s + .13)}{s^3 + 7s^2 + 19.6s + 25.2} \tag{8.24}$$

Values for $\widetilde{k}$ of near 0.5 and 2 yield unstable control systems, which are indicated by very large peaks in the maximum of the sensitivity function.

Braatz and Morari (1992) solved the above control problem, using the *D-K* iteration method (Doyle, 1985) and loop shaping, and obtained the following "μ-optimal" controller:

$$c_\mu(s) = \frac{.0443(9.402s + 1)(-2.697s + 1)(0.4789s + 1)}{(0.216s^2 + 0.861s + 1)(0.118s^2 + 0.369s + 1)}. \tag{8.25}$$

Figure 8.37 shows that the maximum magnitude of the sensitivity function is bounded for both control systems. Since a separate Nyquist analysis of the control systems for a process with a spring constant of $k = 0.8$ shows that each control system is stable for that value of spring constant, we conclude that the control system is stable for all values of spring constant in the uncertainty set. Further, since the maximum of the sensitivity function over all frequencies is smaller for controller $c_1(s)$ than for controller $c_\mu(s)$, we expect that time responses for the control system using $c_1(s)$ are likely to be less oscillatory than that for $c_\mu(s)$, at least for some values of the spring constant. The time responses in Figures 8.38 and 8.39 confirm this expectation.

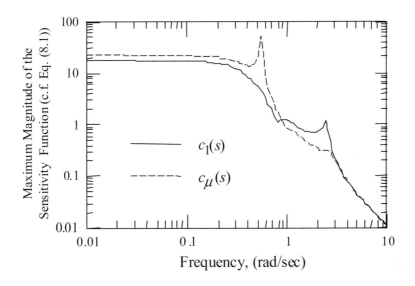

Figure 8.37 Maximum magnitude of the sensitivity function using controllers $c_1(s)$ and $c_\mu(s)$.

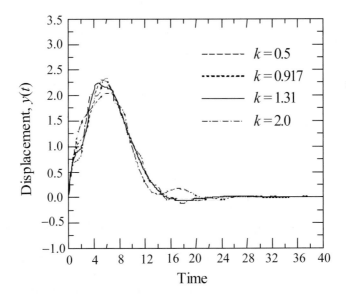

Figure 8.38 Response of $y(t)$ to an impulse disturbance for different spring-constant values using controller $c_1(s)$

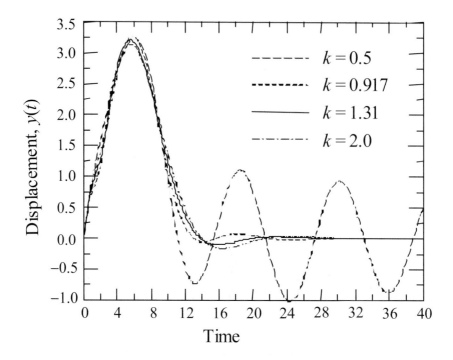

Figure 8.39 Response of $y(t)$ to an impulse disturbance for different spring-constant values using controller $c_\mu(s)$.

8.6 SUMMARY

A 2DF control system can always be designed and tuned to perform at least as well as a 1DF control system, independent of the process uncertainty. By choosing the feedback portion of the controller to be the same as the setpoint portion of the controller, the 2DF controller reduces to that of a 1DF controller when the filter time constants of both portions of the controller are set equal. Otherwise, the speed of the disturbance response can be increased, without degrading the setpoint response, by decreasing the feedback controller filter time constant and possibly increasing the setpoint filter time constant.

Tuning a 2DF controller proceeds in two sequential steps. First, the inner loop is tuned, and then the outer loop. Outer loop tuning uses the same criterion and procedure as in Chapter 7 for 1DF controllers. Different criteria are used for tuning the inner loop, depending on whether the process is stable and overdamped, or underdamped or unstable. In the former case we recommend using a closed-loop transfer function that we call the partial sensitivity function. Tuning using the partial sensitivity function is similar to that using the complementary sensitivity function of Chapter 7. For tuning underdamped or unstable processes, we recommend minimizing the magnitude of the maximum peak of the sensitivity function.

For uncertain, unstable processes, not only the choice of model but also the choice of control structure can have a profound effect on control system performance. In general, the "Algorithmic Internal Model Control" algorithm of Berber and Brosilow (1997, 1999) is preferred over the feedback form of IMC since it can accommodate smaller filter time constants.

Problems

8.1 Compare the one-degree and 2DF disturbance responses for Problems 3.4 and 3.5 of Chapter 3, assuming that the disturbances pass through the process and that the gain and dead time, if any, vary by ± 25%.

8.2 Design and implement controllers for each of the following processes. Assume that the disturbances pass through the process and that the gain and dead time vary by ± 25%.

$$p_d(s) = p(s) = e^{-s}/(s+1)(s^2+.1s+1)$$

$$p_d(s) = p(s) = \frac{e^{-s}}{s(s+1)}$$

$$p_d(s) = p(s) = \frac{2}{2s-1}e^{-0.5s}$$

$$p_d(s) = p(s) = \frac{2(-.5s+1)}{(2s-1)(s+1)}$$

References

Berber, R., and C. Brosilow. 1997. "Internally Stable Linear and Nonlinear Algorithmic Internal Model Control of Unstable Systems." Proceedings of the NATO Advanced Study Institute on Nonlinear Model Based Process Control.

Berber, R., and C. Brosilow. 1999. "Algorithmic Internal Model Control of Unstable Systems." Proceedings of the 7th IEEE Mediterranean Conference on Control and Automation (Med 99)." Haifa, Israel.

Braatz, R. and M. Morari. 1992. "Robust Control for a Noncollocated Springmass System." *Journal of Guidance, Control, and Dynamics*, 15:1103–1109.

Dorato, P., L. Fortuna, and G. Muscato. 1992. *Robust Control for Unstructured Perturbations: An Introduction*, Springer-Verlag, NY.

Doyle, J. C. 1985. "Structured Uncertainty in Control System Design." Proceedings of the 24th IEEE Conference on Decision and Control, 260–265.

Doyle, J.C., B. Francis, and A. Tannenbaum. 1992. *Feedback Control Theory*, Macmillan Publishing Company, NY.

Morari, M., and E. Zafiriou. 1989. *Robust Process Control*, Prentice Hall, NJ.

Wei, B., and D. S. Bernstein. 1990. "A Benchmark Problem for Robust Control Design." *Proceedings of the 1990 American Control Conference*, 961–962.

Wei, B., and D. S. Bernstein. 1992. "Benchmark Problems for Robust Control Design." *Journal of Guidance, Control, and Dynamics*, 15:1057–1059.

Feedforward Control

Objectives of the Chapter

- Describe how to use feedforward control to compensate for measured disturbances.
- Show how to use the material in Chapters 3, 4, 6, and 7 to design and tune combined feedforward/feedback control systems.

Prerequisite Reading

Chapter 3, "One-Degree of Freedom Internal Model Control"
Chapter 4, "Two-Degree of Freedom Internal Model Control"
Chapter 6, "PI and PID Controller Parameters from IMC Designs"
Chapter 7, "Tuning and Synthesis of 1DF IMC Controllers for Uncertain Processes"

9.1   INTRODUCTION

Combined feedforward plus feedback control can significantly improve performance over simple feedback control whenever there is a major disturbance that can be measured before it affects the process output. In the most ideal situation, feedforward control can entirely eliminate the effect of the measured disturbance on the process output. Even when there are modeling errors, feedforward control can often reduce the effect of the measured disturbance on the output better than that achievable by feedback control alone. However, the decision as to whether or not to use feedforward control depends on whether the degree of improvement in the response to the measured disturbance justifies the added costs of implementation and maintenance. The economic benefits of feedforward control can come from lower operating costs and/or increased salability of the product due to its more consistent quality.

Feedforward control is always used along with feedback control because a feedback control system is required to track setpoint changes and to suppress unmeasured disturbances that are always present in any real process.

Figure 9.1a gives the traditional block diagram of a feedforward control system (Seborg et al., 1989). Figure 9.1b shows the same block diagram, but redrawn so as to show clearly that the feedforward part of the control system does not affect the stability of the feedback system and that each system can be designed independently. Figure 9.2 shows a typical application of feedforward control. The continuously stirred tank reactor is under feedback temperature control. Feedforward control is used to rapidly suppress feed flow rate disturbances.

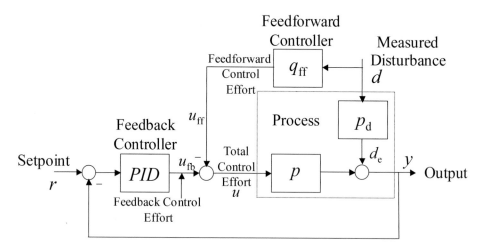

Figure 9.1a Traditional feedforward/feedback control structure.

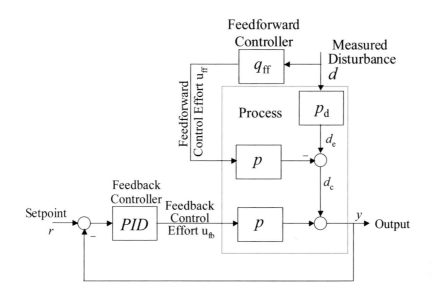

Figure 9.1b Block diagram equivalent to that of Figure 9.1a.

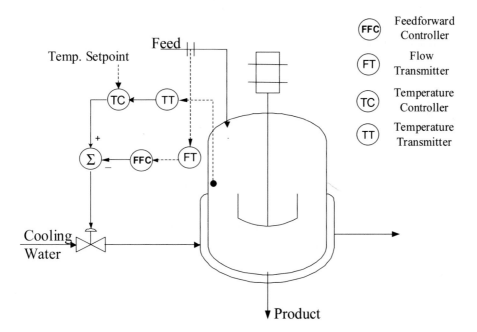

Figure 9.2 Feedforward control on the feed to a continuously stirred tank reactor operating under reactor temperature control.

9.2   CONTROLLER DESIGN WHEN PERFECT COMPENSATION IS POSSIBLE

The transfer function between the process output y and the measured disturbance d from Figure 9.1b is

$$y(s) = \frac{d_c}{1 + PID * p} = \frac{(p_d - pq_{ff})d}{1 + PID * p}.$$ (9.1)

To eliminate the effect of the measured disturbance, we need only choose q_{ff} so that

$$p_d - pq_{ff} = 0.$$ (9.2)

If the dead time and relative order of p_d are both greater than those of p, and p has no right half plane zeros, then q_{ff} can be chosen as

$$q_{ff} = \tilde{p}^{-1}\tilde{p}_d.$$ (9.3)

As in previous chapters, a ~ over a process transfer function indicates that it is a model of the process.

Even in the above case, where the feedforward controller can perfectly compensate the measured disturbance, it may not pay to implement a feedforward controller when the process p is well approximated as a minimum phase system (i.e., one that contains no dead time or right half plane zeros). In this case, it is usually possible to design a 1DF or 2DF PID controller to suppress disturbances well enough so that there is little to be gained by the addition of feedforward control (Morari and Zafiriou, 1989).

Whenever the relative order of $\tilde{p}_d(s)$ is less than or equal to that of $\tilde{p}(s)$, then the noise amplification can be reduced by adding a filter, as in the design of an IMC controller, so that Eq. (9.3) becomes:

$$q_{ff} = \tilde{p}^{-1}\tilde{p}_d f; \quad f \equiv 1/(\varepsilon s + 1)^r$$ (9.4)

The order r of the filter f is either the relative order of $\tilde{p}_d^{-1}\tilde{p}(s)$ or 0 if the relative order of $\tilde{p}_d^{-1}\tilde{p}(s)$ is equal or less than zero. The filter time constant ε is chosen to limit noise amplification, just as was done for IMC controllers for perfect models (see Chapter 3).

The PID controller in Figure 9.1 can be designed and tuned as either a 1DF or 2DF controller using the techniques of Chapters 6, 7, and 8 without regard to the addition of a feedforward controller. Most often, however, the unmeasured disturbances are not severe enough to justify the added complication of a 2DF PID controller.

9.3   CONTROLLER DESIGN WHEN PERFECT COMPENSATION IS NOT POSSIBLE

When the dead time in the disturbance lag p_d is less than that of the control effort lag p, then the feedforward controller q_{ff}, given by Eq. (9.4), is not realizable. In this case, perfect compensation is no longer possible. Much of the literature suggests designing the feedforward controller by simply dropping the unrealizable part of the controller, as is done for a single-degree of freedom IMC controller. However, as we shall soon see by example, the foregoing is by no means the best design. A better feedforward controller can be obtained using a 2DF design. To see why this is, we rewrite the expression for the net effect of the measured disturbance on the output from Eq. (9.1) as

$$d_c = (1 - pq_{ff} p_d^{-1})p_d d = (1 - pq_{ff} p_d^{-1})d_e. \qquad (9.5)$$

Since, by assumption, the term pp_d^{-1} contains a dead time, there is no realizable choice of the feedforward controller q_{ff} that makes the term in brackets in Eq. (9.5) zero. Therefore, there will be a long tail to the response of d_c to a step disturbance d if the lag of p_d is on the order of, or larger than, the lag of p, unless q_{ff} is chosen so that the zeros of $(1 - \widetilde{p}q_{ff} \widetilde{p}_d^{-1})$ cancel the poles of $\widetilde{p}_d$. The procedure for such a choice of q_{ff} is the same as that described in Chapter 4, except that here, the model is $\widetilde{p}\widetilde{p}_d^{-1}$ rather than just $\widetilde{p}$, as in the design of 2DF feedback controllers. Indeed, the 2DF controller design capability of IMCTUNE can be used to determine q_{ff}. To illustrate, we return to Example 8.1, except now the disturbance is measured and there is a dead time of 0.5 time units before the disturbance enters the disturbance lag.

Example 9.1 Comparison of 1DF and 2DF Feedforward Controller Designs

The purpose of this example is to compare the performance of 1DF and 2DF feedforward controllers when the process models are perfect. The process is

$$\widetilde{p}(s) = p(s) = e^{-s}/(s+1); \quad \widetilde{p}_d(s) = p_d(s) = e^{-0.5s}/(4s+1). \qquad (9.6)$$

For the above system, the 1DF and 2DF feedforward controllers from IMCTUNE, and from Equations (9.3) and (9.5) are

$$q_{ff}(s) = (s+1)/(4s+1), \qquad (9.7a)$$

$$q_{ff}(s) = (s+1)(.473s+1)/(4s+1)(0.006s+1). \qquad (9.7b)$$

Notice that the 1DF controller, Eq. (9.7a), does not need a filter, as it is already a lag that does not amplify noise. The 2DF controller, Eq. (9.7b), satisfies the requirement that

$(1 - \widetilde{p}q_{ff}\widetilde{p}_d^{-1})$ cancel the poles of $\widetilde{p}_d$. Only a small filter time constant of 0.006 is needed to reduce the controller noise amplification to 20 because of the relatively large disturbance lag. To obtain Eq. (9.7b) from IMCTUNE, we entered the process model as $\widetilde{p}\widetilde{p}_d^{-1} = (4s+1)e^{-.5s}/(s+1)$ and the model "disturbance" lag as $(4s+1)$. The part of the model to invert is taken as $(4s+1)/(s+1)$, and the order of the controller filter is set to zero. Entering a filter time constant of 0.006 yields Eq. (9.7b).

Figures 9.3 and 9.4 compare the responses of the process output with and without feedback, using the IDF and 2DF controllers given by Equations (9.7a) and (9.7b). The output y, when the feedback loop is open, is the same as the compensated disturbance signal d_c, shown in Figure 9.1b and defined by Eq. (9.5). Notice that the addition of the feedback system improves the response of the output using a 1DF feedforward controller, but degrades the output response using a 2DF feedforward controller. The reason for the degradation in response is that the PID controller attempts to suppress the pulse d_c, created by the feedforward control system. Of course it cannot, and the attempt creates another pulse in the opposite direction that is diminished in amplitude but broader in time. The same effect occurs with the 1DF feedforward controller. However, the response of this controller is long enough with respect to the settling time of the feedback loop that the feedback loop is capable of diminishing the effect of the disturbance on the output.

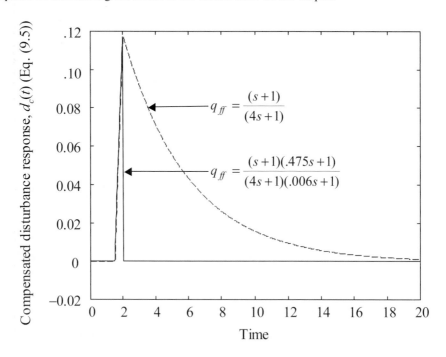

Figure 9.3 Comparison of compensated disturbance responses for Example 9.1.

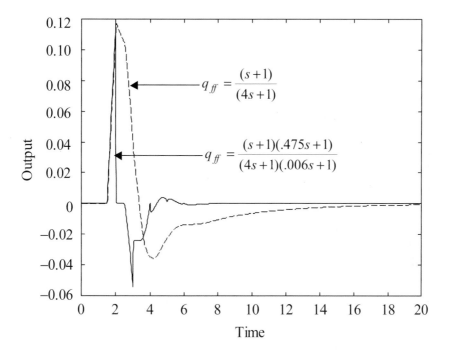

Figure 9.4 Comparison of feedforward plus feedback control system responses to a unit step disturbance for Example 9.1.

The PID controller used to obtain the responses shown in Figure 9.4 was obtained as described in Chapter 6. For a perfect model, an IMC filter time constant of 0.3 yields the following PID controller, which overshoots a step setpoint change by about 10%.

$$PID = 0.610(1 + 1/(1.24s) + .179s/(.090s + 1)).\tag{9.7c}$$

◆

It is possible to prevent the degradation in output response by making the feedback loop blind to the pulse generated by the feedforward controller. This is accomplished by subtracting the pulse from the feedback system, as shown in Figure 9.5.

Based on the responses given in Figures 9.3 and 9.4, the pulse-compensated diagram should be used only when the feedforward control system settling time, with the feedback loop open, is significantly faster than the feedback control system settling time. Thus, for Example 9.1 the pulse compensation of Figure 9.5 improves the response of the 2DF control system, which has a fast response, but degrades the response of the 1DF feedforward control system, which has a slow response relative to that of the feedback system.

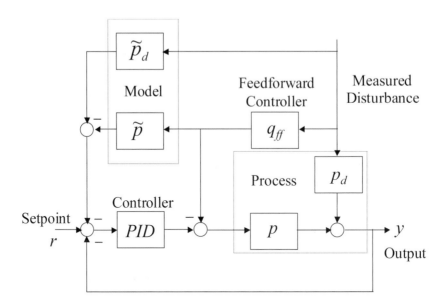

Figure 9.5 Pulse-compensated feedforward control system.

9.4   CONTROLLER DESIGN FOR UNCERTAIN PROCESSES

9.4.1 Gain Variations

Gain variations in either or both $p(s)$ and $p_d(s)$ can result in a nonzero value for the steady state effect of the compensated disturbance d_c (see Figure 9.1b), on the process output. The gain of the feedforward controller K_f should be chosen to minimize either the maximum magnitude of the steady-state compensated disturbance $d_c(\infty)$ or the ratio of the compensated to the uncompensated disturbance $d_c(\infty)/d_e(\infty)$. Mathematically, the problem can be expressed as

$$d_c(\infty)_{opt.} = \min_{K_f} \max_{K_p, K_d} \left| K_d - K_p K_f \right| \qquad (9.8)$$

or

$$(d_c(\infty)/d_e(\infty))_{opt.} = \min_{K_f} \max_{K_p, K_d} \left| 1 - K_p K_f / K_d \right|. \qquad (9.9)$$

The maxima in Equations (9.8) and (9.9) occur at the simultaneous upper bound of K_p and lower bound of K_d, and at the lower bound of K_p and the upper bound of K_d. The values of K_f that minimize these maxima are those values that equalize the values of the two extremes. That is, for Eq. (9.8)

$$(K_f)_{opt} = \left(\frac{\max(K_d) + \min(K_d)}{2} \right) \bigg/ \left(\frac{\max(K_p) + \min(K_p)}{2} \right), \qquad (9.10)$$

$$\equiv (K_d)_{ave} / (K_p)_{ave}$$

and for Eq. (9.9)

$$(K_f)_{opt} = \left(\frac{\max(K_p / K_d) + \min(K_p / K_d)}{2} \right)^{-1}. \qquad (9.11)$$

$$\equiv (K_p / K_d)_{ave}^{-1}$$

The feedforward controller gain given by Eq. (9.11) assures that the magnitude of the compensated disturbance effect $d_c(\infty)$ is always less than the magnitude of the uncompensated disturbance effect $d_e(\infty)$. That is, the action of the feedforward controller always improves the output response over that which would have been achieved without feedforward control. However, this feedforward controller gain also yields the same relative improvement if the ratio of (K_d / K_p) is at its maximum or its minimum, and this weighting tends to amplify the effect of large values of (K_d / K_p) on the output. On the other hand, the feedforward controller gain given by Eq. (9.10) can cause the compensated disturbance to be worse than the uncompensated disturbance in some situations. As we shall see from Example 9.2, the choice between Equations (9.10) and (9.11) depends to a large degree on engineering judgment, and to some degree on the amount of uncertainty in the process and disturbance lag gains.

Example 9.2 Variation in Process and Disturbance Lag Gains

Case 1, Modest Uncertainty:

$$0.8 \leq K_p, K_d \leq 1.2 \,; d = 1$$

From Eq. (9.10), we have $(K_f)_{opt} = 1,$ and from Eq. (9.8), the worst-case compensated effect of the disturbance is

$$\left| d_c(\infty) \right|_{\max.} = \max_{K_p, K_d} \left| K_d - K_p (K_f)_{opt} \right| d(\infty) = \left| K_d - K_p \right|_{K_p = .8, K_d = 1.2} = \left| K_d - K_p \right|_{K_p = 1.2, K_d = .8} = 0.4.$$

The worst case uncompensated effect of the disturbance is $d_e(\infty)_{\max} = (K_d)_{\max} d(\infty) = 1.2.$

From Eq. (9.11) we have $(K_f)_{opt} = (\dfrac{3/2 + 2/3}{2})^{-1} = 12/13,$ and from Eq. (9.9.)

$$\left| d_c(\infty) / d_e(\infty) \right|_{\max} = \max_{K_p, K_d} \left| 1 - K_p K_f / K_d \right| = \left| 1 - \frac{(1.2)(12/13)}{.8} \right| = \left| 1 - \frac{(.8)(12/13)}{1.2} \right| = .385.$$

Since $d_e(\infty)_{\max} = (K_d)_{\max} d(\infty) = 1.2,$ then

$$d_c(\infty)_{\max} = (.385)(\ d_e(\infty)_{\max}) = .46.$$

For most engineering situations the feedforward controller gain from Eq. (9.10) (i.e., $K_f = 1$) would be preferred because this minimizes the maximum effect of the compensated disturbance $\left|d_c(\infty)\right|_{\max}$ on the output (0.4 versus .46).

Case 2, Large Uncertainty:

$$1 \le K_p, K_d \le 3; d = 1$$

From Eq. (9.10) we have: $(K_f)_{opt} = 1$ and from Eq. (9.8), if $K_d = 1$, and $K_p = 3$, then the compensated effect of the disturbance is

$$d_c(\infty) = 1 - 3 = -2.$$

However, the uncompensated effect of the disturbance is only 1.0 since

$$d_e(\infty) = K_d d(\infty) = 1.0$$

From Eq. (9.11) we have $(K_f)_{opt} = (\dfrac{3+1/3}{2})^{-1} = 3/5$, and from Eq. (9.9), if $K_d = 1$, and $K_p = 3$, then the compensated effect of the disturbance is

$$d_c(\infty)/d_e(\infty) = 1 - K_p(K_f)_{opt}/K_d = (1 - (3)(3/5)/1) = -0.8$$

Thus for the situation where $K_d = 1$, and $K_p = 3$, the feedforward gain given by Eq. (9.11) is clearly preferred. However, if the gains of K_d and K_p are reversed so that $K_d = 1$, and $K_p = 3$, then the uncompensated effect of the disturbance is 3.0, and the compensated effect of the disturbance for the feedforward gains given by Eq. (9.10) and Eq. (9.11) are 2 and 2.4 respectively. So, in the first instance the gain given by Eq. (9.11) is preferred, while in the second instance, the gain given by Eq. (9.10) is preferred. "You pays your money and takes your choice." The authors prefer the gain given by Eq. (9.11).

◆

9.4.2 Dead time Variations

This section considers the effect of dead time variations on the choice of feedforward controller dead time. We assume that the process dead time is on the order of, or greater than, the disturbance dead time. Further, the relative order of the process $p(s)$ is the same as that of the disturbance transfer function $p_d(s)$, and all time constants are known exactly. The process description is therefore given by

$$y(s) = Kg(s)e^{-T_p s}u(s) + K_d g_d(s)e^{-T_d s}d(s), \qquad (9.12)$$

where

$$g(0) = g_d(0) = 1, \quad \underline{T}_p \le T_p \le \overline{T}_p, \quad \underline{T}_d \le T_d \le \overline{T}_d, \quad \underline{K} \le K \le \overline{K}, \quad \underline{K}_d \le K_d \le \overline{K}_d$$

The feedforward controller is given by

$$q_f(s) = K_f g_d(s) e^{-\Delta s} / g(s), \tag{9.13}$$

where Δ is a parameter that we need to find.

Substituting Eq. (9.13) into Eq. (9.5) gives

$$d_c(s) = (K_d e^{-T_d s} - K K_f e^{-(T_p + \Delta)s}) g_d(s) d(s). \tag{9.14}$$

When $(T_d > T_p + \Delta)$, the maximum magnitude of the integral of $d_c(t)$ occurs when $(T_d - (T_p + \Delta))$ is a maximum. When $(T_d < (T_p + \Delta))$, the maximum magnitude of the integral of $d_c(t)$ occurs when $((T_p + \Delta) - T_d)$ is a maximum. Choosing Δ so that these two maxima are equal gives

$$\Delta = \frac{(\overline{T}_d + \underline{T}_d) - (\overline{T}_p + \underline{T}_p)}{2} \equiv (T_d)_{ave} - (T_p)_{ave}. \tag{9.15}$$

In Example 9.3 we present the results of simulations on two processes, each of which are similar to Example 9.1. In the first example, the dead time between the measured disturbance and the output is on the order of the process dead time, and hence perfect feedforward compensation is possible for most choices of models in the uncertain set. For this process we use a simple 1DF feedforward controller. In the second example, the dead time between the measured disturbance and the output is smaller than the process dead time, and hence perfect feedforward compensation is not possible. Therefore, for this process we use a 2DF feedforward controller.

Example 9.3 Disturbance and Process Dead Times of Roughly the Same Size

$$p(s) = K e^{-T_p s} / (s + 1); \qquad 0.8 \le K, T_p \le 1.2 \tag{9.16a}$$

$$p_d(s) = K_d e^{-T_d s} / (4s + 1); \quad 0.8 \le K_d, T_d \le 1.2 \tag{9.16b}$$

$$q_f(s) = K_f (4s + 1) / (s + 1) \tag{9.16c}$$

The feedforward controller given by Eq. (9.16c) has a dead time of zero, since the difference between $(T_d)_{ave}$ and $(T_p)_{ave}$ is zero (see Eq. 9.15).

Figures 9.6a and 9.6b show the worst-case feedforward-only responses for the above process, using feedforward controller gains given by Equations (9.10) and (9.11). Notice that for a disturbance lag gain of 1.2 and a process gain of .8 (i.e., $K_d = 1.2$, $K = .8$), the controller gain given by Eq. (9.10) gives the better response. However, when the gains are reversed, (i.e., $K_d = .8$, $K = 1.2$), the controller gain given by Eq. (9.11) gives the better response.

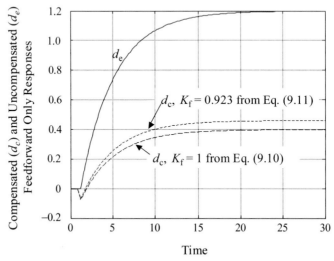

Figure 9.6a Example 9.3 responses of the feedforward-only control system to a unit step disturbance for $K_d = 1.2$, $K_p = .8$, $T_d = 1.2$, $T_p = .8$.

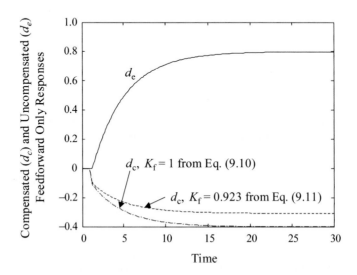

Figure 9.6b Example 9.3 responses of the feedforward-only control system to a unit step disturbance for $K_d = .8$, $K_p = 1.2$, $T_d = 1.2$, $T_p = .8$.

Figures 9.7a and 9.7b show worst-case feedforward plus feedback responses corresponding to the responses in Figures 9.6a and 9.6b. The 1DF PID feedback controller was obtained using Mp tuning (see Chapter 7) and the IMC to PID controller approximation method of Chapter 6. The IMC filter time constant required to achieve an Mp of 1.05 is 1.04, and the resulting PID controller is

$$PID = 0.610(1 + 1/(1.24s) + .179s/(.090s + 1)).\qquad(9.17)$$

The responses in Figures 9.7a and 9.7b should be compared to those of a well-tuned 2DF feedback-only control system, as shown in Figure 9.8. The controller used to generate the responses in Figure 9.8 was obtained using the design and tuning methods described in Chapter 8. Clearly, the performance of the combined feedforward/feedback control system is superior to that of a standalone 2DF feedback control system.

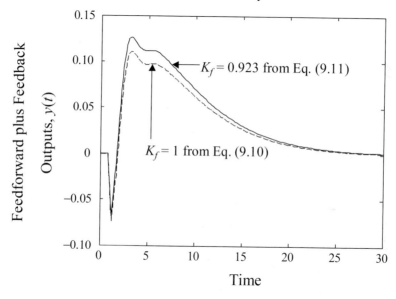

Figure 9.7a Example 9.3 output responses of the feedforward plus feedback control system to a unit step disturbance $K_d = 1.2$, $K = .8$, $T_d = 1.2$, $T_p = .8$.

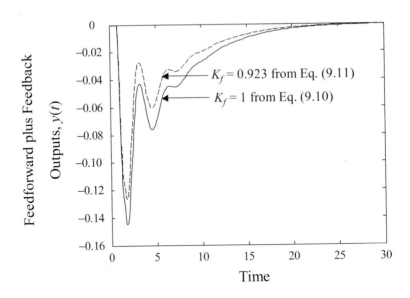

Figure 9.7b Example 9.3 output responses of the feedforward plus feedback control system to a unit step disturbance K_d = .8, K_p = 1.2, T_d = 1.2, T_p = .8.

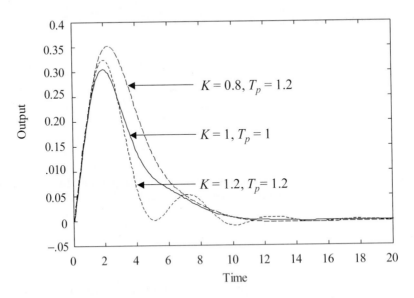

Figure 9.8 Example 9.3 responses of a 2DF feedback control system to a unit step disturbance with feedback controller = $3.52(1+1/(4.04s)+.965s)/(4.17s+1)$.

◆

The next example is the same as Example 9.3, except that the disturbance transfer function dead time is always smaller than the process dead time.

Example 9.4 Disturbance Dead Time < Process Dead Time

$$p(s) = Ke^{-T_p s} /(s+1); \quad 0.8 \le K, T_p \le 1.2, \tag{9.18a}$$

$$p_d(s) = K_d e^{-T_d s} /(4s+1); \quad 0.8 \le K_d \le 1.2, \quad .4 \le T_d \le .6, \tag{9.18b}$$

$$\tilde{p}(s) = e^{-s} /(s+1); \quad \tilde{p}_d(s) = e^{-.5s} /(4s+1). \tag{9.18c}$$

The 1DF feedforward controller is the same as that of Eq. (9.7a):

$$q_{ff1}(s) = (s+1)/(4s+1). \tag{9.18d}$$

The 2DF feedforward controller is the same as that of Eq. (9.7b):

$$q_{ff2}(s) = (s+1)(.473s+1)/(4s+1)(0.006s+1). \tag{9.18e}$$

The PID feedback controller is the same as that of Eq. (9.17):

$$PID = 0.610(1+1/(1.24s)+.179s/(.090s+1)). \tag{9.18f}$$

As before, the following feedforward/feedback responses to a unit step disturbance should be compared with the 2DF feedback-only responses given by Figure 9.8. Figure 9.9 repeats the perfect model responses for the feedback controller of Eq. (9.18f). Notice that the undershoots in Figure 9.9 of the uncompensated controllers are less severe than those in Figure 9.4. This occurs because the feedback controller, Eq. (9.18f), is less aggressive than that of Eq. (9.7c) because Eq. (9.9f) has been tuned to accommodate the specified process uncertainty.

Figures 9.10 and 9.11 show typical worst-case responses of the 1DF and 2DF feedforward/feedback control systems. Unlike Figure 9.8, there is little difference in the responses of the compensated (via Figure 9.5) and uncompensated 2DF controllers, and both controllers perform better than the 1DF controller.

Figure 9.12 shows typical best-case responses that result when the difference between the disturbance dead time and process dead time is 0.2, which is the smallest in the set of uncertain parameters. The poorer behavior of the 2DF controllers relative to the 1DF controller arises from the fact that the 2DF feedforward controller, Eq. (9.18e), was obtained for a difference between the disturbance dead time and process dead time of 0.5.

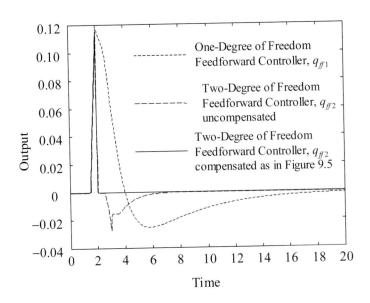

Figure 9.9 Responses of feedforward/feedback control systems to a unit step disturbance at time 1.0 for Example 9.4, $K = K_d = T_p = 1$, and $T_d = 0.5$.

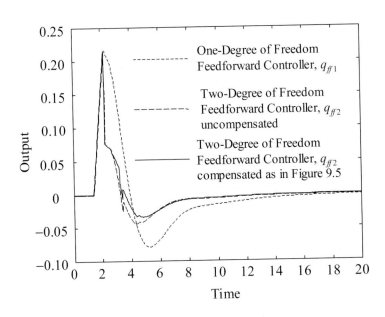

Figure 9.10 Responses of feedforward/feedback control systems to a unit step disturbance at time 1.0 for Example 9.4, and $K = K_d = T_p = 1.2$, and $T_d = 0.4$.

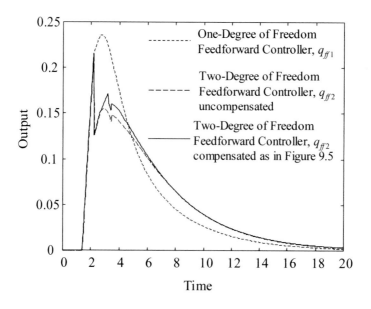

Figure 9.11 Responses of feedforward/feedback control systems to a unit step disturbance at time 1.0 for Example 9.4, $K = .8$, $K_d = T_p = 1.2$, and $T_d = 0.4$.

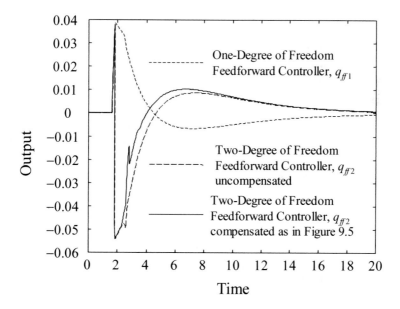

Figure 9.12 Responses of feedforward/feedback control systems to a unit step disturbance at time 1.0 for Example 9.4, $K = K_d = T_p = .8$, and $T_d = 0.6$.

Since the deviations from setpoint in Figure 9.12 are quite small relative to those in Figures 9.10 and 9.11, our conclusion from these figures is that the 2DF feedforward controller design is preferred.

♦

9.5 SUMMARY

A properly designed feedforward/feedback control system will always improve performance over a simple feedback control system, independent of the process uncertainty, provided only that the measured disturbance does not enter directly (i.e., with a unity transfer function) into the process output. However, as one might expect, the greater the process uncertainty, the less the potential improvement in response for processes at the extremes of the uncertainty ranges.

For an uncertain process, the feedforward controller gain should be chosen either as $K_f = (K_d)_{ave} / (K_p)_{ave}$ (see Eq. (9.10)) or as $K_f = (K_p / K_d)_{ave}^{-1}$ (see Eq. (9.11)), depending on the uncertainties in the gains and on control objectives. However, only the latter choice (i.e., Eq. (9.11)) guarantees that the feedforward/feedback controller will perform better than a simple feedback controller for all processes in the set of uncertain processes.

When the dead time between the measured disturbance and the output is greater than the dead time between the control and the output, then a feedforward controller should be designed like a single-degree of freedom IMC controller, except that the difference between the dead times is included in the controller.

When the dead time between the measured disturbance and the output is less than the dead time between the control and the output, then the feedforward controller should be designed as a 2DF IMC controller. If there is also relatively little process uncertainty, better performance can be achieved by modifying the feedforward control system structure to prevent the pulse generated by the feedforward action from being propagated around the feedback control loop.

Problems

Select and tune control systems for each of the following processes. The control objective is to control the measured output $y(t)$, and all disturbances $d_i(t)$ are measured.

9.1 $$y(s) = \frac{2(1 - 2s)e^{-6s}}{(2s + 1)^2} u(s) + \frac{50e^{-4s}d(s)}{(50s + 1)}$$

9.2 $$y(s) = \frac{e^{-10s}}{30s + 1} y_2(s) + d_1(s) \qquad y_2(s) = \frac{e^{-5s}}{4s + 1} u(s) + d_2(s)$$

9.3 $y(s) = \dfrac{Ke^{-4s}}{s(\tau s + 1)}u(s) + \dfrac{K_d e^{-5s}}{s(\tau s + 1)}d(s)$ $1 \le K \le 3 \; ; \; 2 \le \tau \le 5; \; 3 \le K_d \le 4$

9.4 $y(s) = \dfrac{K_1 e^{-Ts}}{5s + 1}y_2(s) + d_1(s)$ $1 \le K_1 \le 3, \;\; 4 \le T \le 6$

 $y_2(s) = \dfrac{K_2 e^{-s}}{3s + 1}u(s) + d_2(s)$ $1 \le K_2 \le 5, \;\; 0 \le u(t) \le 10$

9.5 $y(s) = \dfrac{K_1 e^{-Ts}}{3s + 1}y_2(s) + d_1(s)$ $1 \le K_1 \le 5; \; 2 \le T \le 4$

 $y_2(s) = \dfrac{K_2 e^{-s}}{2s + 1}u(s) + d_2$ $1 \le K_2 \le 5$

9.6 $y(s) = \dfrac{K_1 e^{-Ts}}{3s + 1}y_1(s) + \dfrac{e^{-Ts}d_1(s)}{3s + 1}$ $1 \le K \le 5 \;\; 2 \le T \le 4$

 $y_1(s) = \dfrac{K_2(-2s + 1)}{(2s + 1)^2}u(s) + \dfrac{d_2}{s + 1}$

References

Seborg, D. E., T. F. Edgar, and D. A. Mellichamp. 1989. *Process Dynamics and Control*, John Wiley & Sons, NY.

Cascade Control

Objectives of the Chapter

- To review classical cascade control.
- To present an alternate way of thinking about cascade control that leads to improved performance.
- To introduce controller design methods that accommodate process uncertainty.

Prerequisite Reading

Chapter 3, "One-Degree of Freedom Internal Model Control"
Chapter 4, "Two-Degree of Freedom Internal Model Control"
Chapter 5, "MSF Implementations of IMC Systems"
Chapter 6, "PI and PID Controller Parameters from IMC Design"
Chapter 7, "Tuning and Synthesis of 1DF IMC Controllers for Uncertain Processes"
Chapter 8, "Tuning and Synthesis of 2DF Control Systems"

10.1 INTRODUCTION

Cascade control can improve control system performance over single-loop control whenever either: (1) Disturbances affect a measurable intermediate or secondary process output that directly affects the primary process output that we wish to control; or (2) the gain of the secondary process, including the actuator, is nonlinear. In the first case, a cascade control system can limit the effect of the disturbances entering the secondary variable on the primary output. In the second case, a cascade control system can limit the effect of actuator or secondary process gain variations on the control system performance. Such gain variations usually arise from changes in operating point due to setpoint changes or sustained disturbances.

   A typical candidate for cascade control is the shell and tube heat exchanger of Figure 10.1.

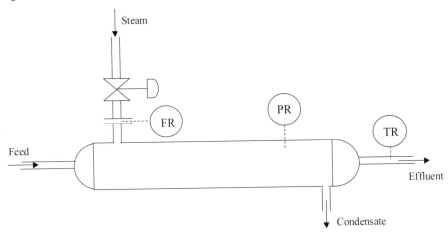

Figure 10.1 A shell and tube heat exchanger.

   The primary process output is the temperature of the tube side effluent stream. There are two possible secondary variables, the flow rate of steam into the exchanger and the steam pressure in the exchanger. The steam flow rate affects the effluent temperature through its effect on the steam pressure in the exchanger. The steam pressure in the exchanger affects the effluent temperature by its effect on the condensation temperature of the steam. Therefore, either the steam flow rate or the steam pressure in the exchanger can be used as the secondary output in a cascade control system. The choice of which to use depends on the disturbances that affect the effluent temperature.

   If the main disturbance is variations in the steam supply pressure, due possibly to variable steam demands of other process units, then controlling the steam flow with the control valve is most likely to be the best choice. Such a controller can greatly diminish the effect of steam supply pressure variations on the effluent temperature. However, it is still necessary to have positive control of the effluent temperature to be able to track effluent temperature setpoint changes and to reject changes in effluent temperature due to feed temperature and flow variation. Since there is only one control effort, the steam valve stem position, traditional cascade control uses the effluent temperature controller to adjust the setpoint of the steam flow controller, as shown in Figure 10.2.

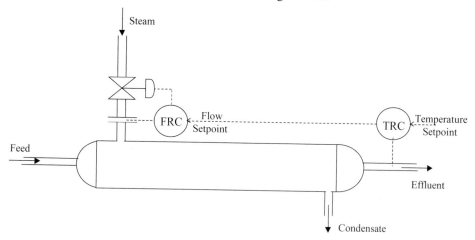

Figure 10.2 Cascade control of effluent temperature via steam flow control.

If feed flow and temperature variations are significant, then these disturbances can be at least partially compensated by using the exchanger pressure rather than the steam flow as the secondary variable in a cascade loop, as shown in Figure 10.3.

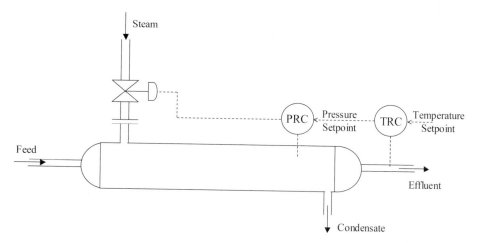

Figure 10.3 Cascade control of effluent temperature via shell side pressure control.

The trade-off in using the configuration of Figure 10.3 rather than that of Figure 10.2 is that the inner control loop from the steam pressure to the valve stem position may not suppress variations in valve gain as well as with an inner loop that uses the valve to control the steam flow rate. This consideration relates to using a cascade control system to suppress the effect of process uncertainty, in this case the valve gain, on the control of the primary process variable, the effluent temperature. We will have a lot more to say about using cascade control systems to suppress process uncertainty in the following sections.

To repeat, cascade control has two objectives. The first is to *suppress the effect of disturbances on the primary process output* via the action of a secondary, or inner control loop around a secondary process measurement. The second is to *reduce the sensitivity of the primary process variable to gain variations* of the part of the process in the inner control loop.

As we shall demonstrate, cascade control can be usefully applied to any process where a measurable secondary variable directly influences the primary controlled variable through some dynamics. We will also demonstrate that despite frequent literature statements to the contrary, inner loop dynamics do not have to be faster than the outer loop dynamics. However, the traditional cascade structure and tuning methods must be modified in order for cascade control to achieve its objectives when the inner loop process has dynamics that are on the order of, or slower than, the primary process dynamics.

10.2 CASCADE STRUCTURES AND CONTROLLER DESIGNS

Figure 10.4 shows the traditional PID cascade control system block diagram (Seborg et al., 1989). This is the cascade structure associated with Figures 10.2 and 10.3. For Figure 10.2, the secondary process variable y_2 is the steam flow rate, while for Figure 10.3, it is the shell-side steam pressure. In both cases, the primary variable y_1 is the effluent temperature.

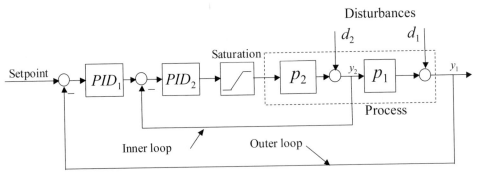

Figure 10.4 Traditional cascade block diagram.

One of the objectives of this section is to present methods for obtaining the parameters of the PID controllers of Figure 10.4 from a well-designed and well-tuned IMC cascade control system, just as we did for single-loop control systems in Chapter 6.

Figure 10.5 shows an IMC cascade block diagram that accomplishes the same objectives as Figure 10.4. There are other equivalent IMC cascade structures to that given by Figure 10.5 (Morari and Zafiriou, 1989). However, the configuration of Figure 10.5 is convenient because it suggests that *controller q_2 should be designed and tuned solely to suppress the effect of the disturbance d_2 on the primary output y_1*, and also convenient because both controller outputs u_1 and u_2 enter directly into the actuator. As we shall see later, this last point facilitates dealing with control effort saturation. However, for the remainder of this section we shall ignore the saturation block in order to study the design and tuning of IMC controllers. These IMC controllers will then be used to obtain the PID controller parameters in Figure 10.4, as was done in Chapter 6 for single-loop control systems.

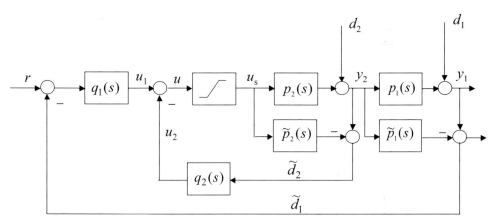

Figure 10.5 IMC cascade structure.

From Figure 10.5, the transfer functions between the inputs to the inner loop, u_1 and d_2, and the secondary process output y_2 are

$$y_2(s) = \frac{p_2(s)u_1(s) + (1 - \tilde{p}_2(s)q_2(s))d_2(s)}{(1 + (p_2(s) - \tilde{p}_2(s))q_2(s))}. \qquad (10.1)$$

The transfer functions between the setpoint and disturbances and the primary process output y_1 are

$$y_1(s) = \frac{p_1 p_2 q_1 r(s) + (1 - \tilde{p}_2 q_2)p_1 d_2(s) + (1 - \tilde{p}_1 p_2 q_1 + (p_2 - \tilde{p}_2)q_2)d_1(s)}{(1 + (p_1 - \tilde{p}_1)p_2 q_1 + (p_2 - \tilde{p}_2)q_2)}. \qquad (10.2)$$

In Eq. (10.2) we have suppressed the dependency of all transfer functions on the Laplace variable s to keep the equation on one line. Based on Equations (10.1) and (10.2) we observe the following:

(1) If the lag time constants of the primary process $p_1(s)$ are large relative to those of the secondary process $p_2(s)$ then the inner loop controller $q_2(s)$ should be chosen so that the zeros of $(1 - \tilde{p}_2(s)q_2(s))$ cancel the small poles (i.e., large time constants) of $\tilde{p}_1(s)$ as outlined in Chapter 4. Otherwise, $q_2(s)$ should simply invert a portion of $\tilde{p}_2(s)$ as described in Chapters 3 and 7.

(2) The outer loop controller $q_1(s)$ should approximately invert the entire process model $\tilde{p}_1\tilde{p}_2(s)$, as described in Chapters 3 and 7.

(3) The IMCTUNE software can be used to design and tune both $q_1(s)$ and $q_2(s)$.

We recommend tuning $q_2(s)$ with the outer loop open, and then tuning $q_1(s)$ with the inner loop closed. That is, first find the filter time constant ε_2 for $q_2(s)$, and then find ε_1 for $q_1(s)$. According to the denominator of Eq. (10.2), the tunings for $q_1(s)$ and $q_2(s)$ interact. Therefore, some adjustment of ε_2 may be necessary after obtaining ε_1.

Having obtained the IMC controllers for Figure 10.5, we would like to use these controllers to obtain the PID controllers in Figure 10.4 in a manner similar to that for single-loop controllers described in Chapter 6. Unfortunately, however, we can do so only very approximately. Figure 10.5 can be rearranged, ignoring the saturation block, as given by Figure 10.6.

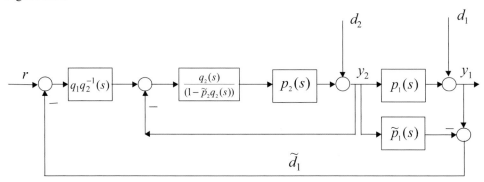

Figure 10.6 IMC cascade control with a simple feedback inner loop.

The controller given by $q_2(s)/(1 - \tilde{p}_2(s)q_2(s))$ can often be well approximated by a PID controller, as described in Chapter 6. Again, IMCTUNE can be used to obtain this controller. However, obtaining PID$_1$ in Figure 10.4 is not so straightforward. Collapsing the feedback loop through $\tilde{p}_1(s)$, while leaving the inner loop alone, yields Figure 10.7.

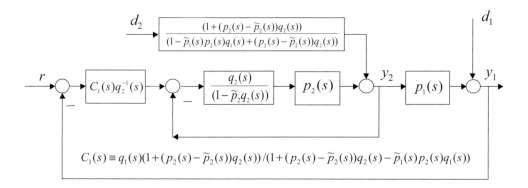

Figure 10.7 Standard feedback form of Figure 10.6.

The controller $C_1(s)$ in Figure 10.7 cannot be realized because it contains the process transfer function $p_2(s)$, which is uncertain and cannot be made part of the controller. We can however approximate $p_2(s)$ with its model $\tilde{p}_2(s)$. In this case $C_1(s)$ becomes

$$C_1(s) \cong q_1(s)/(1-\tilde{p}_1(s)\tilde{p}_2(s)q_1(s)). \qquad (10.3)$$

Another difference between Figures 10.6 and 10.7 is that even if the model $\tilde{p}_1(s)$ is a perfect representation of the process, the pulse created by the inner loop response to the disturbance $d_2(s)$ (i.e., $d_2(s)/(1-\tilde{p}_1(s)\tilde{p}_2(s)q_2(s))$ for a perfect model $\tilde{p}_2(s)$) feeds back around the outer loop of Figure 10.7. Since the primary controller cannot suppress this pulse, it continues around the loop until it dies out.

Even using the approximation given by Eq. (10.3) to obtain a PID controller does not reduce Figure 10.7 to the standard PID cascade diagram of Figure 10.4 because $C_1(s)$ in Figure 10.7 is multiplied by $q_2^{-1}(s)$. If $q_2^{-1}(s)$ is a lead (which will generally occur only if the process description is quite uncertain), then $q_2^{-1}(s)$ can be approximated by a polynomial via a Taylor's series. This polynomial can be multiplied into the PID controller obtained from $C_1(s)$ to obtain a new PID controller after dropping higher order terms. Even if $q_2^{-1}(s)$ is a lag, it may still be possible to approximate the term $C_1(s)q_2^{-1}(s)$ by a PID controller. However, the necessary approximations will have to be carried out by hand, following procedures in Chapter 6, as the current version of IMCTUNE does not carry out the necessary manipulations.

Two rather long examples of cascade control of uncertain processes follow. The individual processes in both examples are first-order lags plus dead time and have significant process uncertainty. In the first example, the secondary process output dynamics are significantly faster than the primary process dynamics. In the second example, the primary and secondary process dynamics have similar dynamic behavior.

Example 10.1 Secondary Process Has Faster Dynamics than the Primary Process

$$p_1(s) = \frac{K_1 e^{-T_1 s}}{\tau_1 s + 1} ; \quad 0.8 \le K_1 \le 1.2, \quad 17.5 \le T_1 \le 22.5, \quad 14 \le \tau_1 \le 16 \quad (10.4a)$$

$$p_2(s) = \frac{K_2 e^{-T_2 s}}{\tau_2 s + 1} ; \quad 0.6 \le K_2 \le 1.8, \quad 2 \le T_2 \le 4, \quad 1 \le \tau_2 \le 3 \quad (10.4b)$$

(a) IMC System Design

Following the suggestions in Chapters 7 and 8, we use the upper-bound gains and dead times and the lower-bound time constants for the process models.

$$\tilde{p}_1(s) = \frac{1.2 e^{-22.5s}}{14s + 1} \quad (10.4c)$$

$$\tilde{p}_2(s) = \frac{1.8 e^{-4s}}{s + 1} \quad (10.4d)$$

Computing the 2DF feedback controller for the inner loop (see Figure 10.5), using IMCTUNE with the outer loop, open gives

$$q_2(s) = \frac{(s+1)(9.05s+1)}{1.8(4.4s+1)^2}. \quad (10.5a)$$

Figure 10.8 shows the tuning curves, while Figure 10.9 shows typical time responses to a step disturbance in the inner loop. Data for both figures was obtained from IMCTUNE.

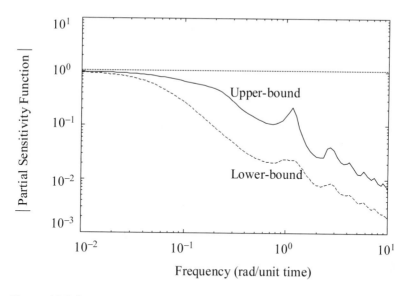

Figure 10.8 Cascade inner loop tuning using controller given by Eq. (10.5a).

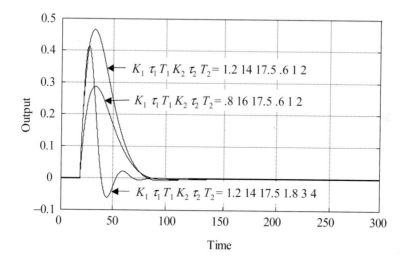

Figure 10.9 Responses to a step inner loop disturbance (d_2) with the outer loop open.

Having obtained the inner loop controller, the outer loop controller can be obtained from the cascade facility of IMCTUNE, and is

$$q_1(s) = \frac{(15s+1)}{2.16(16.87s+1)}.$$ (10.5b)

The tuning curves for the outer loop of the cascade, using Eq. (10.5b), are shown in Figure 10.10. Also in this figure are the closed-loop upper-bound and lower-bound curves for a single-loop controller for a model and controller of

$$\tilde{p}(s) = \frac{2.16\,e^{-26.5s}}{15s + 1} \quad q(s) = \frac{(15s + 1)}{2.16\,(14.3s + 1)}. \tag{10.6}$$

Recall from Equations (10.4a) and (10.4b) that the overall process is

$$p(s) = \frac{Ke^{-Ts}}{(\tau_1 s + 1)(\tau_2 s + 1)} \quad 0.48 \le K \le 2.16, \quad 19.5 \le T \le 26.5, \quad 14 \le \tau_1 \le 16 \quad 1 \le \tau_2 \le 3.$$

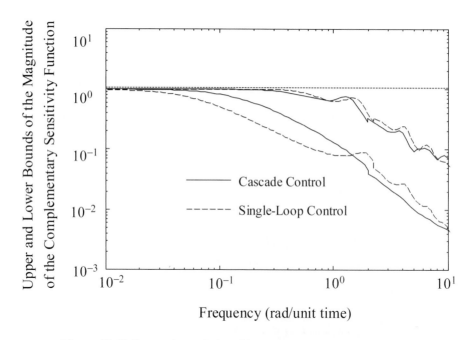

Figure 10.10 Comparison of closed-loop setpoint to output responses.

Based on the closed-loop frequency responses we can conclude that the fastest responses of the single-loop system are slightly faster than those of the cascade system, but more importantly, the slowest responses are significantly slower. Figures 10.11 and 10.12 support these conclusions. Note the different time axes in Figures 10.11 and 10.12.

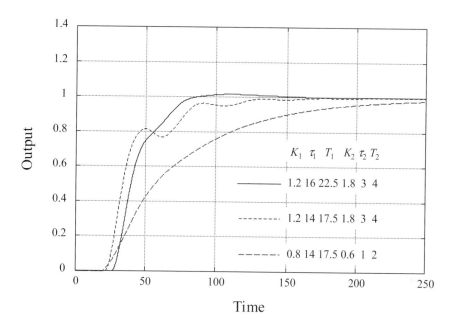

Figure 10.11 Step setpoint responses for the cascade control system of Figure 10.5.

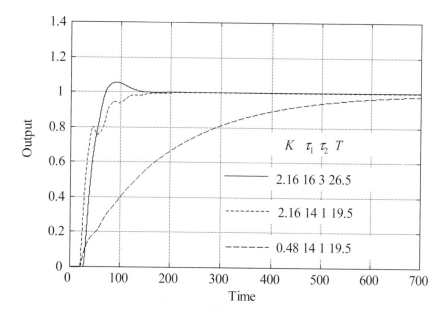

Figure 10.12 Step setpoint responses for the single-loop control system, using Eq. (10.6).

The reason for the improved setpoint response of the cascade system is that the inner loop of the cascade reduces the effect of gain uncertainty in the inner loop process. To show that this is so, Figure 10.13 compares the closed-loop frequency responses of the cascade system with that of a single-loop controller. The process is the same as that given by Equations (10.4a) and 10.4b), except that instead of a lower-bound of 0.48, the lower bounds (*lb*) are 1.1 and 1.44. That is, the single-loop process is

$$p(s) = \frac{Ke^{-Ts}}{(\tau_1 s + 1)(\tau_2 s + 1)} \quad lb \le K \le 2.16, \quad 19.5 \le T \le 26.5, \quad 14 \le \tau_1 \le 16 \quad 1 \le \tau_2 \le 3. \quad (10.7)$$

The model and controller for the process of Eq. (10.7) are the same as given in Eq. (10.6) and are repeated for convenience:

$$\widetilde{p}(s) = \frac{2.16\,e^{-26.5s}}{15s + 1} \quad q(s) = \frac{(15s + 1)}{2.16(14.3s + 1)}$$

A lower-bound gain of 1.44 corresponds to a secondary process (i.e., $\widetilde{p}_2(s)$) with a gain of 1.8 and no gain uncertainty. A lower-bound gain of 1.1 corresponds to a secondary process whose gain varies between 1.375 and 1.8. In other words, the effect on the outer loop of the ratio of the maximum to minimum gain variation of the secondary process has been reduced from a ratio of 3 to a ratio of 1.3. The slowest time responses are compared in Figure 10.14.

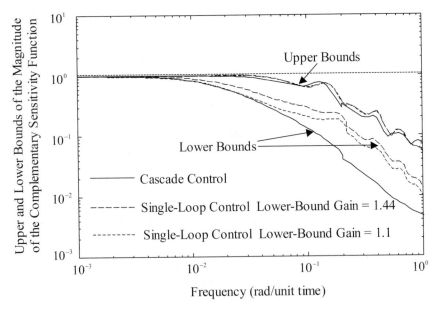

Figure 10.13 Comparison of cascade and single-loop control systems.

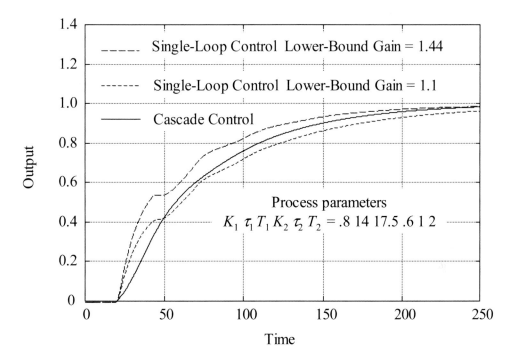

Figure 10.14 Comparison of slowest responses to a step setpoint change.

We now return to the cascade control system responses to a step disturbance to the inner loop, but this time with the outer loop closed. The time responses for the same processes as in Figure 10.9 are shown in Figure 10.15. From this figure we conclude that there is no need to retune the inner loop.

Figure 10.16 shows the effect of using the single-degree of freedom IMC controller given by Eq. (10.8) on the response to a step disturbance in the inner loop. These responses should be compared with those of Figure 10.15.

$$q_2(s) = \frac{(s+1)}{1.8(4.18s+1)}. \tag{10.8}$$

The filter time constant of 4.18 in Eq. (10.8) yields an Mp of 1.05. That is, the controller is tuned so that the worst-case overshoot of the inner loop output y_2 to a step setpoint change to the inner loop is about 10% with the controller q_2 in the forward path. This controller is then used in the feedback path of the inner loop in Figure 10.5.

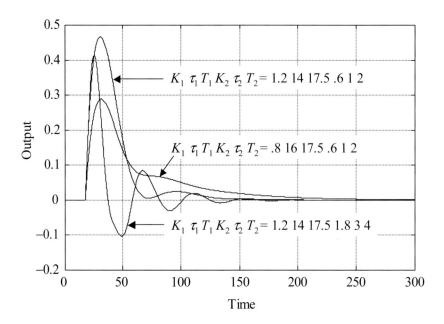

Figure 10.15 Responses to a step inner loop disturbance (d_2) with the outer loop closed.

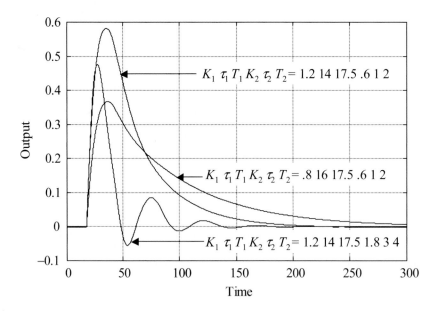

Figure 10.16 Responses to a step inner loop disturbance using the controller given by Eq. (10.8).

While the inner loop disturbance responses using the single-degree of freedom controller Eq. (10.8) are significantly slower than the 2DF controller given by Eq. (10.5a), the responses of the output $y_1(t)$ to setpoint changes to the outer loop are only slightly slower than those given in Figure 10.11.

(b) PID Cascade Controller Designs

Section 10.2 discusses methods for approximating the IMC cascade control system with the traditional cascade system of Figure 10.4. Figure 10.7 shows the IMC equivalent configuration. For convenience, this figure is repeated in Figure 10.17.

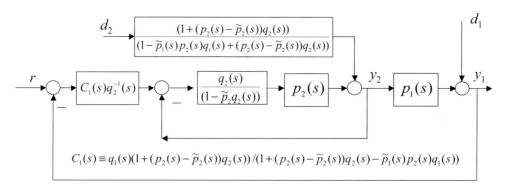

Figure 10.17 Standard feedback form of an IMC cascade control system.

Recall that the controller $C_1(s)$ in Figure 10.7 is not realizable because it contains terms involving the inner loop process $p_2(s)$, which varies within the uncertainty set and cannot be part of the controller. Therefore we suggested replacing $p_2(s)$ with its model, $\tilde{p}_2(s)$. This gives

$$C_1(s) \cong q_1(s)/(1-\tilde{p}_1(s)\tilde{p}_2(s)q_1(s)). \tag{10.9a}$$

IMCTUNE provides the following PID controllers from the IMC controllers obtained previously:

Inner loop: $\dfrac{q_2(s)}{(1-\tilde{p}_2 q_2(s))} \cong PID_2 = 1.79(1+1/(12.05s)+1.68s)/(14.7s+1).$ (10.9b)

Outer loop: $C_1(s) \cong PID_1 = .234(1+1/(23.77s)+5.35s/(.29s+1)),$ (10.9c)

$$q_2^{-1} = 1.8(4.4s+1)^2/((s+1)(9.05s+1)). \tag{10.9d}$$

Figures 10.18 and 10.19 show the disturbance responses for the configuration of Figure 10.17 using the controllers given in equation sets (10.9) and (10.10). Notice that $q_2(s)$ in Eq. (10.9d) is from a 2DF design, and for this reason the responses are labeled Cascade 2.

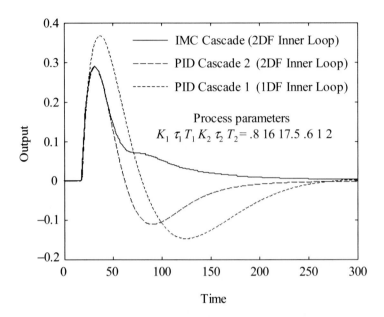

Figure 10.18 Comparison of responses to a step disturbance in the inner loop.

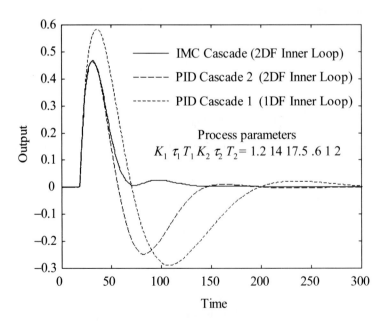

Figure 10.19 Comparison of responses to a step disturbance in the inner loop.

Using the 1DF IMC controller for q_2(s), given by Eq. (10.8) and repeated below, yields the inner loop PID controller given by Eq. (10.10a).

$$q_2(s) = \frac{(s+1)}{1.8(4.18s+1)}$$

Inner loop: $PID_2 = .134(1 + 1/(1.98s) + .319s/(.016s+1)).$ (10.10a)

The outer loop controller remains the same as in Eq. (10.9b) because $q_1(s)$ has not changed. In Figures 10.18 and 10.19 the responses using Eq. (10.10a) are labeled Cascade 1. These responses show the benefits of an IMC outer loop over a PID outer loop. The outer loop PID controller in the responses in Figures 10.18 and 10.19 is cascaded with the term q_2^{-1}. Since q_2^{-1} is a lead, it can be approximated by the Taylor series as the polynomial $1.8(-3.18s^2 + 3.18s + 1)$. Multiplying this polynomial into Eq. (10.10a) and dropping terms higher than second order gives, after some rearrangement,

Outer loop: $PID_1 = .4956(1 + 1/(26.96s) + 7.404s/(.37s+1)).$ (10.10b)

The PID controller given by Eq. (10.10b) can be used in place of Equations (10.8) and (10.10a) and gives virtually identical results. The advantage of Eq. (10.10b) is that it can be used in the traditional cascade configuration of Figure 10.4. Of course, the disturbance response will be that of the 1DF cascade of Figures 10.18 and 10.19.

◆

The purpose of the next example is to demonstrate the advantages of cascade control even when the dynamics of the secondary process are on the same order as the dynamics of the primary process. A common literature fallacy is that the dynamics of the secondary process have to be fast relative to those of the primary process in order to get improved performance from a cascade control system. This fallacy probably arose from the methods used for designing and/or tuning PID cascade control systems. Traditionally, the outer loop controller was designed and tuned assuming that the inner loop is so fast that it can be approximated as a unity gain. When this assumption is not true, the inner and outer loop designs and/or tunings interact, and there existed no good methods of designing and tuning the controller parameters that significantly improved performance over that of a single-loop controller. In an IMC cascade configuration (see Figure 10.5) the inner and outer loops interact mainly by the fact that the inner loop process gain variations are reduced by the action of the inner loop controller. Such interaction is desirable and, as we shall show, does not preclude arriving at controller designs so that cascade performance is significantly better than single-loop performance.

Example 10.2 Primary and Secondary Processes Have Similar Dynamics

The following process was obtained by reducing the time constant and dead time of the primary process of Example 10.1 by a factor of five. This gives the following system:

$$p_1(s) = \frac{K_1 e^{-T_1 s}}{\tau_1 s + 1}; \quad .8 \le K_1 \le 1.2, \quad 3.5 \le T_1 \le 4.5, \quad 2.8 \le \tau_1 \le 3.2. \quad (10.11a)$$

$$p_2(s) = \frac{K_2 e^{-T_2 s}}{\tau_2 s + 1}; \quad .6 \le K_2 \le 1.8, \quad 2.0 \le T_2 \le 4.0, \quad 1.0 \le \tau_2 \le 3.0. \quad (10.11b)$$

(a) IMC System Design

Again following the suggestions in Chapters 7 and 8, we use the upper-bound gains and dead times and the lower-bound time constants for the process models.

$$\tilde{p}_1(s) = \frac{1.2 e^{-4.5s}}{2.8s + 1} \quad \text{and} \quad \tilde{p}_2(s) = \frac{1.8 e^{-4s}}{s + 1}. \quad (10.11c)$$

The controllers associated with the IMC cascade structure of Figure 10.5 are

$$q_2(s) = \frac{(s+1)}{1.8(2.8s + 1)} \quad \text{and} \quad q_1(s) = \frac{(3.8s + 1)}{2.16(5.24s + 1)}. \quad (10.11d)$$

An initial attempt at designing a 2DF controller for the inner loop resulted in the filter time constant reaching the primary process model time constant of 2.8 before achieving an Mp of 1.05 for the partial sensitivity function. In such a situation the inner loop feedback controller is chosen as a 1DF controller with the filter time constant tuned using the partial sensitivity function just as for a 2DF design. This controller is given by Eq. (10.11d). Equation (10.11d) also shows the outer loop controller that achieves an Mp of 1.05 for the complementary sensitivity function. Figures 10.20 and 10.21 show the disturbance and setpoint responses of the IMC cascade control system with models and controllers given by Equations (10.11c) and (10.11d).

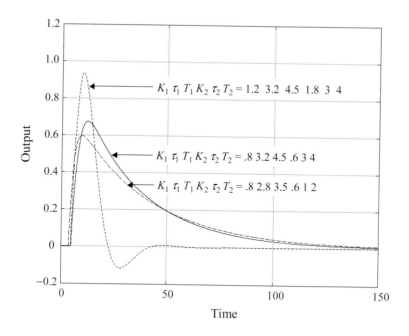

Figure 10.20 Unit step disturbance (d_2) responses for the IMC cascade control system.

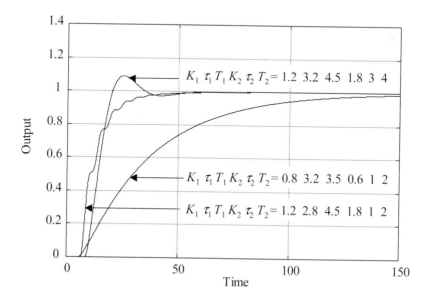

Figure 10.21 Unit step setpoint responses for the IMC cascade control system.

The responses in Figures 10.20 and 10.21 should be compared with those of a well-tuned single-loop control system for the process given by Equations (10.11a) and (10.11b) and rewritten as Equations (10.12a) and (10.12b). Equations (10.12c) and (10.12d) give the associated model and controller:

$$p(s) = \frac{K_1 K_2 e^{-Ts}}{(\tau_1 s + 1)(\tau_2 s + 1)} \ , \tag{10.12a}$$

where $\ .8 \le K_1 \le 1.2, .6 \le K_2 \le 1.8, \ 2.8 \le \tau_1 \le 3.2, \ 1 \le \tau_2 \le 3, \ 5.5 \le T \le 8.5,$

$$p_d(s) = \frac{K_1}{(\tau_1 s + 1)}, \tag{10.12b}$$

$$\widetilde{p}(s) = \frac{2.16 e^{-8.5s}}{(3.8s + 1)}, \tag{10.12c}$$

$$q(s) = \frac{(3.8s + 1)}{2.16(6.31s + 1)}. \tag{10.12d}$$

Notice that Eq. (10.12b) ignores the disturbance dead time since this term changes only the effective arrival time of the disturbance and so cannot be distinguished from the disturbance itself. Also, the model given by Eq. (10.12c) approximates the process lags as a first-order system whose time constant is the sum of the time constants of the two first order process lags. Finally, the controller given by Eq. (10.12d) is a 1DF controller because we are using a single loop controller in spite of the fact that the disturbance, d_2, enters into the primary output through the lag given by Eq. (10.12b).

The single-loop responses given in Figures 10.22 and 10.23 are roughly twice as slow as those of the cascade control system shown in Figures 10.20 and 10.21. Notice that the time scales in Figures 10.22 and 10.23 are from 0 to 300 whereas the time scales in Figures 10.20 and 10.21 are from 0 to 150. Also, the disturbance peak heights in Figure 10.22 are higher than those of the cascade control system in Figure 10.20.

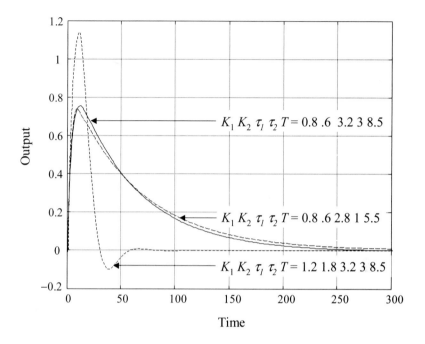

Figure 10.22 Single-loop control system of Eq. (10.12) responses to a step disturbance in d_2.

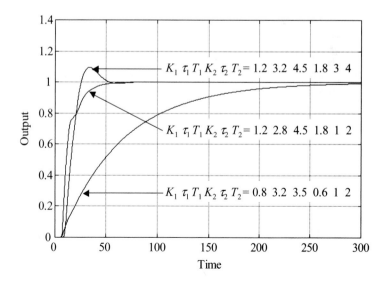

Figure 10.23 Step setpoint responses for the single-loop control system is given by Eq. (10.12).

(b) PID Cascade Controller Designs

Replacing the IMC inner loop with a feedback controller $q_2(s)/(1-\tilde{p}_2 q_2(s))$, as in Figure 10.6, and then approximating the feedback controller with the PID controller given by Eq. (10.13a) does not change the setpoint and disturbance responses of Figures 10.20 and 10.21. That is, there is no degradation of the performance of the mixed IMC-PID cascade control system.

$$PID_2(s) = .178(1+1/2.18s+.456s/(0.0228s+1)). \qquad (10.13a)$$

Approximating the controller C_1 by Eq. (10.3), multiplying it by the Maclaurin series approximation to $q_2^{-1}(s)$, and finally approximating the term $C_1 q_2^{-1}(s)$ as a PID controller, as in Example 10.1, gives

$$PID_1(s) = .485(1+1/8.0s+2.45s/(.122s+1)). \qquad (10.13b)$$

Figure 10.24 shows the inner loop disturbance d_2 response for the traditional cascade configuration of Figure 10.4, using the PID controllers given by Equations (10.13a) and (10.13b). The response for a process with upper-bound parameters is too oscillatory. The step setpoint response for the same process shows a 21% overshoot. The reason for this poorer behavior is probably the interaction between the inner and outer loops.

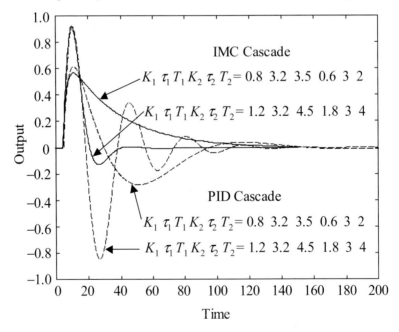

Figure 10.24 Comparison of responses to a step disturbance in the inner loop.

Conceptually, it is possible to extend the cascade feature of IMCTUNE to accommodate the PID cascade configuration of Figure 10.4 and to automatically increase the filter time constant of $q_1(s)$ to tune the outer loop to give a specified Mp. After tuning, IMCTUNE, or any program like it, should be able to provide the PID approximation to the term $C_1 q_2^{-1}(s)$. Unfortunately, such an extension does not yet exist, and the only method that we can suggest to improve the responses in Figure 10.24 is a rather tedious trial-and-error method wherein one increases the filter time constant of $q_1(s)$, re-computes PID$_1$, and then checks the responses of the processes with the upper-bound parameters. This assumes that the worst-case responses will always be those for the upper-bound parameters, which of course may not always be true.

◆

10.3 SATURATION COMPENSATION

10.3.1 IMC Cascade

Figure 10.5 provides the simplest starting point for a discussion of control effort saturation in cascade control systems. For convenience, this figure is reproduced in Figure 10.25.

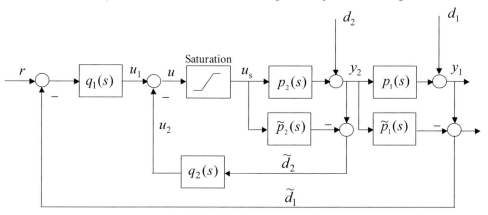

Figure 10.25 IMC cascade structure.

The effect of control effort saturation on the inner loop of Figure 10.25 can be minimized by implementing the inner loop as a model state feedback (MSF) system, as shown in Figure 10.26a.

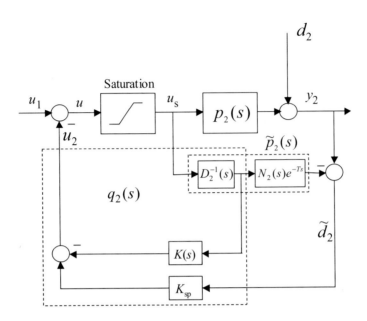

Figure 10.26a An MSF implementation of the inner loop of Figure 10.25.

The only difference between Figure 10.26a and Figure 5.5 of Chapter 5 is that there is no setpoint in Figure 10.26a.

Saturation compensation for the outer loop of Figure 10.25 is not quite so straightforward. One problem is that the outer loop controller is designed to invert portions of both inner loop and outer loop models (i.e., $\tilde{p}_1(s)\tilde{p}_2(s)$). However, there is no such transfer function, since the output of the inner loop model is *not* the input to the outer loop model. One solution is to create a new transfer function, $1/D_1(s)$, where $D_1(s)$ contains the denominator of the transfer function that the controller $q_1(s)$ inverts. Figure 10.26b shows an MSF implementation of the outer loop controller, $q_1(s)$, using this approach. This figure includes inner loop control system of Figure 10.26a, as it is necessary to remove the inner loop control effort u_2 from the signal used to compute the feedback portion of the outer loop control effort u_1.

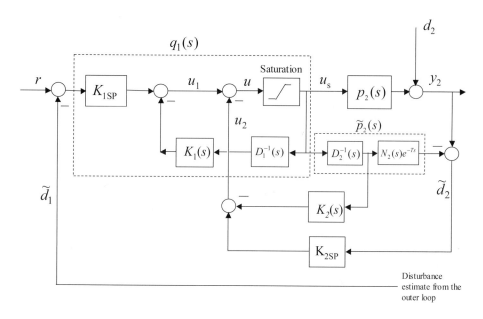

Figure 10.26b MSF implementation of both inner loop and outer loop controllers of Figure 10.25.

10.3.2 IMC/PID Cascade

In the absence of saturation there is usually little performance loss if the inner loop of the IMC cascade structure is replaced by a simple feedback loop, as shown in Figure 10.6. However, the method used in Figure 10.26b to compensate the outer loop for control effort saturation does not readily carry over to the outer loop of the cascade structure in Figure 10.6. The problem is that in the structure of Figure 10.6, there is no explicit calculation of an inner loop control effort, u_2, as there is in Figures 10.5 and 10.26b. For this reason, we recommend implementing the outer loop as shown in Figure 10.27. The limits of saturation block in this figure would ideally be set to the limits of the actual control effort less the contribution of the inner loop control effort u_2 to the total control effort. However, since u_2 is not available without additional calculations, we recommend simply setting the limits to those of the actual control effort. This is, of course, equivalent to assuming that u_2 is zero. Notice that the saturation block in Figure 10.27 is not on the outer loop control effort u_1 but rather only on the input to the inverse of the numerator of $q_1(s)$, which is $D_1(s)$. The reason is that the role of the structure in Figure 10.27 is only to attempt to compensate for saturation in the inner loop, and not to limit the setpoint sent to the inner loop. Finally, we recommend replacing the IMC controller, $q_2(s)/(1 - \tilde{p}_2 q_2(s))$, in Figure 10.6 with a standard anti-reset windup PID controller, as described in Chapter 6.

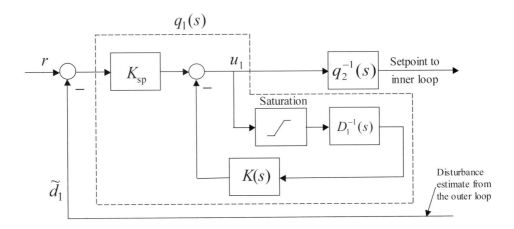

Figure 10.27 Compensating the outer loop of Figure 10.6 for control effort saturation.

10.3.3 PID Cascade

Saturation compensation for the standard PID cascade structure of Figure 10.4 is usually accomplished by either of two methods. The preferred method is to use logic statements that stop the integration in both the inner loop and outer loop PID controllers whenever the control effort reaches a limit, and start it again whenever the error signals change sign or the control effort comes off saturation. The second, and possibly more common, method is to use a standard anti-reset windup controller in the inner loop and implement the integral portion in the outer loop PID controller, as shown in Figure 10.28.

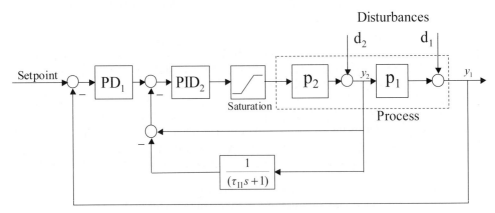

Figure 10.28 Feedback method of outer loop anti-reset windup for PID cascade.

The difficulty with the implementation of Figure 10.28 is that the outer loop integral time constant is not τ_{I1}, but rather a complicated function of τ_{I1} and the parameters of the inner loop transfer function. This complication can make it inadvisable to use the IMC-generated outer loop PID parameters developed in this section. The first method, which simply stops the integration on control effort saturation, does not have the foregoing drawback.

10.4 SUMMARY

To achieve the best disturbance rejection and setpoint tracking the inner loop of the cascade should be designed and tuned as a 2DF controller. The outer loop process lag plays the role of the disturbance lag in the controller design. The outer loop should be implemented as an MSF IMC system. The inner loop can be implemented as either a PID control system or, in the case of very little process uncertainty, in IMC MSF form.

There is no need for the inner loop process to be faster than the outer loop process in order for a well-designed cascade control system to provide significant performance advantages over a single-loop control system.

The techniques of this chapter can often be used to obtain the PID parameters for the traditional cascade structure. However, anti-reset windup for the outer loop should be implemented by stopping integration when the control effort saturates in order to use the calculated integral time constant. If the anti-reset windup for the outer loop is implemented via a lag around the inner loop, then the lag time constant is not necessarily the same as the computed integral time constant. Outer loop setpoint tracking and disturbance rejection is generally better than that achievable with a single-loop control system because the inner loop serves to reduce the apparent gain uncertainty of the inner loop process.

Problems

Design and tune cascade control systems for each of the following processes. The primary output is y, and the measured secondary output is y_2. Also, compare the performance your cascade control system with that of the feedforward control systems found for the problems of Chapter 9. The problems in Chapter 9 are the same as those below except that all the disturbances were considered to be measured whereas now only the process outputs $y(s)$ and $y_2(s)$ are measured.

10.1 $y(s) = \dfrac{e^{-10s}}{30s + 1} y_2(s) + d_1(s)$ $y_2(s) = \dfrac{e^{-5s}}{4s + 1} u(s) + d_2(s)$

10.2 $y(s) = \dfrac{K_1 e^{-Ts}}{5s+1} y_2(s) + d_1(s)$ $1 \le K_1 \le 3,\quad 4 \le T \le 6$

$y_2(s) = \dfrac{K_2 e^{-s}}{3s+1} u(s) + d_2(s)$ $1 \le K_2 \le 5,\quad 0 \le u(t) \le 10$

10.3 $y(s) = \dfrac{K_1 e^{-Ts}}{3s+1} y_2(s) + d(s)$ $1 \le K_1 \le 5,\quad 2 \le T \le 4$

$y_2(s) = \dfrac{K_2 e^{-s}}{2s+1} u(s) + d_2$ $1 \le K_2 \le 5$

10.4 $y(s) = \dfrac{K_1 e^{-Ts}}{3s+1} y_1(s) + \dfrac{e^{-Ts} d_1(s)}{3s+1}$ $1 \le K \le 5,\quad 2 \le T \le 4$

$y_1(s) = \dfrac{K_2(-2s+1)}{(2s+1)^2} u(s) + \dfrac{d_2}{s+1}$

References

Morari, M., and E. Zafiriou, E. 1989. *Robust Process Control*. Prentice Hall, NJ.

Seborg, D. E., T. F. Edgar, and D. A. Mellichamp. 1989. *Process Dynamics and Control*. John Wiley & Sons, NY.

Output Constraint Control (Override Control)

Objectives of the Chapter

- To review classical override control.
- To present an alternate way of thinking about override control that more easily generalizes to multivariable control.
- To show how the material in Chapters 3 through 8 can be used to advantage in designing override control systems that accommodate process uncertainty.

Prerequisite Reading

Chapter 3, "One-Degree of Freedom Internal Model Control"
Chapter 4, "Two-Degree of Freedom Internal Model Control"
Chapter 5, "MSF Implementations of IMC"
Chapter 6, "PI and PID Controller Parameters from IMC Designs"
Chapter 7, "Tuning and Synthesis of 1DF Controllers for Uncertain Processes"
Chapter 8, "Tuning and Synthesis of 2DF Controllers for Uncertain Processes"

11.1 INTRODUCTION

It often occurs that the aim of the control system is to both control several process variables to setpoints while at the same time assuring that other process variables do not exceed specified limits. In general, the number of control efforts is less than the sum of the number of process variables to be held at setpoints plus the number of process variables that must not exceed limits. Typically the term *override control* is used when there is only one control effort to control one process variable to its setpoint while maintaining another process variable between its limits. The discussion in this section focuses on PID and IMC override control systems and a new cascade type constraint controller. Unlike the PID and IMC override controllers, the cascade constraint controller readily extends to general, multivariable, constraint control problems. It can also deal with more general processes than the overdamped processes to which override control is generally applied. However, in this chapter, the cascade constraint controller is used only to treat the same overdamped processes treated by an override control system.

11.2 OVERRIDE AND CASCADE CONSTRAINT CONTROL STRUCTURES

PID and IMC override control systems employ two parallel controllers plus logic to control one variable to a setpoint and keep another below a constraint, as shown in Figures 11.1 and 11.2.

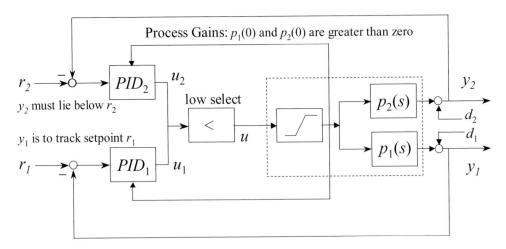

Figure 11.1 PID override control system (y_2 is the constrained output).

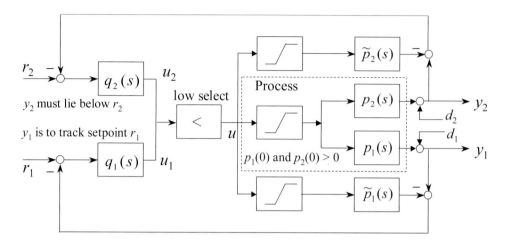

Figure 11.2 IMC override control system (y_2 is the constrained output).

In Figures 11.1 and 11.2 the constrained variable is y_2, and it is constrained to lie at or beneath r_2. The setpoint variable is y_1, and its setpoint is r_1. The steady-state gains between the control effort and each output are assumed to be positive. Each control loop acts to keep its output at its setpoint. That is, PID$_1$ acts to keep y_1 at r_1. PID$_2$ acts to keep y_2 at r_2. The logic block compares the control efforts u_1 and u_2 computed by PID$_1$ and PID$_2$. By selecting the smaller of the two control efforts, the logic block attempts to maintain the output y_1 at its setpoint r_1 while also assuring that output y_2 will not violate its constraint. If the control effort u_1 is greater than the control effort u_2, then the assumption is that it will cause output y_2 to violate its constraint. In this case, the logic selects control effort u_2 to maintain y_2 at its constraint. Of course, y_1 will no longer track its setpoint. Whenever the control effort u_2 drops below u_1, due perhaps to a decrease in disturbance d_2 and/or the setpoint r_1, then the logic block will select control effort u_1 and the output y_1 will again eventually track its setpoint. As we shall show by example, this logic works quite well when the process is overdamped (e.g., can be approximated by a first-order lag plus dead time model). However, it is not clear whether the same logic will work well when either or both processes are underdamped, nonminimum phase, or unstable.

The controllers PID$_1$ and PID$_2$ as well as the controllers IMC$_1$ and IMC$_2$ of Figures 11.1 and 11.2 can be designed independently as 1DF, or even 2DF controllers, following the procedures in Chapters 3 through 8. Note, however, that the selected control effort is used in calculating the outputs from both controllers. This is essential because, in general, both controllers use integral action and must therefore be implemented as anti-reset windup controllers. If the anti-reset windup feature is implemented by positive feedback around a lag whose time constant is the integral time constant, then the input to this lag must be the selected control effort, and not the control effort actually calculated by the respective PID controller. This prevents the inactive control loop from winding up, even though its error signal is not zero. The IMC controllers of Figure 11.2 also will not exhibit windup provided that the input to both process models is the selected control effort, as shown.

Generally, the IMC override control system would not be used unless the process uncertainty is small enough so that the IMC controllers have a significant performance advantage over the PID control system. In such a case the IMC implementation should be via MSF for both controllers, as described in Chapter 5. Again, there is nothing special in the implementations except that the selected control is sent to both models.

11.3 CASCADE CONSTRAINT CONTROL

The cascade constraint control system of Figure 11.3 differs from the PID and IMC override control system in that it does not use a logic block to achieve its control objectives. Rather, there is an inner loop that assures that the constrained variable does not violate its limits and an outer loop that controls the process variable to its setpoint. Constraint violations are prevented by restricting the setpoint to the inner loop to the limits of the constrained variable.

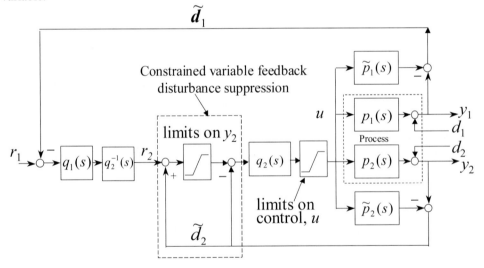

Figure 11.3 Cascade constraint control system.

A unique feature of the cascade constraint control structure is that the inner loop is open when the constrained variable is within its limits. That is, there is no feedback when the constrained variable is within its limits. This is accomplished by having the disturbance estimate added to the setpoint provided by the outer loop, and then subtracted from the output of the limiter, as in a standard IMC configuration. When the sum of these two signals is between the limits, then the signal entering the inner loop controller, $q_2(s)$, is the setpoint signal r_2, and the control effort computed by the inner loop controller is the same as that computed by the outer loop controller $q_1(s)$. In this case, only the outer loop is active. When the sum of the two signals entering the y_2 limiter is outside the limits, then the inner loop is active and the constrained variable is controlled to maintain its limit, and only the inner loop is really under feedback control. Thus the two controllers $q_1(s)$ and $q_2(s)$ can be designed and tuned independently. Also, since there is no logic block, the particular mathematical description of the two processes should not cause any controller design problems beyond those normally encountered for single-loop systems.

The following problem will be solved using all three methods discussed previously. The process is

$$p_i(s) = K_i e^{-T_i s} / (\tau_i s + 1) ; \quad i = 1,2 ; \quad .8 \le K_1 \le 1.2, \quad 14 \le \tau_1 \le 16, \quad 17.5 \le T_1 \le 22.5$$

$$.6 \le K_2 \le 1.8, \quad 1 \le \tau_2 \le 3, \quad 2 \le T_2 \le 4. \tag{11.1}$$

The control objectives are $y_1(t) = 1$, $y_2(t) \le 1.6$. The control effort is limited between zero and 2. The process starts from $y_1(t) = y_2(t) = 0$, and there is a unit step disturbance to y_1 at time = 125.

Example 11.1 PID Override Control

Figure 11.4a gives the SIMULINK implementation of a PID override controller for the above problem. Figure 11.4b shows the details of some of the blocks in Figure 11.4a.

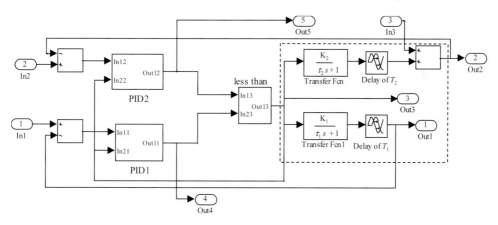

Figure 11.4a SIMULINK diagram for a PID override control system for Eq. (11.1).

In Figure 11.4a, inputs 1 and 2 (i.e., In1 and In2) are steps set point changes of 1 and 1.6 at time zero. Input 3 is a unit step disturbance at time 125. Outputs 1 and 2 (i.e., Out1 and Out2) are y_1 and y_2. Output 3 is the selected control effort. Outputs 4 and 5 are the controls computed by q_1 and q_2.

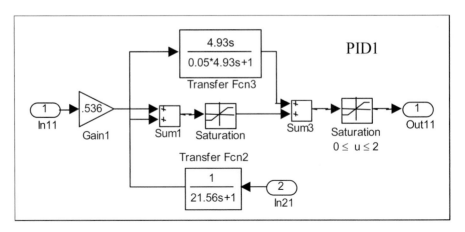

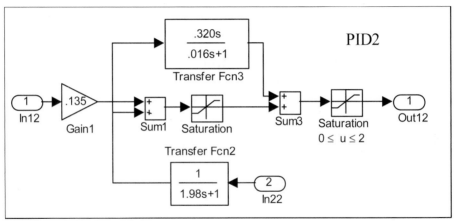

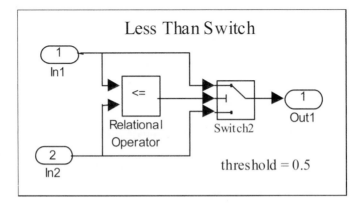

Figure 11.4b Details of selected blocks in Figure 11.4a.

The PID parameters in Figures 11.4a and b were obtained from 1DF IMC controllers using the following model parameters and IMC controllers:

Model Parameters: $\tilde{K}_1 = 1.2,\ \tilde{\tau}_1 = 14,\ \tilde{T}_1 = 22.5;\quad \tilde{K}_2 = 1.8,\ \tilde{\tau}_2 = 1,\ \tilde{T}_2 = 4.$ (11.2a)

IMC Controllers: $q_1(s) = (14s+1)/1.2(11.0s+1),\quad q_2(s) = (s+1)/1.8(4.17s+1).$ (11.2b)

The IMC filter time constants of 11.0 and 4.17 were obtained from IMCTUNE. The associated PID controllers are

$$PID_1 = .536(1 + 1/(21.6s) + 4.93s/(.247s+1))$$ (11.2c)

$$PID_2 = .135(1 + 1/(1.982) + .320s/(.016s+1))$$ (11.2d)

Figures 11.5 and 11.6a and b show the output and control effort responses generated by the SIMULINK diagram of Figure 11.4.

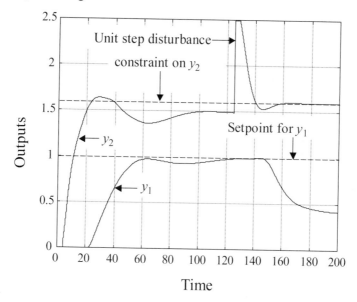

Figure 11.5 Output responses to a unit step setpoint change in y_1 at time 0 and a unit step disturbance in y_2 at time = 125, generated by Figure 11.4.

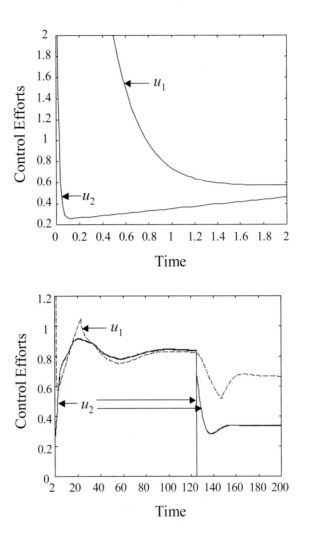

Figure 11.6a Control efforts generated by PID$_1$ and PID$_2$ of Figure 11.4.

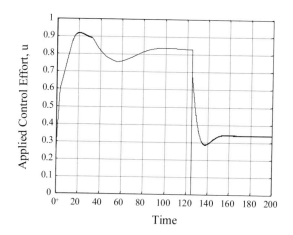

Figure 11.6b Applied control effort that generates the time responses of Figure 11.5.

♦

Example 11.2 IMC Override Control

Figure 11.7 shows the IMC MSF SIMULINK diagram for Equations (11.1) and (11.2).

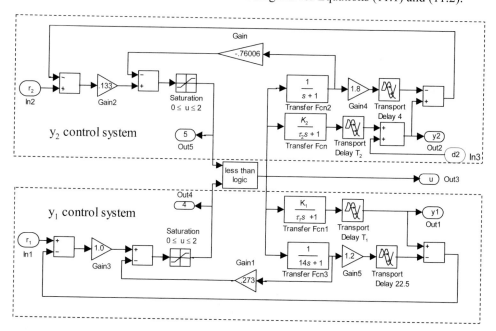

Figure 11.7 IMC MSF SIMULINK diagram for Equations (11.1) and (11.2).

As before, in Figure 11.7, inputs 1 and 2 (i.e., In1 and In2) are steps set point changes of 1 and 1.6 at time zero. Input 3 is a unit step disturbance at time 125. Outputs 1 and 2 (i.e., Out1 and Out2) are y_1 and y_2. Output 3 is the selected control effort. Outputs 4 and 5 are the controls computed by q_1 and q_2.

Figures 11.8 and 11.9 show the output and control effort responses generated by the SIMULINK diagram of Figure 11.7.

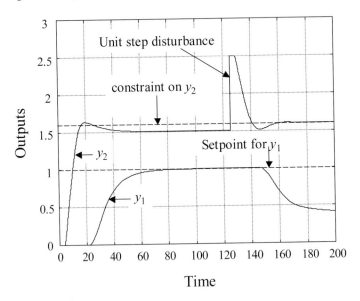

Figure 11.8 Output responses to a unit step setpoint change in y_1 and a unit step disturbance in y_2 at time = 125, generated by Figure 11.7.

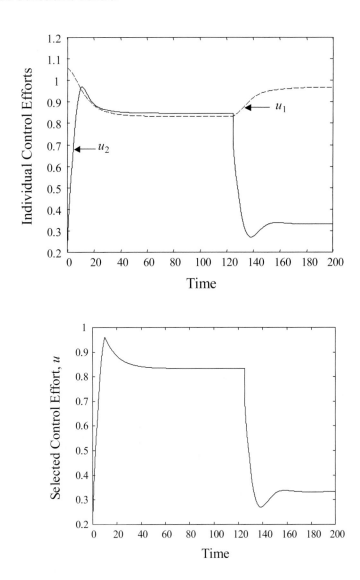

Figure 11.9 Control effort responses generated by Figure 11.7.

The IMC override control responses in Figures 11.8 and 11.9 are not substantially different from the PID override control responses of Figures 11.5 and 11.6.

♦

Example 11.3 Cascade Constraint Control

Figure 11.10 gives the SIMULINK diagram for the cascade constraint control diagram associated with Figure 11.3.

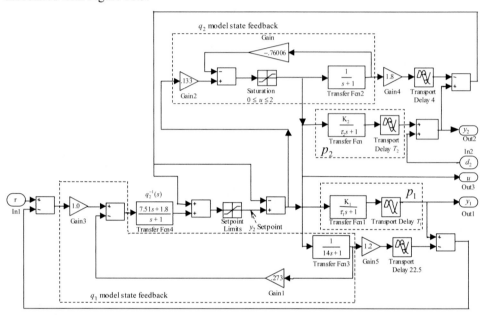

Figure 11.10 SIMULINK diagram associated with Figure 11.3 for MSF implementation of IMC cascade constraint control. Input 1, r, is a unit step at time zero; input 2, d_2, is a unit step at time 125.

The output and control effort responses generated by the SIMULINK diagram in Figure 11.10 are identical to those of Figures 11.8 and 11.9. Thus, for our example problem, there is effectively no performance difference between PID and IMC override control and IMC cascade constraint control. There is also a PID version of cascade constraint control (Magee, 2001), and for the current example, we anticipate the same performance as above. The choice between override (PID or IMC) and cascade constraint (PID or IMC) control is therefore a matter of taste and/or familiarity for simple overdamped processes. As might be expected, we lean towards the cascade constraint control method because of its potentially greater generality and because limiting the setpoint(s) to the control loop for the constrained output variable(s) is conceptually simpler than override logic.

♦

11.4 SUMMARY

The parameters of a PID override control system can be obtained as described in Chapters 6, 7, and 8.

A new cascade constraint control system offers improved control without the need for a logic block. It may therefore apply to processes whose characteristics do not permit application of PID or IMC override control. Extension of the cascade constraint control system to multivariable processes is also much more straightforward (Boyce and Brosilow, 1996).

Problems

11.1 Under what conditions, if any, would you use the following control systems or controller implementations?

    a. Single-degree of freedom PID control
    b. Two-degree of freedom PID control
    c. Lead-lag implementation of IMC
    d. MSF implementation of IMC
    e. Feedforward Control
    f. Cascade Control
    g. Override Control

11.2 Select and tune control systems for each of the following processes. Output y_1 is controlled to a setpoint.

    a. $\quad y(s) = \dfrac{e^{-10s}}{30s + 1}\, u(s) + d_1(s) \quad y_2(s) = \dfrac{e^{-5s}}{4s + 1}\, u(s) + d_2(s) \quad -1 \le y_2 \le 2$

    b. $\quad y_1(s) = \dfrac{K_1 e^{-Ts}}{5s + 1} u(s) + d_1(s) \qquad 1 \le K_1 \le 3, \quad 4 \le T \le 6$

    $\quad y_2(s) = \dfrac{K_2 e^{-s}}{3s + 1} u(s) + d_2(s) \qquad 1 \le K_2 \le 5, \quad 0 \le u(t) \le 10, \quad 0 \le y_2 \le 2$

    c. $\quad y_1(s) = \dfrac{K_1 e^{-Ts}}{3s + 1} u(s) + d_1(s) \qquad 1 \le K_1 \le 5; \quad 2 \le T \le 4$

    $\quad y_2(s) = \dfrac{K_2 e^{-s}}{2s + 1} u(s) + d_2 \quad 1 \le K_2 \le 5 \quad -2 \le y_2$

d. $y(s) = \dfrac{K_1 e^{-Ts}}{3s+1} y_1(s) + \dfrac{e^{-Ts} d_1(s)}{3s+1}$ $1 \leq K \leq 5$ $2 \leq T \leq 4$

$y_1(s) = \dfrac{K_2(-2s+1)}{(2s+1)^2} u(s) + \dfrac{d_2}{s+1}$ $0 \leq y_2 \leq 2$

References

Boyce, J., and C. Brosilow. 1996. "Multivariable Cascade Control for Processes With Output Constraints." Proceedings of European Symposium on Computer Aided Process Engineering-6 (ESCAPE-6).

Magee, L. R. 2001. "Multivariable Cascade Control of Processes with Output Constraints." MS Thesis, Case Western Reserve University.

Single Variable Inferential Control (IC)

Objectives of the Chapter

The purpose of the next two chapters is to present techniques of process control when the primary variable to be controlled is not directly measured or if a significant time lag is associated with this measurement that is detrimental to the control system. We will study control using secondary measurements that carry information about disturbances affecting the primary output variable. Such control is referred to as inferential control (IC) since we are inferring something about the unmeasured variable from the secondary measurements. Issues addressed in this chapter include

- The concept of inferential control.
- Structure and analysis of control using one secondary measurement.
- Inferential control versus classical control strategies.
- Application of inferential control to a distillation column control.

Prerequisite Reading

Chapter 3, "One-Degree of Freedom Internal Model Control"
Chapter 4, "Two-Degree of Freedom Internal Model Control"

12.1 INTRODUCTION

So far, we assumed that the primary variable to be controlled can be measured for feedback correction. There exist a number of processes in which the primary variable to be controlled is difficult to measure or is a sampled measurement with long delay in the sampling and analysis process. In other instances the quantity to be controlled is a calculated variable. In such cases control of the process is usually accomplished by measuring secondary variables (for which sensors are more reliable, cheaper, or more readily available and installed) and setting up a feedback control system using these secondary variables. We define inferential control (IC) as control of a primary, often unmeasured, variable using one or more secondary measurements on the process. A few examples of IC applications are given below.

1. Distillation. In distillation, the primary variables to be regulated are the product compositions (bottoms and distillate purity). Composition analyzers such as gas chromatographs (GC's) are typically used to measure these. Unfortunately, these analyzers are expensive and difficult to calibrate and maintain. They also introduce significant measurement delays because of the time needed to purge the sample line and to heat the sample. Delays of the order of 3 to 15 minutes can be expected. In this case control is often accomplished using temperature measurement on an intermediate tray (see Figure 12.1). Corrections are then made to the temperature setpoint as the product purity measurements become available.

2. Polymerization Reactor. In polymerization reactors primary variables of interest in control are the molecular weight distribution and viscosity of the product. Both quantities are difficult to measure. Control is then accomplished using the secondary measurements, such as temperature and pressure in the reactor.

3. Drying. In drying operations the primary variable is the moisture content of the dried product. Again, it is difficult to sense this quantity using online sensors. Control is accomplished by measuring humidity of the drier exhaust air (dry and wet bulb temperatures), temperature of the solids, and inlet temperatures.

4. Paper-making. In paper-making machines the brightness of the product and the moisture content are of interest in control. However control must usually be accomplished by secondary measurements such as temperature and reflectance.

5. Naphtha Cracking. Naphtha (a product of petroleum crude distillation) is cracked to form high octane products needed in gasoline. The reaction is carried out in large catalytic reactors. The variable of interest here is the octane number of the product. Since this cannot be measured directly, control is accomplished using other variables, such as reactor temperature and reactor pressure.

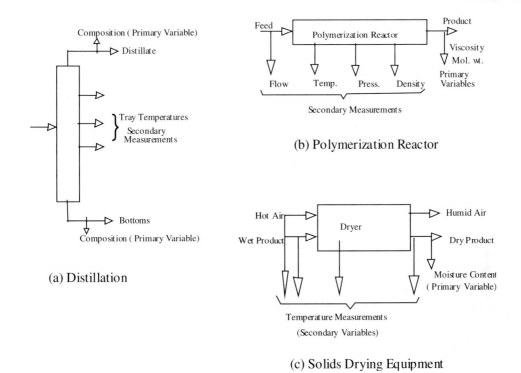

Figure 12.1 Processes where inferential control is needed.

Figure 12.2 shows the general structure of an IC system. The control objective is to keep the primary variable on target in the presence of input disturbances. The objective of this chapter is to develop a design strategy for control using the secondary measurements and process models or IC. Before we get into the IC, let us look at some classical techniques used under similar circumstances which, while being simple to implement, can be costly because of poor controller performance.

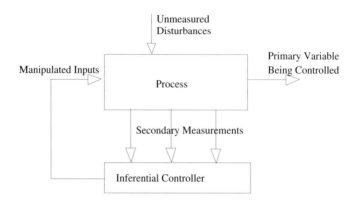

Figure 12.2 IC structure.

12.2 CLASSICAL CONTROL STRATEGIES

12.2.1 Feedback Control of Secondary Variable

This strategy can be used if the primary variable and secondary variable are very closely related. For example, in distillation it is well known that temperature is a very good indicator of product composition. Hence, by maintaining one of the tray temperatures at its setpoint, we can often maintain good control of the product quality.

Consider the simple case of a single disturbance, single primary output and single secondary measurement (we will treat more complicated cases in Chapter 13). The process can be modeled in this case as

$$y(s) = p_y(s) \; u(s) \; + \; p_{dy}(s) \; d(s) : \text{primary output,}$$

$$x(s) = p_x(s) \; u(s) \; + \; p_{dx} \; d(s) : \text{ secondary measurement.} \tag{12.1}$$

The block diagram structure of this model is shown in Figure 12.3.

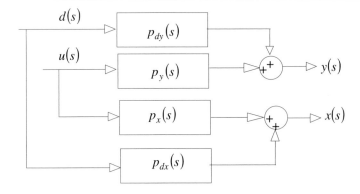

Figure 12.3 Block diagram of a simple process with one primary and one secondary measurement.

Consider direct feedback control of the secondary measurement $x(s)$ using the manipulated variable $u(s)$, as shown in Figure 12.4. For convenience, we drop the Laplace variable s from the following transfer functions.

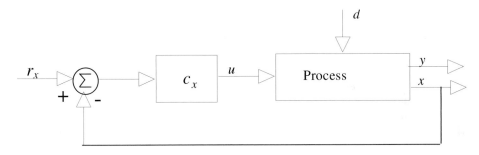

Figure 12.4 Structure of direct feedback control using secondary variable.

Using block diagram reduction techniques we can derive the transfer function for the transfer function between the disturbance and the primary output $y(t)$ as

$$\frac{y}{d} = p_{dx} - \frac{p_y c_x p_{dx}}{1 + p_x c_x}. \tag{12.2}$$

If the controller c_x is integrating, then the steady-state gain of y/d is:

$$\frac{y}{d}(0) = \left(p_{dy}(0) - \frac{p_{dx}p_y(0)}{p_x(0)} \right).$$ (12.3)

The open loop transfer function between the primary output and the disturbance is $p_{dy}(s)$. Thus there is a potential for improvement if the gains of the other process transfer functions are such that

$$\frac{p_{dx}p_y(0)}{p_x(0)} \approx p_{dy}(0)$$ (12.4)

If in addition to the above, the dynamics of $p_{dx}(s)$ and $p_x(s)$ are fast (relative to those of $p_{dy}(s)$ and $p_y(s)$), and $p_{dy}(s)$ and $p_y(s)$ are relatively of the same speed, then feedback control of the secondary measured variable can substantially reduce the impact of disturbances on the desired output. Of course, in general, there will be an offset in the primary output unless the setpoint of the secondary variable is changed to bring the primary variable to its setpoint. This can be done via a cascade control system as described in the next section.

Example 12.1 Composition Control Using Temperature

In a binary distillation column, the temperature and composition have a one-to-one relationship if pressure is fixed. Hence in the steady-state

$$y = \bar{y} + a(T - \bar{T}),$$

where the bar denotes the nominal steady-state value. So, $T = \bar{T}$ and $y = \bar{y}$ in the steady-state. However, if pressure is also variable, then

$$y = \bar{y} + a(T - \bar{T}) + b(P - \bar{P}).$$

In this case, feedback control of T alone is not sufficient. We can keep y constant $(y = \bar{y})$ by setting

$$\bar{y} = \bar{y} + a(T - \bar{T}) + b(P - \bar{P})$$

or

$$T = \bar{T} + \frac{b}{a}(P - \bar{P}).$$

This strategy can be implemented as shown in Figure 12.5. If the distillation is multi-component or if dynamics of T and y are different, the above strategy may not be satisfactory.

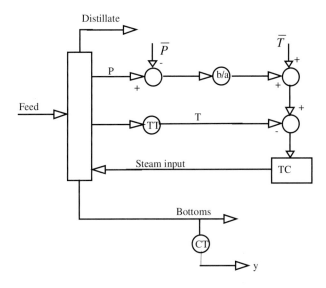

Figure 12.5 Feedback control of temperature with pressure compensation in a distillation column.

♦

12.2.2 Cascade Control on the Secondary Variable

We introduced the classical control strategy in Chapter 10. This strategy is shown in Figure 12.6.

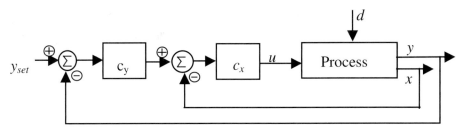

Figure 12.6 Structure of cascade control system.

The inner loop tries to maintain the secondary variable at its setpoint. The outer loop adjusts this setpoint to bring the primary output $y(s)$ back to its setpoint. To implement this strategy, we must have a measurement of the primary variable. Normally, this strategy is shown with a transfer function relating $x(s)$ and $u(s)$ in series with another transfer function relating $x(s)$ and $y(s)$. However, we are interested in a different scenario in which we do not have a direct way to break up the process in the series format. However, the situation is similar enough to think of using a cascade control strategy as discussed in Chapter 10. In the classical implementation, the inner loop should be able to react fast enough to follow constant setpoint changes. If there are significant lags in the inner loop, then the system will not have enough time to settle down, and control system performance will be poor.

The disturbance rejection transfer function for Figure 12.6 is given by (after some algebra)

$$\frac{y}{d}(s) = \frac{\left(p_{dy}(s) - \dfrac{p_{dx}(s)p_y(s)c_x(s)}{(1+c_x(s)p_x(s))} \right)}{\left(1 + \dfrac{c_y(s)c_x(s)p_y(s)}{(1+c_x(s)p_x(s))} \right)}. \tag{12.5}$$

This strategy has the advantage that steady-state error in control of $y(s)$ will be reduced to zero if we use integral action on the outer-loop controller.

The above strategy has no easy extension to the case of multiple secondary measurements. Usually, multiple secondary measurements contain more information about the state of the system. Thus methods that use multiple measurements have an advantage over these multiloop strategies.

12.2.3 Feedforward Control

If *disturbances can be measured* then a feedforward control strategy can and should be used in parallel with the feedback control strategies as shown in Figure 12.7. The feedforward controller output is usually added to the output of the feedback controller in its implementation. The methodology for designing feedforward controllers is covered in Chapter 9. Here we show the simplest form of a feedforward implementation.

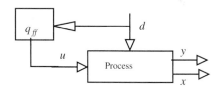

Figure 12.7 Structure of feedforward control.

12.2.4 Feedback Control Using an Output Estimator for *y*

In this strategy we first attempt to build an estimator for the unmeasured output *y*, and then use this estimated value in a feedback control system. If perfect estimation is possible, then this reduces to the classical direct feedback control strategy. Figure 12.8 shows the structure and block diagram of this control strategy.

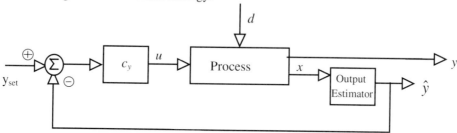

Figure 12.8 Structure of the controller using an output estimator.

If we have a perfect estimator for $y(s)$, then control using the estimator is the same as direct feedback control of *y*. The disturbance rejection transfer function in this case is given by

$$\frac{y(s)}{d(s)} = \frac{P_{dy}}{1 + P_y \cdot c_y}.$$

(12.6)

The advantage with this approach is that an estimate of the unmeasured state $y(s)$ is available through the secondary measurement and this estimate can be displayed to the plant operator(s). This can improve the operator's comfort with the control system, and thereby indirectly improve plant operation. However, if p_{dy} and p_y have large lags, then output estimation can be no better than direct feedback control whose performance is limited by the time lags and time delays present in the feedback loop. When $x(s)$ responds faster to the disturbance and manipulated variable than $y(s)$ then IC is a better option, as we show in the Shell distillation Example 12.2.

12.3 INFERENTIAL CONTROL

12.3.1 Structure

Inferential uses secondary measurements to simultaneously *infer and counter* the effect disturbances on the primary output variable. This approach is fundamentally different from the use estimation techniques to estimate the unmeasured states of a system and then using these estimated states in a feedback control system. Since the effect of the disturbance on the measured secondary variable is frequently faster than its effect on the output of interest, one can often achieve faster disturbance rejection with IC than with the feedback of the estimated output. Later in this section we shall present some comparisons between the different control approaches, using distillation column and reactor examples.

IC can be viewed as an extension of 2DF IMC control system of Figure 4.3, which is repeated below for convenience. In this structure the feedback controller $qq_d(s,\varepsilon)$ computes the feedback portion of the control effort using an estimate $\tilde{d}_e$ of the disturbance effect on the measured output y.

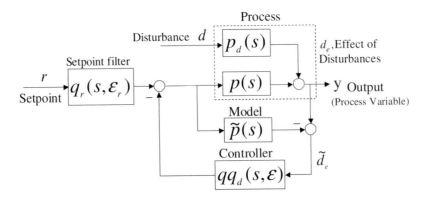

Figure 4.3 The two-degree of freedom internal model controller.

In IC, the controller uses the measured output x to obtain a predicted or delayed estimate of the disturbance effect $\tilde{d}_{ey}^*$ on the unmeasured (or infrequently measured) output of interest y as shown in Figure 12.9 for the process described by Eq. (12.1) (repeated below for convenience). The process has only one unmeasured disturbance, one control effort $u(t)$ a measured output, $x(t)$, and an unmeasured output $y(t)$ that we wish to control.

$$y(s) = p_y u(s) + p_{dy} d(s)$$
$$x(s) = p_x u(s) + p_{dx} d(s).$$
(12.7)

The inferential controller $q_I(s,\varepsilon)$ of Figure 12.9 uses the estimated effect of the disturbance $\tilde{d}_{ey}^*(t)$ on the measured output $x(t)$ to adjust the control effort $u(t)$ to counter the predicted or delayed estimate of the effect of the disturbance $\tilde{d}_{ey}^*(t)$ on the desired output $y(t)$. The setpoint filter $q_r(s,\varepsilon_r)$ adjusts the control effort so that the output $y(t)$ will track the desired setpoint.

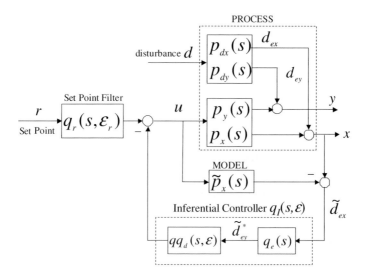

Figure 12.9. Structure of IC of the process of Eq. (12.1).

12.3.2 Design of the Inferential Controller for Perfect Models

The inferential controller, $q_I(s,\varepsilon)$, of Figure 12.9 is the product of three terms $q(s)$, $q_d(s)$, and $q_e(s)$. The design of the controllers $q(s)$ and $q_d(s)$ proceeds in the same manner as the design and tuning of the controllers, $q(s)$, and $q_d(s)$ of the two-degree of freedom IMC control system of Figure 4.2. Similarly, the design of the setpoint controller $q_r(s,\varepsilon_r)$ is the same for both inferential and two-degree of freedom controllers. The special feature of an inferential controller is the design of the estimator-predictor $q_e(s)$. Before embarking on the selection of $q_e(s)$, let us review briefly the design of $q(s)$, $q_d(s)$, and $q_r(s)$.

The perfect model response of $y(s)$ to setpoint and disturbances is:

$$y(s) = p_y(s)q_r(s,\varepsilon_r)r(s) + (1 - p_y p_{dy}^{-1} p_{dx}(s)q_I(s,\varepsilon))p_{dy}(s)d(s), \qquad (12.8)$$

where: $\qquad q_I(s,\varepsilon) \equiv q(s,\varepsilon)q_d(s,\varepsilon)q_e(s). \qquad (12.9)$

According to the design procedure suggested in Chapter 4, both $q_r(s,\varepsilon_r)$, and $q(s,\varepsilon)$ are selected to invert a portion (perhaps all) of the invertible part of the model $p_y(s)$ but possibly with different filter time constants. The portion of the controller labeled $q_d(s,\varepsilon)$ is selected as either unity, or as the ratio of polynomials given by

$$q_d(s, \varepsilon, \alpha) \equiv \frac{\sum_{i=0}^{n} \alpha_i s^i}{(\varepsilon s + 1)^n} \tag{12.10}$$

where the constants α and n are chosen so that the zeros of the term in brackets in Eq. (12.8) (i.e., $(1 - [p_y q(s, \varepsilon) p_{dy}^{-1} p_{dx}(s) q_e(s, \varepsilon)] q_d(s, \varepsilon))$) cancel selected poles in $p_{dy}(s)$. If the term $(1 - [p_y q(s, \varepsilon) p_{dy}^{-1} p_{dx}(s) q_e(s, \varepsilon)])$ has fast dynamics relative to those of $p_{dy}(s)$, then $q_d(s, \varepsilon)$ is usually chosen to be unity as there is little or no advantage in canceling poles in $p_{dy}(s)$.

From the way that we have grouped the transfer functions in the term $(1 - [p_y q(s, \varepsilon) p_{dy}^{-1} p_{dx}(s) q_e(s, \varepsilon)])$, the reader may have guessed that we choose $q_e(s)$ to invert the invertible part of $p_{dy}^{-1} p_{dx}(s)$, and that is indeed the case. However, we also wish to interpret $q_e(s)$ as the estimator part of the inferential controller, $q_I(s, \varepsilon)$. To do so we note that the effect of the disturbance, $d(s)$, on the outputs $x(s)$ and $y(s)$ is given by:

$$\tilde{d}_{ex}(s) = d_{ex}(s) \equiv p_{dx}(s) d(s) = \tilde{p}_{dx}(s) d(s) \tag{12.11}$$

$$\tilde{d}_{ey}(s) = d_{ey}(s) \equiv p_{dy}(s) d(s) = \tilde{p}_{dy}(s) d(s) \tag{12.12}$$

Therefore,

$$\tilde{d}_{ey}(s) = d_{ey}(s) \equiv p_{dy}(s) p_{dx}^{-1}(s) d_{ex}(s) = \tilde{p}_{dy}(s) \tilde{p}_{dx}^{-1}(s) \tilde{d}_{ex}(s) \tag{12.13}$$

The estimate given by Eq. (12.13) can be realized if the term $\tilde{p}_{dy}(s) \tilde{p}_{dx}^{-1}(s)$ is realizable. This requires that the dead time in $p_{dx}(s)$ be equal to or smaller than the dead time in $p_{dy}(s)$, that $p_{dx}(s)$ have no right half plane zeros, and that the relative order of $p_{dy}(s)$ be equal to or greater than that of $p_{dx}(s)$. If the foregoing conditions are not met, then only a delayed or approximate estimate of $d_{ey}(s)$ can be obtained, and this estimate is given by:

$$\tilde{d}_{ey}^*(s) = (\tilde{p}_{dy}(s) \tilde{p}_{dx}^{-1}(s))_m f(s, \varepsilon) \tilde{d}_{ex}(s) \equiv q_e(s) \tilde{d}_{ex}(s), \tag{12.14}$$

where the subscript denotes taking only the realizable and stable portion of the term in brackets. The filter $f(s, \varepsilon)$ has the minimum order required to make $q_e(s)$ realizable. In the case that the relative order of $\tilde{p}_{dy}(s) \tilde{p}_{dx}^{-1}(s)$ is zero or higher, the filter in Eq. (12.14) is unity, and the order of the filter in $q(s)$ is reduced as necessary so that the relative order of $q_I(s, \varepsilon)$ is zero.

There is yet another important case to consider. It often occurs that the dead time in $p_{dy}(s)$ is larger than that in $p_{dx}(s)$. In this case $d_{ey}(s)$ given by Eq. (12.14) contains a dead time, and such a dead time should not be realized. Rather, it is preferable to obtain a prediction of the effect of the disturbance on the desired output so that the control system can suppress the effect of the disturbance on the output more rapidly. To obtain such a prediction, all that is needed is to drop any dead time in $\tilde{p}_{dy}(s)\tilde{p}_{dx}^{-1}(s)$. That is, we choose $q_e(s)$ so that it has no multiplicative dead times, is stable, and has relative order of zero or higher as given below

$$q_e(s) = (\tilde{p}_{dy}(s)\tilde{p}_{dx}^{-1}(s))_m f(s,\varepsilon). \tag{12.15}$$

The following two examples illustrate the design procedures outlined above. In the Shell fractionator example, the dynamics of the secondary measurement are much faster than that of the primary output, and we see the advantage of IC over that of feedback control of the primary output. The reactor example shows the advantage of choosing $q_d(s,\varepsilon)$ according to Eq. (12.10) when the process dead time is greater than the disturbance dead time.

Example 12.2 IC of a Distillation Column

Consider the Shell fractionator system (Prett et al. 1987, 1988) of Figure 12.10. The process model is

$$y(s) = \frac{4.05e^{-27s}}{50s+1}u(s) + \frac{1.44e^{-27s}}{40s+1}d(s) : \text{top end point}$$

$$x(s) = \frac{3.66e^{-2s}}{9s+1}u(s) + \frac{1.27}{6s+1}d(s) \quad : \text{top temperature} \tag{12.16}$$

Here $u(s)$ is the top draw rate and $d(s)$ is the change in heat duty of the top intercooler. Note that this secondary variable responds faster to both the manipulated variable and the disturbance variable. From the process model given in Eq. (12.16),

$$p_y p_{dy}^{-1} p_{dx}(s) = \frac{3.57(40s+1)}{(50s+1)(6s+1)}. \tag{12.17}$$

The above is completely invertible, and therefore we can choose $q_I(s,\varepsilon)$ as,

$$q_I(s,\varepsilon) = \frac{.280(50s+1)(6s+1)}{(40s+1)(\varepsilon s+1)}. \tag{12.18}$$

A filter time constant of .75 (i.e., $\varepsilon = .75$) yields a noise amplification ratio of 10.

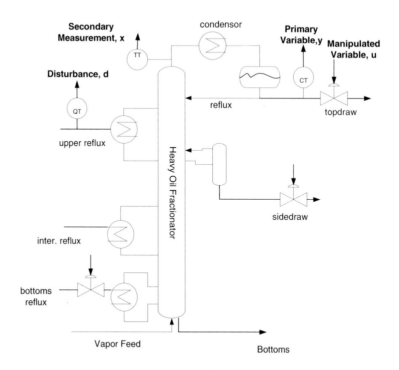

Figure 13.3 Schematic diagram of Shell fractionator.

Now let us repeat the design of the inferential controller, $q_i(s, \varepsilon)$, by designing $q(s)$ and $q_e(s)$ and then multiplying the two terms to get $q_i(s, \varepsilon)$. The controller $q(s)$ inverts the invertible part of $p_y(s)$ and so is

$$q(s, \varepsilon) = \frac{(50s + 1)}{4.05(\varepsilon s + 1)}. \tag{12.19}$$

The estimate of the effect of the disturbance on the top endpoint is

$$\tilde{d}_{ey}(s) = (\tilde{p}_{dy}(s)\tilde{p}_{dx}^{-1}(s))_m \tilde{d}_{ex}(s) = \frac{(1.44e^{-27s})(6s + 1)}{(40s + 1)1.27} \tilde{d}_{ex}(s). \tag{12.20}$$

Therefore to obtain a 27 time unit prediction of $d_{ey}(s)$, we set

$$q_e(s) = \frac{1.44(6s + 1)}{1.27(40s + 1)}. \tag{12.21}$$

Multiplying Eq. (12.19) by Eq. (12.21) yields exactly Eq. (12.18). Figure 12.11 shows the IC system for the Shell fractionator.

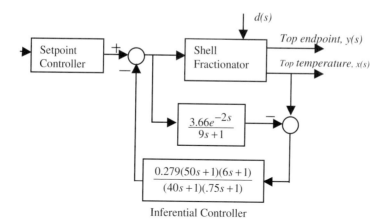

Figure 12.11 IC of top endpoint using top temperature for the Shell fractionator.

Substituting Eq. (12.18) and the various models into Eq. (12.8) gives

$$y(s) = \frac{e^{-27s}}{(\varepsilon_r s + 1)} r(s) + (1 - \frac{1}{(\varepsilon s + 1)}) \frac{1.44 e^{-27s}}{(40s + 1)} d(s). \qquad (12.22a)$$

Choosing both ε and ε_r to give noise amplification ratios of 10 for both $q(s)$ and $q_r(s)$ yields:

$$y(s) = \frac{e^{-27s}}{(5s + 1)} r(s) + \frac{1.08 s e^{-27s}}{(40s + 1)(.75s + 1)} d(s). \qquad (12.22b)$$

Because the term in parentheses in Eq. (12.22a) has very fast dynamics, the residue at the disturbance lag pole of $(-1/40)$ will be small, and $q_d(s)$ can be taken to be one.

Figure 12.12 below compares the step responses of various control configurations. The PID controller for the response labeled feedback control on $y(t)$ is:

$$PID_y = .474(1 + \frac{1}{61.4s} + \frac{.972s}{(.486s + 1)}). \qquad (12.23)$$

The feedback controller for $x(t)$ is (see Figure 12.4):

$$c_x(s) = .913(1 + \frac{1}{9.69s} + \frac{0.642s}{(0.0321s + 1)}). \qquad (12.24)$$

The cascade controller, c_y (see Figure 12.8), is the same as that given by Eq. (12.23).

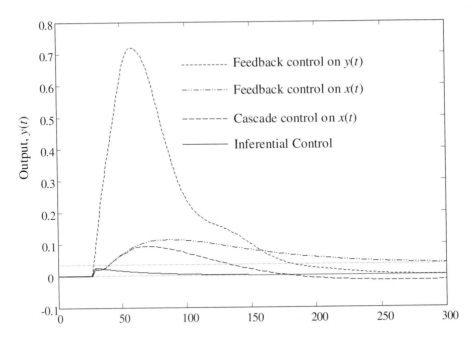

Figure 12.12 Comparison of step disturbance responses of various control systems for the Shell fractionator example.

The long tail in the response of the feedback control system on $y(t)$ could be eliminated by using a two-degree of freedom control system. However, the peak of about 7 units will remain as it is due to the relatively long delay of 27 units in the effect of the control effort on the output. The relatively good performance of the feedback control system on the secondary output $x(t)$ and its cascade enhancement is due to fortuitous relationships between the gains of the various process transfer functions (see Section 12.2.1).

♦

Example 12.3 IC of a Tubular Reactor

Consider the tubular reactor inside furnace as shown in Figure 12.13. The primary variable to be controlled is the product composition y, which is not measured. However a secondary variable, the product temperature x, is readily available. The following transfer functions are available relating these output variables to the fuel flow u, which is the primary manipulated variable and the feed composition d, which is the primary disturbance variable.

$$y(s) = \frac{3e^{-10s}}{10s+1}u(s) + \frac{4e^{-s}}{10s+1}d(s): \text{ primary output}$$

$$x(s) = \frac{e^{-9s}}{5s+1}u(s) - \frac{2e^{-s}}{3s+1}d(s): \quad \text{secondary measurement}$$

(12.25)

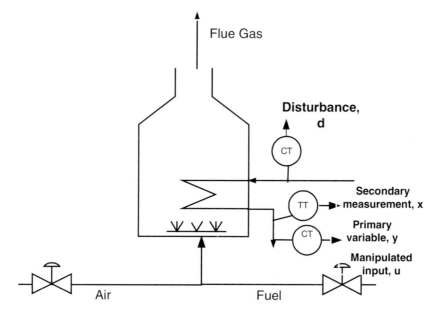

Figure 12.13 A furnace with a tubular reactor.

The primary and secondary variables of Eq. (12.25) have similar dynamics, but very different gains. The mismatch in gains precludes attempting to control the product composition by controlling the product temperature (see Section 12.2.1). Designing an IC system for Eq. (12.25) according to Section 12.3.2 and Equations (12.9) and (12.15) gives:

$$q(s) = \frac{(10s+1)}{3(\varepsilon s+1)}$$

(12.26a)

$$q_e(s) = (\frac{4e^{-s}}{10s+1})(-\frac{3s+1}{2e^{-s}}) = -\frac{2(3s+1)}{(10s+1)}.$$

(12.26b)

So, from Eq. (12.9):

$$q_I(s,\varepsilon) = -\frac{2(3s+1)}{3(\varepsilon s+1)}q_d(s,\varepsilon).$$

(12.26c)

Substituting the above into Eq. (12.8) gives:

$$y(s) = \left(1 - \frac{e^{-10s}}{(\varepsilon s + 1)} q_d(s, \varepsilon)\right) \frac{4e^{-s}}{(10s + 1)} d(s). \tag{12.27}$$

Choosing $q_d(s, \varepsilon)$ with the aid of IMCTUNE so that a zero of the term in brackets cancels the disturbance lag pole at -0.1 for a filter time constant, ε, of 1.0 gives:

$$q_d(s, 1) = \frac{(7.02s + 1)}{(s + 1)} \tag{12.28}$$

Figure 12.14 shows the response of the primary output, $y(t)$, for a two-degree of freedom IC system that chooses $q_d(s, \varepsilon)$ as given by Eq. (12.28) and a one-degree of freedom IC system that chooses $q_d(s, \varepsilon)$ as 1.0. For comparison, Figure 12.14 also includes the response that would be obtained for the primary output from a two-degree of freedom PID control system, if the primary output were measurable. In order to obtain a well-behaved two-degree of freedom PID controller, it was necessary to increase the filter time constant to 4.0, which yields PID parameters of 0.4654, 13.02, and 3.207 for the gain, integral and derivative time constants.

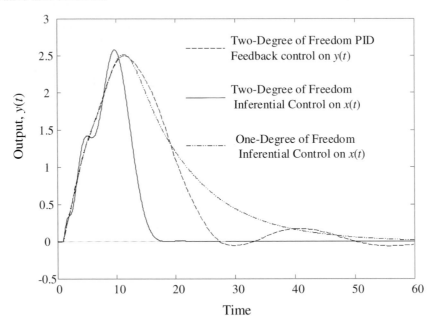

Figure 12.14 Comparison of step disturbance responses of several control systems for Example 12.3.

♦

12.3.3 Combining Inferential and Feedback Control

As pointed out earlier, IC strategy is similar to that of feedforward control. As in feedforward control, modeling errors can lead to an undesirable steady-state error in control. This can be avoided by combining IC strategy with a feedback control strategy as shown in Figure 12.15. A standard PID controller can be used to make corrections and drive $y(s)$ back to its setpoint.

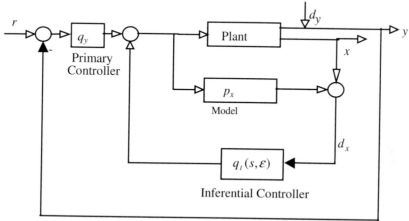

Figure 12.15 Combined inferential and feedback control strategy.

12.3.4 Design of the Inferential Controller for Uncertain Processes

For imperfect models, the closed-loop transfer functions of the measured and unmeasured outputs, $x(s)$, and $y(s)$, of Figure 12.9 are:

$$x(s) = \frac{q_r(s)p_x(s)r(s) + (1 - \tilde{p}_x(s)q_I(s))p_{dx}(s)d(s)}{1 + (p_x(s) - \tilde{p}_x(s))q_I(s)} \tag{12.29}$$

$$y(s) = \frac{q_r(s)p_y(s)r(s)}{1 + (p_x(s) - \tilde{p}_x(s))q_I(s)} + \left(1 - \frac{p_y p_{dy}^{-1} p_{dx}(s)q_I(s)}{1 + (p_x(s) - \tilde{p}_x(s))q_I(s)}\right)p_{dy}(s)d(s). \tag{12.30}$$

The transfer functions between the measured output, $x(s)$, and the setpoint and disturbance are similar in form to those in Chapter 8. However, since $r(s)$ is the setpoint of $y(s)$ rather than $x(s)$, and $q_r(s)$ is designed based on $\tilde{p}_y(s)$, the response of the secondary variable, $x(t)$ to a step setpoint change in y can look quite strange. Similarly, using the inferential controller designs suggested by Eq. (12.9) and Eq. (12.15) can lead to unusual responses of the measured output to disturbances.

The transfer function between the inferred output, $y(s)$, and the disturbance given by Eq. (12.8) is quite different from that used in Chapter 8. Thus the techniques of Chapters 7 and 8 are not directly applicable. We can, however, still tune for robust stability using the setpoint response of Eq. (12.29), which is repeated below in Eq. (12.31) in a slightly modified form.

$$\frac{x(s)}{r(s)} = \left(\frac{q_I(s,\varepsilon)p_x(s)}{1+(p_x(s)-\tilde{p}_x(s))q_I(s,\varepsilon)} \right)(q_r(s,\varepsilon_r)/q_I(s,\varepsilon)). \qquad (12.31)$$

If all the transfer functions in Eq. (12.31) are stable, then the *magnitude of the frequency response of Eq. (12.31) must be finite if the closed-loop system is to be stable* (see Chapter 7). The term in braces in Eq. (12.31) is a complementary sensitivity function with a controller of $q_I(s,\varepsilon)$. However, since in general $q_I(0,\varepsilon) \neq \tilde{p}_x^{-1}(0)$, the low frequency magnitude of this complementary sensitivity function is not unity. Nonetheless we can still use IMCTUNE to compute the upper bound, and then, if desired, normalize the upper bound by its low frequency gain. Values of the normalized peak less than 3 assure that that control system will be stable, and have reasonable robustness.

Example 12.4 IC of the Shell Fractionator with Uncertain Gains

The Shell problem of Example 12.1 was originally posed with uncertain gains as:

$$y(s) = \frac{K_y e^{-27s}}{50s+1}u(s) + \frac{K_{dy}e^{-27s}}{40s+1}d(s): \text{ top end point}$$

$$x(s) = \frac{K_x e^{-2s}}{9s+1}u(s) + \frac{K_{dx}}{6s+1}d(s) \qquad : \text{ top temperature,}$$

$\qquad\qquad\qquad\qquad\qquad\qquad\qquad\qquad\qquad\qquad (12.32)$

where $K_y = 4.05 \pm 2.11$, $K_{dy} = 1.44 \pm .16$, $K_x = 3.66 \pm 2.29$, $K_{dx} = 1.27 \pm .08$.

Using IMCTUNE to check the robustness of the controller design for Example 12.1 as described above yields Figure 12.16 below. A maximum peak of 1.8 indicates that the control system will be adequately robust. The time responses of the primary and secondary outputs of Figures 12.17 and 12.18 lend validity to this conclusion. Notice that there is now an offset from zero in Figure 12.17 because of the modeling errors. However, in the absence of control, the offsets would have been 1.60 and 1.28 for the upper and lower bound gains, respectively. Further, the simple feedback system, whose response to a step disturbance is shown in Figure 12.12, has an Mp of 2.57, and an overshoot of 50% for a step setpoint change when the process gain is 6.16.

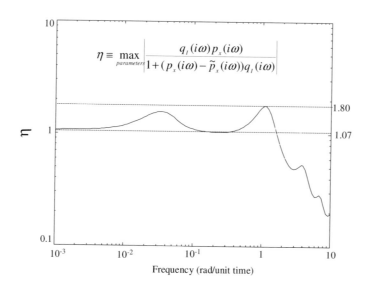

$$\eta \equiv \max_{parameters} \left| \frac{q_l(i\omega)p_x(i\omega)}{1+(p_x(i\omega)-\tilde{p}_x(i\omega))q_l(i\omega)} \right|$$

Figure 12.16 Closed-loop frequency response for the IC system of Example 12.1 ($\varepsilon = 0.75$).

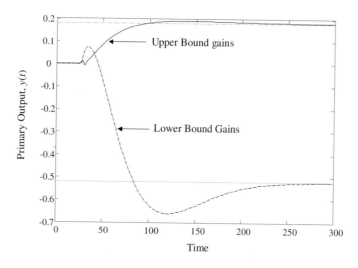

Figure 12.17 Step disturbance responses for the IC system of Example 12.1 ($\varepsilon = 0.75$).

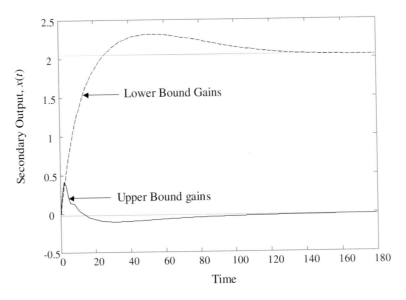

Figure 12.18 Step disturbance responses for the IC system of Example 12.1 ($\varepsilon =$ 0.75).

12.4 SUMMARY

In this chapter we presented a control technique for control of a primary variable using a secondary variable. This strategy attempts to approximate a feedforward control strategy and hence has advantages similar to that of feedforward control such as efficient disturbance rejection. As in the case of feedforward control, we will need models relating the disturbance variable to the outputs. The IC system is subject to modeling errors and one should combine it with direct feedback control of the primary output variable if it can be measured. This strategy is similar to that of a combined feedback/feedforward control strategy. IC is different from cascade control in that the process model is not represented in series format, where a transfer function relating the secondary variable to the primary variable is used. IC should be used when such a direct relationship is not present.

It is important to realize that the above structure is different from an approach that uses the secondary output, $x(t)$, to predict and control the primary output, $y(t)$. Estimating $y(t)$ first and then designing a controller throws away the feedforward nature of the IC structure. If the secondary measurements respond faster to the disturbances, then one can take faster corrective action and hence gets better control system performance by using $x(t)$ in an inferential structure.

Problems

12.1 Consider the reactor process shown in the figure below. It is desired to control the exit composition y of the reactor effluent, using the coolant flow u. The main disturbance is the change in inlet concentration and is not measured. A model is given by

$$y(s) = \frac{1}{5s+1}u(s) + \frac{2}{2s+1}d(s)$$

$$x(s) = \frac{1}{s+1}u(s) + \frac{2}{4s+1}d(s)$$

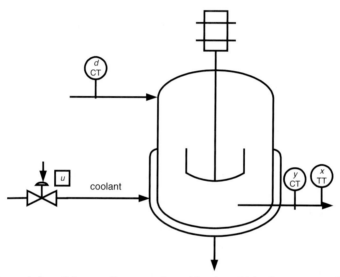

a. Derive an inferential controller to regulate $y(s)$ using $x(s)$ in the presence of disturbances d. Show the block diagram of the IC system.

b. What will be the steady-state error in control of $y(s)$ if direct regulation of $x(s)$ using $u(s)$ is implemented?

c. Suppose the actual process transfer function is given by

$$x(s) = \frac{k}{s+1}u(s) + \frac{4}{2s+1}d(s).$$

d. What values of k will lead to instability in the inferential controller if $\varepsilon = 0.2$ is used?

12.2 A process furnace has two secondary measurements, as shown:

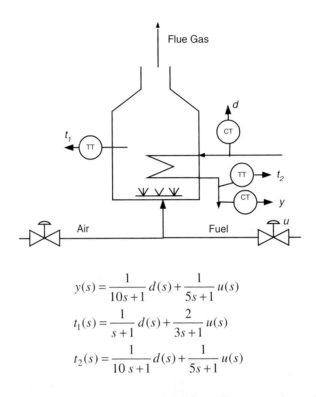

$$y(s) = \frac{1}{10s+1}d(s) + \frac{1}{5s+1}u(s)$$

$$t_1(s) = \frac{1}{s+1}d(s) + \frac{2}{3s+1}u(s)$$

$$t_2(s) = \frac{1}{10\,s+1}d(s) + \frac{1}{5s+1}u(s)$$

Which measurement should be used for the design of an IC system using only one secondary measurement? Explain how you arrived at your conclusion. Compare inferential controller performance with that of a direct feedback control system using y.

12.3 For the Shell distillation column let us say that the transfer function for the output $y(s)$ was erroneously modeled as

$$y(s) = \frac{5e^{-30s}}{40s+1}u(s) + \frac{2e^{-25s}}{30s+1}d(s).$$

Derive a new inferential controller using this model. Implement this controller and report its performance in terms of the frequency response characteristics of disturbance rejection transfer function and the step response. Compare it to the case with no modeling error.

12.4 Derive Eq. (12.2) and Eq. (12.5).

12.5 Derive Eq. (12.8).

12.6 Verify the results shown in Figures 12.13 and 12.14.

12.7 Consider a wet solids drying process, as shown in the figure below. The primary variable to be controlled, y, is the exit moisture content, and this measurement is not very reliable. However, the exit temperature of the air, x, can be used as a secondary measurement. The process model is given by

$$y(s) = \frac{1}{2s+1}u(s) + \frac{1}{3s+1}d(s)$$

$$x(s) = \frac{1}{4s+1}u(s) + \frac{1}{5s+1}d(s).$$

a. Derive an inferential controller. Draw the block diagram structure of the inferential controller.

b. Derive a transfer function between $y(s)$ and $d(s)$ when using the inferential controller scheme.

c. Which would be preferable to use, an IC system or an IMC feedback controller using $y(s)$? Explain your answer.

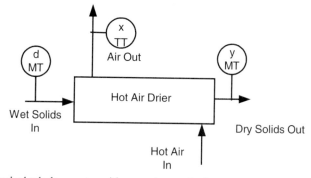

12.8 Consider a packed tubular reactor with secondary and primary measurements given by

$$x(s) = \frac{4e^{-s}}{3s+1}d(s) + \frac{e^{-3s}}{s+1}u(s)$$

$$y(s) = \frac{3e^{-5s}}{4s+1}d(s) + \frac{2e^{-3s}}{2s+1}u(s).$$

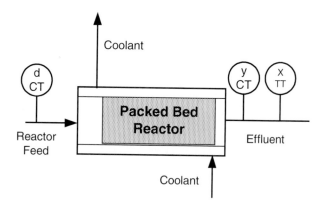

a.   Design an inferential controller for $y(s)$ using x. Draw a block diagram of the IC system showing all transfer functions.

b.   Derive a closed-loop disturbance rejection transfer function $y(s)/d(s)$ and plot its frequency response.

c.   Design and tune a PID controller for direct feedback control of y. Compare the disturbance rejection characteristics of this closed-loop system with that obtained in part b.

References

Preliminary work on control using secondary measurements was reported by Weber and Brosilow (1972). This was followed by a series of papers by Joseph and Brosilow. Brosilow (1982) recognized the model-based nature of IC. Patke et al. (1982) and Luyben and Yu (1987) make comparisons of IC with cascade structures. Budman et al. (1992) and Patke et al. both report on experimental applications of IC. Industrial applications are reported in Ogawa (1999), Montin (1998), and the series of papers by San et al. (1994).

Brosilow, C. B. 1979. "The Structure and Design of Smith Predictors from the Point of View of Inferential Control." Proc. of Joint Automatic Control Conf. Denver, CO.

Brosilow, C. B., and M. Tong. 1978. "The Structure and Dynamics of Inferential Control Systems." *AIChE J.*, *84*, 492.

Chen, C. Y., and C. C. Sun. 1991. "Adaptive Inferential Control of Packed-Bed Reactors." *Chem. Eng. Sci.*, 1041–1054.

Budman, H. M., C. Webb, T. R. Holcomb, and M. Morari. 1992. "Robust Inferential Control of a Packed-Bed Reactor." *Ind. Eng. Chem. Res.*, *31*, 1665.

Haggblom, K. E. 1996. "Combined Internal Model and Inferential Control of a Distillation Column via Closed-Loop Identification." *J. Process Control*, 6(4), 223–232.

Joseph, B., and C. B. Brosilow. 1978. "Construction of Optimal and Sub-Optimal Estimators." *AIChE J.*, *24*, 500.

Joseph, B., and C. B. Brosilow. 1978. "Inferential Control of Processes." *AIChE J.*, *24*, 485.

Kookos, I. and J. D. Perkins. 1999. "A Systematic Method for Optimum Sensor Selection in Inferential Control Systems." *Ind. Eng. Chem. Res.*, *38*(11), 4299–4308.

Lee, J. H. 1991. *Robust Inferential Control: A Methodology for Control Structure Selection and Inferential Control System Design in the Presence of Model/Plant Mismatch.* Ph. D. Thesis. California Inst. Technol., Pasadena, CA, USA.

Montin, T. and A. B. Nynas. 1998. "Online Flash Point Calculation Improves Crude Unit Control. Hydrocarbon Process." *Int. Ed.*, *77*(12), 81–84.

Ogawa, M., M. Ohshima, K. Morinaga, and F. Watanabe. 1999. "Quality Inferential Control of an Industrial High-Density Polyethylene Process." *J. Process Control, 9*(1), 51–59.

Patke, N. G., and P. B. Deshpande. 1982. "Experimental Evaluation of the Inferential System for Distillation Control." *Chem. Eng. Commun, 13*(4–6), 343–359.

Patke, N. G., P. D. Deshpande, and A. C. Chou. 1982. "Evaluation of Inferential and Parallel Cascade Schemes for Distillation Control." *Ind. Eng. Chem. Process Des. Dev., 21*(2), 266–272.

Prett, D. M, and Morari, M. 1987. *The Shell Process Control Workshop.* Butterworths, Stoneham, MA.

Prett, D. M, and Garcia, C. E. 1988. *Fundamental Process Control.* Butterworths, Stoneham, MA.

Quintero-Marmol, E., and W. L. Luyben. 1992. "Inferential Model-Based Control of Multicomponent Batch Distillation." *Chem. Eng. Sci., 47*(4), 887–898.

San, P., K. D. Landells, and D. C. Mackay. 1994. "Inferential Control. Part 1: Crude Unit Advanced Controls Pass Accuracy and Repeatability Tests." *Oil Gas J., 92*(48), 76, 78–80.

San, P., K. D. Landells, and D. C. Mackay. 1994. "Inferential Control. Conclusion: Crude Unit Controls Reduce Quality Giveaway, Increase Profits." *Oil Gas J., 92*(49), 56–61.

Shen, G. C., and W. K. Lee. 1989. "Adaptive Inferential Control for Chemical Processes with Intermittent Measurements." *Ind. Eng. Chem. Res., 28*(5), 55–63.

Shen, G. C., and W. K. Lee. 1988. "Adaptive Inferential Control for Chemical Processes with Perfect Measurements." *Ind. Eng. Chem. Res., 27*(1), 71–81.

Shen, G. C., and W. K. Lee. 1988. "Multivariable Adaptive Inferential Control." *Ind. Eng. Chem. Res., 27*(10), 1863–1872.

Weber, R., and C. B. Brosilow. 1972. "The Use of Secondary Measurements to Improve Control." *AIChE J., 18*, 614.

Yu, C. C., and W. L. Luyben. 1987. "Control of Multicomponent Distillation Columns Using Rigorous Composition Estimators." *I. Ch. E. Symp. Ser., 104*, 29–69.

Yu, C. C., and W. L. Luyben. 1987. "Use of Multiple Temperatures for the Control of Multicomponent Distillation Columns." *Ind. Eng. Chem. Process Des. Dev., 23*(3), 590–597.

Inferential Estimation Using Multiple Measurements

Objectives of the Chapter

The objective of this chapter is to discuss the design and implementation of estimators for inferential controllers using multiple secondary measurements. In this chapter we focus on model-based estimators. Topics covered include

- Design of the steady-state estimator.
- Selection of measurements.
- Adding dynamic compensation.
- Design of optimal estimators.
- Nonlinear estimators.

Prerequisite Reading

Chapter 12, "Single Variable Inferential Control"
Appendix C, "Linear Least-Squares Regression"
Appendix D, "Random Variables and Random Processes"

13.1 INTRODUCTION AND MOTIVATION

In the previous chapter we considered the inferential control using only one secondary variable. In many instances, multiple secondary measurements are available. Using multiple secondary measurements has some advantages and disadvantages. We have more information about the disturbances affecting the process and hence can take more accurate control action. However, this comes at the expense of increased complexity of the estimator and increased susceptibility to measurement errors and sensor failures. In practice one must carefully evaluate the increased benefits of control using multiple measurements versus the increased cost associated with building the estimator and maintaining the system.

This chapter focuses on estimator design when the process model is known. Chapter 19 discusses how to build estimators if no model is available but data are available (called data-driven estimators).

In this chapter we will focus on techniques of estimating the effect of disturbances on the output. We assume that for those significant disturbances that are measurable, one would employ feedforward control. It is the disturbances that are unmeasurable that are of interest in inferential control.

Figure 13.1 shows the structure of an inferential controller using multiple secondary measurements. Note that the design of the controller remains unchanged. The difference is in the estimator, which now has multiple secondary measurements as its input.

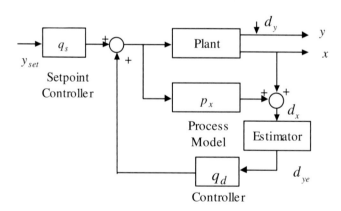

Figure 13.1 Structure of an inferential controller using multiple secondary measurements.

The basic problem in inferential control is how to estimate the effect of the disturbance on y, denoted using d_y, from the measured effect of the disturbance on x, denoted by the variable d_x. The problem can be mathematically defined as follows. Consider a process modeled by the set of equations

$$x(s) = A^T(s)d(s) + p_x(s)u(s),$$
$$y(s) = b^T(s)d(s) + p_y(s)u(s),$$

(13.1)

where

$x(s)$	=	$(p \times 1)$ vector of secondary measurements,
$y(s)$	=	scalar primary output variable to be controlled,
$u(s)$	=	manipulated variable, assumed to be a scalar,
$A^T(s)$	=	$(p \times n)$ matrix of transfer functions,
$a_{ij}(s)$	=	transfer function relating inputs i and output j,
$b^T(s)$	=	$(1 \times n)$ vector of transfer functions relating $y(s)$ and $d(s)$,
$p(s)$	=	transfer function matrix relating $y(s)$ and $u(s)$,
$d(s)$	=	random input disturbances process.

The transposes have been used primarily for notional convenience later on in the text. As an example, the model for a three-disturbance, two-secondary measurement system can be written as

$$x_1(s) = a_{11}(s)d_1(s) + a_{21}(s)d_2(s) + a_{31}(s)d_3(s) + p_1(s)u(s)$$
$$x_2(s) = a_{12}(s)d_1(s) + a_{22}(s)d_2(s) + a_{32}(s)d_3(s) + p_2(s)u(s)$$
$$y(s) = b_1(s)d_1(s) + b_2(s)d_2(s) + b_3(s)d_3(s) + c(s)u(s)$$

Define

$$d_x(s) = x(s) - p_x(s)u(s),$$

(13.2)

$$d_y(s) = y(s) - p_y(s)u(s),$$

(13.3)

then

$$d_x(s) = A^T(s)d(s),$$

(13.4)

$$d_y(s) = b^T(s)d(s).$$

(13.5)

The problem then is to estimate d_y from d_x. To complete the problem statement we must add a model of the disturbances also to the system equations. Since d is a random, unmeasured variable, the model for d will consist of a description of its statistical properties (such as mean and standard deviation).

A formal solution to the problem exists for linear systems. This is known as the *Kalman filter*. However, this filter is rather complex, requiring many parameters related to the disturbance and noise in the measurements to be specified. For this reason, this approach has not found widespread applications in the process industry.

We will first focus on simple, easy-to-implement solutions to the problem. First we consider steady-state estimators. These are usually sufficient in many cases. A design procedure is developed, assuming that the input disturbance does not change rapidly so that the process has time to settle down and operate in one region for a sufficiently long period of time to justify the emphasis on good steady-state estimation. Next we discuss how to add suitable dynamic compensation to the problem. Finally we discuss the formal solution to the problem.

Sometimes a process model of the type described above may not be available. If, on the other hand, we have some operational data on the process, then we can attempt to relate the secondary variables to the primary output. Such estimators are called data-driven estimators. We will discuss the design of linear and nonlinear data-driven estimators in Chapter 19.

13.2 DERIVATION OF THE STEADY-STATE ESTIMATOR

The steady-state form the process model discussed in the previous section can be written as

$$d_x(0) = A^T(0)d(0) \tag{13.6}$$

$$d_y(0) = b^T(0)d(0) \tag{13.7}$$

or in simplified notation, dropping the zeros,

$$d_t = A^T d \tag{13.8}$$

$$d_y = b^T d. \tag{13.9}$$

The steady-state estimation problem is to estimate d_y from d_x. Since both are linearly related to d, we can propose a linear estimator of the form

$$
\begin{aligned}
d_{ye} &= a^T d_x \\
&= \alpha_1 d_{x1} + \alpha_2 d_{x2} + \dots + \alpha_p d_{xp}
\end{aligned}
\tag{13.10}
$$

and determine the coefficients α_i.

If A is square (the number of measurements = number of disturbances), then an exact solution exists. If A is not square, then we can write a least-square solution.

Let us assume that the input disturbances are random variables. A review of the basic theory used to describe these random variables is given in Appendix D.

$$d = \begin{bmatrix} d_1 \\ d_2 \\ . \\ d_n \end{bmatrix} \tag{13.11}$$

then the mean of d is defined as

$$E\{d\} = \begin{bmatrix} E\{d_1\} \\ E\{d_2\} \\ . \\ E\{d_n\} \end{bmatrix} = \bar{d} . \tag{13.12}$$

Without loss of generality, we can assume that the disturbances have zero mean (i.e., $\bar{d} = 0$). The covariance of d is defined as

$$\Phi_{dd} = E\left\{(d - \bar{d})(d - \bar{d})^T\right\}$$

$$= \begin{bmatrix} E\left\{(d_1 - \bar{d}_1)(d_1 - \bar{d}_1)^T\right\} & E\left\{(d_1 - \bar{d}_1)(d_2 - \bar{d}_2)^T\right\} & . & E\left\{(d_1 - \bar{d}_1)(d_n - \bar{d}_n)^T\right\} \\ E\left\{(d_2 - \bar{d}_2)(d_1 - \bar{d}_1)^T\right\} & E\left\{(d_2 - d_2)(d_1 - \bar{d}_2)^T\right\} & . & E\left\{(d_2 - \bar{d}_2)(d_n - \bar{d}_n)^T\right\} \\ . & . & . & . \\ E\left\{(d_n - \bar{d}_n)(d_1 - \bar{d}_1)^T\right\} & E\left\{(d_n - \bar{d}_n)(d_2 - \bar{d}_2)^T\right\} & . & E\left\{(d_n - \bar{d}_n)(d_n - \bar{d}_n)^T\right\} \end{bmatrix} . \tag{13.13}$$

We restrict ourselves to linear estimations of the form $\bar{d}_y = \alpha^T d_x$ for two reasons. First, these estimators are easier to implement online. Second, if the input disturbances are Gaussian distributed, then it can be shown that the linear estimators minimize the expected square of the error in estimation. If we use a linear estimator to estimate y, then the mean square error in estimation is

$$e = \left(b^T - \alpha^T A^T\right)d \tag{13.14}$$

and

$$E\left\{\|e\|^2\right\} = \left(b^T - \alpha^T A^T\right)\Phi_{dd}\left(b - A\alpha\right). \tag{13.15}$$

The minimization of the least-square error can be carried out as follows:

$$\overline{e^2} = b^T \Phi_{dd} b + \alpha^T A^T \Phi_{dd} A\alpha - \alpha^T A^T \Phi_{dd} b - b^T \Phi_{dd} A\alpha \tag{13.16}$$

which can be written as (after completing squares)

$$e^2 = b^T \Phi_{dd} b - b^T \Phi_{dd} A (A^T \Phi_{dd} A)^{-1} A^T \Phi_{dd} b$$
$$+ [\alpha^T - b^T \Phi_{dd} A (A^T \Phi_{dd} A)^{-1}] A^T \Phi_{dd} A [\alpha - (A^T \Phi_{dd} A)^{-1} A^T \varphi_{dd} b] \tag{13.17}$$

The left hand side is minimized by choosing

$$\alpha = (A^T \Phi_{dd} A)^{-1} A^T \Phi_{dd} b \tag{13.18}$$

When $\Phi_{dd} = \mathbf{I}$, this becomes

$$\alpha = (A^T A)^{-1} A^T b \tag{13.19}$$

The error in estimation given by

$$e = (b^T - \alpha^T A^T) d . \tag{13.20}$$

The mean square error in estimation is given by

$$\bar{e}^2 = r^T \Phi_{dd} r . \tag{13.21}$$

where

$$r = b - A\alpha \tag{13.22}$$

is called the residual error in the estimator α. Refer to the tutorial article by Rhodes (1971) for a more complete explanation and derivation of the least-square estimator.

A scaled measure of the error in estimation is the *projection error*, defined as

$$\begin{aligned} \left(\text{Projection Error} \right)^2 &= \frac{E\{e^2\}}{E\{d_y^2\}} \\ &= \frac{r^T \Phi_{dd} r}{b^T \Phi_{dd} b}, \end{aligned} \tag{13.23}$$

for the case $\Phi_{dd} = I$, the projection error is given by

$$Projection\ Error = \sqrt{\frac{r^T r}{b^T b}} . \tag{13.24}$$

We want to keep this as close to zero as possible.

Example 13.1 Steady-state Estimator

Consider the steady-state model

$$d_{x1} = d_1 + 2d_2 + 3d_3$$
$$d_{x1} = 2d_1 + d_2 + d_3$$
$$d_y = 3d_1 + 3d_2 + 4d_3$$

with $\overline{d} = 0$, $\Phi_{dd} = I$.

In this case

$$A = \begin{bmatrix} 1 & 2 \\ 2 & 1 \\ 3 & 1 \end{bmatrix} \quad and \quad b = \begin{bmatrix} 3 \\ 3 \\ 4 \end{bmatrix},$$

and hence

$$\alpha = (A^T A)^{-1} A^T b = \begin{bmatrix} 1 \\ 1 \end{bmatrix}.$$

Hence the estimator is

$$d_y = d_{x1} + d_{x2}$$

◆

The solution will exist if and only if $A^T A$ is a nonsingular matrix. If the matrix is nonsingular, then measurements x_1, x_2, and so on are giving us different information regarding the input disturbances (i.e., the rows of A^T are linearly dependent). If this is not the case, then some measurements in the set are linearly dependent on others and the procedure needs to be modified.

Example 13.2 Collinear Measurements

Suppose for Example 13.1, we have another measurement

$$d_{x3} = -d_1 + d_2 + 2d_3$$

then $(A^T A)$ becomes singular. In this case

$$d_{x3} = d_{x1} - d_{x2}$$

and hence no new information is contained in d_{x3}. However, if the measurements are noisy, then it may be better to include this additional measurement to improve the accuracy of estimation as seen in Example 13.4.

◆

Example 13.3. Error in Estimation

Consider the estimation of

$$d_y = 3d_1 + 3d_2 + 4d_3$$

from

$$d_x = d_1 + 2d_2 + 3d_3$$

Let us assume that $\boldsymbol{\Phi}_{dd} = \boldsymbol{I}$. The least-square estimator is given by

$$\alpha = (A^T A)^{-1} A^T b$$
$$= 3/2$$

and

$$e = \frac{3}{2}d_1 - \frac{1}{2}d_3 = (3/2 \quad 0 \quad -1/2)\begin{pmatrix} d_1 \\ d_2 \\ d_3 \end{pmatrix}$$

$$E\{e^2\} = 5/2$$
$$E\{d_y^2\} = 9 + 9 + 16 = 34$$
$$\text{Projection Error} = \sqrt{\frac{5}{68}} = 0.271 = 27\%$$

Whether this error is acceptable or not depends on the allowable variation in the primary output. Additional measurements should be considered if the estimation error is not satisfactory.

♦

13.2.1 Noise in Measurements

Let us say that the measurements are corrupted by noise

$$d_x = A^T d + n_x, \tag{13.25}$$

where the noise has zero mean:

$$E\{n_x\} = 0, \tag{13.26}$$

and the variance is given by

$$E\left\{n_x \ n_x^T\right\} = \Phi_{nn}.$$ (13.27)

In this case the least-square estimate is given by

$$\alpha = (A^T \Phi_{dd} A + \Phi_{nn})^{-1} A^T \Phi_{dd} b .$$ (13.28)

The error in estimation is increased by the presence of noise in the measurements.

13.3. SELECTION OF SECONDARY MEASUREMENTS

In this section we will address the issue of selecting secondary measurements to use in an inferential feedback control system (e.g., on what trays in a distillation column should temperatures be located to get good control of product composition?). The criteria used for selecting measurements include
1. The ability to get good estimates of the output from the secondary measurements.
2. The cost of installing and maintaining the sensor.
3. The ability of the control system to cope with sensor failures.

We will limit the discussion in this chapter primarily to choosing measurements based on criterion 1. Both the second and third criteria favor using a small number of measurements. The last criterion favors simpler control structures that do not depend on many measurements. As the complexity of the control system increases, it becomes more difficult to cope with temporary sensor failures.

If secondary measurements are used, then we can break down the first criterion into two parts: (1) the ability to estimate the primary output as correctly as possible (the *accuracy or performance* of the estimator), and (2) the ability to cope with modeling errors, (the *sensitivity or robustness* of the estimator).

We will primarily focus on steady-state criteria for selecting measurements. These can be extended to incorporate dynamic considerations, discussed in the next section.

13.3.1 Effect of Measurement Selection on Accuracy of Estimators

In the previous section we derived a measure of the steady-state accuracy of estimation called the projection error. It is defined as

$$\text{Projection Error} = \text{PE} = \sqrt{\frac{E\left\{e^2\right\}}{E\left\{y^2\right\}}}$$ (13.29)

$$= \frac{r^T r}{b^T b} = \frac{\|A\alpha - b\|}{\|b\|}. \tag{13.30}$$

Thus to maximize accuracy of the estimation, we must minimize the projection error associated with the secondary measurement. This can be used as a criterion for selecting measurements (to determine the smallest number needed and the location of these sensors).

Example 13.4 Selection of Temperature Measurements in a Column

Consider the depropanizer column, studied by Tong and Brosilow (1978) with temperature measurements available on five different trays, as shown in Figure 13.2. The steady-state process input/output model is determined as follows:

$$x_1 = -7.99 d_1 - 9.78 d_2 \quad -5.28 d_3 \quad +3.59 d_4 +6.09 d_5$$
$$x_2 = -11.29 d_1 -15.91 d_2 \quad -4.23 d_3 \quad +3.63 d_4 \quad +6.09 d_5$$
$$x_3 = -18.28 d_1 -16.43 d_2 \quad -0.47 d_3 \quad +3.96 d_4 \quad +4.60 d_5$$
$$x_4 = -42.02 d_1 -35.92 d_2 \quad +4.45 d_3 \quad +1.10 d_4 \quad +0.46 d_5$$
$$x_5 = -50.47 d_1 -25.26 d_2 \quad +4.45 d_3 \quad +1.10 d_4 \quad +0.32 d_5$$
$$y = -0.188 d_1 -0.163 d_2 +0.0199 d_3 +0.0043 d_4 +0.002 d_5$$

where y = overhead composition, and d_i are feed composition disturbances. The projection errors for estimators using each of these measurements are

$$PE\ (1) = 61.97\%,$$
$$PE\ (2) = 44.59\%,$$
$$PE\ (3) = 24.34\%,$$
$$PE\ (4) = 0.75\%,$$
$$PE\ (5) = 28.27\%.$$

From this, clearly x_4 alone is sufficient to yield a good estimate of the overhead butane composition.

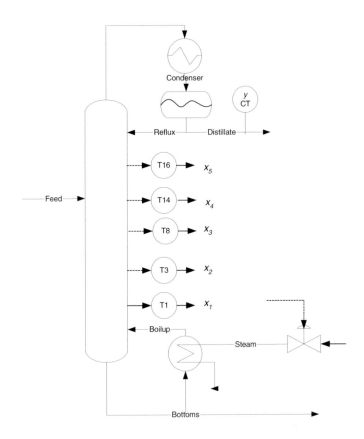

Figure 13.2 Schematic of a Depropanizer.

13.3.2 Robustness of Estimators

By robustness of the estimators, we mean the sensitivity of the estimator to errors in modeling. A robust estimator is one whose parameters do not change significantly when small errors are introduced in the process model.

The discussion in Appendix C on the robustness of least-square estimation shows that the condition number of the A matrix can be used as an indicator of the measure of robustness.

Example 13.5 Sensitivity of Estimator to Model Errors

Consider the estimation problem

$$x_1 = d_1 + 1.1 \ d_2,$$
$$x_2 = d_1 + 1.1001 \ d_2,$$
$$y = d_1 + d_2.$$

The A matrix in this case is given by

$$A = \begin{bmatrix} 1 & 1 \\ 1.1 & 1.1001 \end{bmatrix},$$

with condition number

$$\rho(A) = 44202.$$

The estimator, using both measurements, is

$$y = 1001x_1 - 1000x_2.$$

Suppose there is a small modeling error in x_2:

$$x_2 = d_1 + 1.1002 \ d_2.$$

Then the estimator is

$$y = 501x_1 - 500x_2.$$

Thus a very small error in the model leads to a very large change in the estimator parameters. As detailed in Appendix C, one should use PCA-based least-square solution in this case. Large condition numbers can be caused by two reasons: (1) the columns of A are nearly dependent, or (2) the matrix A has elements which differ significantly in size. The former situation arises, for example, when two measurements are located close to each other and respond to the unknown input disturbances in nearly the same way so that no new information is contained in the second measurement (collinearity of measurements).

♦

Example 13.6 Effect of Adding Noise on Estimator

Consider the estimation of y from two secondary measurements, using the process model

$$x_1 = d_1 + 2d_2 + 3d_3$$
$$x_2 = d_1 + 2d_2 + 3d_3$$
$$y = d_1 + 2d_2 + 3d_3$$

so that the A and b matrices are given by

$$A = \begin{bmatrix} 1 & 1 \\ 2 & 2 \\ 3 & 3 \end{bmatrix} \text{ and } b = \begin{bmatrix} 1 \\ 2 \\ 3 \end{bmatrix}$$

Notice that the two measurements respond identically to disturbances. Clearly, both carry the same information. Now if we attempt to build an estimator, we find that the matrix $A^T A$ is singular and hence noninvertible.

Now suppose that we add random noise to the measurements:

$$x_1 = d_1 + 2d_2 + 3d_3 + n_1$$
$$x_2 = d_1 + 2d_2 + 3d_3 + n_2$$

with

$$E(n) = 0$$

$$E(nn^T) = \Phi_{nn} = \begin{bmatrix} 0.0001 & 0 \\ 0 & 0.001 \end{bmatrix}$$

$$where \ n = (n_1 \ n_2)$$

so there is really very little noise. Now the estimator is given by

$$a = (A^T A + \Phi_{nn})^{-1} A^T b = \begin{bmatrix} 0.5 \\ 0.5 \end{bmatrix},$$

and the estimator is

$$\hat{y} = 0.5x_1 + 0.5x_2$$

Thus the addition of even a small amount of noise has made it possible to construct an estimator, and the result is physically meaningful; one merely takes the average of the two measurements.

However, there is a problem. The matrix to be inverted $(A^T A + \Phi_{nn})$ is still very close to being singular with condition number = 280000. The PCA method described in Examples 13.7 and 13.8 avoid the problem of inverting such nearly singular matrices.

♦

Example 13.7 PCA Estimator

Consider the Shell distillation example from Chapter 12. See Figure 13.3. This column has two temperature measurements that can be used to control the product composition.

The process model equations are

$$d_{x1} = 1.16d_1 + 1.27d_2$$
$$d_{x2} = 1.73d_1 + 1.79d_2$$
$$d_y = 1.20d_1 + 1.44d_2$$

Here, x_1 and x_2 are the temperature measurements, and y is the top product composition. Note that we have not included the effect of manipulated variables on these output variables. The matrices for building the estimator are

$$A^T = \begin{bmatrix} 1.16 & 1.27 \\ 1.73 & 1.79 \end{bmatrix} \text{ and } b = \begin{bmatrix} 1.20 \\ 1.44 \end{bmatrix}.$$

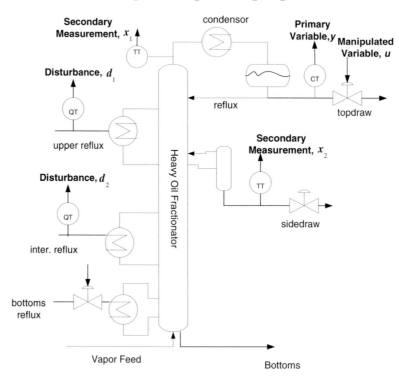

Figure 12.10 Schematic diagram of Shell fractionator.

Using MATLAB, two estimators are constructed:

Estimator 1 (using 1 latent variable): $\hat{d}_y = 0.3514d_{x1} + 0.5087d_{x2}$, $PE = 6.45\%$

Estimator 2 (using 2 latent variables): $\hat{d}_y = 2.8434d_{x1} - 1.2129d_{x2}$, $PE = 0$

In this case, Estimator 1 may be preferred because the projection error is small and it will be less sensitive to noise and model errors.

♦

For further discussion on the effect on collinearity on measurement selection, see Mejdell and Skogestad (1993). Lee and Morari (1991) discuss the effect of measurement selection on robustness of the control system. For an application study involving multiple secondary measurements see Budman et al. (1992).

13.4 ADDING DYNAMIC COMPENSATION TO THE ESTIMATORS

Next we consider the problem of dynamic compensation to the steady-state estimators constructed previously. We begin by discussing the motivation for adding dynamic compensation. Consider an example process with

$$d_x(s) = \frac{1}{s+1}d(s)$$

$$d_y(s) = \frac{1}{10s+1}d(s)$$

In this case perfect estimation is obtained by using the estimator

$$\hat{d}_y = \frac{s+1}{10s+1}d_x(s)$$

Note the lead-lag nature of the dynamic compensator. The object of this section is to discuss how to build such dynamic compensation.

We will first discuss some ad hoc (heuristic) methods for the addition of dynamic compensation to the steady-state estimators. These ad hoc procedures are easy to implement and will be sufficient to meet the needs of many industrial applications. If accurate models of the disturbances are available, then design procedures are available for building estimators that are optimal in a least-square sense.

Example 13.8 Adding Process Dynamics to Estimator

Consider the problem of estimating

$$d_y(s) = \frac{1}{s+1}d_1 + \frac{1}{s+1}d_2$$

from the measurements

$$d_{x1}(s) = \frac{1}{2s+1}d_1 + \frac{2}{2s+1}d_2,$$

$$d_{x2}(s) = \frac{1}{3s+1}d_1 + \frac{2}{3s+1}d_2.$$

The steady-state estimation problem is given by

$$d_y = d_1 + d_2$$
$$d_{x1} = d_1 + 2d_2$$
$$d_{x2} = d_1 + 4d_2$$

with the steady-state estimator given by

$$\hat{d}_y = 1.5d_{x1} - 0.5d_{x2}$$

Comparing the time constants associated with d_x, d_{x1} and d_{x2}, we have,

$$\hat{d}_y(s) = 1.5\frac{2s+1}{s+1}d_{x1} - 0.5\frac{3s+1}{s+1}d_{x2},$$

which results in perfect estimation in the dynamic case:

$$e(s) = d_y(s) - \hat{d}_y(s) = 0.$$

♦

The key factor behind the design procedure just described is that all disturbances affect the measurement with the same time lag. If this were not the case, we would not be able to design perfect compensators. However, we can generalize the concept to the case where the dynamics are nearly the same. Consider the case where

$$d_{x1} = a_{11}\frac{p_{11}s+1}{q_{11}s+1}d_1 + a_{21}\frac{p_{21}s+1}{q_{21}s+1}d_2 \qquad (13.31)$$

If p_{11} and p_{21} are roughly the same and q_{11} and q_{21} are roughly the same, then we can approximate

$$d_{x1} \approx (a_{11}d_1 + a_{21}d_2)\frac{p_1s+1}{q_1s+1} \qquad (13.32)$$

where

$$p_1 \approx \frac{1}{2}(p_{11} + p_{21}), \quad q_1 = \frac{1}{2}(q_{11} + q_{21})$$

We then define

$$d'_{x1} = \frac{q_1 s + 1}{p_1 s + 1} d_{x1} = a_{11} d_1 + a_{21} d_2$$

similarly,

$$d'_{x2} = \frac{q_2 s + 1}{p_2 s + 1} d_{x2} = a_{12} d_1 + a_{22} d_2 \quad \text{and} \quad d'_y = \frac{ts + 1}{rs + 1} d_y = b_1 d_1 + b_2 d_2$$

We can reduce the estimation problem to that of designing a steady-state estimator for $\hat{d}'_y$ using d'_{x1} and d'_{x2} with the solution

$$\hat{d}'_y = \alpha_1 d'_{x1} + \alpha_2 d'_{x2}, \quad \boldsymbol{\alpha} = (\boldsymbol{A}^T \boldsymbol{A})^{-1} \boldsymbol{A}^T \boldsymbol{b}.$$

Substituting the definitions, we can get the estimator in terms of original variables,

$$\hat{d}_y = \alpha_1 \frac{(q_1 s + 1)(rs + 1)}{(p_1 s + 1)(ts + 1)} d_{x1} + \alpha_2 \frac{(q_2 s + 1)(rs + 1)}{(p_2 s + 1)(ts + 1)} d_{x2} \tag{13.33}$$

as the dynamic compensator. The error in estimation is given by

$$e(s) = d_y(s) - d'_y(s)$$

$$= \left(b_1 \frac{r_1 s + 1}{t_1 s + 1} - \alpha_1 a_{11} \frac{(q_1 s + 1)(rs + 1)(p_{11} s + 1)}{(p_1 s + 1)(ts + 1)(q_{11} s + 1)} - \alpha_2 a_{12} \frac{(q_2 s + 1)(rs + 1)(p_{12} s + 1)}{(p_2 s + 1)(ts + 1)(q_{12} s + 1)} \right) d_1(s) \tag{13.34}$$

$$+ \left(b_2 \frac{r_2 s + 1}{t_2 s + 1} - \alpha_1 a_{21} \frac{(q_1 s + 1)(rs + 1)(p_{21} s + 1)}{(p_1 s + 1)(ts + 1)(q_{21} s + 1)} - \alpha_2 a_{22} \frac{(q_2 s + 1)(rs + 1)(p_{22} s + 1)}{(p_2 s + 1)(ts + 1)(q_{22} s + 1)} \right) d_2(s).$$

This will not be zero except in the steady-state (s = 0). We can get an idea of the dynamic error in estimation by looking at the frequency response of the transfer functions multiplying $d_1(s)$ and $d_2(s)$. We shall use this property later on in this section to design suboptimal estimators.

◆

Example 13.9 A Dynamic Estimator for the Shell Column

Consider the shell fractionator with multiple secondary measurements shown in Figure 13.3. We want to control the top endpoint using the temperature measurements.

Top Endpoint:
$$y(s) = \frac{1.20e^{-27s}}{45s+1}d_1(s) + \frac{1.44e^{-27s}}{40s+1}d_2(s) + \frac{4.05e^{-27s}}{27s+1}u(s)$$

Top Temperature:
$$x_1(s) = \frac{1.16}{11s+1}d_1(s) + \frac{1.27}{6s+1}d_2(s) + \frac{3.66e^{-2s}}{9s+1}u(s)$$

Upper Reflux Temperature:
$$x_2(s) = \frac{1.73}{5s+1}d_1(s) + \frac{1.79}{19s+1}d_2(s) + \frac{5.92e^{-11s}}{12s+1}u(s)$$

where

$$d_1 = \text{Intermediate Reflux Duty}$$
$$d_2 = \text{Upper Reflux Duty}$$
$$u = \text{Top Draw Rate}$$

We will focus on the estimation problem. Estimate

$$d_y(s) = y(s) - \frac{4.05e^{-27s}}{27s+1}u(s) = \frac{1.20e^{-27s}}{45s+1}d_1(s) + \frac{1.44e^{-27s}}{40s1}d_2(s)$$

from

$$d_{x1}(s) = x_1(s) - \frac{3.66e^{-2s}}{9s+1} = \frac{1.16}{11s+1}d_1(s) + \frac{1.27}{6s+1}d_2(s)$$

$$d_{x2}(s) = x_2(s) - \frac{5.92e^{-11s}}{12s+1} = \frac{1.73}{5s+1}d_1(s) + \frac{1.79}{19s+1}d_2(s)$$

Solution

First we design a steady-state estimator. The steady-state relationships are

$$d_{x1} = 1.16d_1 + 1.27d_2 = a_1^T d$$
$$d_{x2} = 1.73d_1 + 1.79d_2 = a_2^T d$$
$$d_y = 1.20d_1 + 1.44d_2 = b^T d$$

Therefore

$$a = A^{-1}b = \begin{bmatrix} 2.84 \\ -1.21 \end{bmatrix}$$

The steady-state error in estimation is zero. Next we add the dynamic compensation. Note that we can approximate

$$d_{x1} \approx \frac{1}{8.5s+1}(1.16d_1 + 1.27d_2)$$

where we have chosen the common lag to be the average of the two lags associated with d_1 and d_2. This is a rather coarse approximation.

Similarly,

$$d_{x2} \approx \frac{1}{12s+1}(1.73d_1 + 1.79d_2)$$

$$d_y \approx \frac{e^{-27s}}{42.5s+1}(1.20d_1 + 1.44d_2)$$

This leads to the estimator

$$\hat{d}_y(s) = e^{-27s}\left(2.85\frac{8.5s+1}{42.5s+1}d_{x1} - 1.21\frac{12s+1}{42.5s+1}d_{x2}\right)$$

The error in estimation is given by

$$e(s) = d_y(s) - \hat{d}_y(s) = \frac{1.20e^{-27s}}{45s+1}d_1(s) + \frac{1.44e^{-27s}}{40s+1}d_2(s)$$

$$- e^{-27s}2.85\frac{8.5s+1}{42.5s+1}\left(\frac{1.16}{11s+1}d_1 + \frac{1.27}{6s+1}d_2\right)$$

$$+ 1.21e^{-27s}\frac{12s+1}{42.5s+1}\left(\frac{1.73}{5s+1}d_1 + \frac{1.79}{19s+1}d_2\right)$$

$e(s)$ can be written also as

$$e(s) = q^T(s)d(s).$$

The norm of $q(s)$ is an indication of the effect of $d(s)$ on the error. We can plot $q(s)$ as a function of the frequency to see whether any particular frequencies in the input disturbances are amplified by the estimator.

Figure 13.4 shows the plot of $|q(i\omega)|$ as a function of frequency. This was generated by MATLAB program *ch13ex10.m*. The figure shows a maximum in the error at a frequency of about .1 rad/min. At this frequency, the error in estimation is about 0.5 times the magnitude of the disturbance. Hence disturbances occurring within a period of $(2\pi/0.1)$ or 60 min will result in the max error in estimation (about 50% of the input disturbance in the worst case). On the other hand, disturbances at a frequency of about 0.01 rad/min or a period of above 600 min will be estimated within 10% accuracy. Whether or not this estimator is acceptable will thus depend on the expected variations in the disturbance signal. Some knowledge about the frequency content of the disturbance is needed before making any conclusions regarding the suitability of the estimator.

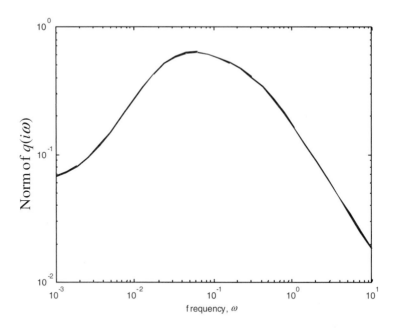

Figure 13.4 The frequency response of the error in estimation.

♦

13.4.2   Design of Suboptimal Compensators

In the above ad hoc design procedure just described, we arbitrarily chose the time lags in the compensator as the averages of the time lags associated with each disturbance. We can choose the time constants in the lead-lag compensators to minimize the prediction error $|e(i\omega)|$ over a range of frequencies at which the disturbances normally appear. This leads to a suboptimal design procedure (suboptimal in the sense that we have restricted the structure of the estimator to be a lead-lag). The error in estimation is given by

$$e(s) = E(s)d(s), \tag{13.35}$$

where $E(s)$ is dependent on the lead-lag compensators associated with the estimator. Let us say that we restrict the estimator to be of the form

$$\alpha(s) = \begin{bmatrix} \alpha_1 \dfrac{p_1 s + 1}{q_1 s + 1} \\ \dots \\ \alpha_n \dfrac{p_n s + 1}{q_n s + 1} \end{bmatrix}, \tag{13.36}$$

where the α_i have been computed from the steady-state gains. The suboptimal estimation problem can be stated as,

Find p_i, q_i such that the maximum value of $|E(i\omega)|$ is minimized over a frequency range, $0 \le \omega \le \omega_{max}$,

$$\underset{p_i, q_i}{Min} \left\{ \underset{0 \le \omega \le \omega_{max}}{Max} |E(i\omega)| \right\}, \tag{13.37}$$

where ω_{max} is based upon the maximum expected frequency of the variation in the random input disturbances. We can always make $|E(i\omega)| \to 0$ *as* $\omega \to \infty$ by adding lags to the estimator, if necessary.

If perfect estimation is possible, then $E(0) = 0$. In this case $|E(i\omega)|$ will exhibit a maximum at a certain frequency. On the other hand, if $E(0) > 0$, then $|E(i\omega)|$ may be monotonically decreasing with frequency. In this case we need to modify the objective function to provide a meaningful optimization problem. We can consider the error to consist of two parts:

$$E(i\omega) = E(0) + (E(i\omega) - E(0)) \tag{13.38}$$

The first part represents the steady-state error, and the second part represents the dynamic error. Then we can design the lead-lag compensator to minimize this dynamic part, since $E(0)$ is not influenced by the dynamic compensator. Thus we have the optimization problem

$$\underset{p_i, q_i}{Min} \left\{ \underset{0 \le \omega \le \omega_{max}}{Max} |E(i\omega) - E(0)| \right\} \tag{13.39}$$

This type of optimization problem can be solved using numerical search techniques such as those available in the Optimization Toolbox of MATLAB. In the following example we illustrate the use of this design procedure for a simple problem.

Example 13.10 Design of a Suboptimal Compensator
for the Shell Column

In the previous ad hoc design procedure, we arbitrarily chose the lags in the estimator:

$$d_y \approx \frac{e^{-27s}}{42.5s+1}(1.20d_1 + 1.44d_2)$$

Suppose instead of choosing the time constant in the lag to be the average, we let

$$d_y \approx \frac{e^{-27s}}{\tau s+1}(1.20d_1 + 1.44d_2)$$

Let us see how τ affects the dynamic error in estimation $|E(i\omega)|$. Figure 13.5 shows a plot of the maximum value attained $|E(i\omega)|$ for various values of τ. This figure was generated by program *ch13ex11.m*.

A minimum is attained when $\tau = 56.0$. The difference between this choice and $\tau = 42.5$ as we guessed earlier, is not significant. Slightly better design of the compensator can be obtained by choosing more of the time constants in the estimator using this minimization technique.

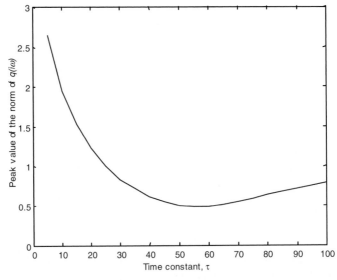

Figure 13.5 The maximum error in estimation as a function of the parameter τ.

♦

13.5 OPTIMAL ESTIMATION

By optimal estimation we mean estimators that are designed to minimize a norm of the error in estimation (e.g., the Euclidean or l_2 norm). In suboptimal estimators designed in the previous sections we restricted the structure of the estimator. The optimal estimation problem can be stated as follows: Let $d(s)$ be a set of random input disturbances entering the process. Assume these random processes can be modeled using their known statistical properties. Then find an estimator $\alpha(s)$ such that the expected value of the square of the error in estimation is minimized in some sense. No restrictions are placed on the structure of $\alpha(s)$.

A formal solution to this problem was first derived by Wiener (Newton et al. 1957). The solution was presented using the frequency spectrum of the measurements involved and proved difficult to compute, especially for multivariable problems. A solution in the time domain was presented by Kalman and Bucy in 1963, and this is now known as the Kalman filter. We will give a brief outline of the Kalman filter and then show how it may be adapted to the problem that we have of estimating d_y from d_t.

The Kalman filter may be expressed in many forms (continuous, discrete, steady-state, dynamic, linear, or nonlinear). Here we present only the most rudimentary form of the filter. No derivation is given. Refer to standard texts on linear system theory, such as Kwakernaak and Sivan (1972), Kailath (1989), Antsaklis and Michel (1997), and Chen (1999).

Consider a linear dynamic system described by

$$dx / dt = Ax + Bu + Gw$$

with measurements given by

$$z = Cx + Du + v.$$

Here w and v are assumed to be process and measurement white noises with

$$E(w) = 0, \ E(v) = 0, \ E(ww^t) = Q, E(vv^T) = R, E(wv^T) = T.$$

R must be positive definite; Q and T are nonnegative definite. Also, the noises are assumed to be not correlated with the initial state of the system. The problem is to estimate the state of the system x from the measurements z. In order to be able to do this, the system must satisfy certain observability conditions, which will ensure that enough information is contained in the output measurements to estimate all the states of the system.

The Kalman filter is another linear dynamic system that is simulated in parallel with the process, driven by the measurements, as its input,

$$\dot{\hat{x}} = A\hat{x} + Bu + LC^T R^{-1}(z - C\hat{x} - Du), \tag{13.40}$$

where L is the Kalman filter gain found by solving the Matrix-Riccatti equation

$$AL + LA^T - (LC^T + GT)R^{-1}(CL + T^T G^T) + GLG^T = 0. \tag{13.41}$$

$\hat{x}$ is an optimal estimate of x in a least-square sense. This is the steady-state version of the Kalman filter for continuous linear systems with correlated white measurement and input noise. The Control System toolbox in MATLAB contains commands to construct the dynamic system describing the filter from the system definition (see MATLAB command `kalman`).

13.5.1 Application of the Kalman Filter to Inferential Estimation Problem

Now let us consider how we can use such a filter for estimating the effect of unmeasured disturbances on y from the measurements z.

First the linear model expressed in the earlier sections in the Laplace domain is converted into state-space form. Typically, the input disturbances in chemical process systems cannot be modeled as white noise. It is best to model them as ARIMA type models (see Appendix D), which in the continuous domain can be modeled as output of a linear system driven by white noise.

Conversion to state-space format will yield the necessary matrices for use with the design of the Kalman filter. We will also need estimates of the measurement and input white noise variances. The matrix L is next computed, and it is then used to construct the state-space model of the filter. It is best implemented in this format in real-time and driven by the actual measurements to produce the state estimates. From the state estimates, the output y can next be calculated. We illustrate the use of MATLAB Control Toolbox to design Kalman filter. A simple example is used. The extension to the Shell column is rather complex and hence not included here.

Example 13.11 Use of Kalman Filter

Consider the problem of estimating

$$y_1(s) = \frac{1}{s+1} d(s)$$

from the measurement

$$y_2(s) = \frac{1}{s+2} d(s) + n(s)$$

where $n(s)$ is the measurement noise. Design a Kalman filter to estimate $y_1(s)$ using $y_2(s)$, making suitable assumptions regarding the statistical nature of the input disturbances and measurement noise.

Solution

First we form a state-space model for the system:

$$dy_1 / dt = -y_1 + d,$$
$$dy_2 / dt = -2y_2 + d + v,$$

where we have assumed v as a modified form of the measurement noise. Let us assume that both d and v can be modeled as white noises to simplify the construction of the Kalman filter. Further, let us assume

$$E(d) = 0; E(dd^T) = 1$$
$$E(v) = 0; E(vv^T) = 1$$

The model can be written in the form

$$dx / dt = Ax + Bu + Gd$$
$$z = Cx + Du + v$$

with

$$x = \begin{bmatrix} y_1 \\ y_2 \end{bmatrix}; A = \begin{bmatrix} -1 & 0 \\ 0 & -2 \end{bmatrix}; B = \begin{bmatrix} 0 \\ 0 \end{bmatrix}; C = \begin{bmatrix} 0 & 1 \end{bmatrix}; D = \begin{bmatrix} 0 \end{bmatrix}; G = \begin{bmatrix} 1 \\ 1 \end{bmatrix}.$$

Substituting these into the Matrix-Riccatti equation, we get

$$\begin{bmatrix} -1 & 0 \\ 0 & -2 \end{bmatrix}\begin{bmatrix} l_{11} & l_{12} \\ l_{21} & l_{22} \end{bmatrix} + \begin{bmatrix} l_{11} & l_{12} \\ l_{21} & l_{22} \end{bmatrix}\begin{bmatrix} -1 & 0 \\ 0 & -2 \end{bmatrix} - \begin{bmatrix} l_{11} & l_{12} \\ l_{21} & l_{22} \end{bmatrix}\begin{bmatrix} 0 \\ 1 \end{bmatrix}\begin{bmatrix} 0 & 1 \end{bmatrix}\begin{bmatrix} l_{11} & l_{12} \\ l_{21} & l_{22} \end{bmatrix} + \begin{bmatrix} 1 \\ 1 \end{bmatrix}[1][1 \quad 1] = 0$$

with the solution given by

$$\begin{bmatrix} l_{11} & l_{12} \\ l_{21} & l_{22} \end{bmatrix} = \begin{bmatrix} \dfrac{5+2\sqrt{5}}{2(6+2\sqrt{5})} & \dfrac{1}{1+\sqrt{5}} \\ \dfrac{1}{1+\sqrt{5}} & \sqrt{5}-2 \end{bmatrix} = \begin{bmatrix} .4522 & .3090 \\ .3090 & .2360 \end{bmatrix}$$

Next we construct the estimator

$$d\hat{x} / dt = A\hat{x} + LC^T(z - C\hat{x})$$
$$= \begin{bmatrix} -1 & 0 \\ 0 & -2 \end{bmatrix}\hat{x} + L\begin{bmatrix} 0 \\ 1 \end{bmatrix}[y_2 - \hat{x}_2]$$

In the Laplace domain this leads to the estimator equation

$$\hat{y}_1(s) = \hat{x}_1(s) = \frac{(s+2+l_{22})l_{12} - l_{12}l_{22}}{(s+1)(s+2+l_{22})} y_2(s)$$

$$= \frac{(s+2)}{(1+\sqrt{5})(s+\sqrt{5})(s+1)} y_2(s)$$

$$= \frac{(.309s+.618)}{(s^2+3.23s+2.23)} y_2(s)$$

The calculations can be automated using the commands `lqe` or `kalman` from the Control System toolbox. This is illustrated in the script file *ch13ex12.m.*

♦

13.5.2 Difficulties with the Kalman Filter

In the previous section the Kalman filter was constructed assuming that the input disturbances were white noise. In reality, it is more likely to be modeled as an output of another dynamic system driven by white noise input. This will require additional specifications on the disturbance.

There is a relationship between the estimator we constructed and the suboptimal estimators that were designed earlier. One can show that if the input disturbances were modeled as integrated white noises and the measurement noise were neglected, then we would get optimal estimators that are identical, at least in the steady-state, to the least-square estimators constructed using the techniques discussed in earlier sections. However, construction of optimal estimators when integrated white noises are present is not as straightforward. See Joseph and Brosilow (1978) for a detailed discussion on this point.

Although the Kalman filter has been widely used in the aerospace industry, it has not found many applications in the process industry. A few applications to simulated and laboratory-scale chemical processes have been reported in academic studies. The major problems seem to be associated with the need for specifying a good linear model of the system along with good models for the disturbances and measurement noise. Thus significant manpower is required to construct the filter. Implementation in real time may be computer-intensive for processes with large numbers of state variables. Some studies have shown that the advantages of optimal estimation over suboptimal estimation are not significant. Perhaps with increased availability of modeling, identification, and control system design tools, we may see some applications in the future.

The Kalman filter idea has been extended to do state estimation of nonlinear systems as well. For a thorough discussion of this topic, see Jawinski (1970). For an application study that compares this extended Kalman filter with other methods, see Jang et al. (1986).

13.6 SUMMARY

Inferential control using multiple secondary measurements is used when the primary output to be controlled is not easily measured or when there are substantial delays associated with the primary measurement. The advantages with using multiple measurements are that (1) multiple measurements carry redundant information and hence there is some built-in immunity to measurement noise, and (2) you can detect disturbances early and take corrective action before the primary output is affected severely by the disturbance. The disadvantage is that the control system design is more complex and is more susceptible to sensor failures unless failure protection is programmed in the estimator. We recommend the design of two or more alternative estimators which can be switched when a sensor failure is detected.

The key problem that we addressed in this chapter was the estimation of the effect of the input disturbance on the primary output variable to be controlled. Thus we defined two new variables:

d_y = Effect of unmeasured input disturbances on the primary output y

d_x = Effect of unmeasured input disturbances on the secondary variables x.

d_x is computed from x by subtracting the effect of the manipulated variables from x. A process model relating x to the manipulated input variables is required for this. Such a model is generally determined using well-designed identification tests on the plant.

The design procedure for *linear model-based estimators* can be summarized as follows:

1. Build a process model relating the manipulated variables u and unmeasured disturbances d to the primary and secondary variables. Ideally, this should be a dynamic process model, although a steady-state model may be employed if process remains at steady-state most of the time or if the dynamics characteristics of the outputs are all very similar. If such a model is not available or is difficult to construct, then consider data-driven regression estimators of the type discussed in Chapter 19.
2. Use PCA-based regression methods to design the steady-state estimator. Check the accuracy of the estimator using the projection error criteria. If it is not low enough, then investigate whether additional measurements can be taken. Check robustness by examining the condition number of the matrix to be inverted.
3. Add dynamic compensation if needed, using the ad hoc procedure outlined in Section 13.4. Lead lag compensation can also be tuned on-line.
4. If the process is nonlinear, then *nonlinear estimators* must be designed. Variable transformation techniques are simple to use and should be tried first. If that fails, then a nonlinear model of the process can be used for good steady-state estimation.

A formal solution to the estimation problem exists and is known as the Kalman filter. However, due to its complexity and the large number of parameters required for its specification, the Kalman filter has not yet found widespread application in the process industry. In this chapter we presented some suboptimal solutions that are easier to construct and implement.

Problems

13.1 Given a process with measurements

$$x_1 = u_1 + 2u_2 + 3u_3 + n_1,$$
$$x_2 = u_1 + 2u_2 + 3u_3 + n_2,$$

with

$$E\{u\} = 0; \ E\{n\} = 0; \ \Phi_{uu} = I; \ \Phi_{nn} = I; \ \Phi_{un} = 0,$$

compute the best linear estimator for

$$y = 3u_1 + 3u_2 + 4u_3$$

a.   Using x_1 alone.

b.   Using both x_1 and x_2.

c.   Estimate the accuracy and sensitivity of both estimators by computing the projection error and the condition number.

d.   Which estimator would you recommend, a or b? Explain your choice.

e.   If the noise in the measurement is reduced to zero, will your answer to d change? Explain.

f.   Under what conditions is it worthwhile to put sensors at the same location?

13.2 Verify the results given in Example 13.4. Find the best estimator using only two measurements. Compute the projection error for this estimator.

13.3 Given

$$y(s) = \frac{1}{2s + 1} \ u(s),$$

$$x(s) = \frac{1}{2s + 1} \ u(s) + n(s) \ ; \ n(s) \ \text{is zero mean white noise}$$

design a Kalman filter for $y(t)$ using $x(t)$.

13.4 Verify the results shown in Fig. 13.4 using MATLAB.

13.5 Verify the results shown in Fig. 13.5 using MATLAB.

13.6 Verify the results shown in Example 13.7 using the PCA procedure described in Appendix D. Repeat, using the PLS toolbox, if that is available.

13.7 Derive Eq. (13.28), which gives the estimator when noise is present in the measurements.

13.8 Consider the packed bed reactor shown in the following figure. The primary variable to be controlled is the exit reactor concentration. The disturbances are changes in the feed concentration. There are two secondary measurements of temperature along the length of the reactor. The following process model is available:

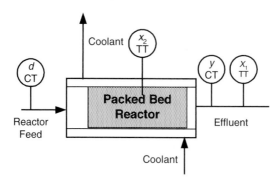

$$x_1(s) = \frac{1}{2s+1}d_1(s) + \frac{3}{3s+1}d_2(s) + \frac{1}{s+1}u(s)$$

$$x_2(s) = \frac{1}{2s+1}d_1(s) + \frac{1}{2s+1}d_2(s) + \frac{1}{2s+1}u(s)$$

$$y(s) = \frac{1}{4s+1}d_1(s) + \frac{4}{5s+1}d_2(s) + \frac{2}{s+1}u(s)$$

a. If it is desired to construct an inferential control system using only one secondary measurement, which measurement should be used? Justify your answer with calculations.

b. If both measurements can be used, derive an estimator for the inferential controller, based on steady-state considerations alone. Will there be any steady-state offset in the inferential control of y?

13.9 A process has secondary measurements given by

$$x_1(s) = \frac{2}{4s+1}d(s) + \frac{2}{3s+1}u(s),$$

$$x_2(s) = \frac{3}{3s+1}d(s) + \frac{2}{3s+1}u(s).$$

It is desired to control the output

$$y(s) = \frac{5}{2s+1}d(s) + \frac{2}{4s+1}u(s).$$

a. Design an inferential estimator using $x_1(s)$.

b. Design an inferential estimator using $x_2(s)$.

c. Design an inferential estimator using both $x_1(s)$ and $x_2(s)$. Which of these three estimators do you recommend, and why?

References

Amirthalingam R., S. W. Sung, and J. H. Lee. 2000. "Two-Step Procedure for Data-Based Modeling for Inferential Control Applications." *AIChE J. 46*(10), 1974–1988.

Antsaklis, P. J., and A. N. Michel. 1997. *Linear Systems*, McGraw Hill, NY.

Bettoni, A., M. Bravi, and A. Chianese. 2000. "Inferential Control of a Side-Stream Distillation Column." *Comput. Chem. Eng. 23*(11–12), 1737–1744.

Budman, H. M.,T. Holcomb, and M. Morari. 1991. "PLS-based robust inferential control of a packed-bed reactor", Proc. of the 1991 American Control Conference, Boston, 1, 256–61.

Chen, C. T. 1999. *Linear System Theory and Design.* 3rd Ed., Oxford University Press, NY.

Chen, C. Y., and B. Joseph. 1987. "On-Line Optimization of a Packed Tubular Reactor." *Ind. and Eng. Chem. Res.*, 26, 1924–1930.

Chen, C. Y., and C. C. Sun. 1991. "Adaptive Inferential Control of Packed-Bed Reactors." *Chem. Eng. Sci.*, *46*(4), 1041–1054.

Clough, D. E., and D. C. Gyure. 1984. "Stochastic Inferential Control of the Fluidized State via Differential Pressure Measurements." *Chem. Eng. Commun.*, *28*(1–3), 59–71.

Doyle, F. J., III. 1998. "Nonlinear Inferential Control for Process Applications." *J. Process Control*, *8*(5-6), 339–353.

Goulding, P. R., B. Lennox, Q. Chen, and D. J. Sandoz. 2000. "Robust Inferential Control Using Kernel Density Methods." *Comput. Chem. Eng.*, *24*(2–7), 835–840.

Guilandoust, M. T., A. J. Morris, and M. T. Tham. 1988. "An Adaptive Estimation Algorithm for Inferential Control." *Ind. Eng. Chem. Res.*, *27*(9), 1658–1664.

Haggblom, K. E. 1996. "Combined Internal Model and Inferential Control of a Distillation Column via Closed-Loop Identification." *J. Process Control*, *6*(4), 223–32.

Hua, X., M. Mangold, A. Kienle, and E. D. Gilles. 1998. "Nonlinear Inferential Control of an Autonomous Periodic Fixed-Bed Reactor." *J. Process Control*, *8*(4), 239–250.

Jang, S.S., H. Mukai, and B. Joseph. 1986. "A Comparison of Two Approaches to On-Line Parameter Estimation and State Estimation of Nonlinear Systems." *Ind. and Eng. Chem. Proc. Des. Dev.*, *25*, 809.

Jang, S. S., H. Mukai, and B. Joseph. 1987. "Control of Constrained Multivariable Nonlinear Processes Using a Two-Phase Approach." *Ind. and Eng. Chem. Proc. Des. Dev.*, *26*(10) 2106–2113.

Jawinski, A. H. 1970. *Stochastic Processes and Filtering Theory*. Academic Press, NY.

Joseph, B., C. B. Brosilow, J. Howell, and N. R. D. Kerr. 1976, (March). "Improve Control of Distillation Columns Using Multiple Temperature Measurements." *Hydrocarbon Processing Journal*.

Joseph, B., and C. B. Brosilow. 1978. "Inferential Control of Processes: Part I. Steady-State Analysis and Design." *AIChE J.*, *24*(3), 485–492.

Joseph, B., and C. B. Brosilow. 1978. "Inferential Control of Processes: Part III. Construction of Optimal and Suboptimal Dynamic Estimators." *AIChE J.*, *24*(3), 500–509.

Kailath, T. 1989. *Linear Systems*, Prentice-Hall, NJ.

Kalman, R. E., and R. S. Bucy. 1961. "New Results in Linear Filtering and Prediction Theory," Tans. ASME, Ser. D, *Journal Basic Eng*, 83, 95–107.

Kano, M., K. Miyazaki, S. Hasebe, and I. Hashimoto. 2000. "Inferential Control System of Distillation Compositions Using Dynamic Partial Least-Squares Regression." *J. Process Control*, *10*(2–3), 157–166.

Kookos, I. K., and J. D. Perkins. 1999. "A Systematic Method for Optimum Sensor Selection in Inferential Control Systems." *Ind. Eng. Chem. Res.*, *38*(11), 4299–4308.

Kresta, J. V., T. E. Marlin, and J. F. MacGregor. 1994. "Development of Inferential Process Models Using PLS." *Comput. Chem. Eng.*, *18*(7), 597–611.

Kwakernaak, H., and R. Sivan. 1972. *Linear Optimal Control Systems*, John Wiley & Sons, NY.

Lee , J., and M. Morari. 1991. "Robust Measurement Selection." *Automatica*, *27*, 519.

Lee, J. H., and A. K. Datta. 1994. "Nonlinear Inferential Control of Pulp Digesters." *AIChE J.*, *40*(1), 50–64.

Lee, J. H., and M. Morari. 1992. "Robust Inferential Control of Multi-Rate Sampled-Data Systems." *Chem. Eng. Sci.*, *47*(4), 865–885.

Mejdell, T., and S. Skogestad. 1993. "Output Estimation Using Multiple Secondary Measurements: High Purity Distillation." *AIChE J.*, *39*(10) 1641.

Morari, M., and A. K. W. Fung. 1982. "Nonlinear Inferential Control." *Comput. Chem. Eng.*, *6*(4), 271–281.

Newton, G. C., L. A. Gould, and J. F. Kaiser. 1957. *Analytical Design of Linear Feedback Controls*. John Wiley & Sons, NY.

Ogawa, M., M. Ohshima, K. Morinaga, and F. Watanabe. 1999. "Quality Inferential Control of an Industrial High Density Polyethylene Process." *J. Process Control*, *9*(1), 51–59.

Ohshima, M., H. Ohno, I. Hashimaoto, M. Sasajima, M. Maejima, K. Tsuto, and T. Ogawa. 1995. "Model Predictive Control with Adaptive Disturbance Prediction and its Application to Fatty Acid Distillation Column Control." *J. Process Control, 5*(1), 41–48.

Oisiovici, R. M., and S. L. Cruz. 2001. "Inferential Control of High-Purity Multicomponent Batch Distillation Columns Using an Extended Kalman Filter." *Ind. Eng. Chem. Res., 40*(12), 2628–2639.

Parrish, J. R., and C. B. Brosilow. "Nonlinear Inferential Control." *AIChE J., 34*(4), 633–644.

Quintero-Marmol, E., and W. L. Luyben. 1992. "Inferential Model-Based Control of Multicomponent Batch Distillation." *Chem. Eng. Sci., 47*(4), 887–898.

Rhodes, I. 1971. "A Tutorial Introduction to Estimation and Filtering." *IEEE Trans. on Aut. Control,* AC–16, 6, 688.

Shen, G. C., and W. K. Lee. 1989. "A Predictive Approach for Adaptive Inferential Control." *Comput. Chem. Eng., 13*(6), 687–701.

Shen, G. C., and W. K. Lee. 1989. "Adaptive Inferential Control for Chemical Processes with Intermittent Measurements." *Ind. Eng. Chem. Res., 28*(5), 557–563.

Shen, G. C., and W. K. Lee. 1988. "Adaptive Inferential Control for Chemical Processes with Perfect Measurements." *Ind. Eng. Chem. Res., 27*(1), 71–81.

Shen, G. C., and W. K. Lee. 1988. "Multivariable Adaptive Inferential Control." *Ind. Eng. Chem. Res., 27*(10), 1863–72.

Soroush, M. 1998. "State and Parameter Estimations and Their Applications in Process Control." *Comput. Chem. Eng., 23*(2), 229–245.

Takamatsu, T., S. Shioya, and Y. Okada. 1986. "Design of Adaptive/Inferential Control System and Its Application to Polymerization Reactors." *Ind. Eng. Chem. Process Des. Dev., 25*(3), 821–828.

Tham, M. T., G. A. Montague, A. J. Morris, and P. A. Lant. 1991. "Soft-Sensors for Process Estimation and Inferential Control." *J. Process Control, 1*(1), 3–14.

Voorakaranam, S., and B. Joseph. 1999. "Model Predictive Inferential Control with Application to a Composites Manufacturing Process." *Ind. Eng. Chem. Res., 38*(2), 433–450.

Yiu, Y. S., and R. K. Wood. 1989. "Multirate Adaptive Inferential Control of Distillation Column Bottoms Composition." *Adv. Instrum. Control, 44*(1), 65–78.

Yu, C. C., and W. L. Luyben. 1984. "Use of Multiple Temperatures for the Control of Multicomponent Distillation Columns." *Ind. Eng. Chem. Process Des. Dev., 23*(3), 590–597.

Discrete-Time Models

Objectives of the Chapter

- Introduce discrete-time models used in computer-controlled systems.
- Introduce the z-transform representation.
- Introduce FIR (finite impulse response) and FSR (finite step response) models.
- Introduce discrete-time versions of the PID controller.

Prerequisite Reading

Chapter 2, "Continuous-Time Models"

We shift our focus from continuous-time models and controllers to discrete-time models and controllers. Application of computer control systems necessitates this representation. The computer is a discrete-time device that can only measure (i.e., collect data from sensors) and take control action at discrete points in time. A typical DCS (distributed computer control system) employed in the process industries has sample times (the time intervals at which measurements and control actions are taken) of 1 to .25 second. Thus the lower level controllers act with this sampling rate. A higher-level controller in the dynamic optimization layer may collect data and take control action every few minutes. The sample time used is usually based on the response time of the process to the control actions taken.

In order to develop model based control strategies for such sampled data systems, we must first have some ways of representing process behavior in the discrete-time (sampled data) domain. The objective of this chapter is to discuss various types of discrete-time model representations.

We start with the development of a mathematical framework for modeling sampled data systems. This leads to a discrete equivalent of the continuous-time Laplace transform called z-transform representation. Z-transforms provide for a compact representation of relationships among discrete-time variables. Next we introduce the discrete equivalent of the impulse response model. The impulse response of a process usually becomes negligibly small after some time, and this allows us to build FIR models and the equivalent FSR models. The latter form is used extensively in industrial implementations of model-based control.

14.1 THE Z-TRANSFORM REPRESENTATION

Consider a continuous function $y(t)$, sampled at equal intervals of time, with a sampling time of T_s as shown in Figure 14.1. The sampled values can be written as the sequence $y(0)$, $y(T_s)$, $y(2T_s)$, $y(nT_s)$.

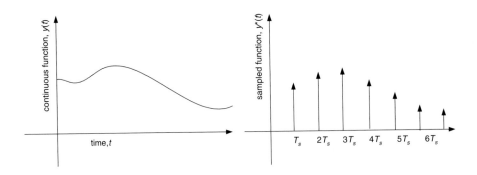

Figure. 14.1. A continuous function and its sampled values.

We define a sampled function $y^*(t)$ as

$$y^*(t) \equiv y(0)\delta(0) + y(T_s)\delta(T_s) + y(2T_s)\delta(2T_s) + \ \dots\ + y(nT_s)\ \delta(nT_s)\ + \ \dots, \quad (14.1)$$

where $\delta(t)$ is the Dirac Delta function. Thus the sampled function $y^*(t)$ is a sequence of impulses with strength given by the value of $y(t)$ at the time of sampling. The Laplace transform of $y^*(t)$ is given by

$$y^*(s) = y(0) + y(T_s)e^{-T_s s} + y(2T_s)e^{-2T_s s} + \dots + y(nT_s)e^{-nT_s s} + \dots \qquad (14.2)$$

We define $z^{-1} = e^{-T_s s}$ (z^{-1} is called the **unit delay operator**).

$$y^*(s) \equiv y(0)\delta(0) + y(T_s)z^{-1} + y(2T_s)z^{-2} + \dots + y(nT_s)z^{-n} + \dots,$$
$$\equiv y(z) \qquad (14.3)$$

$y(z)$ is called the z-transform of $y^*(t)$. We define the operator $Z(\bullet)$ as

$$Z\left\{y^*(t)\right\} = y(z). \qquad (14.4)$$

The operator z^{-1} represents a delay of one sample time. Hence

$$z^{-1} y(z) = Z\left[y^*\left(t - T_s\right)\right]. \qquad (14.5)$$

Frequently, we use the compact notation y_n to represent the sampled function $y(z)$. Using this notation, Equation (14.5) can be written in compact notation as

$$z^{-1}y_n = y_{n-1}.$$

Example 14.1 Discrete Transfer Functions

Consider a system modeled by a difference equation that relates the states of the system at discrete points in time. $x(nT_s)$ represent the state of a system at $t = nT_s$.

$$x(nT_s) = a\ x((n-1)T_s) + b\ u(nT_s).$$

If we apply the z-transform to both sides of this relationship, we get

$$x(z) = z^{-1}x(z) + bu(z).$$

Writing in the compact notation,

$$x_n = az^{-1}x_n + bu_n.$$

Solving for $x(z)$, we get

$$x(z) = \left(\frac{b}{1 - az^{-1}}\right)u(z).$$

The function

$$G(z) = \frac{b}{1 - az^{-1}}$$

is the **discrete transfer function** relating $x(z)$ and $u(z)$.

◆

Example 14.2 Z-Transform of the Unit Step Function

Consider the unit step function given by

$$u(t) = 0 \quad t < 0,$$
$$u(t) = 1 \quad t \geq 0.$$

If it is sampled at $t = 0, 1, 2, ...,$ we get

$$u^*(t) = 1, \; t = 0, 1, 2 \; ...$$

$$
\begin{aligned}
u(z) &= u(0) + u(1)z^{-1} + u(2)z^{-2}... \\
&= 1 + z^{-1} + z^{-2} + ... \; + z^{-n} + \; ... \\
&= \frac{1}{1 - z^{-1}}.
\end{aligned}
$$

◆

Example 14.3 Stability in the Z-Domain

Consider a signal with transfer function

$$y(s) = \frac{1}{s + p}.$$

This has the time domain representation

$$y(t) = e^{-pt},$$

which implies that it is stable (finite for all values of t) only if $p > 0$. The corresponding z-domain representation of the signal (sampled values of $y(t)$) is given by

$$y(z) = y_0 + y_1 z^{-1} + y_2 z^{-2} + \; ...,$$

where

$$y_k = y(kT_s) = e^{-pk T_s}.$$

Hence

$$y(z) = 1 + e^{-pT_s} z^{-1} + e^{-2pT_s} z^{-2} + \ \cdots$$

$$= \frac{1}{1 - e^{-pT_s} z^{-1}}.$$

If a discrete transfer function has the form

$$g(z) = \frac{1}{1 - az^{-1}} = 1 + az^{-1} + a^2 z^{-2} + \ \cdots,$$

then the resulting time domain sequence is finite only if $|a| < 1$, which is true if p is on the RHP. If a is a negative, we will get oscillatory response. In general, if any pole of a z-transfer function lies outside the unit circle, then the system is unstable.

♦

The general form of the discrete transfer function is given by

$$G(z) = \frac{y(z)}{u(z)} = \frac{\left(b_1 z^{-1} + b_2 z^{-2} \cdots + b_m z^{-m} \right)}{\left(a_0 + a_1 z^{-1} + \cdots + a_k z^{-k} \right)}, \qquad (14.6)$$

which corresponds to the discrete-time difference equation model

$$a_0 y_n = -(a_1 y_{n+1} + a_2 y_{n-1} + a_k y_{n-k})$$
$$+ b_1 u_{n-1} + \cdots + b_m u_{n-m}. \qquad (14.7)$$

Table 14.1 lists the z-transform of some common functions.

Table 14.1. Z-Transform of Common Function.

Function	Z-Transform	Laplace Transform
$u(t)$	$\dfrac{1}{1 - z^{-1}}$	$\dfrac{1}{s}$
$e^{-t/\tau}$	$\dfrac{1}{1 - e^{-T_s/\tau} z^{-1}}$	$\dfrac{\tau}{\tau s + 1}$
$\delta(t)$	1	1
$f(t - kT_s)$	$F(z) \cdot z^{-k}$ where $F(z) = Z\{f(t)\}$	$e^{-kT_s s}$
$1 - e^{-t/\tau}$	$\dfrac{z^{-1}(1-a)}{(1 - z^{-1})(1 - az^{-1})}$ $a = e^{-T_s/\tau}$	$\dfrac{1}{s(\tau s + 1)}$

14.2 MODELS OF COMPUTER-CONTROLLED SYSTEMS

In computer control of continuous processes, we are faced with the problem of relating the discrete events taking place in the computer with the continuous events going on in the process. The discrete sequence of control commands emanating from the computer $(u_1, u_2, u_n, ...)$ is converted into a continuous function by means of a device called the zero-order hold or ZOH (higher order holds can also be used, but will not be discussed here). The ZOH generates a continuous input $u(t)$ for controlling the process.

The ZOH keeps the output constant in between sample times; that is,

$$u(t) = u(nT_s) \quad nT_s \leq t \leq (n+1)T_s$$

The transfer function model of the ZOH can be obtained by considering the response of the process to a unit impulse. Consider a unit impulse input to the ZOH device

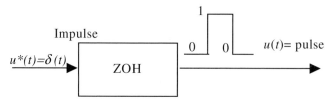

The unit impulse input results in a pulse of height 1. This response has the transfer function (from the figure above)

$$G_{ZOH}(s) = \left(\frac{1}{s} - e^{-T_s s} \frac{1}{s} \right). \tag{14.8}$$

Now consider a process in conjunction with a ZOH and a sampling device

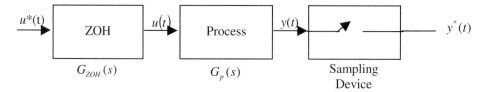

where $y^*(t)$ represents the sampled form of $y(t)$, where the sampler reads the value of the output at every sample time. This is typical of the situation with a computer control system. The Z-transfer function relation between $y^*(t)$ and $u^*(t)$ is given by

$$G(z) = Z\left[G_{ZOH}(s) \cdot G_p(s)\right] .$$

(14.9)

Example 14.4 A First Order Process with ZOH

Consider a first-order process with a ZOH:

The Z-transfer function, after some algebra and using the transforms of Table 14.1, is given by.

$$
\begin{aligned}
\frac{y(z)}{u(z)} &= Z\left[\frac{1}{s}(1-e^{T_s s})\frac{K}{\tau_s+1}\right] \\
&= Z\left\{\frac{K}{s(\tau s+1)}\right\} - z^{-1}Z\left\{\frac{K}{s(\tau s+1)}\right\} \\
&= K\frac{(1-a)z^{-1}}{1-az^{-1}},\ a = e^{-T_s/\tau},
\end{aligned}
$$

(14.10)

or equivalently, converting to the difference form,

$$y_{n+1} = ay_n + K(1-a)\,u_n.$$

(14.11)

Example 14.5 shows an alternate way to convert continuous-time models to discrete-time models. Such conversions may also be done using MATLAB's c2d command.

♦

Example 14.5 Finite Difference Approximations

Another way to obtain discrete models from continuous models is to use the finite difference approximation. Consider the system described by

$$\frac{y(s)}{u(s)} = \frac{K}{\tau s+1}.$$

Converting back to time domain,

$$\tau \dot{y}(t) + y(t) = Ku(t).$$

Discretizing, using finite difference approximation,

$$\tau \frac{(y_{n+1} - y_n)}{T_s} - y_n = Ku_n,$$

which is equivalent to replacing s (i.e., d/dt) with the discrete operation $(z-1)/T_s$. Solving for y_n,

$$
\begin{aligned}
y_{n+1} &= \frac{\left(\dfrac{\tau}{T_s} - 1\right)}{(\tau/T_s)} y_n + \frac{KT_s}{\tau} u_n. \\
&= a_1 y_n + b_1 u_n
\end{aligned}
\tag{14.12}
$$

Problem 14.1 compares the two approaches towards discretization presented in Examples 14.4 and 14.5.

♦

Example 14.6 Discrete-Time First-Order Filter

Consider the transfer function of a first-order transfer function given by

$$g(s) = \frac{1}{\tau s + 1}.$$

The frequency response of this transfer function has a corner frequency at $\omega_c = 1/\tau$, which means that frequencies greater than ω_c contained in the input will be reduced in amplitude (filtered out) by this system. Hence this transfer function can be viewed as a noise filter. A discrete form of the filter can be obtained as follows:

Let us assume that the filter is implemented using a ZOH.

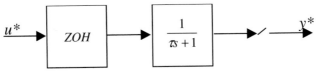

Then

$$y(z) \quad = \quad Z[ZOH(s) \cdot g(s)]u(z)$$

$$= \quad \frac{(1-a)z^{-1}}{1-az^{-1}}u(z); a = e^{-T_s/\tau}$$

or

$$y_{n+1} = ay_n + (1-a)u_n.$$

We can compare the frequency response of the discrete filter with the frequency response of the continuous filter. To do this, recognize that $z^{-1} = e^{-T_s s}$. Hence for the discrete filter, the Laplace transform is given by

$$g_d(s) = \frac{(1-a)e^{-T_s s}}{1-ae^{-T_s s}}.$$

Letting $s = i\omega$, we can compute the frequency response,

$$g_d(i\omega) = \frac{(1-a)e^{-T_s i\omega}}{1-ae^{-T_s i\omega}}.$$

Figure 14.2 compares the continuous and discrete filters for $\tau = 1$, $T_s = 1, .5, .1$. As the sample time gets smaller, the approximation gets better.

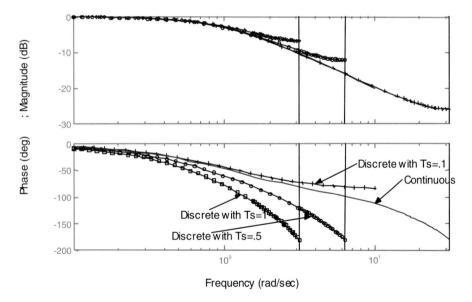

Figure 14.2 Comparison of continuous and discrete forms of a first-order filter.

◆

Example 14.7 Digital Implementation of the PID Controller

Consider the ideal PID controller:

$$m(t) = k_c \left[e(t) + \frac{1}{\tau_I} \int_o^t e(t)dt + \tau_D \frac{de}{dt} \right].$$

A discrete form of the controller can be obtained by using rectangular integration for the integral term and finite difference for the derivative term.

$$m_n = k_c \left[e_n + \frac{1}{\tau_I} \sum_{k=0}^{n} e_k T_s + \tau_D \frac{e_n - e_{n-1}}{T_s} \right],$$

where $e_n = e(nT_s)$ is the error at current time and $m_n = m(nT_s)$ is the current measurement. This is called the *position form* of the PID controller, since it gives the current valve position, m_n, directly. An *incremental (velocity) form* of the controller is obtained by subtracting

$$m_{n-1} = k_c \left[e_{n-1} + \frac{1}{\tau_I} \sum_{k=0}^{n-1} e_k T_s + \tau_D \frac{e_{n-1} - e_{n-2}}{T_s} \right],$$

resulting in the equation

$$\Delta m_n = m_n - m_{n-1} = k_c \left[e_n - e_{n-1} + \frac{1}{\tau_I} e_n T_s + \tau_D \frac{e_n - 2e_{n-1} + e_{n-2}}{T_s} \right].$$

In z-transform notation, this can be written as

$$\Delta m_n(z) = k_c \left[(1-z) + \frac{T_s}{\tau_I} + \frac{\tau_D}{T_s} (1 - 2z^{-1} + z^{-2}) \right] e(z).$$

This is called the velocity form of the control algorithm.

$\blacklozenge$

14.3 DISCRETE-TIME FIR MODELS

Consider a process with the transfer function model

$$y(s) = G(s)u(s). \tag{14.13}$$

Suppose $u(t) = \delta(t) \Rightarrow u(s) = 1$, then the output is given by

$$y(s) = G(s). \tag{14.14}$$

Hence $G(s)$ is the impulse response of the process. Let

$$g(t) = L^{-1}[G(s)], \tag{14.15}$$

where $g(t)$ is the time domain representation of the impulse response. Since

$$y(t) = L^{-1}[G(s)u(s)], \tag{14.16}$$

which by using the convolution property of Laplace transform is

$$y(t) = \int_o^t g(t-\tau)u(\tau)d\tau, \tag{14.17}$$

assuming that $u(t) = 0$ for $t < 0$. Otherwise,

$$y(t) = \int_{-\infty}^t g(t-\tau)u(\tau)d\tau, \tag{14.18}$$

which says that the current value of y is the result of convoluting the past inputs with its impulse response. We will use the latter form when we want to predict the current or future values of the output using past control actions.

Next consider a continuous process with a ZOH and output sampler as we discussed earlier:

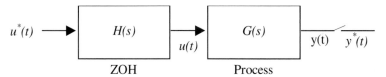

$$u^*(t) \quad \boxed{H(s)} \quad u(t) \quad \boxed{G(s)} \quad y(t) \quad y^*(t)$$

ZOH   Process

If $u^*(t)$ is a unit impulse, then $u(t)$ will be a unit pulse and $y(t)$ will be **pulse** response of the process, $G(s)$. Let us say that the sampled values $y^*(t)$ of this response is given by

$$h_0, h_1, h_2, ..., h_n.$$

The coefficients h_k can be obtained directly from the z-transfer function representation of the system with ZOH. Let

$$G(z) = Z\{H(s)G(s)\}.$$

If $G(z)$ can be expanded as an infinite series of the form

$$G(z) = h_0 + h_1 z^{-1} + h_2 z^{-2}... + h_n z^{-n} + ...,$$

then if $u(z) = 1$ (for an impulse input), output will be the sequence $y_0 = h_0$, $y_1 = h_1$, $y_n = h_n$, and so on.

Next consider the case where there are two input pulses. If the input consisted of a pulse u_0 at $t = 0$ and u_1 and $t = T_s$ (next sample time), then the output will be the sum of the effect of the two pulses (using the linearity property of linear systems):

$$y_n = h_n u_0 + h_{n-1} u_1.$$

Extending this idea, we can express the current y_n in terms of all the past inputs u_n, u_{n-1}, u_{n-2}, and so on.

$$\begin{aligned} y_n &= h_0 u_n + h_1 u_{n-1} + h_2 u_{n-2} \dots \\ &= \sum_{k=0}^{\infty} h_k u_{n-k}. \end{aligned} \tag{14.19}$$

This is the equivalent of the continuous-time impulse response representation Eq. (14.18).

For stable systems, the impulse response decays to zero as $t \to \infty$. Let us say that the coefficients h_n can be neglected after n exceeds some value r.

$$h_n \approx 0, n > r.$$

Under these circumstances, we have a *discrete-time FIR* (finite impulse response) model of the process

$$y_n = \sum_{k=0}^{r} h_k u_{n-k}. \tag{14.20}$$

14.4 DISCRETE-TIME FSR MODELS

Consider the FIR model

$$y_n = \sum_{k=0}^{r} h_k u_{n-k}.$$

To get a unit step response, we set

$$\begin{aligned} \left. \begin{array}{r} y_k = 0 \\ u_k = 0 \end{array} \right\} & \quad k < 0 \\ u_k = 1 & \quad k \geq 0. \end{aligned}$$

Hence

$$
\begin{aligned}
y_0 &= h_0 & &= a_0 \\
y_1 &= h_0 + h_1 & &= a_1 \\
y_r &= h_0 + h_1 + h_2 + \ldots h_r & &= a_r \\
y_{r+1} &= h_0 + h_1 + h_2 + \ldots\ h_r & &= a_r \\
y_{r+1} &= y_{r+2} = \ldots & &= a_r.
\end{aligned}
$$

Hence the unit step response coefficients are given by a_0, a_1, a_2, and so on. Also note that the inverse relationship,

$$
\begin{aligned}
h_k &= a_k - a_{k-1}, \quad k = 1,\ 2,\ \ldots \\
h_0 &= a_0.
\end{aligned}
$$
(14.21)

So we can write

$$
\begin{aligned}
y_n &= \sum_{k=0}^{r} h_k u_{n-k} \\
&= a_0 u_n + (a_1 - a_0)u_{n-1} + \ldots + (a_r - a_{r-1})u_{n-r} \\
&= a_0(u_n - u_{n-1}) + a_1(u_{n-1} - u_{n-2}) + \ldots + a_{r-1}(u_{n-r+1} - u_{n-r}) + a_r u_{n-r} \\
&= a_0 \Delta u_n + a_1 \Delta u_{n-1} + \ldots + a_{r-1} \Delta u_{n-r+1} + a_r u_{n-r}.
\end{aligned}
$$
(14.22)

This is called the FSR (finite step response) model of the process. It is used in some industrial model predictive control algorithms (see Chapter 17). The coefficients a_k are easily obtained by sampling the unit step response of the process. Note that we cannot obtain impulse response coefficients by reading the values of the impulse response at sampled points (see Problem 14.7).

Example 14.8 FIR Model of a Second-Order System

Consider a process with transfer function

$$
y(s) = \frac{1}{s^2 + s + 1}\, u(s)
$$

The impulse response is shown in Figure 14.3a

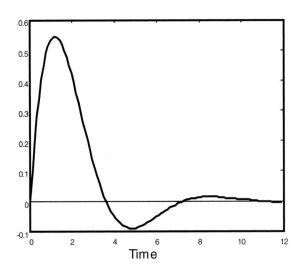

Figure 14.3a Impulse response of the process in Example 14.8.

The response becomes negligible after about 10 units of time. The z-transform of $y(s)$ with a ZOH and a sample time of 0.5 is given by

$$y(z) = \frac{.1044 + .08828\ z^{-1}}{1 - 1.44\ z^{-1} + .6065\ z^{-2}}, \tag{14.23}$$

and the coefficients h_i can be obtained by carrying out the long division suggested in Eq. (14.23). Another way to obtain these coefficients is to generate a unit impulse response of the discrete transfer function. This gives

$h = [0 \quad 0.1044 \quad 0.2359 \quad 0.2702 \quad 0.2389 \quad 0.1739 \quad 0.1010\ 0.0373 \quad -0.0085 \quad -0.0347$
$-0.0439 \quad -0.0410 \quad -0.0313 \quad -0.0194 \ -0.0085 \quad -0.0002 \quad 0.0049 \quad 0.0070 \quad 0.0069$
$0.0056].$

This is shown in Figure 14.3b. Figure 14.4a shows the step response. Figure 14.4b shows the step response coefficients a_i, which are

$a = [0 \quad 0.1044 \quad 0.3403 \quad 0.6105 \quad 0.8494 \quad 1.0234 \quad 1.1244 \quad 1.1616 \quad 1.1531 \quad 1.1184$
$1.0746 \quad 1.0336 \quad 1.0023 \quad 0.9828 \quad 0.9744 \quad 0.9742 \quad 0.9790 \quad 0.9860 \quad 0.9929$
$0.9985].$

Note that there is a one-to-one correspondence between the step response coefficients and the values of the continuous-time step response at sample times. This correspondence is not true for the pulse response coefficients, since the sample time affects the coefficients.

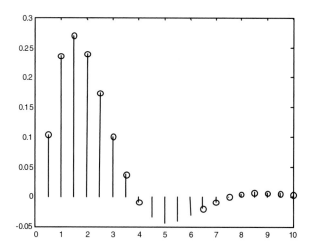

Figure 14.3b FIR coefficients of the process in Example 14.8.

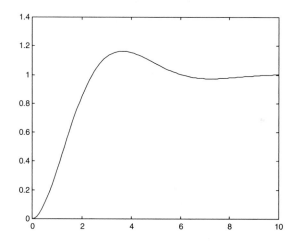

Figure 14.4a Step response of the process shown in Example 14.8.

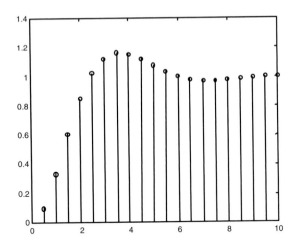

Figure 14.4b Step response coefficents of the process shown in Example 14.8.

◆

14.5 SUMMARY

In this chapter we introduced the concept of a sampled data system. The z-transform is a compact way of representing sampled data systems. For computer-controlled systems, the FIR and FSR models provide an alternate way of representing the process model, even though using a large set of parameters.

When the sampling time is small compared to the response time of the process (i.e., the sampling rate is high compared to the bandwidth of the process), the discrete and continuous systems behave similarly. However, lower sampling rates can introduce additional delay in the loop and adversely affect the stability of the closed-loop system (see Problem 14.4).

Most often, we have to build empirical models from sampled input/output data. For empirical model building, the discrete-time models (z-domain, FIR, or FSR) are the most convenient. Once a model is obtained in the discrete domain, we can generate the continuous-time model, if needed. Chapters 15 and 16 discuss the techniques used to generate process models, given input/output data sets.

Industrial implementations emphasize the finite step or impulse response model of the process. This has the added convenience of not having to make any assumptions regarding the order of the system and the time delays present. The disadvantage is in identifying the large number of model parameters needed for this representation.

Problems

14.1. Consider the process modeled by

$$\frac{dy_1}{dt} + y_1 = u$$

$$\frac{dy_2}{dt} + y_2 = y_1.$$

    a.   Derive a transfer function model between y_2 and u.

    b.   Derive a discrete-time transfer function model using a ZOH and sample time = 0.2.

    c.   Check your answer using MATLAB function *c2d*.

    d.   Repeat b, using a finite difference approximation by replacing s with $\left(1 - z^{-1}\right)/T_s$.

    e.   Another way of obtaining a discrete model is to use the trapezoidal rule for integration. Show that this is equivalent to replacing $s\left(ie.,\ \dfrac{d}{dt}\right)$ by $\dfrac{2}{T}\dfrac{z-1}{z+1}$. Check your answer using the *c2d* function in MATLAB.

    f.   Compare the frequency response of all three discrete approximations with the continuous system frequency response. You can use the dbode function in MATLAB to graph the bode plot of discrete transfer functions.

14.2 Find an FIR model of the system in Problem 14.1, using $T_s = 0.2$ and $r = 10$. Check the first few coefficients, using the ZOH approximation obtained in Problem 14.1. Also, an FSR model.

14.3 Compare the frequency response of the FIR model in Problem 14.2 with the frequency response of the continuous system. Estimate the frequency at which the error approximation becomes greater than 1%.

14.4 Consider the proportional control of a sampled data first-order system, as shown in the figure below.

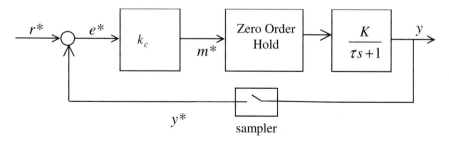

    a.   Obtain a z-transfer function relation between $y(z)$ and $m(z)$.

    b.   Derive the closed loop z-transfer function relation between $y(z)$ and $r(z)$.

Note: The rules for manipulating z-transfer function diagrams are identical to those for s-transfer function diagrams.

c.   A z-transfer function is unstable if the poles of the system lie outside the unit circle (i.e., $\left|z_p\right| > 1$, where z_p is a root of the denominator of the closed-loop transfer function). Determine the smallest value of k_c that will make the closed-loop system unstable. For continuous systems, the system is stable for all values of k_c. However, sampling introduces a delay that causes instability at large values of k_c.

The answer is given by $k_c = \dfrac{1}{K}\left(\dfrac{1+e^{-T_s/\tau}}{1-e^{-T_s/\tau}}\right)$. As T_s gets smaller, this k_c gets larger, as expected.

14.5   Show that the result given in Eq. (14.11) can be obtained by solving the differential equation for the output $y\left[(n+1)T_s\right]$ with

$$\tau \frac{dy}{dt} + y = K\, u(t),$$
$$y(n\,T_s) = y_n,$$
$$u(t) = u_n \quad nT_s < t \le (n+1)T_s.$$

14.6   Show that the FIR coefficients of a continuous process with a ZOH is given by

$$h_k = \int_{(k-1)T_s}^{k\,T_s} g(t)\,dt \ ,$$

where T_s is the sample time and $g(t)$ is the impulse response of the process. These coefficients can be approximated by $h_k = g(kT_s)T_s$.

14.7   Show that discrete representation obtained in Example 14.1, using a finite difference, approximates the discrete model derived using a ZOH as τ/T_s gets large (i.e., T_s is small if compared to τ).

14.8   Obtain the z-transfer function of a PID controller with a first-order noise filter built into the derivative action. Use the first-order finite difference form suggested in Problem 14.1. The transfer function of such a PID controller is given by

$$m(s) = K_c(1 + \frac{1}{\tau_I s} + \frac{\tau_D s}{\alpha\tau_D s+1})e(s).$$

Compare the frequency response of the PID controller $\left(1+\dfrac{1}{s}+\dfrac{s}{.1s+1}\right)$ with its discrete equivalent for $T_s = 0.1, 0.5$.

14.9 Show that the z-transform of $\dfrac{1}{s(\tau s+1)}$ is given by $\dfrac{z^{-1}(1-a)}{(1-z^{-1})(1-az^{-1})}$.

Hint: Use partial fractions to expand the function, and then apply the results of Examples 14.2 and 14.3.

14.10 One way to obtain an approximate discrete form of a continuous system is by using trapezoidal integration. In this approximation the term $z^{-1} = e^{-sT_s}$ is replaced by the (1,1) Padé approximation $z^{-1} = e^{-sT_s} = \dfrac{1-sT_s/2}{1+sT_s/2}$, yielding $s = \dfrac{2}{T_s} = \dfrac{1-z^{-1}}{1+z^{-1}}$. This is called the Tustin approximation. For the first-order system $p(s) = \dfrac{1}{s+1}$, compare the frequency response of the following approximations (for $T_s = 0.1, .5$)

    a. $p(z)$ obtained using a ZOH,
    b. $p(z)$ obtained using Tustin approximation,
    c. $p(z)$ obtained using backward differences $s = (1-z^{-1})/T_s$.

Use the *bode* command in MATLAB to get frequency response. Which method yields the best approximation?

References

Åström, K. J., and B. Wittenmark. 1990. *Computer Controlled Systems: Theory and Design,* 2nd Ed. Prentice Hall, NJ.

Deshpande, P. B., and R. H. Ash. 1988. *Computer Process Control with Advanced Control Applications,* 2nd Ed.. Instrument Society of America, NC.

Franklin, G. F., and J. D. Powell. 1990. *Digital Control of Dynamic Systems,* 2nd Ed. Addison-Wesley, NY.

Luyben, W. L. 1989. *Process Modeling, Simulation, and Control for Chemical Engineers,* 2nd Ed. McGraw Hill, NY.

Ogunnaike, B., and W. H. Ray. 1994. *Process Dynamics Modeling and Control.* Oxford University Press, NY.

Seborg, D. E., T. F. Edgar, and D. A. Mellichamp. 1989. *Process Dynamics and Control*. John Wiley & Sons, NY.

Stephanopoulos, G. 1984. *Chemical Process Control: An Introduction to Theory and Practice*. Prentice Hall, NJ.

William, T. J. 1984. *The Use of Digital Computers in Process Control*. Instrument Society of America, Instrument Learning Modules Series, Research Triangle Park, NC.

CHAPTER **1 5**

Identification: Basic Concepts

Objectives of the Chapter

The overall objective is to present the basic concepts for determining a process model from empirical tests conducted on the plant. Specific topics to be discussed include

- The identification problem.
- The least-squares formulation and solution.
- Practical considerations.

Prerequisite Reading

Chapter 14, "Discrete-Time Models"
Appendix C, "Linear Least-squares Regression"
Appendix D, "Random Variables and Random Processes"

15.1 INTRODUCTION

Identification is used to describe the process of determining a process model from empirical test data collected from a process. In Chapter 2 we discussed the step test method used by process engineers to estimate parameters for an FOPDT (first-order plus dead time) model.

In this chapter we will establish more accurate methods for determining process models. Since the data available is in discrete form, we will focus on identification of discrete-time models first. We will also illustrate how to estimate time domain models from the discrete model.

The topic is discussed in two chapters. In this chapter we will introduce the basic concepts along with some practical considerations. The next chapter will focus on issues such as design of the plant test, filtering of noise, modifications to the basic least-squares algorithm, and application to multiple output systems. A comprehensive case study example is included in the latter.

Due to the difficulty of obtaining accurate dynamic process models from first principles, industry depends on empirically determined models for implementing model-based control. The quality of the model (accuracy of the model) depends on the type and duration of empirical tests conducted and the algorithms used to determine the model parameters. The identification problem, as we define it, is as follows:

> *Perturb the system using suitable input forcing functions. Measure the output response. Determine process models that lead to good control-system performance.*

Although the final objective is to achieve good control, the problem is usually simplified by redefining the objective as follows:

> *Given input/output operational data, determine process models that fit the data, usually in a least-squares sense.*

There are two main aspects to this problem:

1. *The structure of the process model.* We will restrict ourselves to linear models in this text. Because of the sampled nature of the data that is generally available, attention will be focused on discrete-time linear models. Even if we restrict ourselves to linear process models, there is still a question of the representation used. For example, do we use a transfer function model or a discrete finite impulse response (FIR) model? If a transfer function model is used, what order should we use for the numerator and denominator?

2. *The choice of the objective function used to evaluate the fit.* An obvious choice for the objective function to be minimized is the sum of the square of the prediction errors, where

    Prediction Error (PE) = actual output − predicted output

This is called a *least-squares fit*. Variations in the objective function arise from weighting functions applied to the error to enhance or retard specified characteristics of the model fit. For example, we could emphasize that the fit be good in a certain frequency range using a weighted least-squares.

In this chapter we first show that the identification problem can be formulated as a least-squares parameter estimation problem. Next we present the solution to the general least-squares identification problem. The properties of the parameter estimates resulting from a least-squares solution are given. It is shown that the accuracy depends on the quality of the data used, which in turn is dependent on the type of input signals used. The chapter concludes with a discussion on the practical considerations involved.

15.2 LEAST-SQUARES ESTIMATION OF PARAMETERS

As mentioned earlier, the problem of identification can be reduced to the problem of least-squares estimation of parameters. We illustrate this idea using an example and then generalize the idea to linear discrete-time dynamic process model identification.

15.2.1 Formulation of the General SISO Identification Problem (Known Time Delay Case)

In this section we will develop the general procedure for determining a discrete transfer function model from plant data. The continuous transfer function parameters may be recovered from the relationship between the discrete and continuous models that were discussed in Chapter 14. Much of what follows uses *linear least-squares theory*, a brief summary of which is given in Appendix C.

Consider the problem of identifying a transfer function model,

$$G(z) = z^{-nk} \frac{B(z)}{A(z)},$$

$$B(z) = b_1 + b_2 \, z^{-1} + \ldots + b_{nb} \, z^{-nb+1}, \tag{15.1}$$

$$A(z) = 1 + a_1 \, z^{-1} + \ldots a_{na} \, z^{-na},$$

where

na = order of denominator polynomial
nb = order of numerator polynomial
nk = time delay in the process, assumed known for now.

This leads to the discrete-time model

$$y_n + a_1 \, y_{n-1} + a_2 \, y_{n-1} \ldots + a_{na} \, y_{n-na}$$

$$= b_1 \, u_{n-nk} + b_2 \, u_{n-nk-1} + b_{nb} \, u_{n-nk-nb+1}. \tag{15.2}$$

Let e_n represent the error in the measurement. Assume we have the following set of input/output data

n	u_n	y_n
0	u_0	y_1
1	u_1	y_2
...	...	...
N	u_N	y_N

Then we can set up a linear least-squares problem as follows:

$$\left(k > n_a, \, n_k + n_b, \, n = k + 1, \, ... \, N\right)$$

$$-a_1 y_{k-1} - a_2 y_{k-2} ... - a_{na} y_{k-na} + b_1 u_{k-nk} + b_2 u_{k-nk+1} + ... + b_{nb} u_{k-nk-nb+1} = y_k + e_k$$

$$-a_1 u_k - a_2 y_{k-1} ... - a_{na} y_{k+1-na} + b_1 u_{k+1-nk} + b_2 u_{k-nk} + ... + b_{nb} u_{k-nk-nb+2} = y_k + e_{k+1}$$

$$-a_1 y_{N-1} - a_2 y_{N-2} ... - a_{na} y_{N-na} + b_1 u_{N-nk} + b_2 u_{N-nk-1} + ... + b_{nb} u_{N-nk-nb+1} = y_N + e_N,$$

which represents $N - k + 1$ equations in the $na+nb$ unknown parameters $a_1, a_2, ..., a_{na}$, $b_1, b_2, ..., b_{nb}$. This can be written compactly as

$$\mathbf{\Phi} \cdot \mathbf{\theta} = \mathbf{Y} + \mathbf{e}, \tag{15.3}$$

where

$$\mathbf{\Phi} = \begin{bmatrix} -y_{k-1} & -y_{k-2} & \cdot \cdot & y_{k-na} & u_{k-nk} & u_{k-nk-1} & u_{k-nk-nb+1} \\ -y_k & -y_{k-1} & \cdot \cdot & y_{k-na+1} & u_{k-nk+1} & u_{k-nk} & u_{k-nk-nb+2} \\ \cdot & \cdot & \cdot \cdot & \cdot & \cdot & \cdot & \cdot \\ \cdot & \cdot & \cdot \cdot & \cdot & \cdot & \cdot & \cdot \\ -y_{N-1} & -y_{N-2} & \cdot \cdot & y_{N-na} & u_{N-nk} & u_{N-nk-1} & u_{N-nk-nb+1} \end{bmatrix} \tag{15.4}$$

and

$$\mathbf{\theta} = \begin{bmatrix} a_1 \\ a_2 \\ \cdot \\ a_{na} \\ b_1 \\ b_2 \\ b_{nb} \end{bmatrix}; \quad \mathbf{y} = \begin{bmatrix} y_k \\ y_{k+1} \\ \\ \\ \\ \\ y_N \end{bmatrix}; \quad \mathbf{e} = \begin{bmatrix} e_k \\ e_{k+1} \\ \\ \\ \\ \\ e_N \end{bmatrix}.$$

In general, we will need to collect far more data points than we have parameters to reduce the effect of the measurement errors, so we will have more equations than unknowns. If we seek a solution that minimizes the sum of the square of the errors.

$$V(\boldsymbol{\theta}) = Min\|\mathbf{e}\|_2^2 = \sum_{i=k}^{N} e_i^2, \tag{15.5}$$

then the problem can be formulated as

$$\underset{\boldsymbol{\theta}}{Min}\ V(\boldsymbol{\theta}), \tag{15.6}$$

where

$$V(\boldsymbol{\theta}) = \|\boldsymbol{\Phi}\,\boldsymbol{\theta} - \mathbf{y}\|_2^2. \tag{15.7}$$

The least-squares solution is given by

$$\begin{aligned} \boldsymbol{\theta} &= \left(\boldsymbol{\Phi}^\mathrm{T}\boldsymbol{\Phi}\right)^{-1}\boldsymbol{\Phi}^\mathrm{T}\,\mathbf{y} \\ &= \boldsymbol{\Phi}^+\mathbf{y}. \end{aligned} \tag{15.8}$$

The matrix $\boldsymbol{\Phi}^+ = \left(\boldsymbol{\Phi}^\mathrm{T}\boldsymbol{\Phi}\right)^{-1}\boldsymbol{\Phi}^T$ is called the pseudo inverse of $\boldsymbol{\Phi}$. The assumption is that the matrix $\boldsymbol{\Phi}^\mathrm{T}\boldsymbol{\Phi}$ is not singular.

Example 15.1 Identifying a First-Order Process

Consider a first-order process modeled by the equation $y(s) = \dfrac{k}{\tau s + 1}\,u(s)$. For identification purposes, we will use the discrete-time representation of this system. Using the methods described in the previous chapter, we can write this model in discrete-time as (using a zero-order hold, or ZOH method with sample time = T_s)

$$G(z) = \frac{k(1 - e^{-T_s/\tau})z^{-1}}{1 - e^{-T_s/\tau}z^{-1}} = \frac{b_1 z^{-1}}{a + a_1 z^{-1}},$$

where

$$b_1 \equiv k(1 - e^{-T_s/\tau}) \text{ and } a_1 \equiv -e^{-T_s/\tau}.$$

Converting the z-domain relation to discrete-time domain we get

$$-a_1\,y_{n-1} + b_1\,u_{n-1} = y_n.$$

This is the representation we will use for identification. Let us say that we have some data as follows:

n	Input, u_n	Output, y_n
0	0	0
1	1	0.18
2	1	0.33
3	1	0.45
4	1	0.55
5	1	0.67

On applying the model to this data, we get the equations (e_i is the error in the fit caused by model errors and measurement noise)

$$\text{for } n=1: \quad -a_1 \ 0.00 + b_1 \cdot 1 = 0.18 + e_1,$$
$$n=2: \quad -a_1 \ 0.18 + b_1 \cdot 1 = 0.33 + e_2,$$
$$n=3: \quad -a_1 \ 0.33 + b_1 \cdot 1 = 0.45 + e_3,$$
$$n=4: \quad -a_1 \ 0.45 + b_1 \cdot 1 = 0.55 + e_4,$$
$$n=5: \quad -a_1 \ 0.55 + b_1 \cdot 1 = 0.67 + e_5,$$

Here we have a set of linear algebraic equations to solve. These equations can be assembled in the form

$$\mathbf{A}\mathbf{x} = \mathbf{d} + \mathbf{e},$$

where

$$\mathbf{x} = \begin{bmatrix} a_1 \\ b_1 \end{bmatrix}, \ \mathbf{d} = \begin{bmatrix} 0.18 \\ 0.33 \\ 0.45 \\ 0.55 \\ 0.67 \end{bmatrix} \ \mathbf{A} = \begin{bmatrix} 0 & 1 \\ -.18 & 1 \\ -.33 & 1 \\ -.45 & 1 \\ -.55 & 1 \end{bmatrix}.$$

We seek a solution that minimizes the sum of the square of the errors, $\|\mathbf{e}\|_2^2$. This is a standard least-squares problem. Appendix C discusses the solution procedure. The least-squares solution is given by

$$\mathbf{x} = (\mathbf{A}^T\mathbf{A})^{-1}\mathbf{A}^T\mathbf{d} = \begin{bmatrix} -.7414 \\ .1942 \end{bmatrix} \Rightarrow a_1 = -.7141, b_1 = .1942.$$

The data for this example was generated from a first-order process with $\tau = 5, k = 1$, and $T_s = 1$. This corresponds to (see Example 15.4)

$$a_1 = -e^{-T_s/\tau} = -0.87$$
$$b_1 = 1 - e^{-T_s/\tau} = 0.17.$$

Random noise with a standard deviation of 0.01 was added to the data. This results in some error in the parameter estimates.

♦

15.2.2 Determining the Time Delay

Unfortunately, there is no simple, direct way to compute the time delay. Typically, a trial and error procedure is used, varying nk until the prediction error is minimized. This implies that one must solve a multitude of linear least-squares problems and then select the time delay that yields the best fit. In the next chapter we will present an alternative formulation of the least-squares problem using FIR and FSR models that overcomes this problem to some extent. Another approach is to set up a nonlinear least-squares problem with the time delay as one of the parameters to be estimated.

15.2.3 Recovering Continuous-Time Models from Discrete-Time Models

After a discrete-time model has been fitted to the data, we can recover a continuous time model using the relation between z and s domain transfer functions. From Chapter 14, we have the following approximate relations

$$z = 1 + T_s s \qquad : \quad \text{Backward Difference}$$
$$z = \frac{1}{1 - T_s s} \qquad : \quad \text{Forward Difference} \qquad\qquad (15.9)$$
$$z = \frac{1 + T_s s/2}{1 - T_s s/2} \qquad : \quad \text{Trapezoidal rule (Tustin approximation)}$$

One of these substitutions may be used to recover the continuous model from the discrete model. If the discrete data used was generated using a ZOH, then the relation between z-domain and s-domain using the ZOH can be used. This was illustrated in Example 15.1.

15.3 PROPERTIES OF THE LEAST-SQUARES ESTIMATOR

Certain questions arise regarding the accuracy of the parameters estimated as well as the accuracy of the model itself. In this section we discuss these and other useful properties of the least-squares estimator.

15.3.1 Existence of the Parameters

The existence of the parameters depends on the matrix $\Phi^T\Phi$ being nonsingular (i.e., invertible). In practical terms, this requires that the matrix be well conditioned. A matrix is said to be ill-conditioned if it is nearly singular. A good measure of the conditioning of a matrix is the condition number defined by

$$Condition \ \ No. = \|A\|_2 . \|A^{-1}\|_2 = \frac{Max \ \ Singular \ \ value}{Min \ \ \ \ Singular \ \ value} . \tag{15.10}$$

See Appendix C for more details on singular values and how to compute them. The condition number can vary from 1 (for an identity matrix) to ∞ (for a singular matrix). In practice, condition numbers above 100 imply that the matrix is difficult to invert accurately. An ill-conditioned matrix implies that the data is insufficient or redundant. Data may be insufficient if the plant was not sufficiently excited by the input sequence used during the identification test. As an extreme example, if the input remained constant, then we would not get any dynamic information at all from the data and the matrix would be singular if no noise were present. The presence of noise would actually make the matrix nonsingular but ill-conditioned. In this case the answers will be inaccurate.

Example 15.2 Identification with Insufficient Data

Consider the data given in the following table.

n	Input, u_n	Output, y_n
0	0	−0.1274
1	0.01	0.556
2	0.01	−1.094
3	0.01	−0.7268
4	0.01	1.410
5	0.01	−0.6139
6	0.01	0.2441
7	0.01	−1.579
8	0.01	−0.3935
9	0.01	−0.7623
10	0.01	−.5417

The data was obtained from the first-order process modeled in Example 15.1 by perturbing the system with a small step change in the input and adding a random noise with a standard deviation of .8 to the output. The excitation in u is small, and there is a considerable amount of noise in the measurements. The matrix $\Phi^T\Phi$ is ill-conditioned (condition number = 93) and the least-squares estimates obtained are

$$a_1 = 0.33$$

$$b_1 = -41.4075,$$

which are quite far from the correct values. This points to the importance of designing good input-forcing functions.

◆

In the identification literature the requirement for the data to contain enough information is called the need for persistent excitation. Isermann (1980) gives the necessary conditions for persistency of excitation for ARMA (Auto Regressive Moving Average) type processes. However, in practice, considerably more data are needed than is called for by the minimum requirement.

In order to obtain a reasonable excitation, Cutler and Yocum (1991) recommend using a *signal to noise ratio (S/N ratio)* of at least 6. The S/N ratio is defined as the ratio of the variance of the signal to the variance of the noise. To get an estimate of the noise variance, we can obtain some data for a time period in which time the input was maintained constant, and estimate the variance of the output as

$$\sigma_n^2 \approx \sum_1^n (y_i - \bar{y})^2 / (n-1),$$ (15.11)

while keeping u constant.

This must be compared against the variance of y during the identification test

$$\sigma_y^2 = \sum_1^n (y_i - \bar{y})^2 / (n-1),$$ (15.12)

during identification test.

The S/N ratio is then

$$S/N \ ratio = \sigma_y^2 / \sigma_n^2.$$ (15.13)

15.3.2 Accuracy of Model

How good is the model? There are a few statistical measures that we can use to evaluate the goodness of the least-squares fit. A commonly used measure is the correlation coefficient R^2.

$$R^2 = 1 - \frac{SS_E}{SS_{yy}},$$ (15.14)

where

$$SS_E = \|\mathbf{e}\|_2^2 = \sum_{i=1}^{N} e_i^2 = \sum_{i=1}^{N} (y_i - \hat{y}_i)^2, \tag{15.15}$$

$$\hat{y}_i = predicted \quad value \quad of \quad y, \quad y_i = Actual \quad measurement, \tag{15.16}$$

$$SS_y = \sum_{i=1}^{N} (y_i - \bar{y})^2 = \|\mathbf{y} - \bar{\mathbf{y}}\|_2^2. \tag{15.17}$$

For a perfect fit, $SS_E = 0$ and $R^2 = 1$. A high value of R^2 implies that the model fits the data well, but it does not necessarily mean that it is a *good* model. This can be explained using the following analogy. Consider fitting a polynomial through a set of (x, y) points. As we increase the order of the polynomial, we will be able to reduce the error in the fit at the expense of an increased number of parameters. Eventually, the error in the fit can be reduced to zero if a sufficiently high-order polynomial is used. If the data were corrupted by noise, the polynomial that we fitted will perform poorly when applied to a new set of data. The prediction error for the training data is not a good measure of the accuracy of the predictive capabilities of the model (see Problem 15.2). To evaluate the predictive capabilities of the model, the approach shown in Figure 15.1 is used.

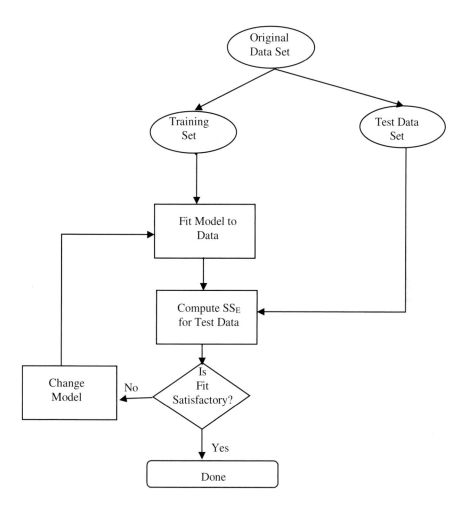

Figure 15.1 Procedure to test predictive capability of model.

The available data is divided into two sets. The first set (called the training data set) is used to compute the least-squares parameters. The resulting model is then used to compute the Prediction Error ($PE = SS_E$) for the second (test) data set. The model that yields the lowest PE for the test data set is chosen as the best. This approach avoids the problem of over-fitting.

Example 15.3 Accuracy of Model

Consider the training data shown in Fig 15.2. Using 0.1 sec sample time, 370 data points were collected. The input is random binary signal with ±1 value.

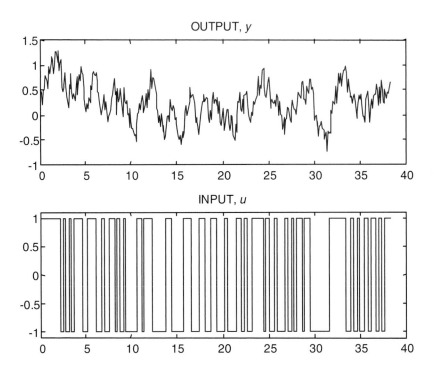

Figure 15.2 Input/output data used for identification.

This data was generated from a discrete system modeled using the equation
$G(z) = \dfrac{0.09516z^{-1}}{1 - 0.9048z^{-1}}$. This was fitted using four different models of the form

$$G(z) = \frac{b_1 \, z^{-nk} \; + \; b_2 \, z^{-nk+1} \; ...+ \; b_{nb} \, z^{-nb-nk+1}}{1 \; + \; a_1 \, z^{-1} \; + \; a_2 \, z^{-2} \; + \; a_3 \, z^{-3}... \; + \; a_{na} \, z^{-na}},$$

using the Identification toolbox. Four different models were fitted. The prediction error of these models for a different test data set were calculated. The results are shown in Figure 15.3. See *ch15ex3.m* on the website associated with this book. This program also shows how to use the commands from the Identification toolbox to generate the models.

Fitted Model [na nb nk]	Prediction Error	Transfer Function
[1 1 1]	1.4997	``` 0.0975 ---------- z - 0.8467```
[2 2 1]	.9566	``` 0.08716 z + 0.0599 ---------------------- z^2 - 0.4309 z - 0.3808```
[3 3 1]	.6140	``` 0.09344 z^2 + 0.07362 z + 0.03982 --------------------------------- z^3 - 0.2487 z^2 - 0.1471 z - 0.374```
[4 4 1]	.6140	``` 0.09678 z^3+.08034 z^2+0.04472 z+0.02261 --- z^4-0.1804z^3-0.1024 z^2-0.3126 z-0.1389```

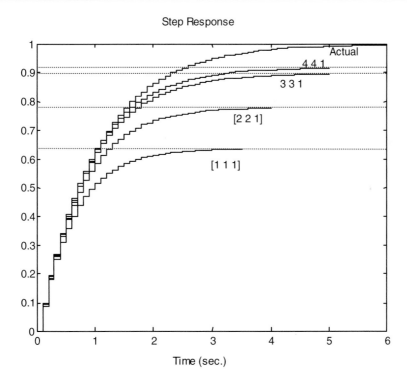

Step Response

Figure 15.3 Comparing step response of actual process and identified models.

Surprisingly, the higher-order models do a better job of fitting the data than the first-order model, although the data was generated using a first-order system. This example shows how difficult it can be to pick out the best model if we do not know the model structure in advance. Ljung (1999) recommends minimizing an information criteria that includes a penalty for higher-order models containing large numbers of parameters.

If the model is good, then the errors should be randomly distributed and there should be no correlation between the input variable u_n and the prediction error e_n. Hence compute the *cross-correlation* function

$$r_{ue}(k) = \sum_{i=1}^{N} u_i \cdot e_{i+k} \tag{15.18}$$

for $k = 0, \pm 1, \pm 2, \pm 3$, and so on. If the model is good, then this should be close to zero for all values of k. Correlation between e and u for negative lags indicates that there is some feedback influence in the system. Correlation at positive lags indicates that the model may not be capturing the system dynamics very well.

The steady-state gain is estimated incorrectly by all models because the input data did not contain long periods of constant input value. This brings up another important point: The input forcing function used to generate the identification data must be chosen to excite the frequencies that we want to estimate accurately. We will have more on this in Chapter 16.

♦

15.3.3 Accuracy of Parameters

First consider the *bias* in the estimators (see Appendix C for an explanation of the term bias).

$$
\begin{aligned}
\mathbf{E}\{\boldsymbol{\theta}\} &= \mathbf{E}\left\{\left(\boldsymbol{\Phi}^T\boldsymbol{\Phi}\right)^{-1}\boldsymbol{\Phi}^T\mathbf{y}\right\}, \\
&= \mathbf{E}\left\{\left(\boldsymbol{\Phi}^T\boldsymbol{\Phi}\right)^{-1}\boldsymbol{\Phi}^T\left(\boldsymbol{\Phi}\,\boldsymbol{\theta}_a + \mathbf{e}\right)\right\}, \\
&= \boldsymbol{\theta}_a + \left(\boldsymbol{\Phi}^T\boldsymbol{\Phi}^T\right)^{-1}\boldsymbol{\Phi}^T\,\mathbf{E}\{\mathbf{e}\},
\end{aligned}
\tag{15.19}
$$

where $\boldsymbol{\theta}_a$ = actual parameters. If we assume that the errors e_i are zero mean, then

$$\mathbf{E}\{\boldsymbol{\theta}\} = \boldsymbol{\theta}_a. \tag{15.20}$$

In this case we say the estimator is *unbiased*. If there are disturbances that affect the output such that their mean is not zero, then we can get biased estimators.

The *variance* of the parameters can be estimated as (Draper & Smith, 1981)

$$\text{var }(\boldsymbol{\theta}) = \left(\boldsymbol{\Phi}^T\boldsymbol{\Phi}\right)^{-1}\boldsymbol{\Phi}^T\,\mathbf{R}\boldsymbol{\Phi}\left[\boldsymbol{\Phi}^T\,\boldsymbol{\Phi}\right]^{-1}, \tag{15.21}$$

where $\mathbf{R}$ is the covariance of the error vector e. If e_i can be characterized as white noise with variance σ_e^2, then the variance in $\boldsymbol{\theta}$ reduces to

$$\text{var} \left(\boldsymbol{\theta} \right) = \sigma_e^2 \left(\boldsymbol{\Phi}^T \boldsymbol{\Phi} \right)^{-1}. \tag{15.22}$$

Hence the variance in $\boldsymbol{\theta}$ can be lowered by reducing the noise in the measurement or by making $\boldsymbol{\Phi}^T \boldsymbol{\Phi}$ larger (i.e., collecting more data and/or increasing the size of the input perturbations). Doubling the number of data points will effectively double the elements of $\boldsymbol{\Phi}^T \boldsymbol{\Phi}$ and reduce the variance by a factor of 2. To reduce the *standard deviation* by a factor of 2, one must *quadruple* the number of data points. Thus the payoff in accuracy diminishes with increasing data points. On the other hand, doubling the size of the inputs will reduce the standard deviation by a factor of 2. The input variations are usually limited by operational constraints. We will discuss such practical aspects of identification testing later.

The noise covariance σ_e^2 can be estimated from the data. First consider

$$SS_E = \left\| \mathbf{e} \right\|^2 = \left\| \mathbf{Y} - \boldsymbol{\Phi} \, \boldsymbol{\theta} \right\|^2 = \mathbf{V}(\boldsymbol{\theta}). \tag{15.23}$$

This residual has N–p degrees of freedom associated with it (p = no. of parameters). Hence the mean square value of SS_E is given by

$$\hat{\sigma}_e^2 = \frac{V(\theta)}{N - p}.$$

This is a model-dependent estimate of σ_e^2, since the estimate was based on the model.

Confidence intervals on the parameter estimates can also be computed (see, for example, Draper & Smith, 1981).

Example 15.4 Properties of the Least-Squares Estimate

Consider the Example 15.1 again. Some new data are available:

n	Input, u_n	Output, y_n	n	Input, u_n	Output, y_n
1	1.0000	-0.1390	11	1.0000	0.8378
2	1.0000	0.2042	12	1.0000	0.7810
3	1.0000	0.3568	13	1.0000	1.1107
4	1.0000	0.4146	14	1.0000	1.1201
5	1.0000	0.6884	15	1.0000	0.7870
6	1.0000	0.5524	16	1.0000	1.1441
7	1.0000	0.6051	17	1.0000	0.8697
8	1.0000	0.7532	18	1.0000	0.9362
9	1.0000	0.8377	19	1.0000	1.0282
10	1.0000	0.7838	20	1.0000	0.9452

Estimate the properties of the least-squares parameters.

Let us first identify a first-order process transfer function using this data. Generate the matrices for least-squares estimation of parameters:

```
A = [ -y(1:19) u(1:19)]; b = y(2:20),
```

where we have used the MATLAB notation to get the arrays. The estimated coefficients of the transfer function are

```
x = A\b =   [ -0.6427    0.3142],
```

where the operator \ gives the psuedo–inverse of the matrix A; x is thus the least-squares solution to the problem. The following are the model coefficients using data which was generated

```
xactual = [ -0.8187    0.1813]
```

There are substantial errors in the parameters due to noise in the data. The correlation coefficient is given by

```
e = y-A*x;Rsquared = 1-(norm(e)/norm(y))^2 = 0.9679.
```

which shows that the data is fitted fairly well. So, fitting the data does not necessarily imply that parameters are accurate. The estimated bias in the estimates,

```
bias = inv(A'*A)*A'*e =   1.0e-015 * [ -0.1387  -0.1965],
```

is very small. The covariance of the noise is estimated as

```
sigma_squared = norm(e)^2 ./(length(e)-length(x)) = 0.0240.
```

The actual noise had a covariance of 0.01. The variance of the estimates is

```
covar_x = sigma_squared*inv(A'*A); =   [  0.0122    0.0076].
```

Hence, the estimate of the standard deviation in the first parameter $x(1)= -.6427$ is about 0.1.

♦

15.4 GENERAL PROCEDURE FOR PROCESS IDENTIFICATION

In practice, identification is one of the most time-consuming and expensive parts of an advanced control project. The paper by Cutler and Yocum (1991) discusses some important issues related to assuring that good data are obtained. They suggest conducting a pretest to estimate time constants involved and to correct any instrumentation problems. The problems they cite include

1. Improper calibration of sensors. Sometimes sensor spans are left at very large values, required for startup/shutdown periods. This will reduce the accuracy of the measurement.
2. Improperly tuned low level controllers. Sometimes the process being identified contains low level control loops which becomes part of the transfer function being identified. For example, low level flow control loops are often cascaded with temperature control loops. If these low level controllers are not properly tuned, the system will exhibit oscillatory response to changes in the outer loop which makes identification difficult. By observing each loop response visually, one could determine if the loop controller requires retuning.
3. Sticky valves and valve hysteresis can introduce limit cycles in the control loop. Adjustments on the control valve should normally take care of this problem.
4. Measurement noise should be removed at the DCS level, where the scan rate is usually very high. Hence a low-pass filter can be implemented without introducing significant lag or attenuation in the signal itself. Data are then sampled at a larger sampling rate, used by the identification algorithm. Unfortunately, this procedure throws away valuable information contained in the original data.
5. The pretest results should be used to determine the smallest possible changes in the manipulated variables to give a S/N ratio of about 6.
6. An engineer should be present at all times to manually record any operator interventions and other unforeseen events that may invalidate segments of the identification data. If an operator intervenes to correct a situation, it will correlate the manipulated variable to the outputs.
7. In a large system many input/output pairs may be uncorrelated to each other. These should be identified, because the step response coefficients will be indistinguishable from the noise. It is best to set these to zero to avoid poor estimation of their transfer function.

Cutler and Yocum suggest simultaneous moves in all manipulated variables and simultaneous identification of all transfer functions. For a 20-inputs, 40-outputs system, this will imply 800 transfer functions. Many of these will be zero. Yet it may be difficult to analyze each one individually to see if the fit is good. Hence good graphical tools that can present the results for visual inspection are essential in screening the models. With such a large number of inputs, moving each input one at a time may be impractical.

The general outline of the procedure used in process identification is shown in Figure 15.4. The main steps in the procedure are:

1. pretesting
2. design of input signal
3. data screening and noise filtering
4. model structure selection
5. least-squares estimation of parameters
6. model validation

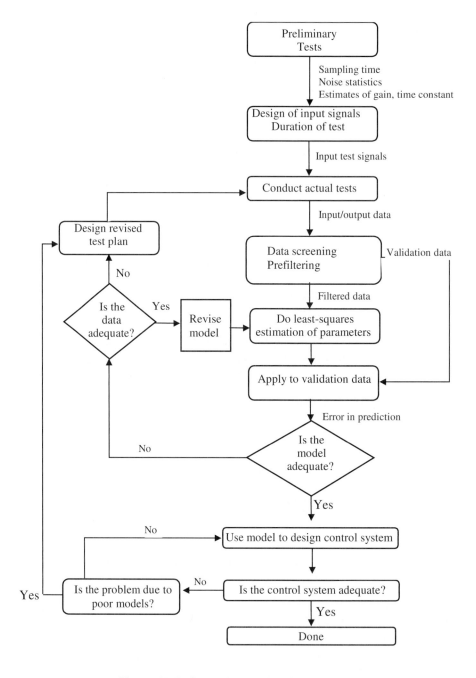

Figure 15.4 General procedure for process identification.

Step 1: Do identification pretesting. The objective of this step is to get an idea of the nature of the system being modeled. Recall that we are primarily concerned with the optimizing and multivariable control layer of the overall control system. This means that the lower level PID regulatory controllers, such as flow-control loops, pressure-control loops and level-control loops (all of which are primarily independent SISO control loops) have already been tuned and are working right. Thus the control engineer should have a clear definition of the manipulated variables and controlled variables that are part of the multivariable control system. Most likely, there may be some high level control loops already installed that are necessary to run the plant. These control loops would have to be placed on manual during the identification process. This requires cooperation and commitment from the plant operators and management. Since the identification has to be carried out without disrupting production, limits of variations in input and output variables should be determined in consultation with the operating personnel. During the pretesting phase, faulty and out-of-calibration instruments, and sensors, improperly sized or installed valves, and poorly tuned low-level PID loops should be identified and repaired. Tests conducted are in the form of step changes up and down in key variables, and the responses are recorded.

A preliminary study of the response plots should give an idea of the response time (order of magnitude estimates) and the process gains. The data can also be used to compute the statistics of the noise in the signals. This includes the standard deviation of the noise as well as a rough idea of the frequency content of the noise.

Step 2: Design of input signal. The objective of this step is to design input-forcing functions (typically a sequence of step changes) to perturb the appropriate frequency bands of interest in the process. The duration of the test will have to be determined based upon the noise that was observed in the output variable. The amplitude of the input changes would be based upon operational constraints and estimated process gains. This aspect is discussed in greater detail in the next chapter.

Step 3: Conduct actual tests. This phase is perhaps the most crucial of all the steps. Here the plant is perturbed from its normal operation, using the test signals designed in steps 1 and 2. Caution is needed here to prevent violation of any process constraints.

There will be some unavoidable external disturbances that will corrupt the data, and such events and their duration should be recorded so that the data segments during such periods can be discarded or modified. If the inputs drive the plant too close to constraint violations, operators will override and bring the plant back to normal. In this instance also, the data is corrupted.

Step 4: Data screening and prefiltering. The data set should be broken up into periods in which abnormal external events did not disrupt the testing process. Each period of uninterrupted test duration will provide one data set for identification. An outlier detection scheme may be employed to smooth any glitches caused by the measurement system.

A filter is next used to smooth the input and output data. The next chapter discusses how to design such noise prefilters.

Step 5:   Select model and do least-squares estimation of parameters. A part of the data is set aside for model validation (test data set). The model parameters are estimated using the remaining (training) data set.

Various types of models are employed in control system design. Many implementations use the FIR or FSR models because their representations are quite general and do not require that a model structure or time delay be specified. Multivariable identification problems can still be treated, using linear least-squares with this model representation. The main disadvantage of such representation is the presence of a large number of correlated parameters in the model. Various techniques have been proposed to overcome this correlation problem, and one of these techniques is discussed in the next chapter. The selection of sample time is also very crucial in this type of model. An alternative is to employ structured transfer function models with unknown coefficients to be estimated. Such representation reduces the number of parameters to be computed, but the disadvantage is that the structure must be specified (usually the order of numerator and denominator polynomials and the time delay). After a model structure is selected, the parameters are computed to minimize the error between model prediction and actual observations. Standard least-squares methods are employed here.

Step 6:   Model validation. The parameters determined using the training data set are used to compute the prediction error for the test data set. Generally, over-parameterized models will perform well on the training data but will not be able to predict the test data set well. If the validation test fails, then either the model structure must be modified or the data set used in training is inadequate.

Finally, the model is used to design the control system. Simulated tests on the control system performance are conducted to see if it is adequate. If problems arise, further refinement of the model may be required.

Most processes change with time. If operating conditions change substantially, control system performance will degrade. Thus control system performance should be constantly monitored to determine if a model update is required. This also implies that one should not spend any extra effort at trying to finetune a model, but rather design control systems that can tolerate a certain level of uncertainty in the model.

If it is known that the process must operate over a wide range of operating conditions, then separate tests must be conducted around each operating point to determine a set of linear models.

15.5 SUMMARY

In this chapter we introduced the basic least-squares identification algorithm to obtain discrete-time process models. We also indicated how continuous time models may be derived from the discrete-time models. The main issues that arise in the application of this methodology are

1. Choice of model representation (transfer function versus FIR or FSR form).
2. Design of input sequences that will facilitate the identification of accurate process models.
3. Planning and execution of the identification experiments in the process plant.

In the next chapter we will address the issue of input test signal design, the pretreatment of data, and modifications to the basic least-squares algorithms to overcome some practical problems.

Problems

15.1 Consider the problem of fitting a line to a set of data points.

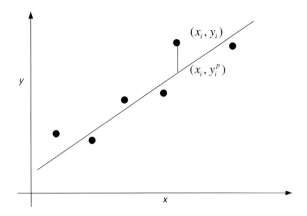

The predicted values of y are given by $y_i^P = a\,x_i + b$. Show that the least-squares solution is given by the equations

$$a = \bar{y} - b\bar{x},$$
$$b = S_{xy}/S_{xx},$$

where

$$\bar{y} = \left(\sum_1^n y_i\right)\Big/n, \quad \bar{x} = \left(\sum_1^n x_i\right)\Big/n$$

$$S_{xx} = \sum_1^n (x_i - \bar{x})^2, \quad S_{xy} = \sum_1^n (x_i - \bar{x})(y_i - \bar{y}).$$

15.2 The time evolution of a variable y was measured as follows:

Time	0	1	2	3	4	5
y	1.4447	3.6154	7.7919	13.9218	21.7382	31.1763

a.   Using MATLAB, fit a polynomial model of the form using the least-squares method.

$$y^P = a_0 + a_1 + + .. + a_n \, t^n$$

b.   for $n = 1, 2, ..., 5$. Also report the sum of the square of the errors (SS_E), defined as $SS_E = \sum \left(y_i^P - y_i \right)^2$, and the R^2 value for each fit. What value of n gives the best fit? Hint: use `polyfit` or `polyval`.

c.   Test how well each of the five models is able to fit the following set of new data, obtained on the same process:

Time	0	1	2	3	4	5
y	0.5674	1.3344	7.1253	13.2877	19.8535	32.1909

d.   Based on this result, which model would you recommend?

15.3   Consider the data given below. The system was at steady state at the beginning of the test.

Time	Input	Output	Time	Input	Output
0.0	2	3.9945	2.1	3	4.7546
0.1	3	4.0195	2.2	3	4.8661
0.2	3	4.0121	2.3	3	4.843
0.3	3	4.0295	2.4	3	4.8428
0.4	3	4.0457	2.5	3	4.8662
0.5	3	4.0023	2.6	3	4.8609
0.6	3	4.1332	2.7	3	4.8825
0.7	3	4.1486	2.8	3	4.8723
0.8	3	4.3072	2.9	3	4.9173
0.9	3	4.3068	3	3	4.944
1.0	3	4.3687	3.1	3	4.9287
1.1	3	4.4888	3.2	3	4.9468
1.2	3	4.5271	3.3	3	4.9101
1.3	3	4.5143	3.4	3	4.933
1.4	3	4.5446	3.5	3	4.9785
1.5	3	4.6715	3.6	3	4.973
1.6	3	4.637	3.7	3	4.9553

Time	Input	Output	Time	Input	Output
1.7	3	4.6787	3.8	3	4.9699
1.8	3	4.7436	3.9	3	4.9960
1.9	3	4.7318	4.0	3	4.9257
2.0	3	4.7738			

a. Graph the data. Fit a continuous FOPDT model using the method discussed in Chapter 2.

b. Use the least-squares method presented in this chapter to fit a discrete FOPDT model. Use a search procedure to determine the delay term that minimizes the prediction error of the model. Plot and compare your model predictions with the data given for this problem. Compare this model with the result from Part a.

15.4 Obtain an estimate of the S/N ratio for the following data set. Fit a second-order transfer function to the data and reevaluate the noise covariance using the model.

Time	Input	Output
0	0	0.0607
0.4000	0	–0.0137
0.8000	0	–0.0067
1.2000	0	–0.0635
1.6000	0	–0.0832
2.0000	0	–0.0352
2.4000	0	0.0140
2.8000	0	–0.0271
3.2000	0	–0.0667
3.6000	0	0.0536
4.0000	1.0000	–0.0356
4.4000	1.0000	0.0610
4.8000	1.0000	0.1912
5.2000	1.0000	0.3249
5.6000	1.0000	0.4949
6.0000	1.0000	0.5808
6.4000	1.0000	0.6084
6.8000	1.0000	0.7175
7.2000	1.0000	0.8410
7.6000	1.0000	0.8115

15.5 Verify the result given for Example 15.3.

15.6 Verify the results given in Example 15.4.

References

Åström, K. J., and B. Wittenmark. 1990. *Computer Controlled Systems: Theory and Design,* 2nd Ed. Prentice Hall, NJ.

Cutler, C.R. and F. H. Yocum. 1991. "Experience with the DMC Inverse for Identification." *Proceedings of the Fourth International Conference on Chemical Process Control,* Padre Island, TX. AIChE, NY. 297–317.

Draper, H. R., and H. Smith. 1981. *Applied Regression Analysis,* 2nd Ed. John Wiley & Sons, NY.

Eykhoff, P. 1974. *System Identification.* John Wiley & Sons, NY.

Isermann, R. 1980. "Practical Aspects of Process Identification," *Automatica,* 16, 575.

Ljung, L. 1999. *System Identification, Theory for the User,* 2nd Ed. Prentice Hall, NJ.

Mathworks. 2000. *System Identification Toolbox for MATLAB.* Mathworks, Inc., Natick, MA.

Identification: Advanced Concepts

Objectives of the Chapter

The overall objective is to present some of the advanced concepts used in process identification. Specific topics to be discussed include

- Design of input-forcing functions.
- Prefiltering of noise in the data.
- Modifications to the basic least-squares method to improve model accuracy.
- Identification of multivariable systems.

The chapter concludes with a comprehensive example that illustrates most of the ideas presented on the subject of identification.

Prerequisite Reading

Chapter 15, "Identification: Basic Concepts"

16.1 DESIGN OF INPUT SIGNALS: PRBS SIGNALS

One way to model the dynamic characteristics of a process is by using frequency domain representation. Given an input/output data set $u(t)$, $y(t)$, the frequency response is given by

$$g(i\omega) = y(i\omega)/u(i\omega), \tag{16.1}$$

where $y(i\omega)$ and $u(i\omega)$ are the Fourier transforms of the output and input sequence respectively. This equation tells us that to characterize the behavior of a process over a given frequency range, the input signal must be rich (contain sufficient power or strength) in that frequency range. If $u(i\omega)$ is small, then the corresponding excitation in y, $y(i\omega)$, will also be small, and hence the calculation indicated in Eq. (16.1) indicates that small errors in y or u will cause large errors in the model at that frequency. We will use this insight in designing input signals used in identification tests.

From frequency domain analysis (see Appendix B), we know that accurate information about gain and phase near the critical frequency region is essential for good control system design. Hence the input-forcing function $u(t)$ should excite the system sufficiently around the mid-frequency region that surrounds the critical frequency. There are two problems with this. One is that we do not know a priori what the critical frequency is. Secondly since most processes are nonlinear the critical frequency would depend on the current operating point. The suggested approach here is to use the preliminary test to get an approximate value of the process delay and time constant at a nominal operating point. This value is then used to design the input signal for identification. If the resulting model is quite different from the preliminary model, then it might be necessary to revise and repeat the identification tests. If the plant operating conditions vary significantly, tests may be repeated at the various levels of operation to determine the range of variation of the process parameters.

One of the classical tests used by industry is the step test, as discussed in Chapter 2. The process is perturbed by a step increase in the input manipulated variable. The time constant, time delay, and process gain are then estimated from the response of the process output. The step input can be represented by

$$u(s) = \frac{1}{s},$$

with the frequency domain representation

$$u(i\omega) = \frac{1}{i\omega}.$$

This has amplitude given by

$$|u(i\omega)| = \frac{1}{\omega},$$

which implies that low frequencies are represented or perturbed very strongly, whereas higher frequencies are not. This implies that the steady-state gain can be estimated well, whereas the time constant estimate will be less accurate.

Generally, the presence of noise in the measurement and the presence of unmeasured disturbances will make it difficult to obtain a model from a single-step test. Figure 16.1 shows the step response of a process with noise. Noise effects can hide the actual response of the system. For this reason, we must consider more sophisticated input signals, which usually consist of multiple step responses.

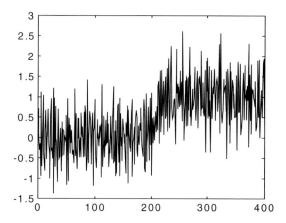

Figure 16.1 Step response of a process with measurement noise. Random noise with a standard deviation of 1.0 added to the signal.

Industry typically employs a sequence of step changes to get a better estimate of the model parameters. According to Cutler and Yocum (1991), each input variable should be stepped up and down over a range of frequencies. Also, "It is essential that on two or three occasions during the test, each of the independent variables be moved and held for a period equal to the time to steady-state." One of the commonly employed input signals is the *PRBS (pseudo random binary signals)*. Figure 16.2 shows a typical PRBS and its associated properties. A PRBS consists of a series of up and down step changes in the input variable. It is a deterministic, two-state binary sequence (i.e., $x(j) = a, j N$). Its properties include (Davies, 1970):

1. It is a periodic sequence with amplitude $\pm a$ and period P. The value of the sequence changes at discrete times (kT_c), where k is an integer and T_c is called the clock tick time.

2. In every period of length P there are T clock ticks $(T = P/T_c)$. There are $T/2$ intervals over which the signal is of length 1 (length 1 means signal is constant over one clock tick), $T/4$ intervals over which the signal is of length 2, $T/8$ of length 3, and so on. This distribution effectively perturbs the plant over a wide variety of frequency ranges. T is given by the equation

$$T = 2^n - 1.$$

3. The power spectrum (energy of signal at various frequencies) of a PRBS is given by

$$\Phi_{uu}(\omega) = \frac{\alpha^2(T+1)}{T}\left[\frac{\sin(\omega T_c/2)}{\omega T_c/2}\right]^2$$

$$\omega = \frac{2\pi k}{TT_c}\ k = 1,2,...T,...,etc.$$

(16.2)

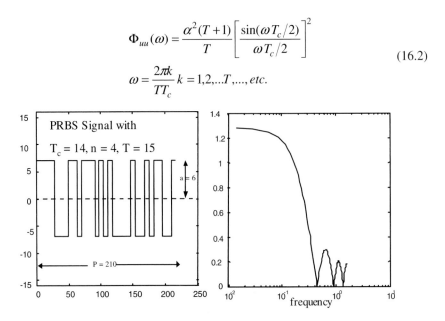

Figure 16.2 A typical PRBS and its power spectrum.

A MATLAB program, *prbs.m*, is provided with this book to facilitate PRBS generation with desired properties. The calling sequence for this function is

$$u = \text{prbs}\ (\omega_{\min},\omega_{\max},\ \text{samp-time, amplitude}),$$

where $\omega_{\min},\omega_{\max}$ are the minimum and maximum frequencies over which we would like to perturb the system.

The advantages of using PRBS include the following.

1. A PRBS perturbs the system around a steady-state and does not result in significant deviations from this value. This is very important in preventing significant upsets to the production schedule during the tests. Since tests can last over a significant period, any disruption in production will not be acceptable to plant management.

2. The frequency content of the PRBS can be adjusted relatively easily to meet the desired specifications on the desired excitation frequency range.

3. To increase accuracy of estimates, we can repeat the PRBS signal many times, provided the operating points do not change significantly due to outside disturbances.

The key information that is needed to design a PRBS signal is the specification of the frequency range of interest. Some guidelines can be developed, as follows. Suppose we have a process with an open-loop dominant time constant τ. A suitable frequency range of interest in control will be

$$\frac{1}{\beta\tau} < \omega < \frac{\alpha}{\tau}. \tag{16.3}$$

Rivera et al. (1995) suggest a value of 3 for β and 2 for α. This assumes that we will be able to design a control system that would respond at least twice as fast in the closed-loop as the open-loop response time. This is in line with the recommendations of other practitioners; see Cutler and Yocum (1991), for example. The PRBS signal has a low frequency amplitude given by (from Eq. (16.2))

$$\Phi_{uu} = a^2 \, (T+1)/T \,, \tag{16.4}$$

and this is reduced to half this value at $\omega = 2.8/T_c$. The lowest frequency perturbed is given by $k = 1$ in Eq. (16.2),

$$\omega_{low} = \frac{2\pi}{TT_c}. \tag{16.5}$$

So, for practical purposes, we can say that the PRBS perturbs the system over

$$\frac{2\pi}{TT_c} < \omega_{PRBS} < \frac{2.8}{T_c}. \tag{16.6}$$

Actually, there will be some perturbation outside this range, but it decays rapidly with frequency. Comparing this to Eq. (16.3) we conclude that

$$T_c = \frac{2.8\tau}{\alpha}, \tag{16.7}$$

and

$$T = \frac{2\pi\alpha\beta}{2.8} \approx 2(\alpha\beta). \tag{16.8}$$

Thus the clock tick time is determined by the highest frequency of excitation desired, and the period of the PRBS is determined by the lowest frequency. Usually, we choose n (the number of shift registers used to generate the PRBS signal) which is closest to

$$T = 2^n - 1. \tag{16.9}$$

16.1.1 Clock Tick Versus Sample Time

In practice (for computer-controlled systems), all signals are sampled, and hence we must also be aware of the sampling rate. According to the sampling theorem, we should sample at a sampling rate that is

$$T_s < \frac{\pi}{\omega_{high}}, \qquad (16.10)$$

where ω_{high} rad/sec is the highest frequency present in the signal. Hence if

$$\omega_{high} = \frac{2.8}{T_c}, \qquad (16.11)$$

then the following is implied:

$$T_s < \frac{\pi}{2.8} T_c. \qquad (16.12)$$

In practice, we will want to sample at a higher frequency and then filter out (using a band-pass filter) the high and low frequencies that are outside the range of interest. Hence a sampling time that is at least 10 times smaller than the clock tick time should be used.

16.1.2 Choice of Amplitude, a

The amplitude of the PRBS is selected based on the range over which the controlled variable is allowed to change and the knowledge about the process gain. For instance, if the controlled variable is allowed to vary by 20% of the sensor transmitter span, and if the process gain is about 2, then the input signal should be varied by 10% of the controller output span. Normally, the higher the amplitude, the larger the S/N to ratio. A S/N ratio of at least 6 is recommended by practitioners. The S/N ratio is estimated from

$$S/N \ ratio \approx \frac{\sigma_s^2}{\sigma_N^2}, \qquad (16.13)$$

where

$$\sigma_s = \text{standard derivation of signal,}$$

$$\sigma_N = \text{standard derivation of noise.}$$

An observation of the output when input is kept constant can be used to estimate the standard deviation of the noise. In practice, further restrictions on the input magnitude may be placed due to operational constraints and safety considerations. This would further compound the identification problem.

16.1.3 Duration of the Test

The duration of the test is primarily determined as a compromise between the cost of conducting the test versus increased accuracy obtained by a longer duration test. The minimum duration of the test is $P = TT_c$, one period of the PRBS signal. This test should be repeated to reduce the covariance of the error in the estimates. As shown in the previous section, to reduce the standard deviation of the estimated parameters by a factor of 2, the number of data sampled should be quadrupled. Most processes are time-varying and nonlinear. Hence one should not put too much time and effort into identifying very accurate linear process models. It is better to get a set of linear models at different operating conditions to gather estimates of the parameter variations and then use the techniques of controller tuning for uncertain processes outlined in Chapters 7 and 8.

Example 16.1 Design of a PRBS for a Process

Consider a process with the following known characteristics:

1) The input manipulated variable can range from 0% to 100% (span of controller output).
2) The output variable also has a range of 0% to 100%, but operational constraints require that this variable be kept within ±10% of its nominal steady-state value during the test runs.
3) A preliminary identification test reveals that a 10% change in input causes a 15% change in the output (i.e., effective gain is around 1.5). However, this is only a rough estimate.
4) A pulse of short duration and a 10% change in input was seen to cause a change in the output a few minutes later. The change would be detected in the output in about 10 minutes, before it was buried in the noise. This indicates that the time constant should be 10 minutes or so.

With this information, we are ready to proceed with the design of an input signal.

Choice of T and T_c. Given that the time constant τ is around 10, we can expect to design a closed-loop system with the frequency band of interest (from Eq. (16.3))

$$0.03 < \omega < .20 \, \text{rad/min},$$

which covers the range from one third of the open-loop response time to about two times the open-loop response time. This leads to the following choice of T and T_c (using Eq. (16.7) and Eq. (16.8))

$$T_c = \frac{2.8\tau}{\alpha} = 14\,\text{min}$$

$$T \approx 12.$$

Choose $n = 4$ so that

$$T = 2^4 - 1 = 15$$

This will require $n = 4$ registers $(2^4 - 1 = 15 = T_{actual})$.

Choice of Sample Time. With a clock tick time of $T_c = 14 \min$, we should use a sample time of 1.4 min or smaller. We will use a sample time of 1 min.

Choice of Amplitude. The gain of the process is 1.5, and output should be restricted to 10% of its nominal value. Hence use an input amplitude of

$$a \approx \frac{10}{15} \approx 6.$$

Final Design. The final design uses the following parameters:

$$
\begin{aligned}
T_c &= 14 \\
n &= 4 \\
T &= 15 \\
\omega_{min} &= 0.02 \\
\omega_{max} &= .20 \\
a &= 6
\end{aligned}
$$

The length of the signal is 210 minutes, and the clock tick time used is 14 minutes. This signal and its frequency content is shown in Figure 16.2.

♦

16.2 NOISE PREFILTERING

Generally, the output signal is corrupted by random noise. These noises come from two sources:

1. High frequency noise is added by the measurement instrumentation. Typically, these appear at frequencies smaller than 1 Hertz (cycles/sec). A low-pass filter is normally employed on the sensor/transmitter to remove noise higher than a set cut-off frequency, such as 10 Hertz or so, but residual noise may still appear in the sampled signal.

2. Medium to high frequency noise is caused by process noise (caused by turbulence, flow fluctuations, poor mixing, etc.).

3. Low to medium frequency noise is caused by unmeasured disturbances entering the system. Since the test duration is necessarily long, any number of external events can cause changes in the output variable. If the disturbances have to travel through the process, then this will typically appear as slow drifts in the output

variable. An example of this would be noise induced by ambient temperature variations which has a dominant 24 hour cycle.

A *prefilter* is a transformation of the input and output signal that removes a portion of the signal corresponding to specified frequency ranges. A *low-pass filter* would remove high frequency components, a *high-pass filter* will remove low frequency components, and a *band-pass* filter will pass a specified frequency band. Low-pass filters are generally used before identification algorithm is applied to the data. These filters are characterized by the *cut-off frequency*, or the frequency at which the gain of the filter drops below $1/\sqrt{2}$. This is considered to be the *bandwidth* of the filter.

Example 16.2 A First-Order Filter

Consider the first-order system

$$G(s) = \frac{1}{\tau s + 1}, \tau = 10.$$

The bode plot of this transfer function is shown in Figure 16.3.

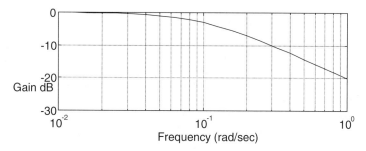

Figure 16.3 Bode plot of 1/(10s+1).

This shows that this system will allow frequencies below $\omega = 0.1$ to pass with little or no attenuation, whereas signal at a frequency higher than $\omega = 0.1$ will be reduced in amplitude. The bandwidth of this filter is $\omega = 0.1$ when the $AR = 1/\sqrt{2} = 0.707$.

The discrete equivalent of the first-order filter is given by discretizing the transfer function. For example, if we use the Tustin approximation (i.e., the trapezoidal integration rule) to discretize the transfer function by substituting (see Chapter 14)

$$s = \frac{2}{T_s} = \frac{z-1}{z+1},$$

$$T_s = \text{sampling time,}$$

we get

$$G(z) \; = \; \frac{1+z^{-1}}{(1-\frac{2}{T_s})z^{-1}+(\frac{2}{T_s}+1)} = \frac{b_0 + b_1 z^{-1}}{a_0 + a_1 z^{-1}}.$$

If we let y_n denote the raw data, then the filtered data is given by

$$\frac{y_{fn}}{y_n} = \frac{b_0 + b_1 z^{-1}}{a_0 + a_1 z^{-1}}$$

$$y_{fn} = \frac{1}{a_0}[b_0 y_n + b_1 y_{n-1} - a_1 y_{fn-1}].$$

This equation is used in the calculation of the filtered data. In filter design, Tustin approximations are preferred to discretization using a zero-order hold (ZOH) because of better agreement with continuous systems in the frequency domain (see Chapter 14).

Let us say we want to design a filter to filter out most of the signal beyond a frequency of 0.1. This would require a filter with $\tau = 10$. Using $T_s=1$ in the above first-order filter, we get the filter

$$G(z) = \frac{1+z^{-1}}{21-19z^{-1}}.$$

Consider a signal with two frequencies present:

$$y(t) = \sin \omega_1 t + \sin \omega_2 t,$$
$$\omega_1 = 0.05, \omega_2 = 1.$$

Since the filter has a cut-off frequency of $1/\tau = 0.1$, only the first sine wave will pass through this filter. This is illustrated in Figure 16.4.

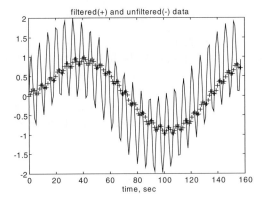

Figure 16.4 Use of low-pass filter.

A high-pass filter can be constructed from a low-pass filter by subtracting it from 1:

$$G_{Hp}(z) = 1 - G(z).$$

For the first-order filter designed in our example,

$$G_{Hp} = \frac{19 - 20z^{-1}}{20 - 19z^{-1}}.$$

♦

Example 16.3 Second-Order Butterworth Filter

This filter is given by the transfer function

$$g(s) = \frac{1}{(s/\omega_B)^2 + 1.42(s/\omega_\beta) + 1},$$

where ω_B = bandwidth of filter.

This is an underdamped second-order system with $\xi = 0.71$. The corresponding discrete domain representation, using the Tustin approximation, is given by

$$b_0 = 1, b_1 = 2, b_2 = 1,$$

$$a_0 = \frac{4}{\omega_B^2 T_s^2} + \frac{2.84}{T\omega_B} + 1, a_1 = 2 - \frac{8}{T_s^2 \omega_B^2},$$

$$a_2 = \frac{4}{\omega_B^2 T_s^2} - \frac{2.84}{T_s \omega_B} + 1.$$

The performance of this filter with $\omega_B = 1, T_s = 1$ on the same data as in Example 16.2 is shown in Figure 16.5. This filter does a much better job of removing the second frequency component in the signal. (Note, however, that the filter introduces a phase lag in the signal. This can be nullified in off-line calculations (see MATLAB Signal Processing toolbox function `filterfilter`). Higher-order filters are generally used to obtain a sharp cut-off and smooth frequency response. See standard books on filtering, such as Franklin et al. (1990).

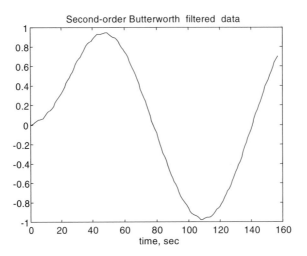

Figure 16.5 The use of a second-order Butterworth filter.

♦

In the previous section on input signal design we studied how to design PRBS signals that excite selected frequency bands in the process. The output measurement will contain (in addition to the actual response to the input-forcing function) high-frequency noise and low-frequency disturbance responses. A band-pass filter is used to reduce the effect of such extraneous noise and to improve the accuracy of the parameter estimates.

16.2.1 Prefiltering Using Noise Models

An alternative approach to prefiltering is to identify a pre-whitening filter simultaneously with the model parameters. Consider the Box-Jenkins (Ljung, 1999) representation of a process model,

$$y = \frac{B(z)}{F(z)} u(z) + \frac{C(z)}{D(z)} e(z), \qquad (16.14a)$$

where the measurement is assumed to be corrupted with a noise generated by white noise passing through a dynamic system $C(z)/D(z)$. This type of noise is called colored noise. We can rewrite the model as

$$\frac{D(z)}{C(z)} y(z) = \frac{B(z)D(z)}{F(z)C(z)} u(z) + e(z) \qquad (16.14b)$$

or

$$y_F = \frac{B(z)}{F(z)} u_F(z) + e(z), \tag{16.15}$$

where

$$y_F = \frac{D(z)}{C(z)} y(z) \tag{16.16}$$

and

$$u_F = \frac{D(z)}{C(z)} u(z) \tag{16.17}$$

can be looked upon as filtered versions of the output and input. The filter $D(z)/C(z)$ is computed to minimize the prediction error of the filtered data. Note that this problem can no longer be posed as a least-squares problem because the unknown coefficients of polynomials B, D, C, and F appear in product form in the model. A nonlinear regression method must be employed in this instance to compute the coefficients. The order of the filter must be predetermined. The program *bj.m* in the Identification toolbox of MATLAB implements a nonlinear regression program to calculate the coefficients of the Box-Jenkins model. Since this model requires nonlinear regression, it is not easily scalable to multivariable systems.

16.3 MODIFICATIONS TO THE BASIC LEAST-SQUARES IDENTIFICATION

FIR (finite impulse response) type models are the most widely used in industrial applications of model-predictive control. As pointed out earlier, due to the large number of parameters in the model, the identification algorithm is sensitive to noise. To overcome this problem, a number of modifications to the DFIR identification algorithm have been suggested in the literature. These include a penalty for either the size or the change in the contiguous FIR coefficients; improved noise prefiltering; partial least-squares, canonical variate analysis; and polynomial smoothing functions. A comparison among these showed the polynomial smoothing to be the most powerful. Hence this method is described in some detail here, and software implementing the algorithm is provided with the book.

The basic idea is to take the FIR coefficients array

$$\underline{h} = [h_0 \ h_1 \ ... \ h_r] \tag{16.18}$$

and say that these can be sampled from a smooth, continuous polynomial function in time, $h_c(t)$:

$$\underline{h} = [h_c(0) \ h_c(T_s) \ ... \ h_c(rT_s)] \tag{16.19}$$

and

$$h_c(t) = a_0 + a_1 t + \ldots\ldots + a_N t^N \quad t_{min} \le t \le t_{max}, \tag{16.20}$$

where t_{min} = estimate of the dead time of the process and t_{max} = estimated time at which the impulse response becomes negligible. In practice, t_{min} may be taken to be 0, if no prior information is given, and t_{max} may be set to be the largest estimate of the time it takes the process to come to a steady-state after a step or impulse input. For computation efficiency, it is better to represent the function $h_c(t)$ in terms of orthogonal basis functions:

$$h_c(t) = \sum_{i=1}^{N+1} a_i \,_i(t). \tag{16.21}$$

Using this model, the identification problem can be reduced to a linear least-squares problem in the unknown parameters $a_1, \ldots .. a_{N+1}$ (Ying & Joseph, 1999). Generally, a value of N, much smaller than r, can be used.

Example 16.4 DFIR Identification Using Polynomial Smoothing Filters

Consider the data used in Chapter 15, Example 15.3 (see Figure 15.2). Using the program POLYID, the DFIR model was identified. The step response resulting from this model is shown in Figure 16.6. This should be compared with the ARX models shown in Figure 15.3. The polynomial smoothing, compared to raw least-squares filtering, does a much better job of determining the model.

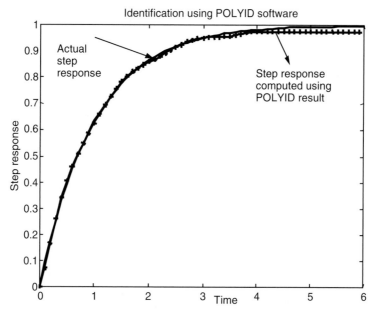

Figure 16.6 Step response of model identified using POLYID compared with the true response.

♦

16.4 MULTIPLE INPUT, MULTIPLE OUTPUT (MIMO) SYSTEMS

MIMO systems can be identified using the SISO techniques described earlier if it is possible to perturb each input one at a time, independently. For example, consider the 2 x 2 system

$$y_1 = g_{11}(z)u_1 + g_{12}(z)u_z,$$
$$y_2 = g_{21}(z)u_1 + g_{22}(z)u_z. \tag{16.22}$$

If u_1 is perturbed while maintaining $u_2 = 0$, we have data on the subsystem

$$y_1 = g_{11}(z)u_{1,}$$
$$y_2 = g_{21}(z)u_1. \tag{16.23}$$

Using this data, we can identify g_{11} and g_{12} separately through SISO identification methods.

It may not always be possible to perturb each input separately. It may be necessary to adjust more than one input at a time to keep the plant from violating operational constraints. Another problem is the total time needed for the identification test. Perturbing each input one at a time essentially multiplies the total identification test time on the plant by the number of inputs. In such cases, it is possible to set up the identification of all transfer functions simultaneously, at the risk of increased computational complexity. This is illustrated next.

Suppose we have a 2 x 2 model of the form

$$y_1 = \frac{B_{11}(z)}{A_{11}(z)} u_1(z) + \frac{B_{12}(z)}{A_{12}(z)} u_2(z),$$

$$y_2 = \frac{B_{21}(z)}{A_{21}(z)} u_1(z) + \frac{B_{22}(z)}{A_{22}(z)} u_2(z).$$

(16.24)

The equation for y_1 can be written as (dropping the z in parentheses for convenience)

$$A_{11} A_{12} y_1 = B_{11} A_{12} u_1 + B_{12} + A_{11} u_2.$$

This equation is nonlinear in the coefficients of the transfer functions. Hence the identification problem becomes a nonlinear least-squares problem.

Note that if one chose to use FIR representation, then the problem remains a linear least-squares problem even when simultaneous moves are made in all the inputs. In this case the model can be written as

$$y_1 = H_{11}(z)u_1(z) + H_{12}(z)u_2(z)$$

$$y_2 = H_{21}(z)u_1(z) + H_{22}(z)u_2(z)$$

(16.25)

In this formulation, there are no non-linear terms in the least-squares identification problem. Hence the problem can be solved efficiently using linear least-squares methods outlined earlier. For large MIMO systems this represents a significant advantage.

16.5 A COMPREHENSIVE EXAMPLE

In this section we present a comprehensive example that considers each of the steps in the identification procedure. The steps needed are
  1. pretesting to estimate time constants and noise characteristics
  2. design of noise filter
  3. design of input signal
  4. estimation of model parameters
  5. testing of the goodness of fit

Example 16.5 Identification of a Model for a Hot Air Blower

Consider the identification of a model relating the heater power (manipulated variable) with the temperature of the exit hot air stream in the process shown in Figure 16.7. Initially, the controller is kept on manual and tests are conducted by manipulating the heater power manually.

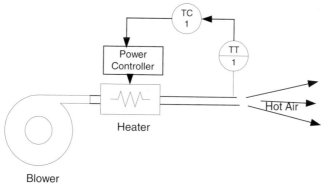

Figure 16.7 Schematic of hot air blower process.

Step 1. Pretesting. The objective of pretesting is to determine approximately the process response time, process gain, noise content and noise spectrum present in the measured output signal. A typical test would be to let the input remain constant for a while and then to step up or down the input signal by a reasonable amount (determined after consultation with operators). The response is recorded. Figure 16.8 shows the step response obtained. Note that the initial half of the response is primarily noise and hence can be used to determine the noise frequencies present. Typically, the output noise will hide the actual response of the process.

The frequency spectrum of the first half of the output response is shown in Figure 16.9. The initial high value in the response corresponds to the step change that was made. The noise seems to have a maximum around a frequency of .4 rad/sec. This noise can be removed by means of a low-pass filter. A low-pass filter with a cut-off at 0.5 was used to generate the filtered step response shown in Figure 16.10.

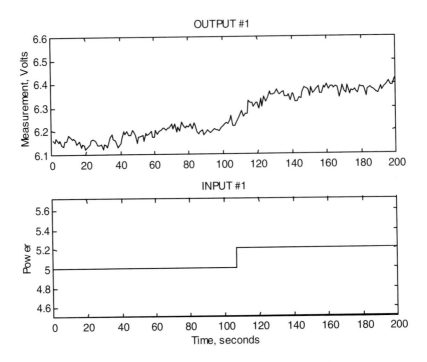

Figure 16.8 Pretesting: Response to a step change in input.

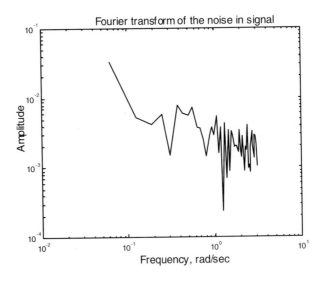

Figure 16.9 Frequency spectrum of output noise.

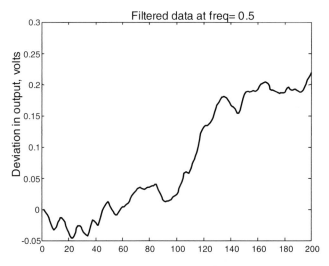

Figure 16.10 Filtered step response.

This shows the step response more clearly, although a significant amount of noise is still present. But the gain of the process can be estimated to be about 1, since the input was changed by unit step. The time constant appears to be around 10 seconds (a rough guess, looking at the step response graph). We will use this to design the actual test signal. If our estimates were badly off, we will have to repeat the input signal design at a later stage. The S/N ratio is of the order of 4. This is actually smaller than the value of 6 recommended in the literature. If possible, we should increase the input signal amplitude to allow for a larger S/N ratio. In this example we have assumed that the changes in input should be between +0.2 and −0.2.

Step 2. Design of input signal. From the results of Example 16.1, we know that for a process with an estimated time constant of 10 seconds, the input should be designed to excite the plant in the frequency range of 0.03 to 0.20 rad/sec. The sample time used is the same as in the pretest phase (1 sec). Using this, we use the m-file *prbs.m* to design our input signal. The result is shown in Figures 16.11a and 16.11b. The input signal is a series of steps of short duration. The frequency spectrum shows that the signal contains frequencies in the range of 0.03 to 0.2 as was desired.

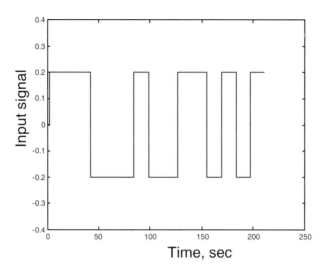

Figure 16.11a PRBS test signal.

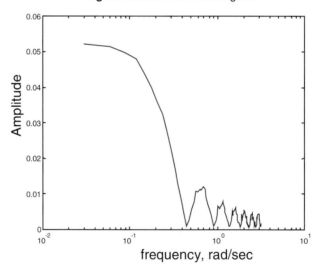

Figure 16.11b Spectrum of signal.

Step 3. Actual identification test. The input signal from step 2 is applied to the plant to generate data for identification. Two cycles of the *prbs* signal were used in the test, resulting in a test duration of about 400 seconds. The result is shown in Figure 16.12. An identical test is used to generate data for validating the model. The output is corrupted by high-frequency noise, and hence a filter is applied to remove this noise. A Butterworth low-pass filter with a cut-off at w = .5 rad/sec is applied to the input and output data. The resulting filtered response is shown in Figure 16.13.

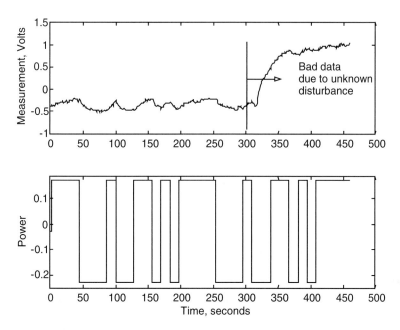

Figure 16.12 Response of plant to PRBS test signal.

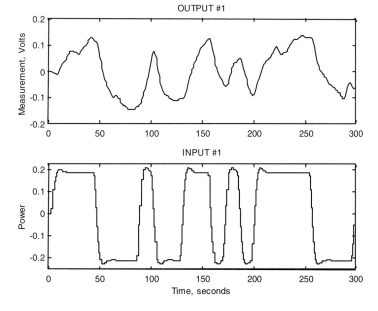

Figure 16.13 Input and output signals after removing bad data points and then applying a low-pass filter.

Steps 4 and 5. Data-fitting and model validation. The next step is to fit this data to a model. We must choose the type of model that should be used. To test how good the model is, we use the validation data and compute the SS_E in prediction. The validation data (after low-pass filtering) is shown in Figure 16.14.

Next we must fit a model to the test data. This work is best done using the System Identification toolbox in MATLAB. We can characterize the ARX model, using the three parameters *na*, *nb*, and *nk* where

na = order of the denominator in the transfer function,
nb = order of the numerator in the transfer function,
nk = the time delay in the response.

A discrete-time (z-transfer function) model of the following form is assumed:

$$y(t) + a_1 y(t-1) + ... + a_{na} y(t-na) = b_1 u(t-nk) + ... + b_{nb} u(t-nk-nb+1).$$

The coefficients a_i and b_i are estimated to minimize the SS_E in prediction. Another approach is to use the polynomial identification method suggested in the previous section. The results of various models and the resulting prediction errors (for the validation data set) are shown in Table 16.1.

From this table, the best fit is obtained using a polynomial FIR model (since it has the lowest Prediction Error). Better fits could be obtained by trying other model types, but since the data is never perfect, fine-tuning a model may not be worth the effort.

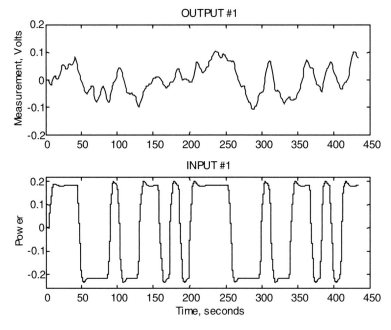

Figure 16.14 Validation data after low-pass filtering.

Table 16.1 Comparison of Fit Using Various Models.

Model Type	Prediction Error	Transfer Function (obtained by converting the discrete model to a continuous model)
[1 1 1] ARX Model	0.7422	$$\frac{0.05542}{s + 0.08466}$$
[2 2 1] ARX Model	0.8763	$$\frac{0.05208\ s + 0.009669}{s^2 + 0.2114\ s + 0.01863}$$
[3 3 1] ARX Model	0.7237	$$\frac{0.04068\ s^2 + 0.02198\ s + 0.01345}{s^3 + 0.3581\ s^2 + 0.2811\ s + 0.02052}$$
POLYID Model	0.6325	$$\frac{0.57231\ \exp(-2*s)}{1 + 9.383*s}\ (\ FOPDT\ model\ fitted)$$

Figure 16.15 compares the model predictions and validation data plotted on the same graph for the best fit. One can see some of the high-frequency noise is still present in the validation data. Most of the error is contributed by this noise rather than by the model error itself. To get a better fit, one would also have to include a noise model. Some techniques to do this are described by Ljung (1999). However, in process industry, noise models are seldom used. The step responses resulting from the four models are shown in Figure 16.16.

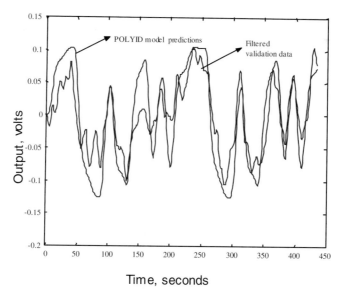

Figure 16.15 Comparison of actual and model-predicted responses for validation data.

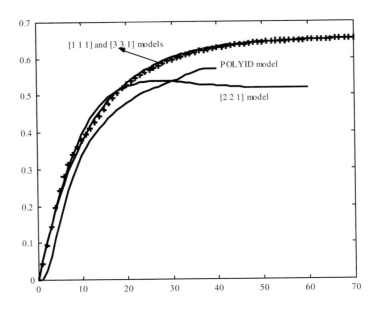

Figure 16.16 Comparison of the step responses obtained by various models.

16.6 EFFECT OF PREFILTER ON PARAMETER ESTIMATES

In this section we consider the rationale for selecting a filter. This section may be skipped without loss of continuity. Consider a system represented by a transfer function relationship

$$y(k) = G(z, \theta)u(k) + e(k), \tag{16.26}$$

where $u(k)$ is the input.

     Here $e(k)$ represents the noise and other unknown disturbances in the output measurement. If the system is modeled by $y_m = G(z, \tilde{\theta}) \cdot u(k)$,

$$y_m = G(z, \tilde{\theta}) \cdot u(k), \tag{16.27}$$

where

$$G(z, \theta) = \frac{b_0 + b_1 z^{-1} + \dots b_{nb} z^{-nb} z^{-nk}}{a_1 + a_1 z^{-1} + \dots + a_{na} z^{-na}}, \tag{16.28}$$

in which $\theta = (\tilde{b}_0, \tilde{b}_1, ..., \tilde{b}_{nb}, a_1, ...\tilde{a}_{na})$ represents unknown model parameters, the prediction error is defined by

$$\varepsilon(k,\theta) = y(k) - y_m(k),$$
$$= [G(z,\theta) - G(z,\theta)]u(k) + e(k). \tag{16.29}$$

In least-square estimation, we seek to minimize

$$\underset{\theta}{Min}\, V(\theta) = \left\| \varepsilon(k,\theta) \right\|_2^2$$
$$= \sum_{k=1}^{N_K} \varepsilon^2(k,\theta), \tag{16.30}$$

where N_k is the number of data points. As $N_k \to \infty$, we can apply Parseval's theorem (see Appendix B)

$$V(\theta) = \frac{1}{2\pi} \int_{-\pi}^{\pi} \Phi_\varepsilon(\omega,\theta)d\omega, \tag{16.31}$$

where $\Phi_\varepsilon(\omega,\theta)$ is the power spectrum of the signal $\varepsilon(k,\theta)$. Now,

$$\Phi_\varepsilon(\omega,\theta) = \left| G(e^{i\omega},\theta) - G(e^{i\omega},\theta) \right|^2 \Phi_u(u) + \Phi_e(\omega)d\omega, \tag{16.32}$$

assuming e and u are uncorrelated. Suppose we used a prefilter $L(z)$ to smooth the input $u(.)$ output $y(.)$. The filtered prediction error is given by

$$\varepsilon_f(k,\theta) = L(z)y(k) - G(z,\tilde{\theta})L(z)u(k). \tag{16.33}$$

Minimizing the filtered prediction error leads to the least-squares problem

$$\underset{\theta}{Min}\, \hat{V}_F(\theta) = \frac{1}{2\pi} \int [|G(i\omega,\theta) - G(i\omega,\theta)|^2 \Phi_u(i\omega) + \Phi_e(\omega)]|L(i\omega)|^2 \, d\omega. \tag{16.34}$$

Thus prefiltering is equivalent to incorporating a weighting function to the minimization objective function in the frequency domain. By adjusting the properties of $L(i\omega)$, we can enhance the fit at selected frequencies at the expense of minimizing the contribution at other frequencies.

The objective of the prefilter should be to suppress the contribution of $\Phi_e(\omega)$ to the objective function $\hat{V}_F(\theta)$. Thus $|L(i\omega)|$ should be small when $\Phi_e(\omega)$ is large. From our discussion in this chapter, a band-pass filter would be most appropriate for process system identification. The frequency bands to pass should be selected based on

1. The frequency bands of interest to the closed-loop system.

2.   The frequencies at which noise will dominate.

A preliminary test on the process should include collecting data over a period in which the input variable is kept constant. A spectral analysis of this data will provide insights into the selection of the pass-band for the prefilter. The time period over which this data is collected should span several time constants of the process to observe the low-frequency disturbances present. If the noise frequency band overlaps with closed-loop response band, then it is not possible to reduce the contribution of $\Phi_e(\omega)$. In this case consider increasing Φ_u in order to get more accurate estimates.

16.7 SOFTWARE FOR IDENTIFICATION

MATLAB provides a rich variety of tools in its Identification toolbox. These tools are meant for discrete systems. The use of these commands is illustrated in the MATLAB file for Example 15.3. In addition, two software modules (MODELBUILDER and POLYID) for identification of SISO systems are included on the website. See Appendix H for a description of the software.

16.8 SUMMARY

In this chapter we explored the following concepts:
- How to design input signals.
- How to selectively suppress the effect of noise.
- How to improve upon the basic least-square algorithm.
- How to identify MIMO systems.

The design of input signals is governed by the desired frequency content of the input signal. Typically, the frequency band of interest is taken to be

$$\frac{1}{\beta\tau} < \omega < \frac{\alpha}{\tau},$$

where τ is the dominant time constant of the process (determined by a pretest). Typical values for α and β are 2 and 3 respectively.

The design of the noise prefilter also plays an important role. If possible, the noise frequency content of the signal must be measured before designing the noise prefilter. Some authors recommend removing the low-frequency noise by formulating the identification problem in terms of incremental changes.

MIMO identification problems can also be formulated as a linear least-squares problem when using DFIR models. However, one difficulty with DFIR models is the large number of model parameters. It is important to use some smoothing techniques to the DFIR coefficients to minimize the sensitivity of the model to noise. One such technique was presented in this chapter.

Problems

16.1 A process with transfer function

$$p(s) = \frac{e^{-s}}{s+1}$$

was perturbed with three input sequences, and the sampled data (corrupted with noise) is given in the files on the website as *ch16pr1a.dat*, *ch16pr1b.dat*, and so on. Fit a FOPDT model to the process (assume delay is known) using each of the three sets. Compare the % error in gain and time constant in each case.

Compute the frequency spectrum of the input signals in each case. What range of frequencies were excited in each case? Discuss your identification results in the context of the input frequency content.

16.2 A process with transfer function

$$p(s) = \frac{e^{-s}}{s+1}$$

was perturbed with a properly designed input PRBS signal. The output corrupted by noise is given in file *ch16pr1a.dat*. Apply a first-order filter with cut-off frequencies at 0.1, 1, and 10 rad/sec. Use filtered data to identify gain and time constant of the process model.

Discuss your results. What cut-off frequency do you recommend? Why?

16.3 Consider the data given in Problem 16.2. Apply the following algorithms to identify an FOPDT model (assume delay is not known):

    a.   ARX program from Identification Toolbox

    b.   MODELBUILDER program

Determine the gain, delay, and time constant. Repeat after applying a second-order filter with cut-off frequency at $\omega = 5 \, \text{rad/sec}$. Comment on your results.

16.4 Consider the process transfer function

$$p(s) = \frac{1}{s^2 + s + 1}.$$

Design a PRBS signal to identify the process. Plot the frequency content of the signal.

References

Amirthalingam, R., and J. H. Lee. 1999. "Subspace Identification-Based Inferential Control Applied to a Continuous Pulp Digester." *J. Process Control, 9*(5), 397–406.

Banerjee, P., S. L. Shah, S. Niu, and D. G. Fisher. 1995. "Identification of Dynamic Models for the Shell Benchmark Problem." *J. Process Control, 5*(2), 85–97.

Braatz, R. D., M. L. Tyler, M. Morari, F. R. Prnackh, and L. Sartor. 1992. "Identification and Cross-Directional Control of Coating Processes." *AIChE J., 38*(9), 1329–1339

Cutler, C. R., and F. H. Yocum. 1991. "Experience with the DMC Inverse for Identification." *Proceedings of the Fourth International Conference on Chemical Process Control,* Padre Island, TX. AIChE, NY. 297–317.

Davies, W. D. T. 1970. *System Identification for Self-Adaptive Control.* John Wiley, London.

Defranoux, C., H. Garnier, and P. Sibille. 2000. "Identification and Control of an Industrial Binary Distillation Column: A Case Study." *Chem. Eng. Technol., 23*(9), 745–750.

Favoreel, W., B. De Moor, and P. Van Overschee. 2000. "Subspace State Space System Identification for Industrial Processes." *J. Process Control, 10*(2–3), 149–155.

Franklin, F. G., and J. D. Powell. 1990. *Digital Control of Dynamic Systems.* Addison-Wesley, Reading, MA.

Koenig, D. 1991. *Control and Analysis of Noisy Processes.* Prentice Hall, NJ.

Koung, C. W., and J. F. MacGregor. 1993. "Design of Identification Experiments for Robust Control: A Geometric Approach for Bivariate Processes." *Ind. Eng. Chem. Res., 32*(8), 1658–66.

Ljung, L. 1990. *System Identification, Theory for the User,* 2nd Ed. Prentice Hall, NJ.

Mathworks, Inc. 2000. MATLAB System Identification Toolbox. Mathworks, MA.

Rivera, D. E., and S. V. Gaikwad. 1995. "Systematic Techniques for Determining Modeling Requirements for SISO and MIMO Feedback Control." *J. Process Control, 5*(4), 213–224.

Sung, S. W., S. Y. Lee, and H. J. Kwak. 2001. "Continuous-Time Subspace System Identification Method." *Ind. Eng. Chem. Res.*, *40*(13), 2886–2896.

Ying, C. Y., and B. Joseph. 1999. "Identification of Stable Linear Systems Using Polynomial Kernels." *Ind. And Eng. Chem. Res.*, *38*, 4712–4728.

Zhu, Y. 1998. "Multivariable Process Identification for MPC: The Asymptotic Method and Its Applications." *J. Process Control*, *8*(2), 101–115.

Basic Model-Predictive Control

Objectives of the Chapter

In this chapter we describe a discrete-time implementation of model-based control algorithm called model-predictive control (MPC). The main issues addressed are

- Reduction of a control problem into an optimization problem that is solved at every sample time.
- Incorporating effect of measured disturbances (feedforward control).
- Control of multivariable systems.

Prerequisite Reading

Chapter 14, "Discrete-Time Models"
Appendix C, "Linear Least-Squares Regression"

17.1 INTRODUCTION

In this chapter we will consider discrete-time (computer-control) systems. The control problem is expressed as a problem of minimizing an objective function subject to constraints. The availability of a model allows us to predict the future effects of control actions taken at the present, and hence we can choose these control actions to meet our control objectives in the best way possible. Reformulating and solving the control problem at every sample time allows us to correct any new disturbances as well as to account for modeling errors. Since this approach tends to base all the control functions on a single-objective function, we will refer to this as single-objective model-predictive control (SOMPC).

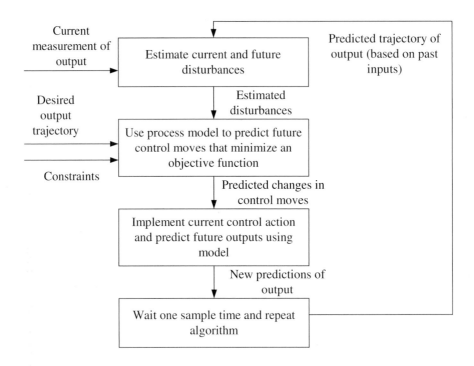

Figure 17.1 Generic form of the SOMPC algorithm.

These single-objective approaches are centralized control approaches because the control of all variables is simultaneously considered. This has some advantages. In the centralized approach, all interactions among the variables are considered, and hence we remove one of the major difficulties associated with an approach that is based on decentralized SISO control loops. Constraints can be taken into account. The centralized approach also allows formulation of control problems with unequal number of inputs and outputs. If we have excess manipulated variables, then we can use the extra degrees of freedom to seek the most economic operating condition as the process disturbances change.

Figure 17.1 shows the generic form of an SOMPC algorithm. The control problem is converted into an optimization problem by defining an objective function, which captures the controller requirements. For example, we can define the control objective as the sum of the square of the errors (difference between output and setpoint):

$$\underset{\Delta u}{Min} \ \{ \left\| \mathbf{y} - \mathbf{y}^r \right\|^2 \} \tag{17.1}$$

Here, y is a vector of predicted future values of the output variable, and $\mathbf{y}^r$ is the desired setpoint of the output. The minimization is carried out over the proposed changes in the manipulated variables. The problem is formulated in the time domain with control moves taken at every discrete point in time. Figure 17.2 shows the typical scenario at an arbitrary sample time.

These single-objective approaches have been used for constrained control of highly coupled unit operations, such as FCC (fluid catalytic cracker) systems in refineries. This control algorithm is implemented at the supervisory level, above the regulatory control (see Chapter 1) layer. Thus slow composition control loops are good candidates to consider for this algorithm. The manipulated variables consist mainly of the setpoints of lower level regulatory controllers (fast flow, level, and pressure-control loops).

The development of these algorithms dates back to the 1970s, when industry started implementing model-based control. Cutler and Ramaker (1980) and Richalet et al. (1978) describe applications that use this approach. Cutler and Ramaker called it the dynamic matrix control (DMC) algorithm and Richalet et al. called it the model algorithmic control (MAC) approach. The DMC algorithm is based on finite step response model (also called the dynamic matrix model) of the process, whereas the MAC approach is based on finite impulse response model. The main advantage of these approaches is that all the control actions are coordinated with each other. The disadvantage is that when there are different priorities associated with various control objectives, the formulation of the objective function may be difficult.

Garcia and Morari (1982) realized the linear nature of the algorithm and coined the term IMC (internal model control) to describe algorithms where the process model is part of the controller and the output of the model is subtracted from the measured output. The DMC algorithm was later extended to constrained linear systems as quadratic dynamic matrix (QDMC) control (Ricker, 1985; Garcia and Morshedi, 1986).

Usually, the active constraints on a process are determined by economic considerations. As process disturbances change, economics might dictate switching from one active constraint to another. In many applications the payoff from the economic optimization is quite significant. To address this issue, some authors suggest a modification to create a two-stage MPC algorithm consisting of a steady-state controller in cascade with the standard MPC controller just described (Morshedi et al., 1985; Brosilow & Zhao, 1988; Yousefi & Tournier, 1991; Harkins, 1991). This is referred to as the LP-MPC or QP-MPC algorithm.

This material is organized in three chapters. This chapter first introduces the SOMPC control algorithm for SISO systems with no constraints on manipulated inputs or controlled outputs. Unconstrained control of multivariable systems are discussed next. Chapter 18 extends this algorithm to consider constraints on input and output variables. We also consider the incorporation of economic objectives, using a two stage cascaded MPC algorithm. The methodology is extended to nonlinear and batch process control. Chapter 19 extends the concept to control using secondary measurements.

17.2 SISO MPC

In this section, we will present the single-objective approach for SISO plant with no constraints. This is the simplest problem, and it is presented mainly for pedagogical reasons to explain the concept and the algorithm. The extensions to constrained, multivariable, and nonlinear control problems follow essentially along the same lines. First we discuss the model representation used.

17.2.1 Dynamic Matrix Model

Consider a process at any sample time $t_k = kT_s$. Let the future predicted values of y be represented by the sequence $y_{k+1}^p y_{k+2}^p, \ldots y_{k+p}^p$. These values capture the effect of all past control actions. Because of the ZOH, the control input look like a series of step changes, with a step change made at every sample time (see Figure 17.1).

Assume that we want to consider $(n+1)$ future control moves i.e., $\Delta u_k, \Delta u_{k+1}, \ldots, \Delta u_{k+n}$ with $\Delta u_{k+n+1} = \Delta u_{k+n+2} = \ldots 0$

Using the step response coefficients we can predict *changes in y* that are caused by the *future-control moves* (see Chapter 14).

$$\Delta y_{k+1} = a_1 \Delta u_k$$
$$\Delta y_{k+2} = a_2 \Delta u_k + a_1 \Delta u_{k+1}$$

(Because of the linearity of the process, the effect of step changes Δu_1 and Δu_2 can be added to yield the net changes in y at t_{k+2}).

Similarly

$$\Delta y_{k+3} = a_3 \Delta u_k + a_2 \Delta_{k+1} + a_1 \Delta u_{k+2}$$

....

$$\Delta y_{k+p} = a_p \Delta u_k + a_{p-1} \Delta u_{k+1} + \ldots + a_{p-n} \Delta u_{k+n}$$

(Note that $\Delta u_{k+n+1} = \Delta u_{k+n+2} = 0$.) This leads to the equation

$$
\begin{bmatrix} \Delta y_{k+1} \\ \Delta y_{k+2} \\ \vdots \\ \Delta y_{k+p} \end{bmatrix} =
\begin{bmatrix}
a_1 & 0 & \ldots & 0 \\
a_2 & a_1 & \ldots & 0 \\
a_n & a_{n-1} & \ldots a_1 & 0 \\
\ldots & \ldots & \ldots & \ldots \\
a_p & a_{p-1} & \ldots & a_{p-n}
\end{bmatrix}
\begin{bmatrix} \Delta u_k \\ \Delta u_{k+1} \\ \\ \Delta u_{k+n} \end{bmatrix}
$$

or in matrix notation

$$\Delta y = A \, \Delta u$$

This representation is referred to as the *dynamic matrix model* of the process.

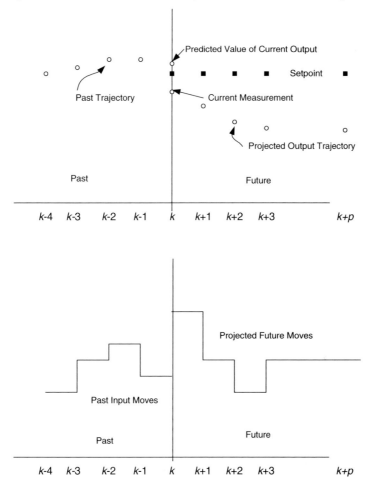

Figure 17.2 A snapshot of input and output variables at any arbitrary sample time, *k*.

17.2.2 Description of the Algorithm

Consider the regulatory control of a single-output variable at its setpoint using a single manipulated input variable. Control is considered in the discrete-time domain with a fixed time at which the output measurement is sampled and a control action is taken. At any arbitrary sample time, denoted by k, the problem can be stated as follows (see Figure 17.2 for notation):

> *Given*
> 1. A predicted trajectory of the output based on past control actions $y_k^p, y_{k+1}^p, ..., y_{k+p}^p$, where k = current time and p = length of output prediction horizon
> 2. A desired output trajectory $y_{k+1}^r, y_{k+2}^r, ..., y_{k+p}^r$,
> 3. Current measurement of output y_k^m,
>
> *compute*
> 1. An estimate of the disturbances $d_{yk}, d_{yk+1}, ..., d_{yk+p}$,
> 2. An estimate of the new desired trajectory,
> 3. The control actions needed to bring the output to the desired trajectory $\Delta u_k, \Delta u_{k+1}, ..., \Delta u_{k+n}$, where n = control move horizon,
> 4. A revised estimate of the output trajectory, assuming that the current control action Δu_k is implemented.

Step 1. Initialization. The algorithm needs a few parameters and variables to be specified at the start. These include

1. *The output horizon length p.* This is the number of sample times into the future over which the output is projected, using a process model. Control moves are computed such that the output follows a specified trajectory over this horizon.
2. *The control move horizon length n.* This is the number of sample times over which the control input will be adjusted. This in effect is the number of changes in the control vector considered. At each sample time, we can change the control input. If we consider n sample times, we have n degrees of freedom to try to bring the process output back to the desired values over the output horizon. In general, the number of control moves considered are smaller than the output horizon p. Generally, the output horizon is chosen long enough to allow the output to reach a steady-state at the end of the horizon.
3. y^p. This is a vector containing the predicted values of output y over the horizon p.

    At the start of the algorithm, this vector is initialized to the current measurement of the output variable. As the algorithm proceeds, this value is updated to include the effect of all past control actions. Computation of future control actions will be based on changes to be made to this projected output trajectory, using future control actions. This vector captures the state of the system.

Step 2. Estimating disturbances. The disturbance estimate is obtained from the difference between the prediction of the current value of y and the current measured value of y.

Let y_k^m denote the current measured value of the output. Then, the current disturbance effect at the output is computed using

$$d_{yk} = y_k^m - y_k^P.$$ (17.2)

Estimating the future values of disturbances requires some assumptions. The simplest assumption is to assume that these are constant:

$$d_{yk+1} = d_{yk+2} = ... = d_{yk}.$$ (17.3)

This assumption seems to work well in most cases. If a model of the disturbance is available, say in the form of an ARIMA type transfer function (see Appendix D), then the future values of d_y can be estimated using the model.

Step 3. Estimate new desired trajectory. In its simplest form, the desired trajectory is made equal to the setpoint:

$$y_{k+1}^r = y_{k+2}^r = ... = y_{k+p}^r = y^{set}.$$ (17.4)

However, sometimes it may be desirable to generate a smooth approach to the set point to prevent large control moves. We will consider such variations later.

Step 4. Compute the control actions. Using the Dynamic matrix model we can express

$$
\begin{aligned}
y_{k+1} &= y_{k+1}^P + a_1 \Delta u_k && + d_{yk+1} \\
y_{k+2} &= y_{k+2}^P + a_2 \Delta u_k + a_1 \Delta u_{k+1} && + d_{yk+2} \\
y_{k+3} &= y_{k+3}^P + a_3 \Delta u_k + a_2 \Delta u_{k+1} + a_1 \Delta u_{k+2} && + d_{yk+3} \\
&\ . \\
&\ . \\
&\ . \\
y_{k+p} &= y_{k+p}^P + a_p \Delta u_k + a_{p-1} \Delta u_{k+1} + ... + a_{p-n} \Delta u_{k+n} && + d_{yk+p},
\end{aligned}
$$ (17.5)

where $a_1, a_2, ..., a_p$ are the unit step response coefficients, and $\Delta u_k, \Delta u_{k+1}, \Delta u_{k+2}, ..., \Delta u_{k+n}$ and so on are the step changes in controller output. In matrix form

$$
\begin{bmatrix}
y_{k+1} \\
y_{k+2} \\
. \\
. \\
y_{k+p}
\end{bmatrix}
=
\begin{bmatrix}
y_{k+1}^P \\
y_{k+2}^P \\
. \\
. \\
y_{k+p}^P
\end{bmatrix}
+
\begin{bmatrix}
a_1 & 0 & . & . & 0 \\
a_2 & a_1 & 0 & . & 0 \\
. & . & . & . & . \\
. & . & . & . & . \\
a_p & a_{p-1} & a_{p-2} & . & a_{p-n}
\end{bmatrix}
\begin{bmatrix}
\Delta u_k \\
\Delta u_{k+1} \\
. \\
. \\
\Delta u_{k+n}
\end{bmatrix}
+
\begin{bmatrix}
d_{k+1} \\
d_{k+2} \\
. \\
. \\
d_{k+p}
\end{bmatrix}
$$ (17.6)

$$y = y^P + A\Delta u + d_y. \tag{17.7}$$

Note that we know y^P, d_y and we want Δu to make y as close to y^r as possible. Hence we can set up an optimization problem with objective function

$$\underset{\Delta u}{Min} \parallel e \parallel^2, \tag{17.8}$$

where

$$\begin{aligned} e &= y - y^r \\ &= y^P + A\Delta u + d_y - y^r. \end{aligned} \tag{17.9}$$

This is a problem of finding the least-squares solution of the set of equations

$$A\Delta u = y^r - d_y - y^P. \tag{17.10}$$

There are p equations in n + 1 unknown $(\Delta u_k, \Delta u_{k+1},..., \Delta u_{k+n})$. Typically, we will have $p \gg n$, which means we will have more equations than unknowns. (We cannot drive all future elements of the vector y to y^r, when using only a few future control moves.) Since we cannot satisfy all the equations exactly, we can attempt to satisfy them in a least-squares sense.

The control moves can be computed using the least-squares solution (see Appendix C):

$$\begin{aligned} \Delta u &= (A^T A)^{-1} A^T (y^r - d_y - y^P) p \\ &= B(y^r - d_y - y^P). \end{aligned} \tag{17.11}$$

Note that since A is known, this calculation can be done ahead of time, and B can be stored online. Thus the computational requirements at every step is reduced. We only need to compute Δu_k, and so only the first row vector of B matrix is needed. Let

$$b^T = \text{first row vector of } B \text{ matrix.}$$

Then

$$\Delta u_k = b_1^T (y^r - d_y - y). \tag{17.12}$$

Step 5. Revise estimates and go to next sample time. At the next sample time, we will want to repeat the whole procedure. We need a new estimate of the output trajectory. We make the assumption that Δu_k will be implemented. $\Delta u_{k+1}, \Delta u_{k+2}, \Delta u_{k+n}$ cannot be implemented yet. Since we will be re-evaluating these based on the situation in the future, we no longer need these values. The reason we incorporated these in the computation was to take into account the fact that requiring only Δu_k to drive y to y^r can lead to large control moves, which may not be feasible or may lead to instability. If Δu_k is implemented, then

$$y^P_{k+1} \leftarrow y^P_{k+1} + a_1 \Delta u_k + d_{yk+1}$$

$$y^P_{k+2} \leftarrow y^P_{k+2} + a_2 \Delta u_k + d_{yk+2}$$

$$\cdot \qquad\qquad\qquad\qquad (17.13)$$

$$\cdot$$

$$y^P_{k+p} \leftarrow y^P_{k+p} + a_p \Delta u_k + d_{yk+p}.$$

This represents our updated prediction of y, taking into account the current control action and current estimate of the disturbances.

Before we go to next sample time, we must update the y^p vector (the first value y^p_k is no longer needed). This is accomplished by moving every element of the vector y up by one index:

$$y^P_{k+i} \leftarrow y^P_{k+i+1}, \qquad\qquad (17.14)$$

thus bumping out the first element of the vector. This leaves a gap in the place of the last entry, which corresponds to our estimate of y one sample time beyond our current output horizon. In the simplest approach this output value is assumed to be the same as the projected value of y at the $k+p^{th}$ sample time:

$$y^P_{k+p+1} = y^P_{k+p} \qquad\qquad (17.15)$$

Likewise, the y^r vector must also have the first element removed and a new element added at the end.

Figure 17.3 shows the block diagram representation of the SOMPC algorithm. Typically, the disturbance model is made part of the algorithm. However, showing it explicitly allows us to understand the extension of this algorithm to cases such as feedforward and inferential control.

17.2.2 Tuning Parameters

1. Sample time T_S and horizon p. The sample time and horizon p determine how far into the future we are going to project the output y.

$$pT_s = \text{horizon time for } y$$

Usually we want to take control action to bring y back to a steady-state value at the setpoint. This leads to the guideline

$$pT_s = \text{time to reach steady-state}$$

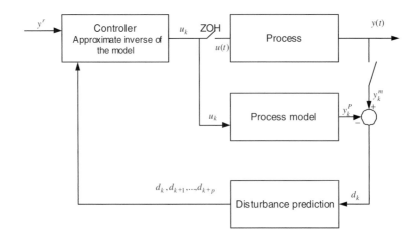

Figure 17.3 Block diagram representation of the SOMPC algorithm.

Since it takes about four time constants for a first-order process, we can determine

$$pT_s = 4\tau,$$

where τ = dominant time constant of the process. The number of samples of y in this horizon will determine the size of the A matrix. Large matrices are more difficult to invert. (They tend to be ill-conditioned and hence sensitive to small errors in modeling.) On the other hand, keeping T_s too large (to reduce p) will mean longer time intervals between control actions, which is also detrimental. Usually, a compromise is sought with

$$p = 20 \text{ to } 30$$

For multivariable systems, p may be kept even smaller to reduce overall matrix dimensions.

Keeping pT_s too small will give a shortsighted nature to the control algorithm. This in turn can lead to instability, because the controller does not see the long-term consequences of the action (consider, for example, an inverse response process).

2. Number of control moves.   Increasing the number of control moves n, increases the degree of freedom and hence tries to make y closer to the setpoint. This is equivalent to requiring the controller to make y follow the setpoint rigidly. Since we have not imposed any restrictions on the control moves, this means large variations in u will be needed to keep $y = y^r$. This is undesirable. In practice, some performance is sacrificed in favor of stable, smooth operation by keeping n small. Keeping n small also reduces the size of the matrix to be inverted ($A^T A$ is an n x n matrix). Typically, n is kept in the order of

$$n = 1 \text{ to } 4.$$

It can be shown that stability of the system is guaranteed by keeping $p \gg n$ (Garcia & Morari, 1983).

3. Move suppression factors β. It has been found that the least-squares solution usually leads to severe control moves to bring the system back to the setpoint. This is because we have not put any constraints on how large Δu can be. This results in significant step changes in u from one time step to the next. There is a need to restrict the size of these moves. This can be done in a number of ways. We can, for example, put constraints on the size of the control moves. However, this will lead to a constrained optimization problem, which is harder to solve online.

Two other approaches have been employed in practice. One approach is to add some additional constraining equations (move suppression factors; see Cutler & Ramaker, 1978) to the least-squares problem. The other approach is to relax the requirement that y approach y^r immediately. Instead, a smooth approach to the desired setpoint is required. This is accomplished by redefining y^r to follow a set reference trajectory.

Consider the approach that uses move suppression factors. Instead of minimizing the sum of the square of the errors

$$\operatorname*{Min}_{\Delta u} \; \left\| y^r - d_y - y^P - A\Delta u \right\|^2, \tag{17.16}$$

we seek to minimize

$$\operatorname*{Min}_{\Delta u} \; \left\| y^r - d_y - y^P - A\Delta u \right\|^2 + \beta^2 \left\| \Delta u \right\|^2, \tag{17.17}$$

where β = more suppression factor, a scalar. This effectively reduces the size of Δu_i in the solution. This is equivalent to adding a new set of equations,

$$\beta \Delta u_k = 0,$$
$$\beta \Delta u_{k+1} = 0,$$
$$\beta \Delta u_{k+2} = 0, \tag{17.18}$$
$$...,$$
$$\beta \Delta u_{k+n} = 0,$$

to the original set of equations,

$$A\Delta u = y^r - d_y - y^P. \tag{17.19}$$

The solution is given by

$$\Delta u = (A^T A + D)^{-1} A^T (y^r - d_y - y^P). \tag{17.20}$$

where D is an $n \times n$ matrix:

$$D = \text{diag}\,[\beta^2 \; \beta^2 \; \beta^2 ... \beta^2\,]. \tag{17.21}$$

The advantage with this approach is that only one additional parameter, β, has to be specified. This parameter is usually treated as an online tuning parameter.

4. Reference trajectory. The second approach taken to suppress large moves in u is to modify the reference trajectory, $\mathbf{y}^r$. Until now, we have treated $\mathbf{y}_i^r$ equal to $\mathbf{y}^{set}$. This is a rather stringent requirement (it is equivalent to saying that we want y to go to the setpoint at the next sample time). We can relax this criterion by making $\mathbf{y}_i^r$ approach $\mathbf{y}^{set}$ gradually, as shown in Figure 17.4.

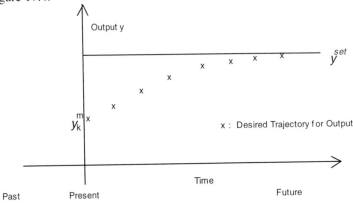

Figure 17.4 Reference trajectory approach.

A reference trajectory, such as the one shown in Figure 17.4, which takes y gradually from the current measurement y_k^m to the desired setpoint, y, can be generated as follows (Richalet et al., 1979):

$$
\begin{aligned}
y_{k+1}^r &= \alpha y_k^m + (1-\alpha)\,y^{set}, \\
y_{k+2}^r &= \alpha y_{k+1}^m + (1-\alpha)\,y^{set}, \\
&\cdots, \\
y_{k+p}^r &= \alpha y_{k+p-1}^m + (1-\alpha)\,y^{set}.
\end{aligned}
\tag{17.22}
$$

The parameter α is related to how fast (or slow) an approach to the setpoint is desired:

$$\alpha = e^{-T_s/T_r}, \tag{17.23}$$

where T_s = sample time and T_r = time constant of approach to the setpoint. To get faster approach, decrease T_r or α. Generally, T_r should be chosen smaller than the open-loop time constant of the process. Keeping T_r too small could lead to severe control actions as the controller attempts to bring the output to the setpoint in a short time.

Example 17.1 MATLAB Implementation of SOMPC

Consider a system modeled by

$$y(s) = 4.05 \ (1+k_e)\frac{(1-13.5s)}{(1+13.5s)(50s+1)}u(s)$$

where k_e is a model uncertainty factor. Assume $k_e = 0$ for the model. The plant may be different. Implement a program in MATLAB to do the following:
1. Generate a finite step response model for the process; that is, construct the dynamic matrix model.
2. Implement the SOMPC algorithm and generate a response to setpoint changes.
3. For various values of the parameter k_e (try $k_e = -.5, 0, .5$) and choices of the tuning parameters β and n (try $\beta = 1, 2, 3$ and $n = 1, 2, 3$), investigate the performance of the control system to setpoint changes.

Solution

A program that implements the algorithm for a SISO process is given in ch17ex1.m. This program is provided primarily to illustrate the mechanics of implementing a simple MPC controller. General implementations are available in the MPC toolbox as well as from control software vendors such as AspenTech. A schematic of the implementation is shown in Figure 17.5.

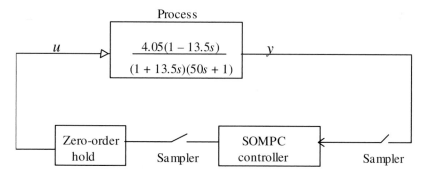

Figure 17.5 Implementation of the SOMPC algorithm for the example process.

The open-loop step response of the process shows that it takes about 300 min to reach steady-state. The presence of the RHP leads to an inverse response behavior of the process.

Since the process takes about 300 min. to reach steady-state, we want $pT_s \approx 300$. For $T_S =$ 5, we find that p should be around 60. For the sake of computational efficiency, we can take a smaller p, or we might consider using a larger sample time.

Keeping $n = 1$ results in a control move that achieves a setpoint response similar to that of the open-loop step response. But $n = 2$ and $n = 3$ (without move suppression) leads to nearly unstable responses due to the inverse response nature of the process. The response with $n = 2$ and $n = 3$ is almost identical. Keeping n larger has a destabilizing effect on the control algorithm. In order to stabilize this, we must use a move suppression factor, which will damp out the large movements in the manipulated input variable.

Addition of the move suppression factor leads to stabilization of the controller by dampening out the controller movements substantially. There is not much difference between the two values $\beta = 2$ and $\beta = 3$, but $\beta = 1$ leads to controller behavior almost identical to that with no move suppression factor at all. It should be remembered that the square of β appears in the equation for computing the control moves. β should be chosen such that it has a substantial impact on the matrix to be inverted, namely, $A^T A$. Thus the square of β should be of the order of the diagonal terms in this matrix to make a significant difference.

Studies using the modeling error (see Figure 17.6) show that the controller is stable with a 50% error in gain, provided a sufficiently large move suppression parameter is used. Reducing the sampling time leads to a shortsighted controller and can lead to instability.

♦

17.2.3 Discussion

The MPC algorithms were primarily developed to treat constrained multivariable systems. The development for an SISO system is given to illustrate the concepts. The advantages of MPC algorithms arise primarily from the ability to incorporate multiple objectives and constraints on input and output variables. These extensions are discussed in Chapters 18 and 19.

The algorithm has disadvantages too. The algorithm is still based on a linear input/output model of the process. This can invalidate its application to systems that exhibit significant nonlinearities. For constrained problems, the computational costs can be high. The initial investment required to build and test the model can be high. Tuning methods for MPC algorithms are still in a state of development. (For a discussion of tuning strategy for MPC see the paper by Shridhar and Cooper, 1998). For these reasons, this approach can only be justified for processes where high economic payback can be obtained by operating close to constraints.

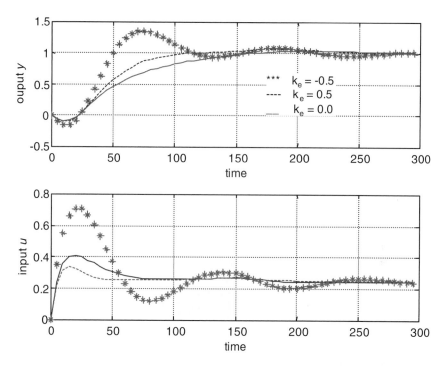

Figure 17.6 Effect of modeling error on the response of the MPC controller to a change in setpoint. Parameters used were $T_s = 5$, p = 30, n = 2, $\beta = 1$.

Example 17.2 Illustrating Use of the MPC Toolbox

This example illustrates the use of SIMULINK blocks provided in the MPC toolbox. Figure 17.7 shows the SIMULINK diagram for implementing control of $y1$ using $u1$ for the Shell fractionation column (see Appendix I). The script file that defines the controller is given in ch17ex2.m. Figure 17.8 shows the disturbance response of this control system.

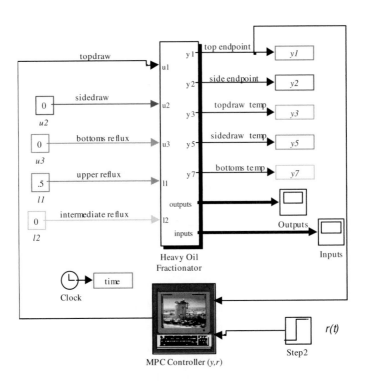

Figure 17.7 SIMULINK diagram of MPC implementation for the Shell column.

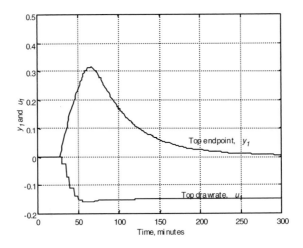

Figure 17.8 Response of y_1 to a step change in disturbance l_1.

17.2.4 Feedforward MPC

Whenever a disturbance can be detected before it enters the process, feedforward MPC can yield better performance than feedback MPC alone. The penalty, of course, is the added cost of measurement and modeling. To incorporate feedforward action to MPC, one must first build a model that describes the effect of the disturbance input on the process output. A step response model can be employed. Consider, for example, the response shown in Figure 17.9, which shows the output behavior when the input disturbance is performed by a unit step increase.

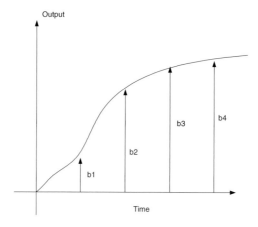

Figure 17.9   Unit step response of y to input disturbance d_m.

The disturbance response model can be written as

$$\Delta y_{k+1} = b_1 \Delta d_{mk},$$
$$\Delta y_{k+2} = b_2 \Delta d_{mk},$$
$$...,$$
$$\Delta y_{k+p} = b_p \Delta d_{mk},$$

(17.24)

or in matrix form

$$\begin{bmatrix} \Delta y_{k+1} \\ \Delta y_{k+2} \\ . \\ \Delta y_{k+p} \end{bmatrix} = \begin{bmatrix} b_1 \\ b_2 \\ . \\ b_p \end{bmatrix} \Delta d_{mk},$$

(17.25)

which can be written as

$$\Delta y = B\Delta d_{mk}.$$
$$\tag{17.26}$$

Combining this with the effect of u, we get

$$y = y^P + A\Delta u + B\Delta d_{mk}.$$
$$\tag{17.27}$$

Using this control move, u can be calculated from

$$\Delta u = (A^T A)^{-1} A^T (y^r - y^P - B\Delta d_{mk} - d).$$
$$\tag{17.28}$$

Incorporating measured input disturbance requires modification of only one step in the control move calculation. The extension to multiple input disturbances is along the same lines. Figure 17.10 shows the block diagram representation of the feedforward MPC.

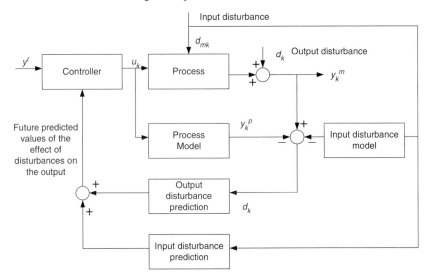

Figure 17.10. Feedback MPC with feedforward correction.

Example 17.3 MPC with Feedforward Applied to the Shell Fractionator Problem

Let us assume that the disturbance l_1 entering the Shell fractionator column is measured. We can use this in a feedforward mode in the MPC algorithm. The implementation of this in SIMULINK, using the MPC toolbox, is shown in Figure 17.11. The script file that implements this is given in program *ch17ex3_script.m*. The result for a step change of 0.5 in l_1 is shown in Figure 17.12. This response should be compared with Figure 17.9, which shows the response with only feedback control.

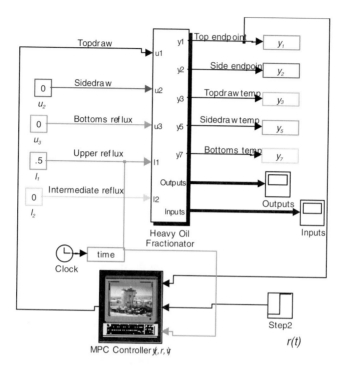

Figure 17.11 SIMULINK diagram of MPC implementation for the Shell column.

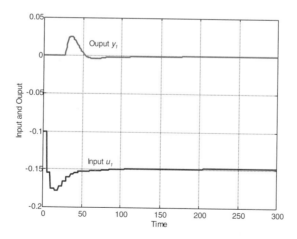

Figure 17.12 MPC with feedforward compensation applied to Shell fractionator. Response of y_1 to a step change in disturbance l_1.

17.3 UNCONSTRAINED MULTIVARIABLE SYSTEMS

17.3.1 Development of the Algorithm for MIMO Case

In this section we will extend the concepts of the single-objective approach to the multivariable case. There is virtually no difference between the SISO and MIMO versions of the algorithm.

Consider the MIMO system shown in Figure 17.13. The relation between y_j and u_i can be modeled as in the SISO case:

$$\Delta y_j = A_{ij} \Delta u_i, \tag{17.29}$$

where A_{ij} is the dynamic matrix relating the changes in the jth output to the ith input. Using these matrices, we can predict the trajectory for the output as

$$
\begin{bmatrix} y_1 \\ y_2 \\ ... \\ y_m \end{bmatrix} = \begin{bmatrix} y_1^p \\ y_2^p \\ ... \\ y_m^p \end{bmatrix} + \begin{bmatrix} A_{11} & A_{12} & ... & A_{in} \\ ... & & & \\ ... & & & \\ A_{m1} & A_{m2} & ... & A_{mm} \end{bmatrix} \begin{bmatrix} \Delta u_1 \\ \Delta u_2 \\ ... \\ \Delta u_m \end{bmatrix} + \begin{bmatrix} d_{y1} \\ d_{y2} \\ ... \\ d_{ym} \end{bmatrix}, \tag{17.30}
$$

where the notation used follows the same style as in the SISO case. Note that y_i^p represents the vector of predicted values of output i into the future. The model can be compactly written as

$$y = y^p + A\Delta u + d_y \tag{17.31}$$

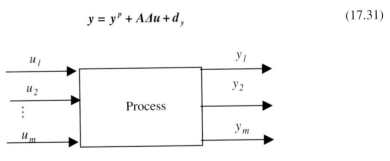

Figure 17.13 A MIMO system.

Step 1. Estimate disturbance vector. At any arbitrary sample time k, the disturbance estimate is obtained as

$$d_{yi} = \begin{bmatrix} d_{yi,k} \\ d_{yi,k+1} \\ d_{yi,k+P} \end{bmatrix},$$
(17.32)

where

$$d_{yi,k} = y^m_{i,k} - y^p_{i,k}$$
(17.33)

and

$$d_{yi,k+1} = d_{yi,k+2...} = d_{yi,k+p} = d_{yi,k}.$$
(17.34)

Step 2. Compute desired trajectory. In the simplest form,

$$y^r_{i,k+1} = y^r_{i,k+2} = ... = y^r_{i,k+p} = y^{set}_i.$$
(17.35)

We can also use a reference trajectory to determine this vector, as explained in the SISO case.

Step 3. Compute control actions.

$$\Delta u = (A^T A + D)A^T (y^r - d_y - y^p)$$
(17.36)

where D is a diagonal matrix of move suppression factors.

Step 4. Revise estimates of y^p.

$$y^p \leftarrow y^p + A\Delta u^1 + d_y,$$
(17.37)

where Δu^1 consists of the vector of control moves including only the move at time t_k :

$$\Delta u^1_i = \begin{bmatrix} \Delta u_{i,1} \\ 0 \\ 0 \\ 0 \\ 0 \end{bmatrix} \text{ etc.}$$
(17.38)

Step 5. Move to the next sample time, t_{k+1}. At this point the values of y^p are updated, as in the SISO case.

In the development of the algorithm for the MIMO case, we assumed that all outputs were given equal weight in the formulation of the control objective. It is possible to weight these outputs differently. In this case the control objective is formulated as

$$Min_{\Delta u}(y^r - y)^T Q(y^r - y) + \Delta u^T R\Delta u, \tag{17.39}$$

where Q and R are positive-definite matrices of weights associated with the outputs and inputs respectively.

17.3.2 Tuning

As in the SISO case, tuning parameters can be incorporated in the form of move suppression factors or reference trajectories. The effects of output horizon p and control horizon n are similar to the SISO case.

Example 17.4 MATLAB Implementation of MIMO-MPC

The MATLAB m-file `ch17ex3.m` illustrates an implementation of the algorithm for the 2 x 2 case. This program is provided primarily to illustrate the mechanics of implementing a multivariable MPC algorithm. The program is set up for a system modeled by the process transfer functions

$$y_1(s) = \frac{4.05(1-13.5s)}{(1+13.5s)(50s+1)}u_1(s) + \frac{1.77(1-14s)}{(14s+1)(60s+1)}u_2(s),$$

$$y_2(s) = \frac{5.39(1-9s)}{(1+9s)(50s+1)}u_1(s) + \frac{5.72(1-7s)}{(1+7s)(60s+1)}u_2(s).$$

This is schematically shown in Figure 17.14.

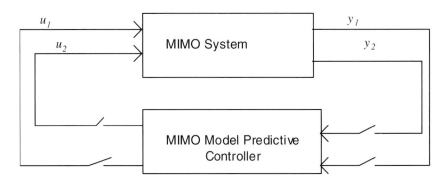

Figure 17.14 Schematic of MIMO problem.

The program sets up the FSR model, using the step response curves. It is possible to introduce a fractional error in each of the gains associated with the process model. This feature can be used to test robustness.

Figure 17.15 shows the response of the system to a step increase in the setpoint of y_1 and y_2. There were no model errors for the response studies shown in this figure. The controller responds well. The response time is similar to that of the open-loop response time of the process. Also, the changes in both manipulated inputs are rather large at the beginning, indicating the emphasis placed on the steady-state response by the controller. Use of a large value for p leads to a control calculation that puts larger emphasis on steady-state behavior.

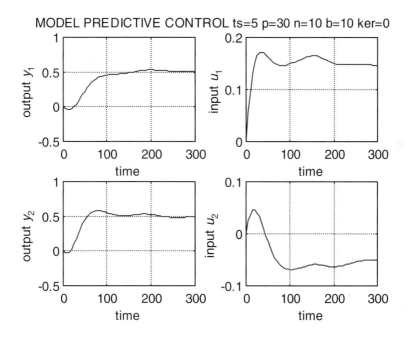

Figure 17.15 Results for Example 17.2. Response of the SOMPC controller simultaneous setpoint change in y_1 and y_2.

17.4 STATE-SPACE FORMULATION OF UNCONSTRAINED SOMPC

In this section we present a formulation of the SOMPC algorithm in state-space form. It may be skipped without loss of continuity. This material is taken from the thesis by Palavajjhala (1994).

Lee et al. (1994) show that the MPC algorithm reduces to an observer-controller formation. We use Lee et al.'s (1994) observer-controller formulation here (see Figure 17.10).

The SOMPC algorithm uses a linear model, in the form of step response coefficients, to predict the influence of past manipulated variable moves on the future controlled variable values. These predicted controlled variable values are then corrected to account for the influence of measured disturbances. Contributions from unmodeled dynamics and unmeasured disturbances, computed by taking the difference between the current measurements and the current predictions, are then used to update the future predictions. The error between the predicted controlled variable values and the desired setpoint values is then computed, and the 2-norm of the error is minimized with or without controlled variable and manipulated variable constraints. The optimization is an unconstrained or constrained least-squares problem. The solution to this optimization problem for the unconstrained case yields a constant-gain state feedback controller.

The observer-controller configuration (shown in Fig 17.16) is described here for a SISO process with no model uncertainty. The formulation can be extended to MIMO processes as well. Equations describing each of the elements shown in Figure 17.16 are given next.

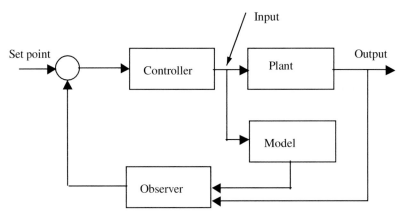

Figure 17.16 Observer-controller configuration of MPC control.

17.4.1 Step Response Model

Let s_i, $i = 1, 2, ..., n$, denote the step response coefficients of a stable process. Then, the step response S is defined as

$$\mathbf{S} = [s_1 \quad s_2 \quad \cdots \quad s_n].\tag{17.40}$$

The step response model that predicts the influence of $u(k-1)$ on the future output variable values, $\mathbf{Y}(k)$, using the linear superposition principle, is

$$Y(k) = MY(k-1) + S\Delta u(k-1), \tag{17.41}$$

where

$$Y(k) = \begin{bmatrix} y(k) & y(k+1) & \cdots & y(k+n-1) \end{bmatrix}^T, \tag{17.42}$$

$$M = \begin{bmatrix} 0 & 1 & 0 & \cdots & 0 \\ 0 & 0 & 1 & \cdots & 0 \\ \vdots & \vdots & \vdots & \ddots & 1 \\ 0 & 0 & 0 & \cdots & 1 \end{bmatrix}, \tag{17.43}$$

and n is the settling time of the step response. The next predicted output value is

$$y(k) = \begin{bmatrix} 1 & 0 & \cdots & 0 \end{bmatrix} Y(k) = NY(k). \tag{17.44}$$

Thus the states of the dynamic system, denoted by $Y(k)$, are the n predicted output values starting from the value at time k (see Eq. 17.42). The output of the dynamic system is the output value at time k (see Eq. 17.44). The dynamic system given by Equations (17.41) and (17.44) describes the plant and its model in state space form.

17.4.2 Prediction

The control action at time k is computed based on the predicted output values. The future output values depend on the past predicted output values $Y(k|k)$, the current and future disturbance $d(k)$, and the future manipulated variable changes $U(k)$. Let the output values over the prediction horizon p, starting from time $k + 1$ based on the measurement at time k, be denoted by $Y(k + 1|k)$. Then,

$$Y(k+1|k) = MY(k|k) + S^d \Delta d(k) + S^u \Delta U(k). \tag{17.45}$$

S^d contains step response coefficients generated from the disturbance model, $d(k)$ is the disturbance vector entering the system at time instant k, and the vector $U(k)$ is the manipulated variable changes vector, consisting of controller moves in the control horizon m:

$$\Delta U = \begin{bmatrix} \Delta u(k) & \Delta u(k+1) & \cdots & \Delta u(k+m-1) \end{bmatrix}^T. \tag{17.46}$$

S^u is called the dynamic matrix and contains the step response coefficients as its elements:

$$S^u = \begin{bmatrix} s_1 & 0 & 0 & 0 \\ s_2 & s_1 & \vdots & \vdots \\ \vdots & \vdots & \ddots & \vdots \\ s_p & s_{p-1} & \cdots & s_{p-m+1} \end{bmatrix}, \tag{17.47}$$

and M is a p x n matrix:

$$M = \begin{bmatrix} 0 & 1 & 0 & 0 & 0 & \cdots & 0 \\ 0 & 0 & 1 & 0 & 0 & \cdots & 0 \\ \vdots & \vdots & \vdots & \ddots & \ddots & \ddots & 0 \\ 0 & 0 & 0 & \cdots & 1 & \cdots & 0 \end{bmatrix}. \tag{17.48}$$

17.4.3 Control Law

The prediction error vector, $E(k+1|k)$, is computed by taking the difference between the set-point trajectory and the predicted output values at each sample time over the prediction horizon. The prediction error computation assumes that no control action is taken between time k and $k + p - 1$. Therefore,

$$E(k+1|k) = \begin{bmatrix} e(k+1|k) & e(k+2|k) & \cdots & e(k+p|k) \end{bmatrix}^T, \tag{17.49a}$$

$$E(k+1|k) = R(k) - (MY(k|k) + S^d \Delta d(k)). \tag{17.49b}$$

$R(k)$ is the vector that prescribes the setpoint trajectory. The controller action is computed by minimizing the 2-norm of the prediction error $E(k+1|k)$. An additional term that penalizes large manipulated variable changes is also introduced in the optimization. Thus the following weighted least-squares problem is solved at each controller sample time to determine the manipulated variable changes vector $U(k)$:

$$\underset{\Delta U(k)}{Min}\, J(k) = \left\| \Gamma[Y(k+1|k) - R(k)] \right\|_2^2 + \left\| \Lambda \Delta U(k) \right\|_2^2, \tag{17.50}$$

where Γ is the output weight and Λ is the move suppression parameter. The solution to the weighted least-squares problem is

$$\Delta U(k) = (S^T \Gamma^T \Gamma S + \Lambda^T \Lambda)^{-1} S^T \Gamma^T \Gamma E(k+1|k) = K_{MPC} E(k+1|k), \tag{17.51}$$

and the control action at time k is

$$\Delta u(k) = \begin{bmatrix} 1 & 0 & \cdots & 0 \end{bmatrix} \Delta U(k). \tag{17.52}$$

K_{MPC} is the controller gain and is a constant if S. Λ and Γ are time-invariant. Thus the control law is essentially a constant-gain state feedback law. Since all the states, $Y(k)$, are not measured at every control sample time, an observer should be used to estimate these states, denoted as $Y(k|k)$, based on the current manipulated and output variable values.

17.4.4 Observer Equation

An observer is a dynamic system whose inputs are the current control action and the current output measurement, and the observer output is a good approximation to the "true" states. The SOMPC algorithm uses a closed-loop observer with a constant gain $\mathbf{I}$,

$$\mathbf{Y}(k \mid k) = \mathbf{M}\mathbf{Y}(k \mid k - 1) + \mathbf{I}(y^m(k) - y(k \mid k - 1)), \tag{17.53}$$

where,

$$\mathbf{I} = \begin{bmatrix} 1 & 1 & \cdots & 1 \end{bmatrix}^T \tag{17.54}$$

and $y^m(k)$ is the output measurement at time k. Replacing $y(k|k-1)=\mathbf{N}(\mathbf{M}\mathbf{Y}(k|k-1) + \mathbf{S} \ u(k-1))$ in Eq. (17.53),

$$\mathbf{Y}(k \mid k) = (\mathbf{M} - \mathbf{I}\mathbf{N}\mathbf{M})\mathbf{Y}(k \mid k - 1) + (\mathbf{I} - \mathbf{I}\mathbf{N})\mathbf{S}\Delta u(k - 1) + \mathbf{I}y^m(k). \tag{17.55}$$

Substituting for $y(k|k-1)$, $\mathbf{Y}(k|k-1)$, and $\mathbf{U}(k-1)$ from equations (17.42), (17.43), and (17.50), respectively, and simplifying, we get

$$\begin{bmatrix} \mathbf{Y}(k) \\ \mathbf{Y}(k \mid k) \end{bmatrix} = \begin{bmatrix} \mathbf{M} & -\mathbf{S}\mathbf{K}_{MPC}\mathbf{M} \\ \mathbf{I}\mathbf{N}\mathbf{M} & \mathbf{M} - \mathbf{I}\mathbf{N}\mathbf{M} - \mathbf{I}\mathbf{S}\mathbf{K}_{MPC}\mathbf{M} \end{bmatrix} \begin{bmatrix} \mathbf{Y}(k-1) \\ \mathbf{Y}(k-1 \mid k - 1) \end{bmatrix} + \begin{bmatrix} \mathbf{S}\mathbf{K}_{MPC} \\ \mathbf{I}\mathbf{S}\mathbf{K}_{MPC} \end{bmatrix} \mathbf{R}(k-1). \tag{17.56}$$

17.4.5 Observer-Error Equation

The observer error $\mathbf{Z}(k) = \mathbf{Y}(k) - \mathbf{Y}(k|k)$ can be computed using Eq. (17.56). The observer-error equation defines the dynamic relationship between the observer errors:

$$\mathbf{Z}(k \mid k) = (\mathbf{M} - \mathbf{I}\mathbf{N}\mathbf{M})\mathbf{Z}(k - 1 \mid k - 1). \tag{17.57}$$

Using similarity transformation, Eq. (17.57) can be rewritten as

$$\begin{bmatrix} \mathbf{Y}(k) \\ \mathbf{Z}(k \mid k) \end{bmatrix} = \begin{bmatrix} \mathbf{M} - \mathbf{S}\mathbf{K}_{MPC}\mathbf{M} & \mathbf{S}\mathbf{K}_{MPC}\mathbf{M} \\ \mathbf{0} & \mathbf{M} - \mathbf{I}\mathbf{N}\mathbf{M} \end{bmatrix} \begin{bmatrix} \mathbf{Y}(k-1) \\ \mathbf{Z}(k-1 \mid k - 1) \end{bmatrix} + \begin{bmatrix} \mathbf{S}\mathbf{K}_{MPC} \\ \mathbf{0} \end{bmatrix} \mathbf{R}(k-1) \tag{17.58}$$

17.4.6 Observer-Controller Closed-Loop Equation

Equation (17.58) can be used to derive the closed-loop observer-controller dynamic equation (see Fig. 17.11). The input to this dynamic system is $\mathbf{R}(k-1)$ and the output is $y(k)$. $\mathbf{Y}(k)$ and $\mathbf{Z}(k|k)$ are the states of this dynamic system. The dynamic system is defined by the following set of equations:

$$\mathbf{X}(k) = \mathbf{A}\mathbf{X}(k - 1) + \mathbf{B}\mathbf{R}(k - 1), \tag{17.59}$$

$$y(k) = CX(k),\tag{17.60}$$

where $X(k) = [Y(k)\ \ Y(k) - Z(k|k)]^T$, $C = [IN\ \ 0]$,

$$A = \begin{bmatrix} M - SK_{MPC}M & SK_{MPC}M \\ 0 & M - INM \end{bmatrix},$$

and

$$B = \begin{bmatrix} SK_{MPC} \\ 0 \end{bmatrix}.$$

The stability of the observer-controller closed-loop system is therefore determined by the eigenvalues of matrix A. Since matrix A is a diagonal matrix, the observer-controller closed-loop configuration is stable if the eigenvalues of $M - SK_{MPC}M$ and the eigenvalues of $M - INM$ lie strictly within the unit circle. The eigenvalues of $M - SK_{MPC}M$ can be adjusted using the controller gain, and the eigenvalues of $M - INM$ depend on the observer gain I, a constant in the MPC formulation. This result is called the *separation theorem* in the control literature, as Eq. (17.58) suggests that the closed-loop design can be decoupled into the controller design and the observer design.

The stability of the observer-error equation (determined by the eigenvalues of $M - INM$) can be interpreted as follows: The observer-error dynamics are described by a discrete-time equation of the form

$$z(k) = Gz(k-1),\tag{17.61}$$

where, $G = M - INM$. The matrix G is independent of the controller parameters and depends only on the process settling time n. For the observer error to decrease with time and finally converge to 0, all eigenvalues of matrix G should strictly lie inside the unit circle. However, if any eigenvalue of G lies outside the unit circle, then the observer error increases with time, and therefore the model predictions diverge from the true value. For example, such a situation arises when the steady-state gain of the model has a sign opposite to that of the true gain. The matrix $G = M + INM$ and has an eigenvalue outside the unit circle, and therefore the observer error diverges with time. The controller that uses predictions from such an observer always computes corrective action in the opposite direction. As a result, the control moves increase with time due to the diverging prediction error, and this causes the closed-loop control system to be unstable.

17.5 SUMMARY

In this chapter we discussed a discrete-time implementation of model-based control called model-predictive control (MPC). In this implementation the control problem is converted into an optimization problem, which requires the minimization of a quadratic objective function at every sample time. For unconstrained problems, this reduces to a linear least-squares problem, which can be solved offline. The primary application of this algorithm is for large constrained multivariable processes. The extension to constrained systems is discussed in Chapter 18.

Problems

17.1 Consider the Wood and Berry (1973) column modeled by

$$\begin{bmatrix} x_D(s) \\ x_B(s) \end{bmatrix} = \begin{bmatrix} \dfrac{12.8e^{-s}}{16.7s+1} & \dfrac{-18.9e^{-3s}}{21s+1} \\ \dfrac{6.6e^{-7s}}{10.9s+1} & \dfrac{-19.4e^{-3s}}{14.4s+1} \end{bmatrix} \begin{bmatrix} R(s) \\ S(s) \end{bmatrix},$$

where R = reflux rate, S = steam rate, x_B = bottoms purity, and x_D = distillate purity. Luyben (1989) gives the following tuning constants for independent PID controllers:

$$x_D - R \ pair: K_{c1} = 0.20, \tau_{I1} = 4.44,$$
$$x_B - S \ pair: K_{c2} = 0.04, \tau_{I2} = 2.67.$$

Tune an MPC controller to control x_D and x_B. Use of MPC toolbox is recommended. Compare the performance of the PID controller with the performance of the MPC controller for setpoint changes in x_D and x_B.

17.2 Exercise the program *ex_17.1.m* to study the effect of varying the parameters on the performance of the system to load disturbances. Try values of n ranging from 1 to 10 and values of p ranging from 10 to 60.

Also study the effect of varying the move suppression parameter β in the objective function on the performance of the controller to maintain the setpoints in the presence of disturbances. Is this influenced by the choice of n and p?

17.3 Consider a process with a transfer function given by

$$G(s) = \frac{1}{4s+1}.$$

Using a sample time of $T_S = 1$, output horizon = 1, and control move horizon = 1,

a. Derive a control equation with and without move suppression factors following the SOMPC approach.

b. Add a constraint on the input variable:

$$0 \le u \le 1.$$

Explain how you will incorporate this into the control algorithm.

17.4 Consider the Shell fractionator problem. Using the SIMULINK model provided on the website associated with this book, develop and implement two PID controllers to control y_1 and y_2 using u_1 and u_2 respectively. Tune these controllers, using IMC tuning methods. Check their performance to the prototype test cases 1 and 2 as outlined in the problem statement given in Appendix I. Are any constraints violated?

Develop an MPC solution to the Shell problem, using the SIMULINK MPC block from the MPC toolbox. Test your implementation using prototype test cases 1 and 2. Compare your results.

References

Cutler, C. R., and B. L. Ramaker. 1980. "Dynamic Matrix Control: A Computer Control Algorithm." Proc. Joint Auto Control Conf., Paper WP5-B, San Francisco, CA.

Garcia, C. E., and A. M. Morshedi. 1986. "Quadratic Programming Solution to Dynamic Matrix Control (QDMC)." *Chem. Eng. Comm.*, *46*, 73.

Garcia, C. E., and M. Morari. 1982. "Internal Model Control, 1: A Unifying Review and Some New Results." *IEC Proc. Des. Dev.*, *21*, 308.

Harkins, B. L. 1991. "The DMC Controller." Instrument Society of America. Paper #91-0427, 853.

Lee, J. H., M. Morari, and C. E. Garcia. 1994. "State Space Interpretation of Model-Predictive Control." *Automatica*, *30*, 707–718.

Lee, J. H., and B. Cooley. 1997. "Recent Advances in Model-Predictive Control and Other Related Areas." *CPC-V*, January 7-12, 1996, Tahoe City, CA, *CACHE & AIChE*, 201–216.

Luyben, W.L. 1989. *Process Modeling, Simulation, and Control for Chemical Engineers,* 2nd Ed., McGraw-Hill, NY.

Morshedi, A. M., C. R. Cutler, and T. A. Skrovanek. 1985. "Optimal Solution of Dynamic Matrix Control with Linear Programming Techniques (LDMC)." Presented at the *ACC85*, Boston, MA, 199–208.

Palavajjhala, S. 1994. "Studies in Model-Predictive Control with Application of Wavelet Transform." D.Sc. Thesis, Washington University.

Prett, D. M., and M. Morari. 1986. The Shell Process Control Workshop. Butterworths, MA.

Prett, D. M., and R. D. Gillette. 1979. "Optimization and Constrained Multivariable Control of a Catalytic Cracking Unit." #51c, AICHE Annual meeting, Houston, Texas.

Prett, D. M., C. E. Garcia, and B. L. Ramaker. 1990. The Second Shell Process Control Workshop. Butterworths, MA.

Prett, D. M., and C. E. Garcia. 1988. *Fundamental Process Control*, Butterworths, MA.

Richalet, J., A. Rault, J. L. Testud, and J. Papon. 1978. "Model-Predictive Heuristic Control: Applications to Industrial Processes." *Automatica*, *14*(5), 413–428.

Ricker, N. L. 1985. "Use of Quadratic Programming for Constrained Internal Model Control." *IEC Proc. Des. Dev.*, 24, 925.

Shridhar, R., and D. J. Cooper. 1998. "A Tuning Strategy for Unconstrained Multivariable Model-Predictive Control." *Ind. Eng. Chem. Res., 37*, 4003–4016.

Wood, R. K., and M. W. Berry. 1973. "Terminal Composition Control of a Binary Distillation Column." *Chem. Eng. Sci., 28*, 1707.

Yousfi, C., and R. Tournier. 1991. "Steady-State Optimization Inside Model-Predictive Control." Presented at *ACC'91*, Boston, MA, 1866–1870.

Advanced Model-Predictive Control

Objectives of the Chapter

In this chapter we discuss some advanced topics related to model-predictive control (MPC). The main issues addressed are

- Incorporating constraints in the control calculation.
- Incorporating local economic objectives in addition to regulatory control objectives (LP-MPC and QP-MPC algorithms) when there are extra degrees of freedom in control.
- Extension to nonlinear systems.
- Extension to batch process control.

Prerequisite Reading

Chapter 17, "Basic Model-Predictive Control"

18.1 INCORPORATING CONSTRAINTS

The dynamic matrix control method of Cutler and Ramaker (1980) was extended to constrained multivariable systems by Ricker (1985) and Garcia and Morshedi (1986). The method given here is based on these two papers. These articles also give some application examples. As we mentioned earlier, the main advantage of these MPC algorithms comes with the ability to deal with constraints in a systematic way. In the past, constraint control was accomplished through clever multiloop strategies unique to the problem at hand. These were often limited in scope due to their complexity. In this section we shall introduce how constraints arise in process control systems and show how constraint handling can be integrated into the single-objective model-predictive control (SOMPC) strategy.

18.1.1 Constraint Types

18.1.1.1 Manipulated Variable Constraints

Often constraints are placed on how fast we can change the manipulated variable. Such constraints are referred to as move size limitations. The manipulated variable may be a setpoint for a lower level controller and in this case fast changes on the setpoints may be difficult to implement. If the manipulated variable is a valve position then the move size may be limited by how fast the valve can respond. Another reason to limit the move size is to reduce wear and tear on the control valve. Move size limitations can be expressed in the form

$$\left| \Delta u_{k+i} \right| \leq \Delta u_{\max} . \tag{18.1}$$

Note that we can implicitly incorporate move size constraints by using move suppression factors, or by using a slow reference trajectory.

Another type of constraint on manipulated variables arises from limits on the upper and lower values achievable by u_k. A valve cannot go more than full open. Likewise, it cannot go beyond full shut position. We can express this type of constraint using

$$u_{k+i} = u_{k-1} + \sum_{j=0}^{i} \Delta u_{k+j} \tag{18.2}$$

$$u_{\min} \leq u_{k+i} \leq u_{\max} \qquad \text{for i = 0, 1, 2, ..., n.}$$

Thus there are constraints placed on the value of u at each sample time, n values into the future. Sometimes there may be equality constraints placed on a manipulated variable. For example, a valve might get stuck, in which case this manipulated variable is essentially lost and all future control actions must take this into account.

If an optimization layer is implemented on top of the control layer, then this layer might suggest optimum operating values for the manipulated inputs. Thus, while the manipulated variable may be used in the short term to correct disturbances, it may be desirable to bring the input back to the optimum value eventually. This type of situation can arise in non-square systems where there are more manipulated variables than regulated outputs. The additional degrees of freedom allow some manipulated variables to assume their optimum steady-state values. These types of control objectives can be handled by incorporating an additional term in the objective function being minimized. For example,

$$\text{Objective function} = (y^r - y)^T A(y^r - y) + \Delta u^T R \Delta u + (u^r - u)^T S(u^r - u),$$

where u^r is the desired value of the input variable. $Q, R,$ and S are weighting factors that determine the relative importance of regulating the output versus bringing the input back to the desired value. The advantage of this treatment (a penalty function approach to constrained optimization) is that it does not further restrict the choice of Δu in solving the optimization problem. This enlarges the size of the feasible region over which one must search for the optimum. Putting too many hard constraints can limit the range of feasible values of the control moves and sometimes leads to infeasible optimization problems.

The disadvantage of this formulation is that unless there are sufficient degrees for freedom to drive the objective function to zero, we will have a steady-state in which all of the y values do not reach the setpoint. It is best to set up the targets for y and u in such a way that the regulatory goals are met. Later on in this chapter we shall introduce extensions to the basic MPC (LP-MPC and QP-MPC), which achieves this goal.

18.1.1.2 Constraints on Output Variables

Constraints on output variables are incorporated in MPC by forecasting the effect of control moves on the output variables and then adding constraints on the forecasted values to the optimization problem used for computing the input control moves. For example, let us say that we want

$$y_{min} \le y \le y_{max}. \tag{18.3}$$

Since the model gives

$$y = A\Delta u + y^P + d, \tag{18.4}$$

we can incorporate new constraints in the optimization problem as

$$y_{min} \le A\Delta u + y^P + d \le y_{max}. \tag{18.5}$$

18.1.1.3 Constraints on Associated Process Outputs

These are constraints placed on outputs other than those considered as regulated variables. There is no attempt to drive these outputs to any setpoint, but rather only to keep these outputs away from a constraint. An example of such a constraint is the upper limit on the differential pressure across a column. Due to flooding possibility, this pressure drop must be kept below an upper limit. To implement constraints of this type, one must model the effect of manipulated variables on these outputs. Let us say that the associated output variable can be predicted using

$$z = B\Delta u + z^p + d_z,$$
(18.6)

where z = associated output variable, z^p, = effect of past inputs on z, and d_z = estimated disturbance effect on z. The quantities B, z^p and d_z are computed in a manner identical to those used for output variable y. The constraints are expressed as

$$z_{min} \le z \le z_{max}.$$
(18.7)

These constraints are enforced using the same procedure described for output variables.

18.1.2 Effect of Input Constraints on Steady-State Regulation

Constraints on manipulated inputs can severely restrict the ability to regulate a process in the presence of disturbances. The effect of input constraints on regulation of SISO systems is discussed in Appendix A. Here we extend those concepts to multivariable systems.

Example 18.1 Steady-State Controllability of a Multivariable Process

Consider the control of the heat exchange system shown in Figure 18.1.

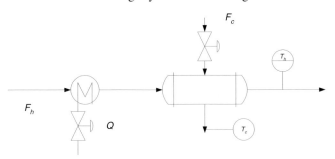

Figure 18.1 Heat exchange system.

The control objective is to maintain the exit temperature of the hot and cold streams near their target valves. The manipulated variables are the heat Q and the flow rate of the cold stream. The following steady-state values were obtained

Condition	F_c Manipulated Variable	Q Manipulated Variables	F_h Disturbance Variable	T_h Controlled Output	T_c Controlled Output
Range	0.05 – .15 lpm	400 – 600 watts	0.05 – .15 lpm	20 – 40 °C	20 – 40 °C
Base Case	0.10	500	0.10	28.8	30.15
Perturb Q	0.10	600	0.10	30.51	32.14
Perturb F_c	0.12	500	0.10	28.2	28.89
Perturb F_n	0.10	500	0.12	28.35	28.92

From this data, we can get the process gains in terms of *scaled, dimensionless* variables (following the procedure outlined in Appendix A) as follows

$$\begin{bmatrix} t_h \\ t_c \end{bmatrix} = \begin{bmatrix} .171 & -.15 \\ .199 & -.315 \end{bmatrix} \begin{bmatrix} q \\ f_c \end{bmatrix} + \begin{bmatrix} -.112 \\ -.3225 \end{bmatrix} [f_n].$$

To evaluate the feasibility of steady-state disturbance suppression, we have graphed the feasible range of t_h and t_c assuming there are no disturbances and that:

$$-0.5 < q < 0.5,$$
$$-0.5 < f_c < 0.5.$$

Figure 18.2 shows the region of space reachable by the output variables (at steady-state) in the absence of any disturbances. The range of variability of t_h is severely limited if t_c is controlled at its nominal setpoint of zero, and vice versa. In this case it appears that both t_h and t_c are closely related to each other, and hence independent control of both variables at their setpoints may not be feasible. For example, if the disturbance f_h is 0.5, then it will not be possible to reach the steady-state, with $t_h = 0$, $t_c = 0$, with the above constraints on the manipulated variables.

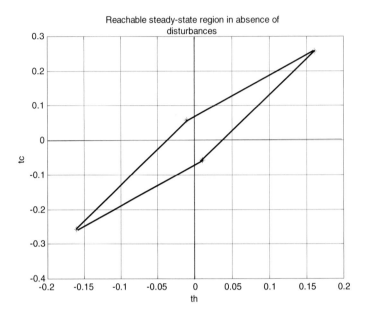

Figure 18.2 Reachable space of the controlled variables, given input constraints.

This example shows that for multivariable systems, the input constraints can put severe restrictions on the feasible set of output setpoints. As the number of variables increases, this raises a concern as to whether the setpoints we have are indeed reachable, given the current set of sustained disturbances. We will address the problem of determining feasible setpoints in Section 18.2, where we introduce a modification to the basic MPC algorithm, called LP-MPC.

One way to evaluate the size of the reachable space for the output is to look at the singular value decomposition (see Appendix C) of the steady state gain matrix. Small singular values would indicate that independent control of all the output variables may be difficult to achieve.

◆

18.1.3 Formulation of the Constrained MPC Algorithm

Now we formulate the MPC algorithm, incorporating constraints. Recall that the unconstrained form computed the control action by solving the optimization problem:

$$\underset{\Delta u}{Min} \quad \left\| y^r - d - y^P - A\Delta u \right\|_Q^2 + \left\| \Delta u \right\|_R^2 . \tag{18.8}$$

The objective function is quadratic in the independent variables. Now consider the constraints on the ith input, u_i. We want

$$u_{i,\min} \leq u_{ik} + \sum_{j=0}^{m} \Delta u_{i,k+j} \leq u_{i,\max}, \tag{18.9}$$

for m = 1, 2 ..., n,

where $u_{i,k}$ is the current value of u_i. This can be expressed in matrix form as

$$
\begin{bmatrix}
1 & 0 & 0 & \dots & 0 \\
1 & 1 & 0 & \dots & 0 \\
1 & 1 & 1 & \dots & 0 \\
. & . & . & . & 0 \\
1 & 1 & 1 & 1 & 1
\end{bmatrix}
\begin{bmatrix}
\Delta u_{i,k} \\
\Delta u_{i,k+1} \\
\Delta u_{i,k+2} \\
. \\
\Delta u_{i,k+n}
\end{bmatrix}
\geq
\begin{bmatrix}
1 \\
1 \\
1 \\
. \\
1
\end{bmatrix}
(u_{i,min} - u_{ik}) \tag{18.10}
$$

and

$$
\begin{bmatrix}
1 & 0 & 0 & \dots & 0 \\
1 & 1 & 0 & \dots & 0 \\
1 & 1 & 1 & \dots & 0 \\
. & . & . & . & 0 \\
1 & 1 & 1 & 1 & 1
\end{bmatrix}
\begin{bmatrix}
\Delta u_{i,k} \\
\Delta u_{i,k+1} \\
\Delta u_{i,k+2} \\
. \\
\Delta u_{i,k+n}
\end{bmatrix}
\leq
\begin{bmatrix}
1 \\
1 \\
1 \\
. \\
1
\end{bmatrix}
(u_{i,max} - u_{ik}) \; . \tag{18.11}
$$

The constraints on outputs may be stated as

$$A\Delta u + y^P + d \leq y_{max},$$
$$A\Delta u + y^P + d \geq y_{min}. \tag{18.12}$$

The quadratic objective function along with these linear constraints constitute the optimization problem, which must be solved at every sample time. This is classified as a *quadratic programming (QP) problem* in optimization because the objective is quadratic and constraints are linear. Several efficient, finite-step algorithms are available for solving quadratic programming problems. The MATLAB Optimization toolbox contains a command, *qp*, for solving quadratic programming problems. We will make use of this command to solve the quadratic programming problem for the example problem.

18.1.3.1 Constraint Softening

The above formulation can lead to situations in which there is no feasible solution to the QP generated. This is undesirable from a practical standpoint. Since output constraints generally are specified with some margin for error, it is preferable to allow some small constraint violation rather than declare infeasibility and shut the control system down, in which case the operator has to take over manual control of the system.

This problem can be avoided by a technique called constraint softening. Consider the following alternative formulation for meeting an output constraint. Given the constraint

$$\Delta u + y^p + d \leq y_{max},$$
(18.13)

we create a new variable, $y_c \geq 0$,

$$\Delta u + y^p + d \leq y_{max} + y_c,$$
(18.14)

and modify the objective function,

$$\underset{\Delta u, \, y_c}{Min} \quad \left\| y - y^r \right\|_Q^2 + \left\| \Delta u \right\|_R^2 + \left\| y_c \right\|_S^2$$
(18.15)

The weighting matrices, **Q, R,** and **S** can be chosen to achieve the desired degree of control over the constraint violation. The controller continuously strives to minimize the constraint violations embodied in the y_c vector.

Example 18.2 Constrained Control of a Column

Consider the Shell distillation column control problem. A part of the problem is schematically shown in Figure 18.3.

The control objectives are
1. Maintain y_1 and y_2 at specified setpoints (0.0 with a tolerance of .005 at steady-state).
2. Keep u_3 as close to -.5 as possible (this maximizes the steam made in the bottom reflux condenser).
3. Reject disturbances l_1 and l_2 entering the column.

The control constraints are
1. All control input move sizes must be kept below 0.05 per minute.
2. The bottom reflux draw temperature y_7 must be kept above −0.5.

The website accompanying this text contains three MATLAB programs that implement a single-objective model-predictive controller for the constrained control of this column. The program is written in two parts:
   i. Part 1 (program *ch18ex2a.m*) constructs a MATLAB model of the column for simulation purposes.
   ii. Part 2 (program *ch18ex2c.m*) implements the constrained model-predictive controller. It calls program *ch18ex2b.m* to construct the necessary finite step response model of the process for the MPC algorithm. The setpoints and disturbances are

specified at the beginning of this program along with the parameters needed to specify the algorithm. This program automatically invokes the first two programs to construct the simulation model and the step response model. The program uses the quadratic program solver from the Optimization toolbox. In the following paragraphs we describe how the algorithm is constructed.

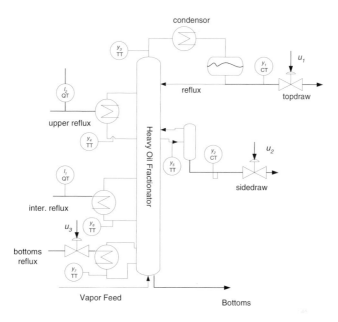

Figure 18.3 Schematic of the distillation column, showing the inputs and outputs.

The process model transfer functions are given by (*for convenience we have labeled the bottoms reflux temperature as* y_3).

$$\begin{bmatrix} y_1(s) \\ y_2(s) \\ y_3(s) \end{bmatrix} = \begin{bmatrix} \dfrac{4.05e^{-27s}}{50s+1} & \dfrac{1.77e^{-28s}}{60s+1} & \dfrac{5.88e^{-27s}}{50s+1} \\ \dfrac{5.39e^{-18s}}{50s+1} & \dfrac{5.72e^{-14s}}{60s+1} & \dfrac{6.90e^{-15s}}{40s+1} \\ \dfrac{4.38e^{-20s}}{33s+1} & \dfrac{4.42e^{-22s}}{44s+1} & \dfrac{7.20}{19s+1} \end{bmatrix} \begin{bmatrix} u_1(s) \\ u_2(s) \\ u_3(s) \end{bmatrix} + \begin{bmatrix} \dfrac{1.20e^{-27s}}{45s+1} & \dfrac{1.44e^{-27s}}{40s+1} \\ \dfrac{1.52e^{-15s}}{25s+1} & \dfrac{1.83e^{-15s}}{25s+1} \\ \dfrac{1.14}{27s+1} & \dfrac{1.26}{32s+1} \end{bmatrix} \begin{bmatrix} l_1(s) \\ l_2(s) \end{bmatrix}.$$

The step response model of the process may be written as

$$y_1 = A_{11}\Delta u_1 + A_{12}\Delta u_2 + A_{13}\Delta u_3 + y_1^P + d_1,$$
$$y_2 = A_{21}\Delta u_1 + A_{22}\Delta u_2 + A_{23}\Delta u_3 + y_2^P + d_2,$$
$$y_3 = A_{31}\Delta u_1 + A_{32}\Delta u_2 + A_{33}\Delta u_3 + y_3^P + d_3.$$

This may be written in short form as

$$y_1 = B_1\Delta u + y_1^P + d_1,$$
$$y_2 = B_2\Delta u + y_2^P + d_2,$$
$$y_3 = B_3\Delta u + y_3^P + d_3,$$

where

$$\Delta u = \begin{bmatrix} \Delta u_1 \\ \Delta u_2 \\ \Delta u_3 \end{bmatrix}$$

and so on. The control objective may be stated as

$$\underset{\Delta u}{Min}\ \Phi = \frac{1}{2}\left\{ \left\| y_1 - y_1^r \right\|^2 + \left\| y_2 - y_2^r \right\|^2 + \alpha\left\| u_3 - u_3^r \right\|^2 + \beta\left\| \Delta u \right\|^2 \right\},$$

where we have included these objectives:
1. Keep y_1 close to its setpoint.
2. Keep y_2 close to its setpoint.
3. Keep u_3 close to its desired value.
4. Keep the move sizes small.

Here α and β are weighting factors that may be tuned online. Note that we have a tradeoff between keeping u_3 close to its setpoint and the outputs at their setpoints. It is possible that with constraints present, we may not be able to reach these setpoints in the steady-state. The offsets may be reduced by changing the weighting factor on u_3.

Next we must convert this objective function to the *standard form* used by the quadratic program solver. This is of the type

$$Min\ \frac{1}{2}\Delta u^T H\Delta\ u + c^T \Delta u$$
$$\Delta u$$
$$Subject\ to\quad A\Delta u \le b.$$

Consider the first term in the objective function

$$\left\| y_1 - y_1^r \right\|^2 = \left\| B_1\Delta u + y_1^P - y_1^r \right\|^2$$
$$= \Delta u^T B_1^T B_1\Delta u + 2k_1^T B_1\Delta u + terms\ independent\ of\ \Delta u,$$

where $k_1 = y_1^P + d_1 - y_1^r$. Similarly, the fourth term in the objective function can be written as $\left\| \Delta u \right\|^2 = \Delta u^T I \Delta u$, where I is an identity matrix of size $3n$ x $3n$ and n is the control move horizon.

The third term is a little harder to convert, since it is given in terms of u rather than Δu. We have the predicted value of u_3 given as:

$$u_{3,k+i} = u_{3,k-1}^P + \Delta u_{3,k} + \Delta u_{3,k+1} + ... + \Delta u_{3,k+i}$$

or

$$\begin{bmatrix} u_{3,k} \\ u_{3,k+1} \\ . \\ u_{3,k+n} \end{bmatrix} = \begin{bmatrix} u_{3,k-1}^P \\ u_{3,k-1}^P \\ . \\ u_{3,k-1}^P \end{bmatrix} + \begin{bmatrix} 1 & 0 & . & 0 \\ 1 & 1 & . & 0 \\ 1 & 1 & . & . \\ 1 & 1 & 1 & 1 \end{bmatrix} \begin{bmatrix} \Delta u_{3,k} \\ \Delta u_{3,k+1} \\ . \\ \Delta u_{3,k+n} \end{bmatrix}.$$

This may be written compactly as

$$u_3 = u_3^P + B_4 \Delta u_3 .$$

Using this notation, we can simplify the fourth term in the objective function as

$$\left\| u_3 - u_3^P \right\|^2 = \left\| u_3^P + B_4 \Delta u_3 - u_3^r \right\|^2$$

$$= \left\| B_4 \Delta u_3 + k_4 \right\|^2$$

$$= \Delta u_3^T B_4^T B_4 \Delta u_3 + 2k_4^T B_4 \Delta u_3.$$

Define

$$B_{44} = \begin{bmatrix} \boldsymbol{0} & \boldsymbol{0} & \boldsymbol{0} \\ \boldsymbol{0} & \boldsymbol{0} & \boldsymbol{0} \\ \boldsymbol{0} & \boldsymbol{0} & B_4^T B_4 \end{bmatrix}, \ k_{44} = \begin{bmatrix} \boldsymbol{0} \\ \boldsymbol{0} \\ B_4^T k_4 \end{bmatrix},$$

$$\left\| u_3 - u_3^r \right\|^2 = \Delta u^T B_{44} \Delta u + 2k_{44} \Delta u .$$

Substituting these into the objective function, we get the standard form

$$\Phi = \frac{1}{2} \left\{ \Delta u^T \left[B_1^T B_1 + B_2^T B_2 + \alpha B_{44} + \beta I \right] \Delta u \right\} +$$

$$\left[k_1^T B_1 + k_2^T B_2 + k_{44} \right] \Delta u.$$

Finally we consider formulation of the constraint equations in the standard form. The constraint on the output can be written as

$$-\boldsymbol{B}_3\boldsymbol{\varDelta u}-\boldsymbol{k}_3 \leq -\boldsymbol{y}_{3,\,min} \; ,$$

which is in the standard form already. The constraint on the control move sizes can be written as

$$\boldsymbol{I}\boldsymbol{\varDelta u} \leq 0.2\boldsymbol{I},$$

$$\boldsymbol{I}\boldsymbol{\varDelta u} \geq -0.2\boldsymbol{I}.$$

Some results, using this controller, are given in Figure 18.4, which shows the response of the controller to a step change in the setpoint of y_1. The result shows that the controller is effective in taking the output y_1 to the new setpoint without much overshoot. The manipulated variable u_3 is slowly moved to its desired value. The final value reached by y_3 is around -0.5.

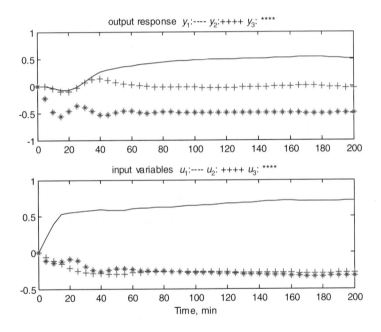

Figure 18.4 Response of the controller to simultaneous changes in both setpoints. The parameters used by the controller were Ts=5, p=30, n=2, beta=4, and alpha=5.

Implementation Using MPC Toolbox

The problem can be solved much more efficiently using the MPC toolbox. A script file containing the necessary setup commands is given in *ch18ex2_script.m*. The corresponding SIMULINK file is given in *ch18ex2.mdl*. This implementation is shown in Figure 18.5. An example of a result from the SIMULINK implementation is shown in Figure 18.6.

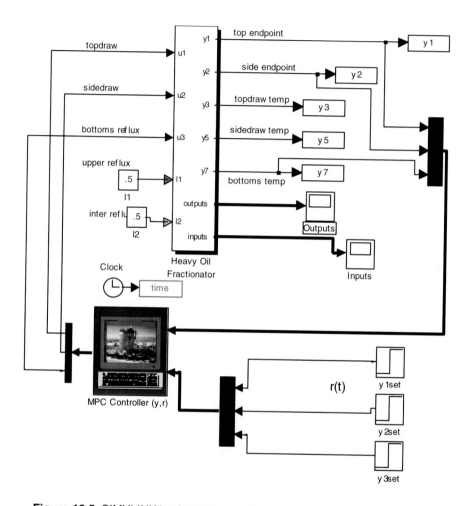

Figure 18.5 SIMULINK implementation of constrained, multivariable control of the Shell fractionator column.

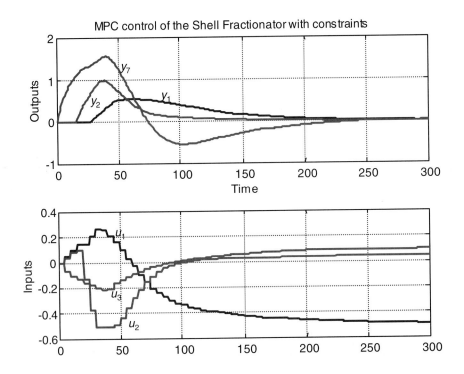

Figure 18.6 Results from the SIMULINK implementation of constrained control. Note that the y_7 target was arbitrarily selected as 0.0. Simultaneous changes in l_1 and l_2 of 0.5 were made to obtain this plot.

In this implementation we put constraints on inputs and outputs as specified in the problem statement. The constraint to minimize u_3 was not implemented. The latter is left as an exercise for the reader.

♦

18.1.4 Stability of the Constrained MPC Algorithm

For the unconstrained MPC, the feedback control can be formulated as a state-space, linear quadratic control problem with an observer for estimating the states from the output measurement. In this case the stability can be determined by examining the poles of the closed loop system as shown in Chapter 17.

For the constrained case, closed form expressions for the control system cannot be derived. However, the stability of the system can still be examined under certain conditions. Rawlings and Muske (1995) considered the stability problem and derived conditions for a stable MPC controller for infinite horizon formulations. Zheng and Morari (1995) later extended this stability analysis. Unfortunately, none of this theory is very helpful in determining the best tuning constants, and industry still relies on experience in selecting the tuning parameters.

18.1.5 Discussion

In this section we presented one method for treating constraints in the process. There are other possible approaches to deal with the constraints. For example, we could have formulated the objective function as a linear function of the independent variables by using the sum of the absolute values of the deviations from the setpoints. In this case we would have a linear programming problem to solve at every sample time. Such problems are easier to solve than QP problems and the results would have been similar, but the basic problem with temporary constraint violations remains.

Whenever constraints are imposed, the feasible region of the optimization problem is restricted. Adding too many constraints can severely restrict the range of available control moves, or worse yet, may preclude any feasible solution to the problem. This can happen, for example, if a constraint is placed on the value of an output at the next sample time, but no control action can drive this variable within the constraint. Control moves take time to take effect due to the time delays and time lags. Thus constraints placed on outputs in the near term may not be met by any control action taken at that point in time. Situations like this should be avoided in real-time implementation because we have to implement some sort of control action at every sample time.

Three approaches can be taken to remedy this situation. The first is to avoid placing constraints on outputs that cannot be met. Hence, if it takes three to four sample times before the effect of an input move can be felt on an output, then constraints on this output should be imposed only after three to four projected values into the future.

The second approach is to add penalty terms to the objective function whenever constraint violations are predicted. This means additional logic in the software package to suitably modify the objective function whenever constraint violations are detected.

The third approach is to use the constraint softening method outlined in Section 18.1.3.1. This approach would allow some constraint violation but the extent of the violation can be kept small by suitable choice of the weighting matrices used in the objective function.

18.2 INCORPORATING ECONOMIC OBJECTIVES: THE LP-MPC ALGORITHM

The initial papers on MPC include the output error $\|y - y_{set}\|^2$ and input move suppression $\|\Delta u\|^2$ terms in the objective function (Cutler & Ramaker, 1980). There are no input error terms $\|u - u_{set}\|^2$ in the objective function. Historically, linear optimal control strategies have incorporated a cost penalty for the input variables in the objective function (Kwakernaak & Sivan, 1972; Kailath, 1980). Inclusion of such a term in MPC formulation leads to steady-state offsets. Cutler and Ramaker (1979) recognized this and chose to pose the optimization problem in terms of changes in the control moves rather than the actual control variable, thus automatically incorporating integral action in the controller and eliminating steady-state offsets.

In many applications, when the number of inputs exceeds the number of controlled variables, it is desirable (from an economic point of view) to also try to achieve some setpoints on the manipulated input variables. Typically, a real-time optimizer is used to compute the economic target values for both output and some input variables, usually in a period of a few hours to a few days (Forbes & Marlin, 1994). One way to achieve these setpoints is to include a cost term $\|u - u_{set}\|^2$ in the objective function used by the MPC algorithm. However, this has two drawbacks: First, the presence of active constraints can lead to loss of degrees of freedom, which in turn can result in steady-state offsets in the systems. This is because the MPC algorithm will distribute the final steady-state errors between the target values for both u and y. Second, the presence of external disturbances can cause the economic optimum to shift, for example, from one set of constraints to another, thus changing the target values for input variables. To address this issue, some authors (Morshedi et al., 1985; Brosilow & Zhao, 1988; Yousfi & Tournier, 1991; Harkins, 1991) suggested a two-stage MPC, which consists of a steady-state controller in cascade with the MPC controller. This outer controller continuously updates the setpoints used by the MPC controller. It can be a linear program (LP-MPC) or a quadratic program (QP-MPC).

Note that as shown in Figures 18.7a and 18.7b, this two-stage MPC is not a substitute for real time optimization (RTO; see Chapter 1). Rather, the two-stage MPC complements and allows the control system to track changes in the optimum caused by disturbances. The second stage uses the disturbance estimate computed by the lower stage MPC at every sample time and then determines the optimum setpoint. As we will demonstrate later (using the Shell case study problem), this approach permits dynamic tracking of the optimum, which is not achievable with a steady-state RTO used in conjunction with a single stage MPC.

The MPC formulation focuses on regulatory control of constrained, multivariable systems. The method is applicable to systems with extra degrees of freedom, for example, if there are more manipulated variables than controlled outputs. This situation occurs frequently and provides an opportunity to incorporate an economic objective simultaneously with the regulation of the process. Thus we can use the extra degree of freedom to move one or more of the manipulated inputs to a point that is economically the best, given the current set of disturbances. One way to achieve this is to use multiloop control as illustrated in the following example.

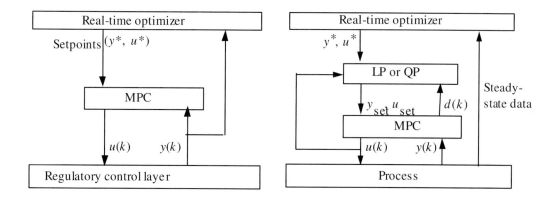

Figure 18.7a MPC in the control system hierarchy.

Figure 18.7b Two-stage MPC system hierarchy.

Example 18.3 A Non-Square System

Consider the process shown in Figure 18.8. The process has two manipulated inputs and one controlled output.

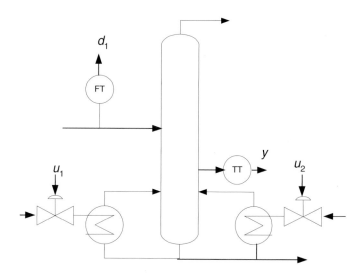

Figure 18.8 Control of a distillation column with two reboilers.

The control objectives for this system are as follows:

Regulatory objective:	Maintain y_1 close to 0.5 (setpoint)
Economic objective:	Minimize u_2 (the more costly input)
Constraints:	$0 \leq u_1 \leq 1$
	$0 \leq u_2 \leq 1$
Model:	$y(s) = \dfrac{u_1(s)}{5s+1} + \dfrac{u_2(s)}{s+1} + d(s)$

Note that y responds faster to u_2. Hence it is desirable to use u_2 in the short run to get better control and u_1 in the long run to achieve better economic performance. Normally, $d = 0$. However, occasionally d can become as low as -1.0. Design a control system that meets the regulatory and economic objectives.

Solution

Let us do a steady-state analysis first. The steady-state model of the process is given by

$$y = u_1 + u_2 + d \ .$$

Under normal circumstances $(d = 0)$ we can keep $u_2 = 0$ and $u_1 = 0.5$. Suppose d becomes -1 at some point. In that case

$$y = u_1 + u_2 - 1 = 0.5 \ .$$

Hence

$$u_1 + u_2 = 1.5 \ .$$

Therefore we must use $u_1 = 1, u_2 = 0.5$ to maintain regulatory control and to optimize the process. One way to achieve this would be to use a split-range controller, as shown in Figure 18.9.

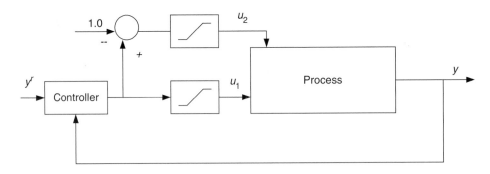

Figure 18.9 Split-range control.

This allows u_1 to respond as long as the controller output is in the 0-1 range. u_2 will remain at its lower constraint of 0. As the disturbance drives controller output above 1, u_2 will respond as needed.

Now consider the design of an MPC controller to achieve the same objectives. One possibility is to enforce this optimization by embedding it into the objective function of the MPC as follows:

$$\underset{\Delta u_1, \Delta u_2}{Min} \left\| y - y^r \right\|^2 + \left\| u_2 - u_2^r \right\|^2 .$$

with $y^r = 0.5$ and $u_2^r = 0$. Unfortunately, this can lead to solutions that violate the steady-state requirement that y return to its setpoint of 0.5, since it is not explicitly included as a requirement. Under normal conditions ($d = 0$), the MPC solution reaches a steady-state value of $u_2 = 0$, $y = 0$, as desired. However, if $d = -1$ the steady-state solution is given by

$$y = 0.25, \ u_{2s} = 0.25, \ u_{1s} = 1.0 .$$

This problem may be alleviated to some extent by assigning different weights to y and u in the objective function

$$\underset{\Delta u_1, \Delta u_2}{Min} \ w_y \left\| y - y^r \right\|^2 + w_u \left\| u_2 - u_2^r \right\|^2 .$$

By making $w_y > w_u$, we can force y to be closer to y^r. However, this may not always be acceptable.

Can we force y_2 to 0.5 by adding it as a constraint in the optimization problem? y must be allowed to deviate around 0.5 in the dynamic sense. Hence enforcing a strict constraint that $y = 0.5$ at all times will lead to infeasible QP problems. We could add an end-point constraint that y_2 return to its target at the end of the output horizon u:

$$y_{k+p} = y^r .$$

This approach will work for simple problems where we know the economic optimum target values of inputs and outputs. In the general situation, these targets may shift, depending on the type of disturbances entering the process. The two-state, cascaded LP-MPC and QP-MPC discussed next provide a powerful way to achieve economic objectives and control objectives at the same time.

♦

18.2.1 General Formulation of the QP-MPC (LP-MPC) Algorithm

In this formulation two optimization problems are solved at every sample time, using a cascade structure (see Figure18.10). The outer layer takes the current estimates of the disturbance variable and calculates the target values of y and u, which will maximize the economic objective *in the steady-state,* subject to the process constraints. The inner layer solves the *dynamic* optimization problem that will regulate the outputs at their target values while using the extra degrees of freedom to drive the input variables to their target values.

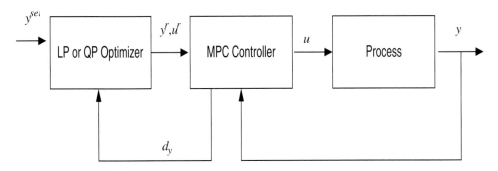

Figure 18.10 LP-MPC (QP-MPC) structure.

Consider a MIMO process modeled in the steady-state by

$$y_s = A_s u_s + d_s,$$ (18.16)

where y_s is the output vector, u_s is the input vector, and d_s is the current estimate of the disturbance variable. Let (y^{set}, u^{set}) be the desired set of operating conditions as computed by the steady-state RTO. The outer QP problem to compute the target values of y^r and u^r is set up as follows:

$$Min_{u^r, y^r} \left\| y^r - y^{set} \right\|^2_{W_y} + \left\| u^r - u^{set} \right\|^2_{W_u},$$ (18.17)

subject to

$$y_s = A_s u_s + d_s$$
$$y^{min} \le y^r \le y^{max}$$
$$u^{min} \le u^r \le u^{max},$$

where W_y = cost weight given to y-variables and W_u = cost weight given to u-variables. If one or more of the outputs have strict equality constraints, then these variables can be ignored in the objective function. W_u and W_y are chosen based on experience (relative costs of each variable) and/or results from the RTO calculation, which can assign relative costs to each variable.

The results of the QP are the setpoints for the inner MPC control calculation. Because both inner and outer calculations use the same model, the solutions computed will be consistent with each other. The inner MPC is formulated as before. Given the plant model

$$y = A\Delta u + d + y^p, \tag{18.18}$$

where y is the vector of output predictions over the output horizon and Δu is the vector of input moves over the input horizon, we want to solve the QP:

$$\underset{\Delta u}{Min} \ \left\| y^r - y \right\|_Q^2 + \left\| \Delta u \right\|_R^2 + \left\| u^r - u \right\|_S^2, \tag{18.19}$$

subject to

$$\begin{aligned}
y &= A\Delta u + d + y^p \\
u_{min} &\leq u \leq u_{max} \\
\Delta u_{min} &\leq \Delta u \leq \Delta u_{max} \\
y^{min} &\leq y \leq y^{max},
\end{aligned} \tag{18.20}$$

where Q = weight given to output variables, R = move suppression weight, and S = weight given to input variables.

Example 18.4 Formulation of LP-MPC for a Simple Problem

Consider the problem given in Example 18.3. The outer layer LP optimization is set up as (given d_1)

$$\underset{y^r, u_1^r, u_2^r}{Min} \ \{u_2\},$$

subject to

$$\begin{aligned}
y &= u_1 + u_2 + d_1 \\
y_1 &= 0.5 \\
0 &\leq u_1 \leq 1 \\
0 &\leq u_2 \leq 1.
\end{aligned}$$

This also can be formulated as a QP with objective function $Min\,(u_2 - u_2^{set})^2$. Under normal operating conditions $(d_1 = 0)$, we will have the solution

$$u_2^r = 0, u_1^r = 0.5, y^r = 0.5.$$

If $d_1 = -1$, then the solution becomes

$$u_1^r = 1, \; u_2^r = 0.5, \; y^r = 0.5.$$

In the inner MPC layer, the objective becomes

$$Min \quad w_y \left\| \left(y - y^r \right) \right\|^2 + w_{u1} \left\| u_1 - u_1^r \right\|^2 + w_{u2} \left\| u_2 - u_2^r \right\|^2 ,$$
$$\Delta u_1, \; \Delta u_2$$

where the weights w_y, w_{u2}, w_{u1} are chosen based on the relative importance of each of these variables. If u_1 is unimportant, then its weight can be set to zero. This formulation brings with it another advantage: We have at our disposal changes in both u_1 and u_2 to regulate y. Since y responds faster to u_2, this variable is used to bring y back to the setpoint in the short term, and u_2 will return to its optimum value in the long run.

18.2.2 Discussion

QP-MPC and LP-MPC is widely used in industry, since most practical problems involve non-square MIMO systems with operational economics that depend on the current set of disturbances affecting the plant. With LP-MPC, we can move the plant to the optimal point continuously and smoothly (as opposed to with abrupt changes in setpoints or active constraints). In fact, much of the incentive for implementing advanced control derives from the ability to track the economic optimum. Regulation alone cannot be used to justify the cost of designing and maintaining advanced control systems.

One might ask, why not rely upon the RTO to get the economic benefit? There are a number of reasons. RTO does provide economic benefit, but is implemented only at steady-state and infrequently. The disturbances to the plant may vary during the periods between executions of the RTO, and in this case the plant will be operating suboptimally. There are other more subtle issues that favor a two-stage implementation. These are discussed next.

Due to the presence of modeling errors, the setpoints $\left(y^{set}, u^{set} \right)$ computed by the RTO (using a detailed steady-state model of the process) may not be consistent with the linear model used in the MPC. (Recall that the MPC model is linear and is therefore consistent with the measurements at each sample time, using the disturbance variable as a correction factor). Thus one should reevaluate a set of target values for y and u that is consistent with the current operating state of the plant. This is precisely what the outer QP layer does. It is possible that under some circumstances, the setpoints computed by the RTO $[y^{set}, u^{set}]$ are not even in the feasible range of operating conditions for the plant. Ignoring this can lead to problems and even instability.

It can be shown that the outer layer of the two-stage MPC acts like a feedforward correction to the setpoint (Ying & Joseph, 1999). Hence the stability of the inner MPC algorithm is not affected by the incorporation of the two-stage algorithm. This stability is preserved even if constraints are present in either inner or outer layers of the algorithm.

The two-stage MPC is also beneficial when considering global setpoint transfers imposed by RTO. Suppose we have two sequential processes: a reactor system and a distillation system (see Figure 18.11). If single-stage MPC is used, a problem can occur if RTO is implemented. When processing conditions change, RTO will give new setpoints for both the reactor and the distillation column. As the setpoints are passed to the MPC controllers, the distillation column controller will try to move the controlled variables to the new setpoints immediately, assuming reactor output has reached new levels. However, because of the delay and lag in the reactor, the reactor output will lag behind. As pointed out in Example 18.1, under some conditions the new setpoints in the distillation column may not even be reachable until the reactor output has reached the new setpoint. Thus, moving of the reactor and column to the new global steady-state optimum must be done while keeping in mind the dynamic lags present in the system.

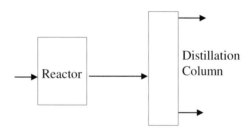

Figure 18.11 Reactor-distillation column system.

In the case of two-stage MPC, the RTO can transfer the new target values for the input and output along with their economic significance. The LP or QP can in turn compute new setpoints for the lower level MPC controller that are consistent with the current plant operating conditions because of the feedback correction present, thus making a successful global setpoint transfer consistent among the units involved.

Example 18.5 Application to the Shell Fractionator Problem

This problem was stated in Example 18.2. The outer-layer LP problem can be formulated as follows:

$$\underset{u_1^r,u_2^r,u_3^r,y_1^r,y_2^r}{Min} \quad \{u_3^r\}, \quad \text{subject to}$$

$$-0.005 \le y_1^r \le 0.005$$

$$-0.005 \le y_2^r \le 0.00$$

$$-0.5 \le y_3^r$$

$$-0.5 \le u_i^r \le 0.5 \qquad \text{where } A_s = \begin{bmatrix} 4.05 & 1.77 & 5.88 \\ 5.39 & 5.72 & 6.90 \\ 4.38 & 4.42 & 7.20 \end{bmatrix}.$$

$$y^r \le A_s u^r + d_{yk}$$

A_s is the matrix of steady-state gains and d_{yk} represents the effect of the current disturbances on y. d_{yk} is calculated at every sample time by the MPC layer, and this is passed to the LP layer. The LP layer then computes the setpoints $y_1^r, y_2^r, y_3^r, u_1^r, u_2^r$, and u_3^r, which are passed on to the MP layer.

Figure 18.12 (Ying & Joseph, 1999) compares the performance of a single-stage MPC with a cascaded LP-MPC implementation for a test disturbance (from $l_1 = l_2 = 0.4$ to $l_1 = l_2 = 0.0$). The single stage MPC is unable to recover the optimum in the steady-state in this case (recall that the economic objective is to keep u_3 as small as possible).

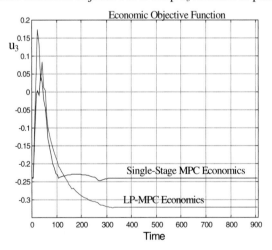

Figure 18.12 The ability of LP-MPC to track the optimum dynamically.

18.3 Extension to Nonlinear Systems

In this section we extend the concepts presented previously to systems modeled using nonlinear differential equations. A two-phase MPC called TP-MPC is discussed.

Linear model-based control is adequate for the majority of systems encountered in the process industry. However, there often arise, particularly in batch process control, processes where the nonlinearity in the process renders linear controls inadequate. We will discuss some practical approaches to this problem without getting into the theoretical issues.

We start with a formulation of the general nonlinear control problem and present a nonlinear MPC algorithm that is an extension of the linear SOMPC strategy presented in Chapter 17.

18.3.1 The Two-phase MPC (TP-MPC) Algorithm

Consider a system described by a nonlinear model,

$$\frac{dx}{dt} = f(x, u, p, d, t) \tag{18.21}$$

$$y = g(x, u, p, d, t)$$

$$x(0) = x_0, \tag{18.22}$$

where

x: state variables,
u: input manipulated variables,
d: unknown disturbances,
p: model parameters,
y: process measurements.

The control objective is taken as

$$Min \int_0^\infty \phi(x, u, p, d, t)dt, \tag{18.23}$$

subject to constraints. The MPC problem can be formulated using a two-phase approach. The *estimation* phase attempts to reconcile the plant measurements with the model predictions. In the *control move calculation* phase, the current control action is computed.

Figure 18.13 shows the schematic of this two-phase nonlinear MPC algorithm (TP-NLMPC). A detailed description of each of the two phases is given next.

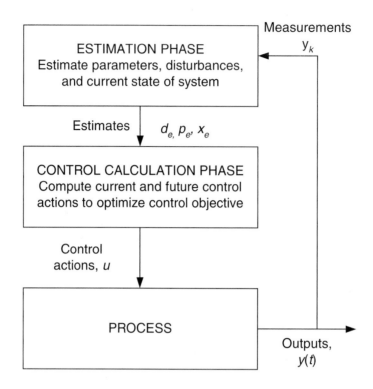

Figure 18.13 The two phase approach for nonlinear MPC.

Phase I: Estimation Phase

Consider any arbitrary point in time, $t = t_k$. We consider an immediate past interval, $t_{k-q} \leq t \leq t_k$, over which it is reasonable to assume that both p and d are constant. (This is a simple form; more complex formulations would consider that p and d are varying with time.) Under these assumptions, the problem is to determine p_e and d_e such that the model predictions are close to the actual measurements. Mathematically, we may formulate the problem as

$$\underset{x_0, p_e, d_e}{Min} \sum_{i=0}^{q} \left[y(t_{k-q+i}) - y_{k-q+i}^m \right]^2 \tag{18.24}$$

subject to

$$x(t-q) = x_0$$

$$\frac{dx}{dt} = f(x, u, p_e, d_e, t) \tag{18.25}$$

$$y \quad = g(x, u, p_e, d_e, t)$$

$$y^m_{k-q+i} \quad = \text{measurement of } y \text{ at } t = t_{k-q+i}$$

where $u(t)$ is known for $t < t_k$.

This is a nonlinear programming problem (NLP), which can be solved using readily available codes. After this problem is solved, the current state of the system, x_e, is evaluated from the model equation.

It is important that the output y show sufficient variation during this time interval so that the problem is well posed. If there was not enough perturbation on y, then it may not be possible to solve for the disturbances and state of the system. Hence if the system has been nearly steady for a while, then the estimation phase will not be effective and will have to be turned off temporarily. No clear guidelines are yet available on how to determine if the estimation problem is well posed.

Phase II: Control Move Calculation Phase

The objective of this phase is to calculate the current control action u_k, However, as in the linear case, we calculate the current as well as a few future control moves so that our control objective function is minimized over an output horizon $t_k \leq t \leq t_{k+p}$. The control move calculation is formulated as another nonlinear programming problem:

$$\min_{\Delta u_k, \Delta u_{k+1}, \dots \Delta u_{k+n}} \int_{t_k}^{t_{k+p}} \phi(x, u, p_e, d_e, t) dt, \tag{18.26}$$

subject to

$$\frac{dx}{dt} = f(x, u, p_e, d_e, t)$$

$$x(t_k) = x_e \tag{18.27}$$

$$h(x, u, p_e, d_e, t) \leq 0.$$

This formulation attempts to minimize the control objective over a moving horizon.

18.3.2 Discussion

For a case study involving the application of NL-MPC, see the article by Jang et al. (1987a). This article presents an application of the concept to control a nonlinear continuous stirred tank reactor (CSTR) process. An application of the concept of a distillation column using reduced-order nonlinear models is presented in Jang et al. (1987a, 1988). Jang et al. (1987b) compares different estimation methods for state and parameter estimation of nonlinear dynamic systems. For a review of more recent work on the subject, see Mayne (1996).

18.4 EXTENSION TO BATCH PROCESSES

In this section the extension of MPC to control of batch process is discussed. The main difference is that batch processes have a limited time horizon of operation. Hence the control horizon is limited in length and shrinks with time. One possible control objective is to achieve an end product quality target. Another possible objective is to maximize an integral of an economic performance measure. Here we present a variation of the nonlinear MPC for batch processes called shrinking horizon MPC (SH-MPC).

18.4.1 The Shrinking Horizon MPC (SH-MPC) Algorithm

Consider a batch process with control objective

$$Max \ \Phi = \int_0^{t_f} \phi(x,u,p,d,t)dt \ + \psi(x_f,t_f), \tag{18.28}$$

subject to constraints

$$h(x,u,p,d,t) \le 0 \ . \tag{18.29}$$

Let the process model be represented as

$$\frac{dx}{dt} = f(x,u,p,d,t)$$
$$y = g(x,u,p,d,t) \tag{18.30}$$
$$x(0) = x_0,$$

We can formulate the SH-MPC as follows. First divide the control u into a set of actions we can take while the batch is in progress:

$$u = \begin{bmatrix} u_1 \\ u_2 \\ \vdots \\ u_n \end{bmatrix}. \tag{18.31}$$

Unlike the continuous control case, each element of u in Eq. (18.31) represents a possible control action that can be taken during the middle of the batch. For example,

u_1 = amount of catalyst to be added at $t = 0$,

u_2 = hold temperaure during the first 20 minutes,

u_3 = temperature ramp during the second 20 minutes,

u_4 = fill level of the reactor,

u_5 = cooling time,

u_6 = batch duration,

and so on.

We can arrange these control actions in such a way that the earliest (first) control action to be taken is placed first and so on. The last item represents the last possible control action. Note that any time $t = t_k$, we can divide the vector u into two parts:

$$u = \begin{bmatrix} u_p \\ u_f \end{bmatrix} \tag{18.32}$$

where

u_p = vector of all past control actions,

u_f = vector of all future control actions.

Similarly, let

$$y = \begin{bmatrix} y_1 \\ y_2 \\ \vdots \\ y_n \end{bmatrix} \tag{18.33}$$

represent the vector of all possible measurements. For example,

y_1 = initial concentration reactant,

y_2 = concentration after 15 minutes,

y_3 = initial temperature of reactor,

y_4 = maximum temperature in the reactor during the first 15 minutes,

y_5 = slope of temperature decrease after 30 minutes,

and so on.

Let y be arranged in the sequence in which it is available. We can divide y also into two parts at any time:

$$y = \begin{bmatrix} y_p \\ y_f \end{bmatrix}, \tag{18.34}$$

where

y_p = vector of all past measurements,

y_f = vector of all future measurements.

At any time, we have the scenario shown in Figure 18.14.

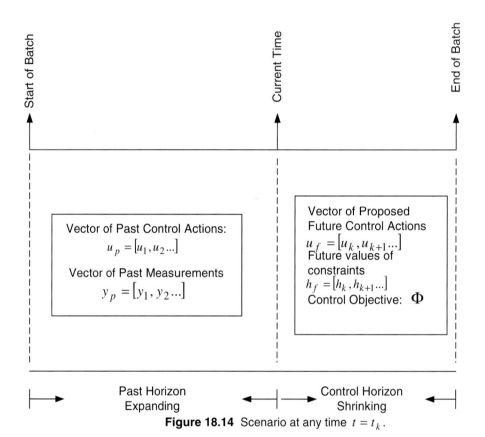

Figure 18.14 Scenario at any time $t = t_k$.

We can state the control problem as, *determine the future control actions u_f so that the control objective is maximized while the future constraints are met*. If future constraints cannot be met within a reasonable tolerance, we may choose to abort the batch run (if, for example, we cannot meet end product quality specifications).

The control calculation at time t_k is formulated in two phases, as in the NL-MPC case. First we get an estimate of the state of the system by solving the estimation problem:

$$\underset{x_0, P_e, d_e}{Min} \left\| y_p - y_p^m \right\|, \tag{18.35}$$

subject to

$$\frac{dx}{dt} = f(x, u, p_e, d_e, t)$$
$$x(0) = x_0 \tag{18.36}$$
$$y(t) = g(x, y, p_e, d_e, t),$$

where

y_p^m = measured value of y_p,

y_p = computed value of measure quantity.

Next these estimates are used to compute the remaining control moves, u_f.

$$\begin{matrix} Max \quad \Phi, \\ u_f \end{matrix} \tag{18.37}$$

subject to

$$\frac{dx}{dt} = f(x, u, p_e, d_e, t)$$
$$x(0) = x_0 \tag{18.38}$$
$$h_f(x, u, p_e, d_e, t) \le 0.$$

The horizon available for control action shrinks as the batch progresses. There may be a point in time beyond which no further control actions are possible (the set u_f is empty). In this case one may still continue to the estimation phase to predict the outcome of the batch and still have the option of aborting the batch if the predicted quality is way off target.

18.4.2 Discussion

For a more detailed discussion of the SH-MPC and application studies see the articles by Wang-Hanratty and Joseph (1993) and Thomas et al. (1997). In these articles, the method is applied to batch control of an autoclave curing process for making polymeric composites. Artificial neural network models were employed to model the process. Thomas et al. also discusses the results of applying this strategy on a bench-scale experimental autoclave process. Tsen et al. (1996) present an application of this strategy to an experimental batch polymerization unit.

18.5 SUMMARY

In this chapter we discussed extensions to the basic MPC algorithm. Constraints on manipulated inputs and process outputs are incorporated by modifying the optimization problem to be solved at every sample time. Unlike the unconstrained case, we can no longer derive closed form expressions for the controller. Hence a quadratic programming problem must be solved at every sample time. With the availability of fast, online computers this is no longer an issue. Care must be exercised in formulating the quadratic programming problem to ensure that the problem has feasible solutions. The presence of input constraints can limit the ability to reject disturbances, more so than for SISO systems due to the dependency among the output variables.

When we have non-square systems with more inputs that outputs, the possibility of using the extra degrees of freedom to optimize the process operation exists. By cascading a steady state LP or QP problem with the MPC algorithm one can optimize the process while meeting control objectives. This LP-MPC and QP-MPC formulation is used to dynamically adjust the setpoints in response to changing process disturbances. A significant portion of the economic payback from using advanced control strategies can be attributed to this ability to incorporate economic objectives into the control algorithm.

Finally we discussed extensions of the MPC algorithm to nonlinear and batch chemical processes. For nonlinear systems a two phase MPC (TP-MPC) consisting of an estimation phase and an optimization phase was presented. For batch processes a shrinking horizon MPC (SH-MPC) in which the control horizon is shrinking as the batch progresses can be used.

Problems

18.1 Consider the Shell problem outlined in Appendix I.

(a)   Using the SIMULINK model provided on the book website, develop and implement two PID controllers to control y_1 and y_2, using u_1 and u_2 respectively. Tune these controllers using IMC tuning methods. Check their performance to the prototype test cases 1 and 2, as outlined in the problem statement given in Appendix I. Are any constraints violated?

(b) Add an override (constraint) controller that maintains y_7 above –0.5, using u_3 as the manipulated variable. Tune the constraint controller. u_3 should be kept at its minimum of –0.5 if the constraint on y_7 is not violated. Check this system using prototype test cases 1 and 2. Are any constraints stated in the problem statement violated?

(c) Develop an MPC solution to the Shell problem using the SIMULINK MPC block from the MPC toolbox. Test your implementation using prototype test cases 1 and 2.

(d) Compare your results for the decentralized solution developed in parts (a) and (b).

18.2 Using the MPC toolbox software, implement the LP-MPC control scheme on the Shell fractionator column. Test its performance for the prototype cases suggested in the problem statement.

18.3 Steady-state regulation of a distillation column.

Consider the following column for separating a mixture of methanol/water. The nominal steady-state operating conditions are shown. Given the following steady-state observations, determine the feasibility of controlling the bottom product purity using (1) the reboiler heat duty and (2) the distillate flow rate when the feed can vary from 95 to 105 lb moles/hr.

Condition	Feed Rate lb moles/hr F	Distillate Rate lb mole/hr D	Reboiler Duty MM Btu/hr Q	Top Purity Mole Fraction x_T	Bottom Purity Mole Fraction x_B
Nominal Steady-state	100	45	2.2	.0057	.0955
Perturb D	100	46	2.2	.0063	.0794
Perturb Q	100	45	2.3	.0047	.0948
Perturb F	105	45	2.2	.0058	.1292

Assume that D can be manipulated between 0 and 100 lb mole/hr and that Q can be manipulated between 0 and 4 MM Btu/hr. The top analyzer has a span of 0-0.02 and the bottom analyzer has a span of 0-0.2. The feed flow sensor has a span of 0-200 lb mole/hr.

    a.   Derive a scaled dimensionless matrix of gains relating the inputs and outputs.

    b.   Determine the feasibility of regulating both the top and bottom purity, assuming that the feed flow can vary between 75 and 125 lb moles/hr in the steady-state.

References

In addition to the referenced papers in the chapter, we have included a few references that deal with the issues related to model-predictive control. The article by Lee and Cooley (1997) contains a larger collection of literature references on the topic.

Bequette, B. W. 1991. "Nonlinear Predictive Control Using Multi-Rate Sampling." *Can. J. Chem. Eng.*, *69*(1), 136–143.

Biegler, L. T. 1998. "Advances in Nonlinear Programming Concepts for Process Control." *J. Process Control, 8*(5-6), 301–311.

Brosilow, C., and G. Q. Zhao. 1986. "A Linear Programming Approach to Constrained Multivariable Process Control." *Control Dyn. Syst.*, *27*(3), 141–181.

Cutler, C. R., and B. L. Ramaker. 1980. "Dynamic Matrix Control: A Computer Control Algorithm." Proc. Joint Auto Control Conf, Paper WP5-B, San Francisco.

Eaton, J. W., and J. B. Rawlings. 1990. "Feedback Control of Chemical Processes Using On-Line Optimization Techniques." *Comput. Chem. Eng.*, *14*(4–5), 469–479.

Forbes, J., and T. Marlin. 1994. "Model Accuracy for Economic Optimizing Controllers: The Bias Update Case." *Ind. Eng. Chem. Res.* 33, 1919–1929.

Garcia, C. E., and A. M. Morshedi. 1986. "Quadratic Programming Solution to Dynamic Matrix Control (QDMC)." *Chem. Eng. Comm.*, 46, 73.

Garcia, C. E., and M. Morari. 1982. "Internal Model Control, 1: A Unifying Review and Some New Results." *I & EC Proc. Des. Dev.*, 21, 308.

Harkins, B. L. 1991. "The DMC Controller." *ISA*, paper #91–0427, 853.

Jang, S. S., H. Mukai, and B. Joseph. 1987a. "On-Line Optimization of Constrained Multivariable Processes." *AIChE Journal*, 33, 26.

Jang, S. S., H. Mukai, and B. Joseph. 1987b. "Control of Constrained Multivariable Nonlinear Processes Using a Two-Phase Approach." *Ind. and Eng. Chem. Proc. Des. Dev.*, 26, 10, 2106 –2113.

Jang, S. S., B. Joseph, and H. Mukai. 1988. "Integrated Model-Based Control of Multivariable Nonlinear Systems." Proceedings of the IFAC Workshop on Model-based Process Control, American Automatic Control Council. Atlanta, GA.

Jang, S. S., H. Mukai, and B. Joseph. 1986. "A Comparison of Two Approaches to On-Line Parameter Estimation and State Estimation of Nonlinear Systems." *Ind. and Eng. Chem. Proc. Des. Dev.*, 25, 809.

Kailath, Thomas. 1980. *Linear Systems*, Prentice Hall, NJ.

Kassmann, D. E., T. A. Badgwell, and R. B. Hawkins. 2000. "Robust Steady-State Target Calculation for Model-Predictive Control." *AIChE J.*, 46(5), 1007–1025.

Kozub, D. J., and J. F. Macgregor. 1992. "Feedback Control of Polymer Quality in Semibatch Copolymerization Reactors." *Chem. Eng. Sci.*, 47(4), 929–942.

Kwakernaak, H., and R. Sivan. 1972. *Linear Optimal Control System.* Wiley Interscience, NY.

Lee, J. H., and B. Cooley. 1997. "Recent Advances in Model-predictive Control and Other Related Areas." *Proceedings of CPC V,* AIChE, New York, NY.

Lid, T., and S. Strand. 1997. "Real-Time Optimization of A Cat Cracker Unit." *Comput. Chem. Eng.*, 21(Suppl.), S887–S892.

Marlin, T. E., and A. N. Hrymak. 1996. "Real-Time Operations Optimization of Continuous Processes." *Proceedings of CPC V*, AIChE, NY.

Mayne, D. Q. 1996. "Nonlinear Model-Predictive Control." *Proceedings of CPC V*, Tahoe City, CA.

Moro, L. F. L., and D. Odloak. 1995. "Constrained Multivariable Control of Fluid Catalytic Cracking Converters." *J. of Process Control*, 5(1), 29–39.

Morshedi, A. M., C. R. Cutler, and T. A. Skrovanek. 1985. "Solution of Dynamic Matrix Control with Linear Programming Techniques (LDMC)." Presented at the *ACC'85*, Boston, MA, 199–208.

Nagy, Z., and S. Agachi. 1997. "Model-Predictive Control of a PVC Batch Reactor." *Comput. Chem. Eng.*, *21*(6), 571–591.

Palavajjhala, S. 1994. "Studies in Model-Predictive Control with Application of Wavelet Transform." D.Sc. Thesis, Washington University.

Prett, D. M., and R. D. Gillette. 1979. "Optimization and Constrained Multivariable Control of a Catalytic Cracking Unit." #51c, *AIChE Annual Meeting*, Houston, TX.

Prett, D. M., C. E. Garcia, and B. L. Ramaker. 1990. *The Second Shell Process Control Workshop*, Butterworths, MA.

Rao, C. V., and J. B. Rawlings. 2000. "Linear Programming and Model-Predictive Control." *J. Process Control*, *10*(2–3), 283–289.

Rawlings, J. B. and K. R. Muske. 1993. "The Stability of Constrained Receding Horizon Control." *IEEE Trans. Auto. Con. 38*(10), 1512–1515.

Rawlings, J. B., N. F. Jerome, J. W. Hamer, and T. M. Breummer. 1990. "Endpoint Control in Semi-Batch Chemical Reactors." IFAC Symp. Ser, 7(Dyn. Control Chem. React., Distill. Columns Batch Processes), 339–44.

Regnier, N., G. Defaye, L. Caralp, and C. Vidal. 1996. "Software Sensor Based Control of Exothermic Batch Reactors." *Chem. Eng. Sci.*, *51*(23), 5125–5136.

Richalet, J., A. Rault, J. L. Testud, and J. Papon. 1978. "Model-Predictive Heuristic Control: Applications to Industrial Processes." *Automatica*, *14*(5), 413–428.

Ricker, N. L. 1985. "Use of Quadratic Programming for Constrained Internal Model Control." *IEC Proc Des Dev,* 24, 925.

Thomas, M. M., B. Joseph, and J. L. Kardos. 1997. "Batch Process Quality Control Applied to Curing of Composite Materials." *AIChE J.*, *43*(10), 2535–2545.

Tsen, A. Y. D., S. S. Jang, D. S. H. Wong, and B. Joseph. 1996. "Predictive Control of Quality in Batch Polymerization Using a Hybrid Artificial Neural Network Model," *AIChE Journal*, *42*(2), 455–465.

Van Antwerp, J. G., and R. D. Braatz. 2000. "Model-Predictive Control of Large Scale Processes." *J. Process Control*, *10*(1), 1–8.

Vargas-Villamil, F. D., and D. E. Rivera. 2000. "Multilayer Optimization and Scheduling Using Model-Predictive Control: Application to Reentrant Semiconductor Manufacturing Lines." *Comput. Chem. Eng.*, *24*(8), 2009–2021.

Voorakaranam, S., and B. Joseph. 1999. "Model-Predictive Inferential Control with Application to Composites Manufacturing Process." *Ind. Eng. Chem. Res.*, 38, 433–450.

Wang-Hanratty, F. H., and B. Joseph. "Predictive Control of Quality in Batch Manufacturing Process Using Artificial Neural Network Models." *Ind. Eng. Chem. Res.*, 32, 1951–1961.

Ying, C. M., and B. Joseph. 1999. "Performance and Stability Analysis of LP-MPC and QP-MPC Cascade Control Systems." *AIChE J.*, (45)7, 1521–1534.

Yousfi, C., and R. Tournier. 1991. "Steady-State Optimization Inside Model-Predictive Control." *Proceedings of the American Control Conference*, Boston, MA, 1866–1870.

Zheng, A., and M. Morari. 1995. "Stability of Model-Predictive Control with Mixed Constraints," *IEEE Transaction on Automatic Control*, 40(10), 1818–1823.

Inferential Model-Predictive Control

Objectives of the Chapter

In Chapter 12 we discussed the design and implementation of inferential controllers. In this chapter we discuss how to incorporate these concepts in a MPC framework. Topics discussed include

- Inferential MPC algorithm.
- Steady-state estimator development using steady-state data.
- Dynamic estimators using time series data.
- Nonlinear estimators using artificial neural networks.

Prerequisite Reading

Chapter 12, "Single Variable Inferential Control"
Chapter 17, "Basic Model-Predictive Control"

19.1 INFERENTIAL MODEL-PREDICTIVE CONTROL (IMPC)

In this section we will consider how to incorporate secondary measurements in discrete MPC schemes such as the algorithms discussed in Chapter 17. Incorporating secondary measurements into the MPC framework is similar to adding feedforward control based on a measured disturbance. Only a slight modification is required for this. The algorithms discussed in this section are equally applicable to inferential control using either single or multiple secondary measurements.

Figure 19.1 shows the block diagram representation of the inferential MPC algorithm. Note the similarity to MPC combined with *feedforward control*. The main difference here is that the disturbance is not directly measured; rather, the effect of the disturbance on the primary output is predicted using the secondary measurements in conjunction with an inferential estimator. We discussed methods for designing the inferential estimator in Chapter 13. This chapter will focus on using this estimator in an MPC framework.

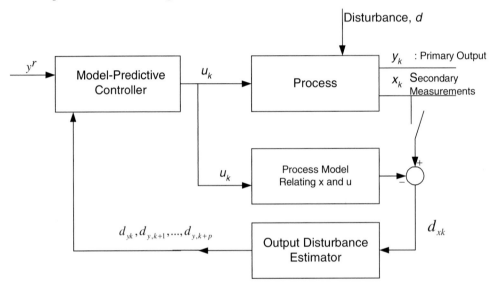

Figure 19.1 Block diagram representation of IMPC.

19.1.1 Derivation of the Inferential MPC Algorithm

Let us review MPC briefly. We started with the following variables. Given:

1. A predicted trajectory of the output based on a process model and past control actions, $y_k^p, y_{k+1}^p, ..., y_{k+p}^p$, where k = current time and p = length of output prediction horizon,

2. A desired output trajectory $y_k^r, y_{k+1}^r, ..., y_{k+p}^r$,

3. Current measurement of output y_k^m,

the first step in the MPC algorithm is to estimate the disturbance on the output variable. The disturbance estimate is obtained from the difference between the prediction of the current value of y and the current measured value of y. Then the current disturbance effect at the output is computed, using

$$d_{y,k} = y_k^m - y_k^P$$

Estimating the future values of disturbances requires some assumptions. The simplest assumption is that these are constant:

$$d_{y,k+1} = d_{y,k+2} = ... = d_{y,k}$$

If a model of the disturbance is available, say in the form of an ARIMA type transfer function (see Appendix D), then the future values of d_y can be estimated using the model. Now suppose y is not measured directly, but a secondary measurement x is available. Then we need a way to estimate $d_{yk}, d_{y,k+1}, d_{y,k+2}, ..., d_{y,k+p}$. We might view the estimator as a dynamic system which is driven by input d_x and produces estimate d_y. See Figure 19.2.

$$d_y(s) = \alpha(s)d_x(s)$$

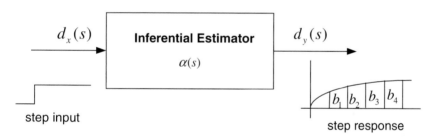

Figure 19.2 The estimator as a dynamic system and its step response model.

At any time, we can calculate d_x from the secondary measurement and then estimate d_y into the future using this estimator. In Chapter 13 we discuss ways of constructing $\alpha(s)$ for process models. It is convenient to convert the estimator into discrete form using the step response coefficients and using the same sampling time as for the MPC algorithm. This can be obtained as shown in Figure 19.2.

A step response model relating d_y and d_x can also be obtained from plant data collected at times when the manipulated variables are kept constant and disturbances are activated. This assumes that the operator can deliberately introduce the common types of disturbances that are present. If that is not possible, we may have to use a rigorous dynamic model of the process to generate this data. Once the data are obtained, then the step or impulse response model relating d_x and d_y can be obtained using the least-squares identification schemes outlined in Appendix C.

Let the step response coefficients be given by b_1, b_2, b_3, ..., b_p. Then we can estimate $d_{y,k}$, ..., $d_{y,k+p}$ as follows:

$$d_{y,k+1} = b_1 d_{xk},$$
$$d_{y,k+2} = b_2 d_{xk},$$
$$....,$$
$$d_{y,k+p} = b_p d_{xk}, \tag{19.1}$$

where d_{xk} is the current estimated value of the disturbance effect on x. The next question is how to obtain d_{xk}. Recall that d_x is the effect of the disturbance on the secondary measurement x:

$$d_x(s) = x(s) - p_x(s)u(s). \tag{19.2}$$

We can represent $p_x(s)$ also in discrete form using the step response coefficients

$$\Delta x_{k+1} = c_1 \Delta u_k,$$
$$\Delta x_{k+2} = c_2 \Delta u_k,$$
$$...,$$
$$\Delta x_{k+p} = c_p \Delta u_k. \tag{19.3}$$

Let us say we have a predicted trajectory for the secondary variable x. That is, we have estimates $x_k^P, x_{k+1}^P, ..., x_{k+p}^P$ over the output horizon. Then the disturbance estimate on x can be calculated as

$$d_{xk} = x_k^m - x_k^P.$$

After the current control action Δu_k has been calculated, it is necessary to update the predicted trajectory of x. Taking into account both the effect of the control effort applied at that sample time and the disturbance effect estimated from the current measurement (which is assumed to be constant as a step disturbance),

$$x_{k+1}^P = x_{k+1}^P + c_1 \Delta u_k + d_{xk},$$
$$x_{k+2}^P = x_{k+2}^P + c_2 \Delta u_k + d_{xk},$$
$$...,$$
$$x_{k+p}^P = x_{k+p}^P + c_p \Delta u_k + d_{xk}. \tag{19.4}$$

Before going to the next sample time, $k+1$, it is necessary to update the x^P vector to include the prediction at $k+p+1$ and discard the prediction at k (recall that we have a moving horizon and when we get to the next sample time, we have to move the trajectory by one sample time forward into the future). This is accomplished by replacing

$$x_{k+i}^p(\text{new}) \leftarrow x_{k+i+1}^p(\text{old}), i = 0,1,..., p-1 \qquad (19.5)$$

and replacing $x_{k+p}^p(\text{new})$ with $x_{k+p}^p(\text{old})$, the last predicted value of x. Figure 19.3 shows the schematic of IMPC.

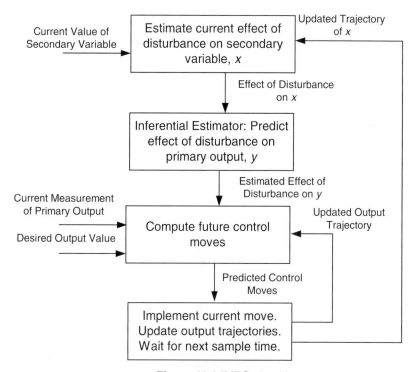

Figure 19.3 IMPC algorithm.

Example 19.1 IMPC of the Shell Fractionation Column

Consider the Shell distillation control problem. The process model is

$$y(s) = \frac{4.05e^{-27s}}{50s+1}u(s) + \frac{1.44e^{-27s}}{40s+1}d(s),$$

$$x(s) = \frac{3.66e^{-2s}}{9s+1}u(s) + \frac{1.27}{6s+1}d(s).$$

Note that this secondary variable responds a lot faster to both the manipulated variable and the disturbance variable. In Chapter 12 we derived an estimator for d_y:

$$d_{ye} = \frac{1.44e^{-27s}}{40s+1} \frac{(6s+1)}{1.27} d_x.$$

Figure 19.4 shows the implementation of this IMPC on the Shell fractionator problem. Here we have explicitly subtracted off the effect of u_1 on y_3 to get d_t. This signal is input to the MPC as an indirect measurement of the disturbance entering the process. The MPC controller uses the estimator transfer function above to estimate how d_t affects the output. The script file *ch17ex1_script.m* is used to define the transfer functions and parameters for the MPC controller.

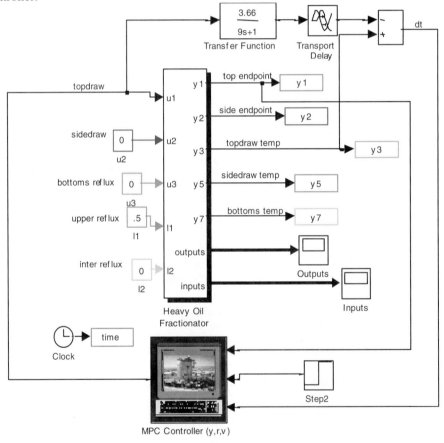

Figure 19.4 IMPC of $y1$ using $u1$ and secondary measurement $y3$.

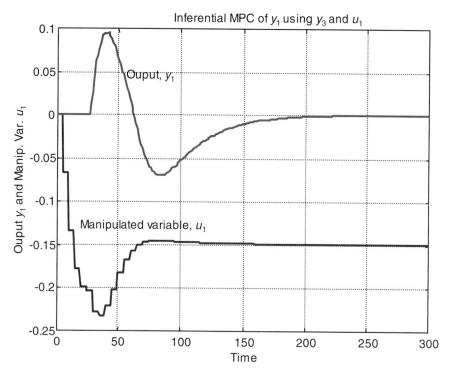

Figure 19.5 The result of IMPC. This response should be compared against direct MPC, as shown in Figure 17.8.

♦

19.2 SIMPLE REGRESSION ESTIMATORS

In this section we discuss the construction of inferential estimators from empirical data collected from the process. Such estimators are called data-driven estimators. Let us restate the inferential estimation problem:

Given a process plant with secondary measurements, x, and primary variable to be controlled, y, generate estimates of the disturbance effect on y (denoted using d_y) from the computed estimates of the effect of the disturbance on x (denoted using d_x).

We want to emphasize the fact that we are not estimating y directly, but d_y. To estimate d_x, we need to subtract the effect of the manipulated variables on x. This will require a model that relates these two variables. This is accomplished by conducting identification experiments on the plant to relate the manipulated variables to the measurements. If all the disturbances could be measured and perturbed independently, then we could have identified the disturbance model also directly from plant tests. However, the assumption is that these disturbances cannot all be measured or manipulated independently, so direct identification through plant tests is not always feasible.

An alternative approach is to collect data on the plant operational characteristics over a long period of time and then try to relate the two quantities d_x and d_y through regression on the data. This problem may be stated as follows:

Given a set of operating data in the form of a table that tabulates y, x, and u for various steady-states attained by the plant, find a linear relationship between d_x and d_y where

$$
\begin{aligned}
d_x &= x - p_x u, \\
d_y &= y - p_y u,
\end{aligned}
\tag{19.6}
$$

where p_x and p_y are the previously determined models relating u to x and y respectively. In this section we limit our discussion to steady-state estimators, which may be sufficient in some cases. The construction of dynamic estimators is discussed in the next section.

The method can be stated in general form as follows: We start with identification experiments to determine the steady-state gains of the manipulated variables to the outputs. This determines the values of p_x and p_y used in Eq. (19.6). Next, a set of steady-state observations of x and y are taken. From the data set, we can calculate d_x and d_y for each of the data sets. This leads to two matrices defined as follows:

$$
X = \begin{bmatrix} dx_{11} & dx_{21} & . & dx_{p1} \\ dx_{12} & dx_{22} & . & dx_{p2} \\ . & . & . & . \\ dx_{1N} & dx_{2N} & . & dx_{pN} \end{bmatrix} \quad \text{and} \quad z = \begin{bmatrix} dy_1 \\ dy_2 \\ . \\ dy_N \end{bmatrix}.
$$

Each row of the X matrix represents one set of steady-state observations on the variable d_x, and each element of the vector z represents the corresponding value of d_y. N is the number of observations. The objective is to build an estimator of the form

$$
d_y = \alpha_1 d_{x1} + \alpha_2 d_{x2} + \ldots + \alpha_p d_{xp}.
\tag{19.7}
$$

The solution is given by

$$
\alpha = (X^T X)^{-1} X^T z
\tag{19.8}
$$

We will call this the regression estimator.

Example 19.2 Regression Estimator for a Distillation Column

Table 19.1 shows data obtained from a steady-state simulation of a distillation column. Shown are 20 data points with two temperatures, the top endpoint, and the manipulated variable, topdraw rate. These data were generated assuming no noise present in the measurement system. The nominal steady-state values of the variables are shown in the first row.

Table 19.1 Steady-State Data Relating Temperatures and Composition Measurements on the Distillation Column.

Data Set Number	Manipulated Input, Reflux Flow, U	Temperature on Tray 24 Deg C, X1	Temperature on Tray 15 Deg C, X2	Distillate purity, Y %,
Nominal Steady-State Value	100.00	200.0	100.0	90.0
1	100.36	203.02	104.6	93.3
2	100.5485	202.7525	104.3372	93.0174
3	100.2618	202.4201	103.7017	92.6071
4	100.5973	203.0652	104.7819	93.4075
5	100.0493	201.5041	102.2101	91.6367
6	100.5711	202.5043	103.9923	92.7488
7	100.7009	203.6734	105.7926	93.9961
8	100.9623	204.256	106.7632	94.6909
9	100.7505	204.0555	106.3521	94.4449
10	100.74	203.3724	105.3218	93.7437
11	100.4319	203.1488	104.8008	93.486
12	100.6343	203.6567	105.6842	94.0253
13	100.803	203.8608	106.0943	94.2473
14	100.0839	201.5668	102.3256	91.7033
15	100.9455	204.9341	107.7328	95.429
16	100.9159	203.6889	105.9075	94.0781
17	100.602	204.2987	106.6025	94.7104
18	100.2536	203.0176	104.5179	93.3092
19	100.8735	204.8709	107.6055	95.3444
20	100.5134	202.8609	104.5007	93.0986

From identification tests, the steady-state gains relating the manipulated variable u to the temperatures x and output composition y were determined as follows (in terms of deviation variables):

$$x_1 = 3.66u,$$
$$x_2 = 5.92u,$$
$$y = 4.05u.$$

Determine a regression estimator relating d_x and d_y.

Solution

Applying regression to the variables

$$d_{x1} = (X_1 - 200) - 3.66(U - 100),$$
$$d_{x2} = (X_2 - 100) - 5.92(U - 100),$$
$$d_y = (Y - 90) - 4.05(U - 100),$$

We get the estimator

$$d_y = 2.84 d_{x1} - 1.2 d_{x2}.$$

Adding noise to the data changes the estimators, as shown in Example 19.3.

♦

Example 19.3 Effect of Measurement Noise

Let us see the effect of adding some random measurement noise to the data set shown above. For comparing different estimators, we use the correlation coefficient, which is defined as

$$R^2 = 1 - \frac{\Sigma(y - y^p)^2}{\Sigma(y - \bar{y})^2}.$$ (19.9)

The y_p is the predicted value of y using the regression estimator.

The estimator coefficients calculated with varying levels of noise in the data are shown in the following table:

Noise Std. Dev.	α_1	α_2	Σe^2	R^2
0	2.84	−1.21	0.00	1.00
0.01	1.81	−0.50	0.102	.9998
0.1	0.45	0.445	0.46	.995
0.5	0.30	0.63	2.83	.772

The regression estimator is sensitive to the noise present in the measurement. The condition number is large for the regression matrix generated from this data set. The use of principal component regression (PCR) can partly overcome this problem. These methods are discussed next (see Appendix C for a discussion on PCR).

♦

Example 19.4 PCR Based Estimators

Application of PCR to the previous data yields the following set of estimators

Noise Std. Dev.	α_1	α_2	Σe^2	R^2
No noise	.3485	.5059	.1356	.9998
.1	.345	.5095	.7426	.9889
.5	.376	.5765	3.3	.72

The estimator, using only one latent variable, is rather insensitive to the noise level in the measurements. However, the performance is slightly worse when using only one latent variable in the PCR estimator. In general, one should use PCR estimators if the secondary measurements used are highly correlated with each other.

♦

The steady-state regression estimators are easy to construct. The problem is generally with acquiring sufficiently rich data that covers the entire possible operating range. All modes of the input disturbances must be present in the data. The estimator will only be valid over the data set covered.

19.2.1 Scaling

Scaling is important in the construction of regression estimators. There are two reasons for scaling variables. First, during regression, larger variables are given more weight automatically since the square of the error is being minimized. Second, if variables differ much in their order of magnitude, then the resulting regression matrix will be ill-conditioned. Scaling is particularly important if the variables involved are of different dimensional units. Suggestions for scaling are given in Appendix A.

19.3 DATA-DRIVEN DYNAMIC ESTIMATORS

Next we consider the problem of building dynamic estimators from process data. This problem is the same as that of identifying a process model using dynamic input/output data, addressed in more detail in Chapters 15 and 16.

Consider the continuous-time lead-lag estimator of the type we designed in Chapter 13. The estimator, using a single measurement, can be written as

$$\hat{d}_y(s) = \alpha \frac{ps+1}{qs+1} d_x(s). \tag{19.10}$$

This can be converted to discrete z-domain as

$$\hat{d}_y(z) = \frac{a_0 + a_1 z^{-1}}{1 + b_1 z^{-1}} d_x(z), \tag{19.11}$$

or in discrete-time-domain as

$$d_y(n) + b_1 d_y(n-1) = a_0 d_x(n) + a_1 d_x(n-1). \tag{19.12}$$

Note that this implies a linear relationship between the sampled values of d_y and d_x. The problem of constructing the estimator in this case is reduced to that of computing the parameters a_0, a_1, and b_1 from a set of sampled values of d_y and d_x at discrete points in time. Once these model parameters are determined, we can calculate the model in any other format, as needed, by the control algorithms being used.

This problem is the same as that of identifying a transfer function model for a process plant from input/output data. Hence we can use the same methodologies that have been developed in the chapter on identification.

Let us say that we have collected sampled values of observations on the primary and secondary variable over a period of time. By keeping the manipulated variable constant or by subtracting out the effect of the manipulated variable on the outputs, we can create data sets of the form

$d_x(0), d_x(1), d_x(2), ..., d_x(p)$: Set of sampled values of secondary measurement with the effect of the manipulated variable removed (i.e., these values represent the effect of the external disturbances on the secondary variable,

$d_y(0), d_y(1), d_y(2), ..., d_y(p)$: Set of sampled values of the primary measurement with the effect of the manipulated inputs removed.

Then we have the following relationships, implied by the proposed lead-lag estimator:

$$d_y(1) + b_1 d_y(0) = a_0 d_x(1) + a_1 d_x(0),$$
$$d_y(2) + b_1 d_y(1) = a_0 d_x(2) + a_1 d_x(1),$$
$$..., \tag{19.13}$$
$$d_y(p) + b_1 d_y(p-1) = a_0 d_x(p) + a_1 d_x(p-1)$$

This can be written as a set of p equations in three unknowns, as follows:

$$\begin{bmatrix} d_y(0) & -d_x(1) & -d_x(0) \\ d_y(1) & -d_x(2) & -d_x(1) \\ . & . & . \\ d_y(p-1) & -d_x(p) & -d_x(p-1) \end{bmatrix} \begin{bmatrix} b_1 \\ a_0 \\ a_1 \end{bmatrix} = \begin{bmatrix} -d_y(1) \\ -d_y(2) \\ . \\ -d_y(p) \end{bmatrix}, \tag{19.14}$$

which can be written as

$$Xa = z. \tag{19.15}$$

α can be solved using the various means (least-squares or PCR) that were discussed earlier. This method is easily extended to the case of multiple secondary measurements.

Example 19.5 Design of a Dynamic Estimator for the Shell Column

The data files *ch19ex5_test_data* and *ch19ex5_reg_data* contain sampled values of temperature and composition generated from an open-loop simulation of the Shell column model. During the simulation, the manipulated variables were kept constant, and hence these data may be interpreted as a table of d_x and d_y, sampled 3 minutes apart. This data does not contain any measurement noise. The program *ch19ex5.m* contains the commands used to generate an estimator from this data. Since the time delay is unknown, various values were tried, and the time delay that yielded the lowest prediction error for the test data was selected for the estimator. Figure 19.6a shows the prediction error as a function of the time delay. This shows that a delay of 6 is the best. The data was regressed to obtain the following estimator:

$$d_y(z) = \left[\frac{0.6436z^{-6}}{1 - .8961z^{-1}} \right] d_x(z).$$

Figure 19.6b shows a comparison between the predicted observed values using this estimator. The results show that the estimator is doing a reasonably good job of tracking the variations in d_y as a function of time.

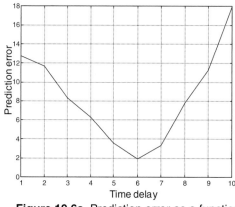

Figure 19.6a. Prediction error as a function of the delay used in the model.

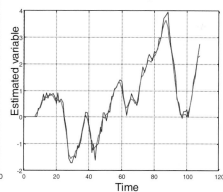

Figure 19.6b Comparison of predicted and observed d_y for the test data set.

◆

19.4 Nonlinear Data-Driven Estimators

If there are large errors associated with the linear estimator, then it is worthwhile to investigate nonlinear estimators of the form

$$\hat{y} = f(x_1, x_2, ..., x_n).$$
(19.16)

The problem here of course is determining the functional form of f. First, one should investigate if some information is available regarding the functional relationship from the physics and chemistry of the process. If this fails, one could try a quadratic regression. The problem with quadratic estimators is that the number of parameters to be regressed is of the order of $n(n + 1)$, where n is the number of secondary measurements. For two measurements, the quadratic estimator is of the form

$$\hat{y} = \alpha_1 x_1 + \alpha_2 x_2 + \alpha_{11} x_1^2 + \alpha_{12} x_1 x_2 + \alpha_{22} x_2^2.$$
(19.17)

Artificial neural networks (ANN) have been suggested for modeling of arbitrary nonlinear input/output mapping. A brief description of ANN is given here. Details are given in many books (for example, Deshpande et al., 1999).

The ANN was inspired by the research done on understanding of biological nervous systems. Numerous structures have been proposed in the literature for building ANN models. Among these, the multilayer perceptron (Rummelhart et al., 1986) has emerged as a model suitable for nonlinear regression modeling. This model, sometimes referred to as feedforward networks, are made up of individual perceptrons (also referred to as neurons), as illustrated in Figure 19.7 and given by Eq. (19.18).

$$y = f(\sum_{i=0}^{N} w_i x_i + b)$$
(19.18)

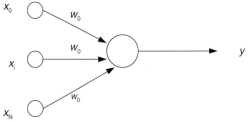

Figure 19.7 A perceptron, which is the building block of a neural network.

Each perceptron has many inputs, and the output is a weighted sum passing through a nonlinear function. The b in Eq. (19.18) is called a bias term. Usually, a sigmoid function is employed:

$$f(\alpha) = \frac{1}{1 + e^{-(\alpha - \theta)}},\qquad (19.19)$$

where θ is a threshold value associated with each node. An ANN is made up of multiple perceptrons arranged in the form of input nodes, one or more hidden layers of nodes and output nodes. Figure 19.8 shows a three-layer ANN with three input nodes and one hidden layer of two neurons, and a single output node.

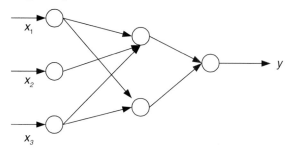

Figure 19.8 Three-layer perceptron.

This three-layer ANN attempts to approximate the mapping

$$y = g\,(x_1, x_2, x_3).$$

The weights w associated with each link, and the bias b associated with each node, are adjustable. These weights and biases are adjusted so that the network can fit a "training set" of data in the form of quadruplets (x_1, x_2, x_3, y). Neural networks with a sufficient number of nodes and hidden layers have been shown to be capable of learning to map arbitrary functions. There are two phases to the use of an ANN:

Training phase: Data sets $(x_1, x_2, ..., x_n, y)$ are repeatedly presented to the network. The weights are adjusted so that error between the value of y predicted by the ANN and the value in the training set is minimized in a least-squares sense. Standard non-linear minimization techniques are employed here.

Production phase: In this phase, given new input data sets $(x_1, ..., x_n)$, the network is used to predict the value of y.

19.3.1 Discussion

The ANN has emerged as a useful regression tool for nonlinear inferential estimation. For an application of inferential estimation using ANN models, see Wang-Hanratty et al. (1995) and Thomas and Joseph (1996). Unger et al. (1997) reviews a number of applications of inferential ANN estimators.

19.5 SUMMARY

In this chapter we extended the concepts of MPC to control using secondary measurements. A variation of the basic MPC called IMPC was introduced. The main component of the IMPC algorithm is an estimator for estimating the effect of input disturbances on the controlled primary output variable. This estimator may be obtained from process models using the procedures discussed on Chapter 13. If process models are not available, data-driven regression estimators may be used in the inferential control system. For this we need process data when disturbances are present. If only steady-state data are available, it can be used to regress a steady-state estimator for d_y using d_x. If the control effort u were changing during the data collection period, then its effect on the outputs must be removed first. Dynamic estimators in the discrete-time domain can be generated using the techniques of process identification.

If a linear model is not sufficient, then nonlinear estimates can be built using the ANN structure. These types of nonlinear estimates are finding many applications in industry and commercial software is available to facilitate the construction and implementation of nonlinear inferential estimators that use ANN-based algorithms.

An important issue to consider is the quality of the data. It is important to perturb all disturbance variables of interest during the data collection period. Data-driven estimators are valid only over the conditions under which the data were collected.

Problems

19.1 A process has two measurements, x and y. The disturbance d to the process was perturbed using a *prbs* signal. The resulting data is saved in file *ch19pr1_t_d_x_y.dat* on the website. The variables are stored in the order t, d, x, y, where t is the time variable.

(a)   Develop a transfer function model relating x to u and y to u. Note the sample time. Use the Identification toolbox in MATLAB. (see file *ch15ex3.m* for commands to use). Describe how you arrived at your choice of the time delay and order of numerator and denominator polynomials.

(b)   Develop an estimator for y using x from the result of (a).

(c)   Develop an estimator relating y and x directly, without using the column vector u.

(d)   Develop a dynamic matrix model relating y and x. Give the elements of the matrix.

(e)   Is the input *prbs* signal designed correctly? Explain with an analysis.

19.2. Consider the data in the following table (see file *ch19pr1.dat* on the website).

Data Set No	Temp 1 Span:50-150	Temp 2 Span:100-200	Pressure Span: 9000-11000	Purity Span: 0-0.1
	T1	*T2*	*P*	*Y*
1	101.2187	149.9846	10094	0.0050
2	9.9877	150.5370	10002	0.0040
3	9.5285	148.8337	9914	0.0032
4	101.6514	149.3900	10001	0.0050
5	99.1673	151.3295	10007	0.0031
6	102.1753	149.6008	9950	0.0061
7	100.6097	150.2827	10018	0.0044
8	102.7978	148.6685	10032	0.0067
9	100.3857	149.6650	10124	0.0044
10	99.3928	149.1823	9702	0.0036
11	98.9388	149.6302	9875	0.0032
12	99.8033	151.0189	10099	0.0045
13	102.2643	151.1753	10035	0.0062
14	96.7270	150.8684	10014	0.0010
15	100.5919	150.1897	10061	0.0048
16	99.1204	150.4312	9916	0.0035
17	100.3411	148.9629	9906	0.0037
18	98.3446	149.3632	9974	0.0021
19	98.6835	151.1431	10185	0.0023
20	102.7317	151.1578	10076	0.0068

(a)   Fit a model to predict the purity, using the other three secondary variables. Calculate the R-squared correlation coefficient.

(b)   Now scale the variables in a suitable way. Repeat the regression fit. Compare with the result from Part a.

References

Kano, M., K. Miyazaki, S. Hasebe, and I. Hashimoto. 2000. "Inferential Control System of Distillation Compositions Using Dynamic Partial Least Squares Regression." *J. Process Control*, *10*(2-3), 157–166.

Kresta, J. V., T. E. Marlin, and J. F. MacGregor. 1994. "Development of Inferential Process Models Using PLS." *Comput. Chem. Eng.*, *18*(7), 597–611.

Lee, J. H., and A. K. Datta. 1994. "Nonlinear Inferential Control of Pulp Digesters." *AIChE J.*, *40*(1), 50–64.

Ponton, J. W., and J. Klemes. 1993. "Alternatives to Neural Networks for Inferential Measurement." *Comput. Chem. Eng.*, *17*(10), 991–1000.

Ogawa, M., M. Ohshima, K. Morinaga, and F. Watanabe. 1999. "Quality Inferential Control of an Industrial High Density Polyethylene Process."   *J. Process Control*, *9*(1), 51–59.

Shen, G. C., and W. K. Lee. 1989. "A Predictive Approach for Adaptive Inferential Control."   *Comput. Chem. Eng.*, *13*(6), 687–701.

Song, J. J., and S. Park. 1993. "Neural MPC for Nonlinear Chemical Processes." *J. Chem. Eng. Jpn*, *26*(4), 347–354.

Tham, M. T., G. A. Montague, A. J. Morris, and P. A. Lant. 1991. "Soft-Sensors for Process Estimation and Inferential Control."   *J. Process Control, 1*(1), 3–14.

Tsen, A. Y. D., B. Joseph, S. S. Jang, and D. S. H. Wong. 1996. "Predictive Control of Quality in Batch Polymerization Using a Hybrid Artificial Neural Network Model." *AIChE J.*, *42*(2), 455–465.

Unger, L., E. H. Hartman, J. D. Keeler, and G. D. Martin. 1996. "Process Modeling and Control Using Neural Networks." Proceedings of the Intelligent Systems in Process Engineering Conference, AIChE Symposium Series, *92*(312), AIChE, NY.

Voorakaranam, S., and B. Joseph. 1999. "Model-predictive Inferential Control with Application to Composites Manufacturing Process." *Ind. Eng. Chem. Res.*, *38*, 433–450.

Voorakaranam, S., J. L. Kardos, and B. Joseph. 1999. "Modeling and Control of an Injection Pultrusion Process." *Journal of Composite Materials*, *33*(13),1173–1205.

Wang-Hanratty, F. H., and B. Joseph. 1993. "Predictive Control of Quality in Batch Manufacturing Process Using Artificial Neural Network Models." *Ind. Eng. Chem. Res.*, *32*, 1951–1961.

Wang-Hanratty, F. H., B. Joseph, and J. L. Kardos. 1995. "Model-Based Control of Voids and Product Thickness during Autoclave Curing of Carbon/Epoxy Composite Laminates." *Journal of Composite Materials*, *29*(8), 1000–1025.

Review of Basic Concepts

Objectives

The objective of this appendix is to review the basic concepts usually covered in an undergraduate course in process control. Concepts reviewed include:

- Block diagram representation of control systems.
- Laplace transform and transfer functions.
- Analysis of block diagrams.
- P, PI, and PID controllers.
- Stability of feedback control systems.

A.1 BLOCK DIAGRAMS

The block diagram notation is extensively used to describe and analyze control systems. The simplest form is shown in Figure A.1. The diagram shows the cause (a change in the input variable) and effect (a change in the output variable) relationship between the input and output variables. We define input and output variables as deviations from an initial nominal steady-state of the system. The box represents the system and the relationship between the input and output. Arrows connecting blocks represent information flow. The block diagram notation is used to represent the control strategy implemented in a process.

Figure A.1   Basic block diagram.

Low-level control strategies (including measurement sensors and actuators) are typically designed and communicated using a process instrumentation diagram. This shows the information flow using dotted lines (for electronic signal lines) or lines such as —⫽—— , which represent pneumatic (air) transmission lines. Using PID, one can show the connectivity between sensors and actuators.

Figures A.2 and A.3 illustrate the difference between a control instrumentation diagram and a block diagram. The instrumentation diagram shows the actual control hardware used, such as sensors, control valves, and controllers. Even though the control block may be a piece of software residing in the control computer, it is shown explicitly as a device to clearly indicate the decision-making process. Figure A.3 is the standard structure of a feedback control loop. Note the terminology used. The terminology is defined in Table A.1.

In Figure A.3 the controller acts on the error between the *setpoint* and the measured value. If the controller sees a difference, then the controller takes action. Changing the controller output changes the manipulated variable (in this case, the coolant flow) and this in turn causes a change in the tank temperature. Finally, the controller sees the effect of its action through the information provided by the temperature transmitter. The effect of the control action is finally *fed back* to the controller through the transmitter. A feedback loop is formed as a result. The feedback of information allows the controller to correct itself, if necessary, to bring the measured value close to the *setpoint*. The majority of the control loops found in industry use this structure. Later on we will introduce controllers that differ from the fundamental structure.

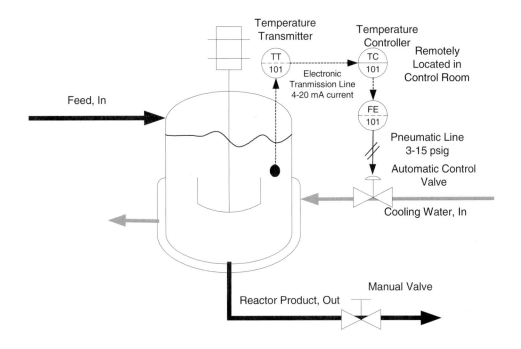

Figure A.2 Process instrumentation diagram of an exothermic reactor temperature control system.

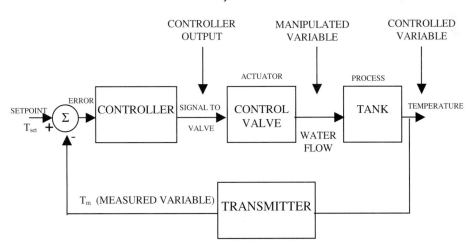

Figure A.3 Block diagram of temperature control system.

Table A.1   Terminology Used.

Term	Abbreviation	Description
Setpoint	SP	Desired value of controlled output
Measured Variable	MV or MS	Measured value of process variable being controlled
Control Valve	CV	The actuator used to control flow of a fluid
Controller Output	CO or OV	Signal going out from the controller to the actuator
Transmitter	T	Device used to send signal to the control room
Disturbance	D	External variables that could affect the controlled output
Error	E	The difference between the setpoint and the measurement
Status	MAN (OFF)	Indicates that controller output is manually set by the operator
	AUTO (ON)	Controller is active and the output to the actuator is determined by the control algorithm
	REMOTE	Controller setpoint is adjusted by another device
Action	DIR	Direct acting, i.e., controller output increases when measurement increases
	REV	Reverse acting, i.e., controller output decreases when measurement increases

Other common instrumentation symbols include

P: Pressure
T: Temperature
F: Flow
C: Concentration
L: Level
D: Density
A: Analyzer
I: Indicator
H: Hand-operated
V: Valve
C: Controller

A.2 LAPLACE TRANSFORM AND TRANSFER FUNCTIONS

The Laplace transform is often used to represent the dynamic relationship contained in a block diagram. It is defined as follows: Given a function of time $f(t)$, $t \geq 0$, we can define a new function $F(s)$ as

$$F(s) = \int_0^\infty f(t)e^{-st}dt. \tag{A.1}$$

This represents a unique one-to-one mapping. $F(s)$ is called the Laplace transform of $f(t)$, and $f(t)$ is the inverse Laplace transform of $F(s)$. This is represented using the notation

$$L[f(t)] = F(s), \tag{A.2a}$$

$$L^{-1}[F(s)] = f(t). \tag{A.2b}$$

s is called the Laplace variable. Table A.2 gives the Laplace transform of some common functions. Some useful properties of the Laplace transform are given in Table A.3.

Table A.2. Laplace Transforms of Common Functions.

Name	Function, $f(t)$	Laplace transform, $F(s)$
Unit step function	$u(t)=0 \quad t<0$ $u(t)=1 \quad t\geq0$	$\dfrac{1}{s}$
Unit impulse function (Dirac Delta function)	$\delta(t) = \infty \quad t = 0$ $\delta(t)=0 \quad t\neq0$	1
Ramp function	$r(t)=0 \quad t<0$ $r(t)=kt \quad t\geq0$	$\dfrac{k}{s^2}$
Sine function	$x(t)=\sin \omega t$	$\dfrac{\omega}{s^2+\omega^2}$

Table A.3. Properties of the Laplace Transform.

Property	Description
Linearity Property	$L[f_1(t) + f_2(t)] = L[f_1(t)] + L[f_2(t)]$ $L[k\,f_1(t)] = k\,L[f_1(t)]$ where k = constant
Time delay	$L[f(t-\theta)] = e^{-\theta s}\,L[f(t)]$ $= e^{-\theta s}\,F(s)$
Differentiation	$L[\dfrac{d\,f(t)}{dt}] = s\,F(s) - f(0)$
Integration	$L\{\int_o^t f(t)dt\} = \dfrac{1}{s}\,F(s)$
Final value theorem	$\lim\limits_{t\to\infty} f(t) = \lim\limits_{s\to 0} s\,F(s)$ provided LHS limit exists.

Consider a system modeled by a first-order differential equation:

$$a_1 \frac{dx(t)}{dt} + a_2 x(t) = b_1 u(t), x(0) = 0, \qquad\qquad (A.3)$$

where $u(t)$ is the input variable and $x(t)$ is the output variable. Notice that we have taken the initial condition to be zero, since both x and u are taken as deviations from the nominal initial steady-state values. Taking Laplace transform of both sides (we will informally use $x(s)$ to denote Laplace transform of $x(t)$),

$$a_1[sx(s) - x(0)] + a_2 x(s) = b_1 u(s).$$

Since $x(0) = 0$,

$$
\begin{aligned}
x(s) &= (\frac{b_1}{a_1 s + a_2})u(s) \\
&= \frac{K}{\tau s + 1}\cdot u(s),
\end{aligned}
\qquad\qquad (A.4)
$$

where $K = b_1/a_2$ is called the *gain* of the system and $\tau = a_1/a_2$ is called the *time constant* of the system. The function

$$G(s) = \frac{K}{\tau s + 1} \qquad\qquad (A.5)$$

is called the transfer function relating $x(s)$ and $u(s)$. The transfer function captures the dynamic relationship between the input and output variable in a compact, convenient, and algebraic (as opposed to the differential equation) form. $G(s)$ is an example of a *first-order transfer function,* since the denominator is first-order in s. Other commonly encountered transfer functions are given below. In general, $G(s)$ can be expressed as a ratio of polynomials:

$$G(s) = \frac{N(s)}{D(s)}, \tag{A.6a}$$

$$x(s) = G(s)u(s). \tag{A.6b}$$

If the transfer function given by Eq. (A.6a) is stable, then the gain of the transfer function is defined as the steady-state change in $x(t)$ for a unit step change in $u(t)$, where $x(t)$ and $u(t)$ are the inverse Laplace transforms of $x(s)$ and $u(s)$. The final value theorem applied to Eq. (A.6b) shows that, by the foregoing definition, the gain of $G(s)$ is $G(0)$.

A.2.1 Common Transfer Functions

1. Second-Order Transfer Functions
The system modeled using the differential equation

$$\tau^2 \frac{d^2 x}{dt^2} + 2\tau\xi \frac{dx}{dt} + x = Ku, x(0) = 0, x'(0) = 0 \tag{A.7a}$$

can be represented using the transfer function

$$x(s) = \frac{\kappa}{\tau^2 s^2 + 2\tau\xi s + 1} \cdot u(s), \tag{A.7b}$$

where $\tau = $ *time constant* and $\xi = $ *damping coefficient.* If we can factor the denominator

$$\tau^2 s^2 + 2\tau\xi s + 1 = (\tau_1 s + 1)(\tau_2 s + 1),$$

where τ_1 and τ_2 are both real (this can be done if $\xi > 1$), then we may write

$$x(s) = \frac{K}{(\tau_1 s + 1)} \cdot (\frac{1}{\tau_2 s + 1}) u(s). \tag{A.8}$$

That is, the second-order system can be viewed as two first-order systems in series. Alternatively, we can think of the second-order differential equation being written as two equivalent, coupled, first-order differential equations.

2. Time Delay or Dead Time: $e^{-\theta s}$

This occurs usually because of the finite time involved in fluid transportation through pipes. The output is the same as the input delayed by θ units of time. Often this is approximated using one of the *Padé approximations*, such as

$$e^{-\theta s} \approx \frac{1 - \theta/2s}{1 + \theta/2s}.\tag{A.9}$$

3. Lead-Lag

$$x(s) = K\frac{(\tau_1 s + 1)}{(\tau_2 s + 1)}u(s)\tag{A.10}$$

τ_1 is the lead time and τ_2 is the lag time.

4. First-Order Plus Dead Time (FOPDT)

$$x(s) = \frac{Ke^{-\theta s}u(s)}{\tau s + 1}.\tag{A.11}$$

This is perhaps the most frequently encountered transfer function in process control. Many process systems are approximated using this transfer function. Note the abbreviation FOPDT, used frequently in this text.

A.3 P, PI, AND PID CONTROLLER TRANSFER FUNCTIONS

In this section we will discuss the transfer functions associated with commonly used controllers. These transfer functions have some parameters associated with them, which must be determined to yield good control system performance. The latter process of setting these controller constants is called controller tuning.

1. Proportional Control (P Control)

The control law is represented by

$$m(t) = K_c e(t) + \bar{m}.\tag{A.12}$$

In the Laplace domain (we omit the bias term since we are dealing with deviation variables),

$$m(s) = K_c \cdot e(s),\tag{A.13}$$

where m is the controller output and $e(t)$ is the error in the control defined by

$$e(t) = \text{setpoint} - \text{measurement}. \tag{A.14}$$

K_c is the *controller gain*. Note the bias term $\overline{m}$ (the steady-state controller output). This is set by the operator or initialized to the current output value when the controller is turned on. One difficulty with this controller is the choice of $\overline{m}$. Since this is not known a priori, it is usually not possible to drive $e(t)$ to zero to eliminate steady-state offset in the controlled variable.

2. Proportional-Integral Control (PI Control)
This control law is given by

$$m(t) = K_c[e(t) + \frac{1}{\tau_I} \int e(t)\, dt]. \tag{A.15}$$

The *integral* or *reset* action allows the controller to eliminate steady-state offset. There is no need for a bias term, as the controller output will change as needed to drive $e(t)$ to zero. The controller transfer function is given by

$$m(s) = K_c[1 + \frac{1}{\tau_I s}]e(s). \tag{A.16}$$

τ_I is called the *integral time constant* and determines the weight given to the integral action relative to the proportional part.

3. Proportional-Integral-Derivative (PID) Control
The *ideal* form of the three-mode PID controller is

$$m(t) = K_c[e(t) + \frac{1}{\tau_I} \int e\, dt + \tau_D \frac{de}{dt}]. \tag{A.17}$$

This adds another parameter τ_D, called the *derivative time constant*. The transfer function of the ideal PID controller is given by

$$m(s) = K_c(1 + \frac{1}{\tau_I s} + \tau_D s)e(s). \tag{A.18a}$$

However, because of the difficulty with differentiating noisy signals, actual industrial controller implementations vary from the ideal form. This is done by adding a noise filter (typically a first-order lag transfer function) before derivative action is applied. A typical implementation of a *real* PID controller would be as follows:

$$m(s) = K_c(1 + \frac{1}{\tau_I s} + \frac{\tau_D s}{\alpha\tau_D s + 1})e(s), \tag{A.18b}$$

where α is typically chosen as a number between .05 and 0.2, depending on the manufacturer of the controller. In some digital control systems the user has control over the value of α. Larger values of α lead to stronger filtering of the derivative action of the controller.

Another form of PID controller still found in older industrial applications is the industrial form of a PID controller, so called because prior to digital control computers it was the most common form of PID controller. The industrial form of a PID controller is given by

$$m(s) = K_c (1 + \frac{1}{\tau_I^* s}) \frac{(\tau_D^* s + 1)}{(\alpha \tau_D^* s + 1)} e(s) \tag{A.18c}$$

The controller given by Eq. (A.18c) is not as general as that given by Eq. (A.18b) since the zeros of Eq. (A.18c) are necessarily real, while those of Eq. (A.18b) may be either real or complex. In the case where the zeros of Eq. (A.18b) are real, then the relationship between the parameters of Equations (A.18b) and (A.18c) are

$$\tau_I \tau_D = \tau_I^* \tau_D^*, \tag{A.18d}$$

$$\tau_I = \tau_I^* + \tau_D^*. \tag{A.18e}$$

Solving for τ_D in terms of τ_I^* and τ_D^* gives

$$\tau_D = \tau_I^* \tau_D^* / (\tau_I^* + \tau_D^*). \tag{A.18f}$$

From the equation series (A.18), changing either τ_I^* or τ_D^* changes both τ_I and τ_D. Therefore, the controller given by Eq. (A.18e) is often called an *interacting* form of the PID controller.

There are many other versions for the real PID controllers, but we will not discuss them here. All of the previously mentioned controllers focus on controlling a *single output* variable using a *single* manipulated *input* variable. Hence they are known as *SISO* control systems.

A.4 STABILITY OF SYSTEMS

The addition of a feedback controller to a process can alter the stability characteristics of the system. We define a system as stable if the output of the system is bounded (finite) for bounded inputs. Since stability is an important characteristic of control systems, we need some tools to analyze the stability of control systems and to evaluate the effect of controller parameters on stability.

Consider a system containing the transfer function

$$y(s) = \frac{K}{(s-p)}x(s). \qquad (A.19)$$

If the system is forced by an impulse input given by the Dirac Delta function

$$x(t) = \delta(t), x(s) = 1, \qquad (A.20)$$

then

$$y(s) = \frac{K}{s-p},$$

which, when inverted back into the time domain, yields,

$$y(t) = Ke^{pt}. \qquad (A.21)$$

Let us consider the behavior of $y(t)$ as a function of p, the *pole* of the system. In general, p can be a complex number given by (see Fig. A.4)

$$p = a + ib,$$

where a = Real part, b = Imaginary part, and $i = \sqrt{-1}$

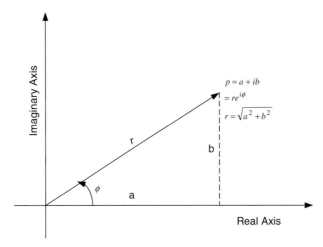

Figure A.4 Location of pole in complex plane, $p = a + ib = re^{i\phi}$.

Complex poles always occur in conjugate pairs in a real system. The response $y(t)$ is given by

$$\begin{aligned} y &= Ke^{pt}, \\ &= Ke^{at}(\cos bt + i \sin bt). \end{aligned} \qquad (A.22)$$

The imaginary part of the output will be cancelled by the conjugate pole. Let us examine the asymptotic behavior of $y(t)$ as $t \to \infty$.

1. If b = 0, response is not oscillatory,
   If b ≠ 0, response is oscillatory
2. If a > 0 $y(t)$ grows unbounded as $t \to \infty$,
   If a < 0, $y(t)$ decays to zero as $t \to \infty$

Systems whose outputs grow unbounded as $t \to \infty$ for any bounded input-forcing function are called *unstable. A necessary condition for stability is that all the poles of the system have negative real parts.* A typical transfer function can be expressed as a ratio of polynomials,

$$G(s) = \frac{N(s)}{D(s)}, \tag{A.23}$$

where $N(s)$ = numerator polynomial in s, $D(s)$ = denominator polynomial in s. Let

$$D(s) = (s - p_1)(s - p_2)...(s - p_n), \tag{A.24}$$

where $p_1, p_2, ..., p_n$ are the roots of the polynomial $D(s)$. These are called the *poles* of the system. We can expand $G(s)$ in a partial fraction expansion as

$$
\begin{aligned}
G(s) &= \frac{N(s)}{D(s)}, \\
&= \frac{N(s)}{(s - p_1)(s - p_2),........(s - p_n)}, \\
&= \frac{K_1}{s - p_1} + \frac{K_2}{s - p_2} + ...+ \frac{K_n}{s - p_n},
\end{aligned}
\tag{A.25}
$$

provided p_1, p_2, and p_n are all unique. Otherwise, the expansion must be modified slightly. If this process is perturbed by an impulse input, then

$$
\begin{aligned}
y(s) &= G(s) \cdot x(s), \\
&= G(s), \\
&= \frac{K_1}{s - p_1} + ... + \frac{K_n}{s - p_n}.
\end{aligned}
$$

Inverting,

$$y(t) = K_1 e^{p_1 t} + ... + K_n e^{p_n t}. \tag{A.26}$$

Thus the behavior of $y(t)$ is characterized by its poles. If any pole has a real part that is positive, the output will grow unbounded. This leads to the following conclusion:

For a transfer function to be stable, all its poles must lie to the left of the imaginary axis in the complex plane; that is, in the LHP.

Figure A.5 shows the behavior of the solution at different pole locations in the complex plane. Note that all poles on the RHP lead to exponential growth. If the pole has an imaginary component, we get oscillations. A pole on the LHP leads to exponential decay (with oscillations if the pole has an imaginary component, we get oscillatory decay). A pole on the imaginary axis leads to sustained oscillatory response (except at $s = 0$).

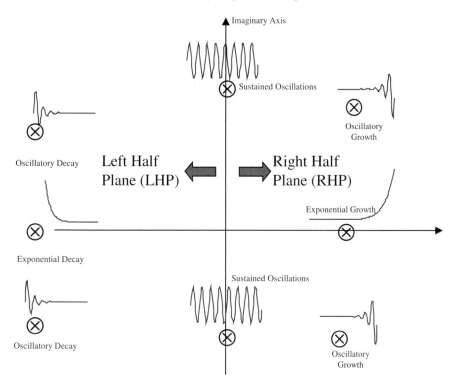

Figure A.5 Response characteristic at different pole locations.

A.5 STABILITY OF CLOSED-LOOP SYSTEMS

In order to analyze the stability of closed-loop systems, we need to develop the transfer functions associated with the closed-loop system. Block diagrams offer a convenient way to do this. The block diagram of a feedback control, shown in Figure A.6 shows four blocks. Let us define the following transfer functions:

Process:
$$G_p(s) = \frac{y(s)}{q(s)}. \tag{A.27}$$

Transmitter:
$$G_T(s) = \frac{y_m(s)}{y(s)}. \tag{A.28}$$

Actuator:
$$G_v(s) = \frac{q(s)}{m(s)}. \tag{A.29}$$

Controller:
$$G_c(s) = \frac{m(s)}{e(s)}. \tag{A.30}$$

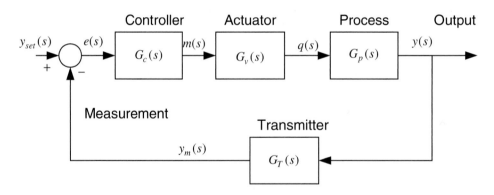

Figure A.6   Block diagram of feedback control system.

If we write down the relationships implied by the diagram, we get

$$
\begin{aligned}
y &= G_p \cdot q \\
  &= G_p \cdot G_v \cdot m \\
  &= G_p G_v \cdot G_c \cdot e \\
  &= G_p G_v G_c (y_{set} - y_m) \\
  &= G_p G_v G_c (y_{set} - G_T \cdot y).
\end{aligned}
$$

Solving for y,

$$y = \frac{G_p G_v G_c}{1 + G_p G_v G_c G_T} y_{set},$$ (A.31)

$$= G_{CL}(s) \cdot y_{set},$$ (A.32)

where $G_{CL}(s)$ is called the *closed-loop transfer function*. Note the following:

1. The stability of the closed-loop is determined by the roots of the polynomial

$$1 + G_p G_c G_v G_T = 0.$$ (A.33)

   This is called the *characteristic equation* of the closed-loop system.
2. The controller transfer function will alter the poles of the closed-loop system. The *roots* of this equation (also referred to as *zeros* of the polynomial) are the *poles* of the closed-loop system.
3. A rule that is useful to remember is that for negative feedback, the closed-loop transfer function can be written as

$$\frac{\left[\begin{array}{l} \text{product of transfer functions in blocks on the} \\ \text{forward path from the setpoint to the output} \end{array} \right]}{1 + \text{product of transfer functions in the loop}}$$

If there is a load or disturbance variable that can also affect the output, as shown in Figure A.7 the transfer function will be

$$y = \frac{G_d(s)}{1 + G_p G_v G_c G_T} d(s).$$ (A.34)

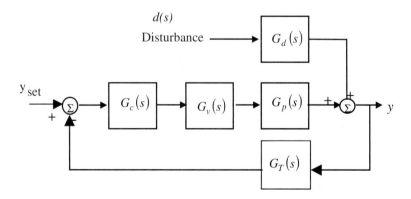

Figure A.7 Block diagram of control loop with disturbance input.

Example A.1 First-Order with PI Control

Let

$$G_p(s) \; = \; \frac{K_p}{(\tau s + 1)},$$

$$G_c(s) \; = \; K_c(1 + \frac{1}{\tau_I s}),$$

$$G_v(s) \; = \; G_T(s) = 1.$$

This leads to the closed loop transfer function

$$G_{CL}(s) = \frac{y}{y_{set}} = \frac{K_c K_p (\tau_1 s + 1)}{(\tau_1 \tau s^2 + \tau_1(1 + K_c K_p)s + K_c K_p)}, \qquad (A.35)$$

with a steady-state gain $G_{CL}(0) = 1$.   Hence no steady-state error (offset) will be present for a step change in the setpoint.

♦

Example A.2 Third-Order Process with P Control

For a process with transfer functions

$$G_p(s) = \frac{1}{(s+1)(s+2)(s+3)}, G_c(s) = K_c, G_v(s) = G_T(s) = 1, \qquad (A.36)$$

the closed-loop response is given by the equation

$$\frac{y}{y_{set}} = \frac{K_c}{(s+1)(s+2)(s+3) + K_c}. \qquad (A.37)$$

The characteristic equation is given by

$$(s+1)(s+2)(s+3) + K_c = 0. \qquad (A.38)$$

It is interesting to study the effect of changing K_c on the poles of the closed-loop system. Table A.4 shows how the zeros of the characteristic equation (system poles) change with the controller gain K_c. These poles are plotted in Figure A.8. As K_c changes, these poles trace a path on the complex plane to generate the *root-locus*.

Table A.4 Table of Roots of the Characteristic Equation for Various Values of K_c.

K_c	root1	root2	root3
0.1	−3.0467	−1.8990	−1.0544
0.2	−3.0880	−1.7909	−1.1211
0.39	−3.1564	−1.4218 − 0.0542i	−1.4218 + 0.0542i
0.6	−3.2212	−1.3894 − 0.3442i	−1.3894 + 0.3442i
1.0	−3.3247	−1.3376 − 0.5623i	−1.3376 + 0.5623i
10.0	−4.3089	−0.8455 − 1.7316i	−0.8455 + 1.7316i
20.0	−4.8371	−0.5814 − 2.2443i	−0.5814 + 2.2443i
30.0	−5.2145	−0.3928 − 2.5980i	−0.3928 + 2.5980i
60.0	−6.0000	0 − 3.3166i	0 + 3.3166i
100.0	−6.7134	0.3567 − 3.9575i	0.3567 + 3.9575i

Since the dynamic behavior of the system is characterized by its poles, we can conclude the following:

1. The system response is nonoscillatory for values of $K_c < 0.39$, since the roots are all real.

2. As K_c approaches 60, the poles move closer to the RHP, and at $K_c = 60$, the system has two poles on the imaginary axis leading to sustained oscillatory response.

3. The system has two poles on the RHP for $K_c > 60$ and is hence unstable for these controller gains.

$K_c = 60$ is called the *stability limit* of the system, also called the *ultimate gain*, K_u. The frequency of oscillation at this gain is called the *ultimate frequency*. From the calculations shown in Table A.4, the ultimate frequency is given by the pole on the imaginary axis $(\omega_u = 3.32/\text{sec})$. The ultimate period of oscillation is $P_u = 2\pi/\omega_u$.

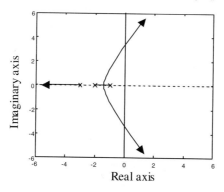

Figure A.8 Root locus for Example A.4.

A.6 CONTROLLER TUNING

PID controllers are used in the majority of SISO feedback controllers because of their effectiveness and simplicity. The transfer function of a typical PID controller is given by

$$m(s) = K_c(1 + \frac{1}{\tau_I s} + \frac{1}{(\alpha s + 1)}\tau_D s)e(s). \tag{A.39}$$

Controller tuning refers to determining the appropriate values of K_c, τ_I, τ_D and α in the controller transfer function. While there is no single "right" way to choose PID parameters, the reader should be aware that the literature is full of wrong ways. Until relatively recently, all controller tuning algorithms were based on the local behavior of the process at the time of tuning. However, local behavior of a process usually changes as operating conditions change due to disturbances, setpoint changes, and process aging. Since most real processes are nonlinear in their global behavior, their local behavior changes as the operating point changes. Thus, a well-tuned controller at one operating point might behave very badly at another. The concept of process uncertainty is one way of attempting to mathematically describe the variations of local process behavior over an operating region in a way that is useful for controller tuning. While it is usually not easy to obtain accurate uncertainty descriptions, it is often possible to obtain reasonable estimates that allow much better tuning than would otherwise be possible. Chapters 7 and 8 present relatively simple methods for including process uncertainty in the tuning of one-degree and two-degree of freedom IMC controllers. The tuned IMC controllers can then be converted into PI or PID controllers, using the techniques of Chapter 6.

In the next section we review the traditional controller tuning methods that are based on local process behavior. Our reasons for doing so are that these methods are still used, even though better methods are available (see Chapters 6, 7, and 8), and it is important for the reader to be familiar with these methods in order to be able to communicate with plant engineers and possibly convince them to convert to better methods.

A.6.1 Tuning Correlations

A number of authors have published empirical correlations for tuning PID controllers based on the selected performance criteria and assumed process model structure. The FOPDT model is used most often. Generally, correlations are provided in terms of the process gain, the time constant, and time delay in the FOPDT model. Separate correlations are provided for servo control and regulatory control. Different criteria are used to evaluate the performance, including

  1.   Integral Squared Error (ISE)

$$ISE = \int_0^\infty e^2 dt \tag{A.40}$$

This criterion is widely used in optimal control theory. It is also the criterion of choice in many industrial implementations of model-predictive control (MPC).

2. Integral Absolute Error (IAE)

$$IAE = \int_{o}^{\infty} |e| \, dt \tag{A.41}$$

3. Integral Time Average Error (ITAE)

$$ITAE = \int_{o}^{\infty} |e| \, t dt \tag{A.42}$$

A few of the well-known correlations follow.

1. Reaction Curve Method (Ziegler-Nichols, 1942)

For a system modeled using an FOPDT transfer function,

$$G(s) = \frac{Ke^{-\theta s}}{\tau s + 1} \tag{A.43}$$

the Ziegler-Nichols tuning constants are listed in Table A.5.

Table A.5   Ziegler-Nichols Tuning Correlation.

Controller Type	Gain, K_c	Integral Time, τ_I	Derivative Time, τ_D
P	$\dfrac{1}{K}\left(\dfrac{\tau}{\theta}\right)$		
PI	$\dfrac{0.9}{K}\left(\dfrac{\tau}{\theta}\right)$	3.33θ	
PID	$\dfrac{1.2}{K}\left(\dfrac{\tau}{\theta}\right)$	2.00θ	0.5θ

2. C-C Tuning (Cohen-Coon, 1953)

These tuning rules are listed in Table A.6.

Table A.6   Cohen-Coon Tuning Correlation.

Con-troller Type	Gain, K_c	Integral Time, τ_I	Derivative Time, τ_D
P	$\dfrac{1}{K}\left(\dfrac{\tau}{\theta}\right)\left(1 + \dfrac{1}{3}\dfrac{\theta}{\tau}\right)$		
PI	$\dfrac{1}{K}\left(\dfrac{\tau}{\theta}\right)\left(0.9 + \dfrac{1}{12}\dfrac{\theta}{\tau}\right)$	$\theta\,\dfrac{30 + 3\,\theta/\tau}{9 + 20\,\theta/\tau}$	
PID	$\dfrac{1}{K}\left(\dfrac{\tau}{\theta}\right)\left(\dfrac{4}{3} + \dfrac{1}{4}\dfrac{\theta}{\tau}\right)$	$\theta\left(\dfrac{32 + 6\,\theta/\tau}{13 + 8\,\theta/\tau}\right)$	$\theta\left(\dfrac{4}{11 + 2\,\theta/\tau}\right)$

Typically the tuning methods presented in this appendix attempt to optimize the closed-loop response to a unit step change in setpoint (see Figure A.9). This response is very similar to the unit step response of a second-order underdamped system modeled by the transfer function

$$G(s) = \frac{1}{\tau^2 s^2 + 2\tau\xi s + 1}. \tag{A.44}$$

For such a system,

$$
\begin{aligned}
\text{Overshoot} &= e^{-\pi\xi/\sqrt{1-\xi^2}} \\
\text{Decay Ratio} &= e^{-2\pi\xi/\sqrt{1-\xi^2}} \\
\text{Period} &= 2\pi\tau/\sqrt{1-\xi^2} \\
\text{Rise Time} &\approx \text{Period}/4
\end{aligned}
\tag{A.45}
$$

Hence τ tells us how fast the process responds and ξ tells us how fast the oscillations decay and how large the overshoot will be.

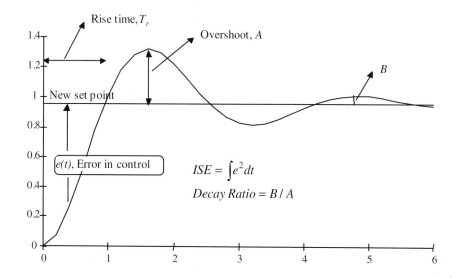

Figure A.9 Response of a control system to a step change in setpoint (servo response).

A.6.2 Tuning of Integrating Processes

Integrating processes frequently arise in level control systems. Since an integrating process is an unstable process, it is best controlled by a two-degree of freedom controller, as discussed in Chapters 4, 6, and 8. However, here we discuss tuning single-degree of freedom PID controllers for such processes using the older literature methods, as we did for stable processes in Section A.6.1 above.

Consider an integrating process with a transfer function given by

$$G(s) = \frac{K}{s},$$ (A.46)

(where K is the slope of the output response for a unit step input).

If we apply a PI controller to this process, we get the block diagram shown in Figure A.10.

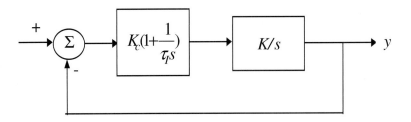

Figure A.10 PI control of an integrating process.

The closed-loop transfer function is given by

$$\frac{y}{y_{set}} = \frac{\tau_I s + 1}{(\tau_I s^2 / KK_c) + \tau_I s + 1}. \tag{A.47}$$

Hence, the closed-loop system will respond as a second-order process. Theoretically, it is stable for all values of $K_c, \tau_1 > 0$. The damping coefficient ξ is given by

$$\xi = 1/2\sqrt{KK_c \tau_I}. \tag{A.48}$$

For values of $\xi < 1$ (i.e., small values of K_c and τ_I) it will be oscillating. Typically, we want to keep ξ close to 0.5 (to get about 20% overshoot). This will yield

$$KK_c \tau_I = 1. \tag{A.49}$$

A reasonable choice for K_c is 5 %/% where the percentages are taken with respect to the span of the variables

$$K_c = \frac{5\% \text{ change in controller output}}{1\% \text{ change in measurement}}$$

$$= 5 \times \frac{\text{controller output span}}{\text{measurement span}} \tag{A.50}$$

τ_I can then be computed from the equation

$$\tau_I = 1/(KK_c). \tag{A.51}$$

A.7 REGULATORY ISSUES INTRODUCED BY CONSTRAINTS

In this section we examine regulatory issues that can be derived from the process model. *The focus is on steady-state suppression of disturbances for SISO systems.* We examine the dynamic suppression issues in Appendix B. For convenience in the analysis we consider first the representation of models using dimensionless variables.

A.7.1 Scaling and Nondimensionalization of Process Variables

Control engineers have long practiced the scaling of variables using the span of the instrument used to measure that variable (the maximum and minimum measurable values of that variable). For a measurement y, the scaling is defined as

$$y = \frac{y_{meas} - \overline{y}_{meas}}{y_{max} - y_{min}},$$
(A.52)

where $y_{max} - y_{min}$ is the span of the measuring instrument (a field adjustable quantity). The smaller the span, the greater the accuracy of the instrument. During startup and shutdown phases, a greater range of the measurement is expected and the span may be kept large. $\overline{y}_{meas}$ is the normal steady-state value of the variable. Ideally, this should be set to the middle of the span to yield a range of

$$-0.5 \le y \le 0.5.$$

A similar definition is used to scale manipulated input variables. With control valves, the nominal steady-state valve is kept closer to 1 (say in the 65% to 75% open range) to reduce pressure losses.

A.7.2 Steady-State Suppression of Disturbances for SISO Systems

Consider SISO system, as shown in Figure A.11. The transfer function model is

$$y(s) = p(s)u(s) + d(s).$$
(A.53)

The control objective is to maintain y close to its nominal steady-state value of 0 in presence of disturbances. First we consider the steady-state case. The problem reduces to

$$y(s) = p(o)u(o) + d(o).$$
(A.54)

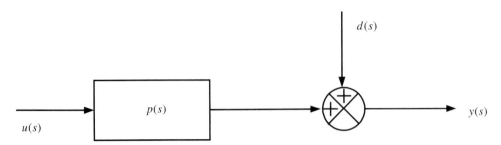

Figure A.11   Process with disturbance

The ability to regulate the process can be determined by the graph shown in Figure A.12.

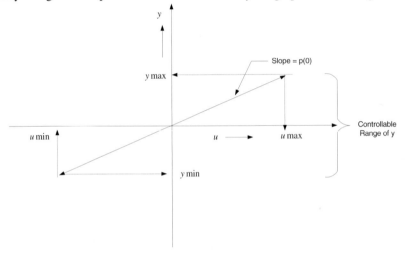

Figure A.12   The effect of input constraints on output.

The figure shows that for a low gain system in which $p(0)$ is small, the "reachable space" of y is limited or small in size. If the effect of the disturbance exceeds this limit, then y cannot be regulated; that is, we cannot reach the desired setpoint in the steady-state. A large value of $p(0)$ implies that the process is very sensitive to small changes in the input u. Usually, there is an uncertainty or error associated with implementing a control action. In this case the process will amplify that uncertainty and this could lead to difficulty in regulation. Ideally, therefore, a gain closer to 1 will usually be a compromise between sensitivity to input uncertainty and ability to regulate in presence of disturbances.

Example A.3 Steady-State Disturbance Suppression in a Distillation Column

Consider the column for separating a mixture of methanol/water. The nominal steady-state operating conditions are shown. Given the following steady-state observations, determine the steady-state regulation of the bottom product purity using the reboiler heat duty and the distillate flow rate when the feed can vary from 95 to 105 lb moles/hr.

Condition	Feed Rate F	Distillation Rate D	Reboiler Duty MM Btu/hr Q	Top Purity X_T Mole Fraction	Bottom Purity X_B Mole Fraction
Span or Range of Variable	0-200 lb moles/hr	0 -100 lb mole/hr	0-4 MMBtu/hr	0-0.02	0-0.2
Nominal Steady-state	100	45	2.2	.0057	.0955
Perturb D	100	46	2.2	.0063	.0794
Perturb Q	100	45	2.3	.0047	.0948
Perturb F	105	45	2.2	.0058	.1292

Solution

We compute the steady-state gains of the process in non-dimensional, scaled variables:

$$\begin{bmatrix} x_T \\ x_B \end{bmatrix} = \begin{bmatrix} k_{11} & k_{12} \\ k_{21} & k_{22} \end{bmatrix} \begin{bmatrix} d \\ q \end{bmatrix} + \begin{bmatrix} k_{f1} \\ k_{f2} \end{bmatrix} f,$$

$x_T = (X_T - \overline{X}_T)/(0.02)$, $x_B = (X_B - \overline{X}_B)/0.2$, $q = (Q - \overline{Q})/4$, $d = (D - \overline{D})/100$, $f = \overline{F}/200$.

From the data given

$$k_{12} = \frac{(.0047 - .0057)/(0.02)}{(2.3 - 2.2)/4.0} = -2, \ k_{11} = 3.0, \ k_{21} = 8.05, \ k_{22} = -.14,$$

yielding

$$K = \begin{bmatrix} 3.0 & -2 \\ 8.05 & -.14 \end{bmatrix}, k_f = \begin{bmatrix} 0.4 \\ 6.74 \end{bmatrix}.$$

The variations in x_B caused by the feed flow disturbance can be computed as $\pm 6.74 * 5/200 = \pm 0.1685$. Using q, the maximum variations achievable in x_B is given by $\pm(0.14) * 50/100 = \pm .07$. *Hence we conclude that it is not possible to regulate x_B using q because of the small gain.*

However, the variations in x_B that can be achieved using d are $\pm 8.05 * (4-23)/4 = \pm 3.4$. *Hence x_B can be regulated easily, using d.* Note that this result is counterintuitive, since our first instinct would be to choose to regulate x_B using q (due to the physical proximity of the two variables).

◆

Problems

A.1 Control system design and block diagram construction. Consider the control of pH in a wastewater treatment facility. Contaminated water from the plant flows into a large pH stabilization tank where it is mixed with caustic solution (NaOH solution) to neutralize the water and bring its pH to 7. The caustic flows into the stabilization tank from another storage tank. The treated water from the stabilizer tank then flows into a lake about 1000 yards away.

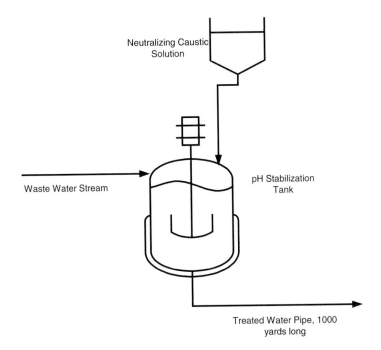

a. Design a feedback control system to control the pH of the water being discharged to the lake. Show the process instrumentation diagram using standard symbols. Identify and label all sensors, transmitters, and control valves.

b. Draw a block diagram of the pH control system, identifying all variables used.

c. Discuss briefly the pros and cons of the following options from the point of view of the performance of the control system. (i) Place the pH sensor in the pipe through which treated water is leaving the stabilization tank. Sensor is located as close to the tank as possible.(ii) Place the pH sensor at the point where the wastewater enters the lake. (iii) Place the pH sensor inside the stabilization tank. (iv) Place the pH sensor in the pipe that brings the water into the stabilization tank.

A.2 Feedback versus feedforward control. A process fluid is heated in a furnace, as shown in the figure. The available instrumentation on the system is also shown. It is desired to control the temperature of the process fluid at the exit at a given setpoint by manipulating the fuel input flow into the furnace.

d. Show the instrumentation diagram of a feedback control strategy to achieve the objective stated. Label and clearly identify the instrumentation used.

e. Show the instrumentation diagram of a feedforward controller that will be used in conjunction with the feedback controller of Part a. This feedforward controller should compensate for variations in the input disturbance, namely, the process fluid flow.

f. Draw a block diagram of the feedforward/feedback control system. Label and identify all blocks and arrows used in the block diagram.

g. It is important to maintain a constant ratio of air to fuel flow to the furnace. Show the instrumentation necessary (transmitters, controllers, etc.) to maintain this ratio constant.

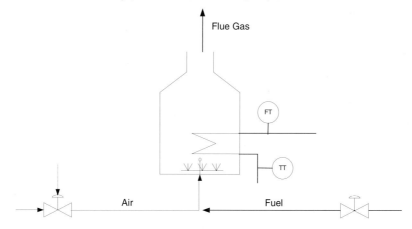

A.3 Step response of a first-order system. You have been assigned the task of measuring the time constant of a temperature sensor being used in a control loop. The following experiment was performed. The sensor was initially at room temperature of 25 °C. It was suddenly inserted into hot water in a cup. You read the temperature as 40 °C after 1 minute. The reading reached a steady-state value of 45°C after a few minutes. Assume the sensor can be modeled as a first-order system.

    a.   Estimate the time constant of the temperature sensor.

    b.   Write the transfer function for the temperature sensor. Use your result from Part a. Show that the gain is 1.

    c.   If the temperature can be read accurately to ± 1 °C on the sensor, what is the accuracy of your time constant estimate in minutes?

A.4 Step response of a first-order system. A bare thermocouple, when suddenly moved from a steady temperature of 80°F into a stream of hot air at a constant temperature of 100°F, reads 90°F after 1 sec. Assume the thermocouple responds as a first-order system.

    a.   Sketch the behavior of the temperature surrounding the thermocouple as a function of time.

    b.   Sketch the expected behavior of the measured temperature as a function of time (qualitatively).

    c.   What will be the final (steady-state) temperature measured by the thermocouple?

    d.   Compute the measured temperature at 2 sec.

A.5 Pulse response of a first-order system. The time-dependent behavior of the outlet concentration of a stirred tank to changes in inlet concentration is given by

$$\tau \frac{dy}{dt} + y = x,$$

where y = outlet tank concentration, x = inlet concentration, t = time in minutes.

Suppose at t = 0, y = 0. The inlet concentration is suddenly increased from 0 to 1 and maintained at 1 for the duration of 1 minute, at which time it is brought back to zero. This is called a unit pulse input.

    a.   Sketch $x(t)$ as a function of time.

    b.   Write an equation for $x(t)$ in terms of unit step functions.

    c.   What is the Laplace transform of $x(t)$?

    d.   What is the Laplace transform of $y(t)$?

    e.   Solve for $y(t)$ by inverting the result of Part d.

    f.   If $\tau = 1$, sketch the response $y(t)$.

A.6 Inverting Laplace transforms. Sketch the behavior of $f(t)$ if its Laplace transform is given by

$$F(s) = \frac{1}{s} + \frac{e^{-s} - e^{-2s}}{s^2} - \frac{e^{-3s}}{s}.$$

A.7 Using dynamic response characteristics in design. A holding tank of uniform cross section has been designed to smooth out the flow from a reactor to a purification column. The reactor has to be shut down periodically for cleaning. Given the following data, estimate the maximum rate of change of the flow into the column. Assume flow out of the holding tank is proportional to the height of liquid in the holding tank.

Data: The tank has a cross-sectional area of 10 ft^2 and a height of 10 ft. Normal (steady-state level) in the tank is 5 ft. The normal flow from the reactor is 5 ft^3/hr and the reactor is periodically shut down for 2 hours.

Hint: First write the model of the holding tank. Then derive a transfer function relating changes in inlet flow to the outlet flow. It is preferable to use deviation variables. This is a first-order transfer function. Shutting down the reactor reduces the inlet flow to 0 suddenly. Hence it is a step change. The outlet flow responds to this step change. Sketch its response. When is the slope of this curve a maximum?

A.8 Closed-loop stability. An open-loop unstable process can be represented by the following transfer function:

$$G(s) = \frac{K}{(4s-1)(\tau_v s+1)(\tau_T s+1)}.$$

where τ_v and τ_T are, respectively, the time constants of the control valve and the transmitter. Assuming proportional control with gain K_c is used, find the range of the loop gain KK_c for which the loop is stable if

    a.   The valve and transmitter time constants are negligible ($\tau_v = \tau_T = 0$).

    b.   The valve time constant is negligible, and $\tau_T = 1.0$ min.

A.9 Response of second-order systems. The temperature of a reactor responds to the feed variations as a second-order system. Let

$T(t)$ = deviation of reactor temperature from steady-state,

$F(t)$ = deviation of reactor feed flow from steady-state,

$$T(s)/F(s) = \frac{(15°\,F/gpm)}{4s^2 + 2s + 2}.$$

The reactor is operating normally at a feed rate of 70 gpm and at 300°F. Previous experience shows that if the reactor temperature exceeds 350°F, the catalyst will be deactivated. The feed is subject to sudden step changes. What is the maximum permissible step change in the feed rate, that will keep the reactor temperature below its specified maximum of 350°F?

A.10 Control valve design. Control valves play an important role in the performance of the control system. This problem considers a typical case where a control valve is installed in line with a process unit, in this case a heat exchanger. The working design equation for the control valve is

$$q = C_v f(l) \sqrt{\frac{\Delta P_v}{g_s}}$$

$$f(l) = R^{l-1}, R = 20$$

$$l = \text{valve position } (0-1)$$

$$C_v = \text{valve size}$$

$$\Delta P_v = \text{pressure drop across valve, psi}$$

$$g_s = \text{specific gravity}$$

$$q = \text{gal/min (gpm)}$$

This is called an equal percentage valve, due to the nature of the dependence of $f(l)$ (the area available for flow) on the valve position.

    a.   Design a control valve (i.e., compute C_v) for a fluid with $g_s = 0.8$. The normal design flow is 250 gpm. The normal pressure drop across the valve is 10 psi and the normal pressure drop across the line is 10 psi. The total pressure drop (pressure drop across valve + pressure drop across line) is always constant at 20 psi. The pressure drop across the line is variable and proportional to the square of the flow rate. The valve should be designed so that it is half open (l = 0.5) at the design flow.

    b.   What will be the flow through the valve when the valve is fully open? What is the pressure drop across the valve at this flow rate? (Hint: It is not 500 gpm.)

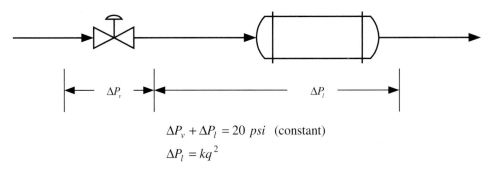

$$\Delta P_v + \Delta P_l = 20 \ psi \ \text{(constant)}$$

$$\Delta P_l = kq^2$$

A.11 Troubleshooting a control valve problem. You have been asked to troubleshoot a problem with a temperature control loop, which controls the outlet temperature of a process seen by adjusting the cooling water flow. The system is not performing well. One suspect is the valve. You collect the following data: (i) The normal flow of cooling water is 100gpm. (ii) The pressure drop across the valve at normal conditions is 5 psi. (iii) The pressure drop across the heat exchanger is 20 psi, the difference between pressures of cooling water entering and leaving the heat exchanger. (iv) You note that a linear valve has been used. For a linear valve, the flow through the valve is given by the equation

$$q = C_v f(l) \sqrt{\frac{\Delta P_v}{g_s}},$$

$$f(l) = l,$$

$$l = \text{valve position } (0-1).$$

The valve size C according to the specs, is 100. Valve is air-to-open (as the signal to the valve increases, the valve opens more).

    a.   What is the maximum achievable flow through the cooling water valve? Assume that the pressure drop across the heat exchanger is proportional to the square of the flow rate. The total pressure drop (the sum of the pressure drops across the valve and the heat exchanger) is a constant. (See figure in Problem A.10.)

    b.   Based on your calculations, could the valve be the cause of the problem with the control loop? Explain why or why not. If the valve were the cause of the problem, what remedial action would you recommend?

References

Cohen G. H., and G. A. Coon. 1953. "Theoretical Considerations of Retarded Control." Trans. ASME, 75, 827.

Luyben, W. L. 1990. Process Modeling, Simulation and Control for Chemical Engineers. 2nd Ed. McGraw-Hill, NY.

Marlin T. E. 1995. Process Control: Designing Processes and Control Systems for Dynamic Performance. McGraw-Hill, NY.

Ogunnaike, B. A., and W. H. Ray. 1994. Process Dynamics, Modeling and Control. Oxford University Press, NY.

Seborg, D. E., T. F. Edgar, and D.A. Mellichamp. 1989. Process Dynamics and Control. John Wiley & Sons, NY.

Smith, C. A., and A. B. Corripio. 1985. Principles and Practice of Automatic Process Control. John Wiley & Sons, NY.

Stephanopoulos, G. 1984. Chemical Process Control. Prentice Hall, NJ.

Ziegler J. G., and N. B. Nichols. 1942. "Optimum Settings for Automatic Controllers." Trans. ASME, 64, 759.

Review of Frequency Response Analysis

Objectives

The objective of this appendix is to review the basic concepts behind frequency response analysis and its use in control system design. Topics discussed include

- Bode and Nyquist plots.
- Nyquist stability theorem.
- Closed-loop response characteristics.
- Controller performance and design criteria.

B.1 INTRODUCTION

By *frequency response* we mean the response characteristics of the system when subject to sinusoidal inputs. The input frequency is varied, and the output characteristics are computed or represented as a function of the frequency. Frequency response analysis provides useful insights into stability and performance characteristics of the control system.

Figure B.1 shows the hypothetical experiment that is conducted.

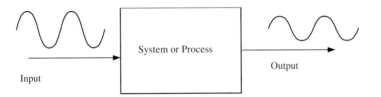

Figure B.1 How frequency response is defined.

The system is subject to an input of the form

$$x(t) = A \ \sin(\omega t) \quad t > 0. \tag{B.1}$$

After some initial transient period, the output settles down to a sine wave of the form

$$y(t) = B \sin(\omega t + \phi), t \quad \gg 0. \tag{B.2}$$

The amplitude and phase are changed by the system, but the frequency remains the same. This is shown in Figure B.2.

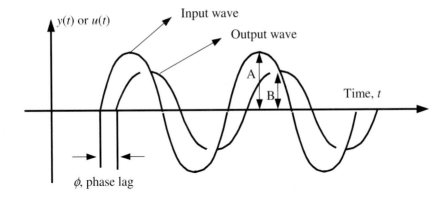

Figure B.2 Frequency response.

Note that the output wave lags behind the input. ϕ is defined as the *phase lag* (usually expressed in degrees or radians). The output amplitude is different from the input, and we can define a ratio:

$$\text{Amplitude Ratio}(AR) = B \,/\, A. \tag{B.3}$$

Now let us examine the effect of changing the frequency of the input. Consider the response of the level in the tank (see Figure B.3) to sinusoidal changes in the input flow.

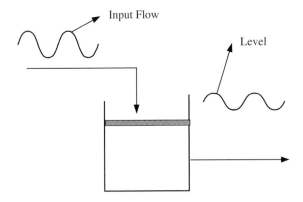

Figure B.3 Self-regulating level response to inlet flow variations.

Let us say that the tank is self-regulating: as the level changes, the outlet flow changes due to the change in hydraulic pressure until, at steady-state, the time average of the inlet flow matches the outlet flow.

If the inlet flow changes sinusoidally, the level will respond likewise. At low frequencies, the level will have plenty of time to keep pace with the inlet flow changes. At very high frequencies, the tank will not have time to respond to the flow variations and the amplitude of level changes will be small. The tank, in effect, will average out the inlet flow fluctuations. The peak in level will occur sometime after the inlet flow has peaked; that is, the changes in level will lag behind the changes in inlet flow.

There are a number of ways to represent the frequency response of a process. We will use two of these representations: *Bode plots* and *Nyquist plots.*

A Bode plot is a plot of the amplitude ratio (AR) and the phase lag as a function of the frequency of the input line wave (which is the same as the frequency of the output wave). Logarithmic scales are used for the frequency axis. The y-axis is often plotted using the units of decibels, which is 20 log (AR). Figure B.4 shows the Bode plot for a first-order process. In this example both the AR and decrease as the frequency increases. At low frequencies, the output is able to respond to the slow varying input disturbances with only a small attenuation (AR close to 1). However, at higher frequencies, the AR decreases rapidly, approaching an asymptote with a slope of –1 in the log-log graph shown for the first-order self-regulating process of the tank. Note that this system acts as a *low pass filter,* that removes the high-frequency inputs. A first-order system has decreasing phase angle, which approaches –90 asymptotically at higher frequencies. This implies that the output will lag behind the input (hence the name first-order lag).

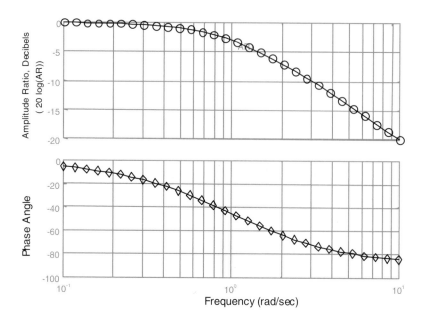

Figure B.4 Bode plot of a first-order process with $G(s)=1/(s+1)$.

The phase angle ϕ is usually negative. In this text we use the convention that phase lag is the negative of the phase angle. As ω increases, ϕ becomes more negative (i.e., the phase lag increases). This again represents the fact that at higher frequencies, the output will peak later than the input.

Typically, most processes exhibit a low AR at high frequencies. Hence, any low frequencies present in the input signal is passed through the process, whereas high-frequency components of the input signal are reduced significantly in amplitude as they pass through the process. We can view such a process as a low pass filter, which allows low frequencies to pass through without attenuation.

Any periodic signal can be viewed as a composite sum of various frequency components (obtained via a Fourier transform of the signal). Likewise, a system can be viewed as a filter that attenuates the input signal according to the frequency contained in the signal.

A Nyquist plot is another way to show the frequency response. Here $G(i\omega)$ is plotted in the complex plane. An example of a Nyquist plot is shown in Figure B.5. Note that there are two parts to the curve: one that shows the plot for ω varying from 0 to ∞ and another that shows the curve for ω varying from $-\infty$ to 0. (This is true only for transfer functions with the degree of the denominator polynomial higher than the degree of the numerator.)

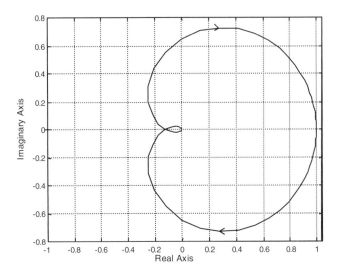

Figure B.5 Nyquist plot of $G(s)=1/(s+1)^3$.

B.2 FREQUENCY RESPONSE FROM TRANSFER FUNCTIONS

The frequency response can be derived from the transfer function using the following theorem: *Consider a process with transfer function $G(s)$. Then the frequency response is given by*

$$AR = |G(i\omega)|,$$
$$\phi = \angle G(i\omega).$$

(B.4)

The proof of the theorem can be found in many undergraduate texts (see Luyben, 1989, for example).

Example B.1 Frequency response of a first-order process

Consider

$$G(s) = \frac{K}{\tau s + 1},$$
$$G(i\omega) = \frac{K}{\tau(i\omega) + 1}.$$

Applying the theorem, we get

$$AR = \left| G\left(i\omega\right) \right| = \frac{K}{\left(1 + \omega^2\,\tau^2\right)^{1/2}},$$

$$\phi = \underline{\big|\, G\left(i\omega\right)} = \tan^{-1}\left(-\omega\tau\right).$$

This is plotted in Figure B.4 for $\tau = 1$ and $K = 1$. As ω increases, the AR gets smaller and smaller. At $\omega_b = \tau$, the $AR = 1/\sqrt{2}$ and $\phi = -\pi/4$ rad $= -45°$. This is called the *break frequency*. (Break frequency is the frequency at which the low frequency and high frequency asymptotes intersect.) The maximum phase lag of $-90°$ is reached as $\omega \to \infty$. At low frequencies, $AR \to K$, which means that the output amplitude is the input amplitude multiplied by the process gain.

♦

The following corollary to the theorem in Example B.1 is useful in computing the frequency response of transfer functions in series.

Corollary: *The frequency response of two transfer functions in series is given by*

$$G(s) = G_1(s)\,G_2(s), \tag{B.5a}$$

$$AR\,(G) = AR\,(G_1)\,AR\,(G_2), \tag{B.5b}$$

$$\phi\left(G\right) = \phi\left(G_1\right) + \phi\left(G_2\right). \tag{B.5c}$$

Thus ARs are multiplied together, whereas the phase angles are additive. For example, we can obtain the frequency response of a third-order transfer function using the result of equation series (B.5).

$$G(s) \;=\; \frac{1}{(s+1)\,(s+1)\,(s+1)},$$

$$AR\,(G) \;=\; AR\left(\frac{1}{s+1}\right)\cdot AR\left(\frac{1}{s+1}\right)\cdot AR\left(\frac{1}{s+1}\right),$$

$$\;=\; \frac{1}{\left(1 + \omega^2\right)^{3/2}},$$

$$\phi\,(G) = 3\,\tan^{-1}\left(-\omega\right).$$

The result is plotted in Figure B.6.

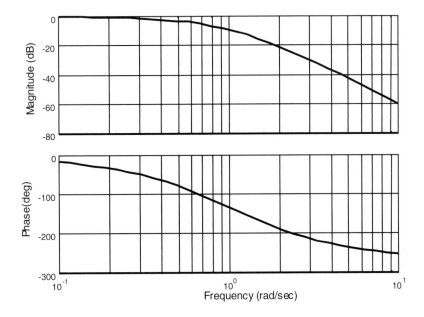

Figure B.6 Bode plot of $1/(s+1)^3$.

Example B.2 Time Delay

Consider

$$G(s) = e^{-Ds}.$$

Then

$$AR(G) = \left| e^{-Di\omega} \right| = 1,$$

$$\phi = -D\omega \text{ radians}.$$

The delay contributes to a phase lag, which increases with frequency.

♦

Example B.3 Frequency Response of a PID Controller

Consider the PID controller

$$G_c(s) = (1 + \frac{1}{\tau_I s} + \tau_D s)K_c.$$

Figure B.7 shows its frequency response for $K_c = 1, \tau_I = 1, \tau_D = 1$. Note that this controller has a high gain at both low and high frequencies. Also shown in the diagram is the frequency response

$$G_c(s) = (+\frac{1}{\tau_I s} + \frac{\tau_D s}{0.05\tau_D s + 1})K_c.$$

which includes a high-frequency noise filter to suppress the high-frequency gain. The addition of a first-order lag to the derivative term reduces the high-frequency gain to 20. Since measurement noise is usually of a high-frequency nature, this filter suppresses noise amplification.

Note the high gain of the controller for low frequencies. This is necessary to eliminate steady-state error when low-frequency disturbances are present. This high gain is caused by the integral term in the controller. Later in this section we justify the need for high gain at low frequencies.

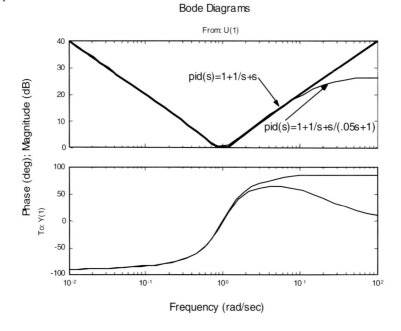

Figure B.7 Bode plots of ideal and real PID controllers.

B.3 DISTURBANCE SUPPRESSION IN SISO SYSTEMS: EFFECT OF CONSTRAINTS

We can examine the capability of a system to suppress disturbances using frequency response methods. Selecting $s = i\omega$, we get

$$y(i\omega) = p(i\omega)u(i\omega) + d(i\omega). \tag{B.6}$$

We can suppress the disturbance effect on $y(s)$ perfectly only if $p(s)$ is invertible. Even then the constraints of u usually prevent us from achieving perfect compensation. Hence at any frequency ω, for perfect compensation,

$$p(i\omega)u(i\omega) = -d(i\omega) \tag{B.7a}$$

or

$$\left| p(i\omega) \right| \left| u(i\omega) \right| = \left| d(i\omega) \right|. \tag{B.7b}$$

The ability to suppress the disturbance depends on the value of $\left| p(i\omega) \right|$ given the constraints on $\left| u(i\omega) \right|$. $\left| p(i\omega) \right|$ is called the *dynamic gain* of the system. The steady-state analysis on the effect of input constraints discussed in Appendix A extends to the dynamic case, with the steady-state gain being replaced by the dynamic gain. Typically, the magnitude of the disturbances, $\left| d(i\omega) \right|$ becomes small at high frequencies. $\left| p(i\omega) \right|$ also tend to become small at high frequency. Hence the question is important in the intermediate frequency range, where $\left| d(i\omega) \right|$ can be significant. If $\left| p(i\omega) \right|$ is small at these ranges of frequency, then we will have a problem with disturbance suppression.

Example B.4 Selection of Manipulated Variables in a Heat Exchange System

Consider the heat exchanger shown in Figure B.8. There are two possible manipulated variables: the bypass flow rate u_2 and the cooling water flow rate u_2. The following transfer functions were obtained. All variables have been scaled and normalized using the procedure outlined in Appendix A.

$$\begin{aligned} y(s) &= p_1(s)u_1(s) + p_2(s)u_2(s) + d(s) \\ &= \frac{10}{s+1}u_1(s) + \frac{2}{.05s+1}u_2(s) + d(s) \end{aligned} \qquad \begin{aligned} -0.5 &< u_1 < 0.5 \\ -0.5 &< u_2 < 0.5, \end{aligned}$$

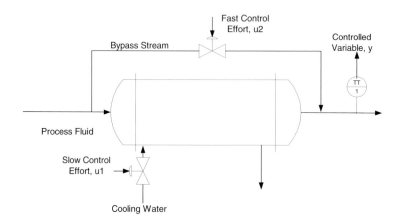

Figure B.8 Instrumentation diagram for a heat exchanger.

where $d(s)$ is the disturbance caused by feed flow variations. $d(t)$ was determined to vary with an amplitude of 3.0 around a frequency of 1.0 rad with amplitude of 1.5 near a frequency of 10 rad/time. Which manipulated variable should be used in a single-loop feedback control of y to suppress the disturbances?

Figure B.9 shows the frequency response of the transfer functions. From this it is seen that at $\omega = 1.0$ and the AR of $p_1(s)$ and $p_2(s)$ are 7.0 and 2.0 respectively. Given that u_2 must lie between –0.5 and 0.5, the maximum amplitude that can be achieved in y using u_2 at this frequency is only 1.0. Hence at this frequency, we can control y using u_1 but not using u_2. At a frequency of $\omega = 10$ rad/time, the ARs of $p_1(s)$ and $p_2(s)$ are 1.0 and 1.78 respectively. In this case we can control y using u_2 but not u_1. Hence both inputs should be used if control is desired over the frequency range of 1.0 to 10 rad/time.

Note that this result is intuitive. We must use fast-responding control effort to suppress fast changing disturbances. However, the slower control effort with a larger gain is useful in suppressing slow disturbances of higher magnitude. *Ideally, we want both large steady-state gain, fast response to control efforts, and large bounds on the control effort.* Chapter 18, "Advanced Model-Predictive Control," discusses how to use both control efforts efficiently. See also the paper by Brosilow et al. (1986).

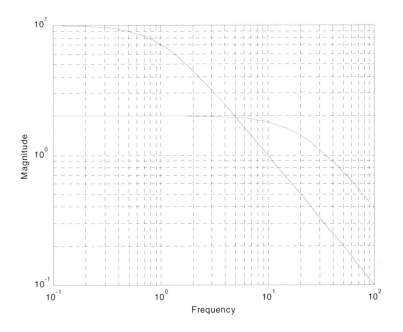

Figure B.9 Dynamic gain of the two control efforts for the heat exchanger.

B.4 STABILITY IN THE FREQUENCY DOMAIN

Nyquist derived conditions for stability of a system from the frequency response characteristics. This result is known as the *Nyquist stability theorem*:

The number of zeros, Z, of the characteristic equation of the closed-loop system shown in Figure B.10, inside a closed contour, D, in the complex plane, is equal to the number of poles of the characteristic equation inside of the closed contour D plus the number of clockwise (counterclockwise) encirclements of the point $(-1/k, 0)$ by the phasor g(s) as s moves in a clockwise (counterclockwise) direction around the closed contour D.

That is, $Z = N + P$,

where   Z = # of zeros of $(1 + kg(s))$ inside D.

N = # of encirclements of $(-1/k, 0)$ point by g(s) as s moves once around D.

P = # of poles of $(1 + kg(s))$ inside D.

A control system is stable if, and only if, the contour D encloses the entire right half of the s plane and the number of zeros Z, as calculated above, is zero. (An intuitive proof of the Nyquist stability theorem can be found in Appendix IV of Vegte, 1986).

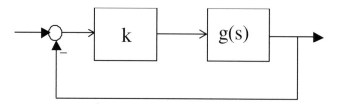

Characteristic equation = 1 + kg(s)

Figure B.10 Feedback diagram for the Nyquist stability criterion.

If g(s) has a pole at s = 0, then the contour D is usually taken, as shown in Figure B.11, and the radius δ is made to approach zero so that the contour encloses the entire right half of the s-plane.

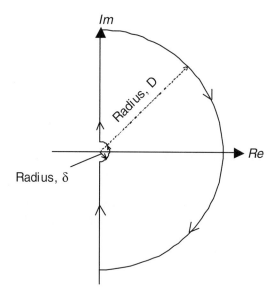

Figure B.11 Nyquist D contour when there is a pole at the origin.

Example B.5. Application of Nyquist Stability Theorem

Figure B.12 shows the Nyquist plots of $G(s)$ for

$$G(s) = \frac{1}{(s+1)(s+2)(s+3)}.$$

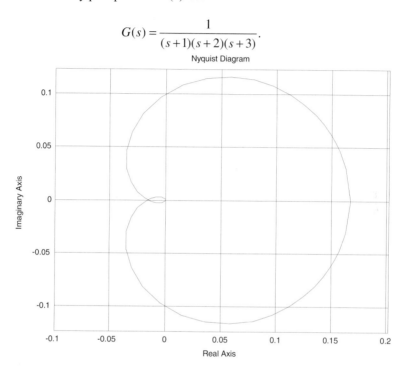

Figure B.12a Nyquist plot of $1/(s+1)(s+2)(s+3)$ is shown. Figure B.12b is a zoomed version near the origin.

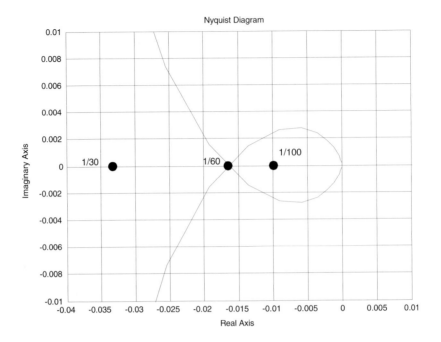

Figure B.12b Nyquist plot of $1/(s+1)(s+2)(s+3)$ near the origin.

Consider proportional feedback control of this system using $K_c = 30$ and $K_c = 100$. For $K_c = 30$, the Nyquist plot of $K_cG(i\omega)$ does not encircle the $(-1/30,0)$ point. Hence $N = 0$. Also, $G(s)$ is open-loop stable, so that $P = 0$. Hence, according to the Nyquist stability theorem, $Z = 0$ (i.e., the closed–loop system should have no unstable poles). This is verified by the root locus plot shown in Table B.1. For $K_c = 100$, there are two encirclements of the $(-.01,0)$ point and hence $Z = 2 + 0 = 2$. Thus the closed-loop system should have two unstable poles. This is also verified by Table B.1.

The ultimate gain, the gain at which the system just becomes unstable, is also shown. This occurs for $K_c = 1/.0167 = 60$. In this case, the system will be oscillatory with the ultimate frequency. The advantage of the Nyquist stability theorem is that we can graphically determine the closed-loop characteristics from the open-loop Nyquist plot. Table B.1 shows the frequency response of the process transfer function $p(s)$. Adding a proportional controller $c(s) = K_c$ will not change the phase angle but will multiply the magnitude by K_c. The ultimate frequency with such a proportional controller is when the phase angle of the open-loop process becomes -180, which occurs at around $\omega = \sqrt{11}$. The magnitude of $p(i\omega)$ at this frequency is given by $1/60$. Hence using a controller gain of 60 will make the magnitude of pc

= 1, causing the Nyquist diagram to go through the (–1,0) point. This is the ultimate gain for this process.

Table B.1 Frequency Response of p(s) = 1/(s+1)(s+2)(s+3).

Frequency rad/sec	Magnitude	Phase Angle, degrees
0.1000	0.1655	– 10.4822
0.1600	0.1638	– 16.7156
0.2560	0.1596	– 26.5264
0.4942	0.1431	– 49.5302
0.7906	0.1176	– 74.6628
1.2649	0.0805	–106.8415
2.0236	0.0430	–143.0395
3.2375	0.0176	–178.3086
3.5565	0.0143	–184.7952
6.8665	0.0027	–221.8740
10.000	0.0009	–236.2802

♦

B.5 CLOSED-LOOP FREQUENCY RESPONSE CHARACTERISTICS

Consider the control system shown in Figure B.13. For this closed-loop system, we can define two transfer functions: $S(s)$ and $CS(s)$. We have, using block diagram algebra,

$$\frac{y(s)}{y_{set}(s) - n(s)} = \frac{p(s)c(s)}{1 + p(s)c(s)} = CS(s) \qquad \text{(B.8a)}$$

$$\frac{y(s)}{d(s)} = \frac{1}{1 + p(s)c(s)} = S(s) \qquad \text{(B.8b)}$$

where

$S(s)$ = sensitivity function,

$CS(s)$ = complementary sensitivity function.

Note that

$$CS(s) + S(s) = 1$$

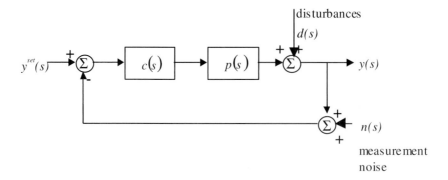

Figure B.13 Block diagram of closed-loop system.

Recall that $|p(i\omega)|$ generally decreases in value rapidly as ω gets large. It shows a downward trend with increasing frequency. This implies that $CS(s)$ will also decrease at large frequencies, assuming $|c(s)|$ does not become extremely large as ω increases. Figure B.14 shows the typical behavior of CS and S.

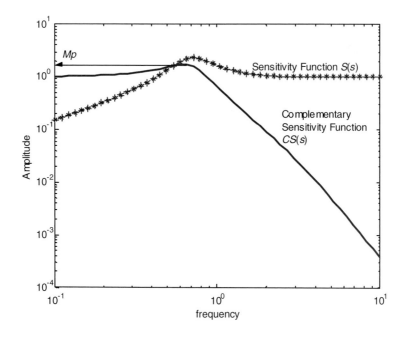

Figure B.14 A typical sensitivity and complementary sensitivity function.

Ideally, we want $S(s)$ to be as small as possible over a wide range of frequencies. Similarly, we want $CS(s)$ to be close to 1 over a wide range of frequencies. The frequency at which $CS(s)$ drops to $1/\sqrt{2}$ is called the *bandwidth* of the system. Beyond that frequency, the controller will not be able to follow the setpoint well. $CS(s)$ may exhibit a maximum.

Now let us consider how to choose the controller response characteristics. Our goal will be to pick $c(s)$ such that the control objectives set in the time domain or frequency domain can be met. Some objectives and their consequences are discussed next.

1. No steady-state error. Using final value theorem, steady-state corresponds to low-frequency behavior. If we have $CS(i\omega) \to 1$ as $\omega \to 0$, then we will have no steady-state error. Since $p(i\omega)$ approaches a finite value, we must have $c(i\omega) \to \infty$ as $\omega \to 0$. The integral term $(1/i\omega)$ in a controller is used to ensure this criterion.

2. Limit measurement noise amplification. Generally, noise in measurement is significant at high frequencies. Noise transmission is determined by $|CS(s)|$, and this is ensured by keeping $|CS(i\omega)| \approx 0$ at higher frequencies. We can limit high-frequency noise amplification by limiting the controller gain at high frequencies, as the process gain usually becomes very small at high frequencies. Typically, we try to keep

$$\lim |c(i\omega)| < 20$$

$$\omega \to \infty.$$

This explains why ideal derivative action is never used in the real world. Ideal derivative action requires the following term in the controller.

$$c(s) = \tau_D s,$$

$$c(i\omega) = \tau_D i\omega.$$

As ω gets large, $|c(i\omega)|$ will get infinitely large. Typically, derivative action is implemented using a lead-lag instead:

$$c(s) = \frac{\tau_D\, s}{\alpha \tau_D s + 1},$$

with $\alpha = 0.05$. At high frequencies,

$$\lim_{\omega \to \infty} |c(i\omega)| \approx \frac{|\tau D i \omega|}{|\alpha \tau_D i \omega|} = \frac{1}{\alpha} = 20.$$

3. Provide good disturbance rejection at as wide a frequency range as possible. Recall that disturbance rejection is given by the sensitivity function

$$S = \frac{1}{1 + pc}.$$

To keep $|S|$ small, we must keep $|1 + pc|$ large. At low frequencies, this is accomplished by the integral action. In the medium range of frequencies, where disturbances usually dominate, we will have to keep $|c(i\omega)|$ large. The controller gain $|c(i\omega)|$ in the medium frequency range will be limited by stability considerations, as increasing the controller gain may cause encirclement of the $(-1,0)$ point in the Nyquist diagram.

4. Provide a high bandwidth. The ultimate frequency, ω_u, will determine the region around which $|CS|$ will peak and start to decrease. The inverse of this frequency determines the dominant time constant and hence the speed of response of the control system. To increase bandwidth, we must make the frequency of which the phase lag approaches $= 180°$ as large as possible. One way to do this is to add phase lead to the controller. This derivative term has the property of adding a phase lead near the intermediate frequency range. Adding the phase lead allows us to push the frequency to the right and hence get a higher bandwidth.

5. Preserve stability in presence of model errors. The maximum value of $CS(i\omega)$ carries a special significance in controller design. According to the Nyquist stability theorem, in order for the closed-loop system to be stable, we must examine the encirclements of the $(-1,0)$ point in the Nyquist plane. Consider a stable closedloop system whose open-loop Nyquist diagram is shown in Figure B.15. According to this diagram, the function

$$CS = \frac{pc}{1 + pc} \tag{B.9}$$

is a measure of how close to $(-1/K,0)$ (with $K = 1$) the Nyquist curve of pc lies. In actual applications, there will be a region of uncertainty surrounding p, as the process transfer function can be different from what was used in the controller design (due to modeling errors and process variations). For an inherently stable process, we can interpret the minimum distance $|(1 + pc)|$ as shown in Figure B.15 as maximum change allowed in pc before an encirclement occurs, resulting in instability of the closed-loop. Hence the quantity

$$l = \left| \frac{1 + pc}{pc} \right| \tag{B.10}$$

represents the maximum fractional change allowed in pc before instability occurs. We want to keep this as large as possible to get a large *stability margin*. This is achieved by keeping

$$M_p = \max_\omega \left| \frac{pc}{1 + pc} \right| = \max_\omega |CS| \tag{B.11}$$

as small as possible. M_p is called the *maximum closed-loop modulus*. If we let $M_p < 1$ then this allows the process gain to increase by 100% before the closed-loop system becomes unstable. Since we know that $|CS| \to 1$ as $\omega \to 0$, a slightly relaxed criteria,

$$M_p < 1.05, \tag{B.12}$$

allows the CS to be close to 1 over a wider frequency range. This criterion is used in this text for tuning controllers. A large value of M_p implies that the closed-loop system is on the verge of instability.

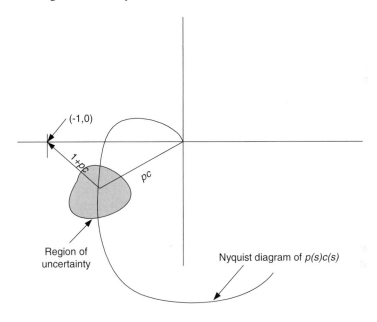

Figure B.15 Effect of uncertainty in process transfer function of stability.

This criterion conflicts with the desire to respond rapidly to setpoint changes and disturbances, which requires the high-frequency controller gain to be as large as possible. This is referred to as the tradeoff between robustness (insensitivity to model errors and plant variations) and performance (speed of response and disturbance suppression).

Problems

B.1   Consider the following FOPDT system

$$G_p(s) = \frac{3e^{-2s}}{5s+10}.$$

    a.   Using frequency response, find the ultimate gain and ultimate period of the process with transfer function (Hint: Assume a proportional controller is used. Obtain the phase angle and AR of G_p. Determine the value of K_c that causes the Nyquist plot of $K_c G_p$ to pass through the $(-1,0)$ point which corresponds to a phase angle of $-\pi$ rad and AR of 1.)

    b.   What are the recommended settings for a PID controller using the Ziegler-Nichols controller tuning method?

    c.   One suggested approximation for the time delay is the (1,1) Padé approximation

$$e^{-Ds} \approx \frac{1 - \dfrac{D}{2}s}{1 + \dfrac{D}{2}s}.$$

       Using this approximation in the transfer function, determine the characteristic equation of the system under proportional feedback control. Use the root locus method to determine the ultimate gain. Compare with Part a.

B.2   Gasoline is blended in-line by mixing two streams. The octane number of the gasoline is controlled by adjusting the flow rate of stream A. The process transfer function can be represented by

$$y(s) = 5e^{-2s}x(s),$$

where y is the octane number and x is the flow of stream B.

    a.   What is the ultimate controller gain if proportional action alone is used?

    b.   If a controller gain equal to one half of the ultimate gain is used, plot the response of y to a unit step change in setpoint. Show scale of x and y axes and the steady-state value reached by y.

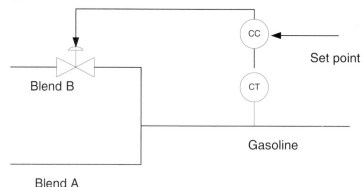

B.3   Consider the process with transfer function

$$G(s) = \frac{1}{(s+1)(2s+1)(3s+1)} = \frac{1}{6s^3 + 11s^2 + 6s + 1}.$$

a. Sketch the Bode and Nyquist plot of the system.

b. Using the frequency domain stability criteria, determine the ultimate gain and ultimate frequency when a proportional feedback controller is used with this process. (Hint: MATLAB has some functions to facilitate the construction of Bode diagrams. Use the help command to obtain additional information about these commands.)

```
» num= [1]; % defines the numerator of the transfer function
» den= [6   11   6   1]; % defines the denominator coefficients
» system=tf (num,den)    % defines the system
» bode (system) % creates the bode plot
» nyquist ( system) % creates the nyquist plot
» grid on
```

c. Suppose a proportional controller with a gain of $Kc = 5$ is used to control this process. Derive the sensitivity and complementary sensitivity functions for the closed-loop system. Plot these functions using MATLAB. What is M_p for the closed-loop system?

B.4 Distillation processes have a large number of time constants. Consider a distillation process response, which can be approximated by 10 first-order transfer functions in series, each with a time constant of 5 seconds. Estimate the ultimate period of this process.

B.5 The temperature in a reactor is measured using a thermocouple with a time constant of 5 seconds. The gain is 1. The normal reactor temperature is 300°F. The temperature high alarm is set at 325°F in the control computer.

Occasionally, the reactor temperature goes into oscillations with a period of 100 seconds and amplitude of 30°F above and below the normal steady-state. We can model these deviations as a sine wave with amplitude of 30°F. Will these oscillations cause the temperature alarm to sound? Explain how you arrived at your conclusion.

B.6 The objective of this exercise is to look at the design characteristics of PID controllers from a frequency response point of view. Use MATLAB and/or SIMULINK to do the calculations.

a. Consider the process with transfer function

$$p(s) = \frac{1}{(10s+1)^3}.$$

What is the ultimate, K_u? Ultimate frequency P_u?

b. Compute tuning constants for a PID controller using the Ziegler-Nichols tuning method. Use a real PID controller with the equation

$$c(s) = K_c (1 + \frac{1}{\tau_I s} + \frac{\tau_D s}{\alpha \tau_D s + 1}).$$

c. Graph the frequency response of this controller.

d. Plot step response to setpoint changes. Discuss your result. Is the response satisfactory?

e. Plot the sensitivity and complementary sensitivity function. What is the M_p for this closed-loop system?

B.7 Consider a first-order plus delay process with transfer function

$$p(s) = \frac{10e^{-Ts}}{s+1}.$$

a. Derive the AR and phase angle as a function of frequency.

b. Write an equation to compute the ultimate frequency. Solve for the case T = .5.

c. Show that the ultimate frequency decreases as T increases.

d. Write an equation to compute the ultimate gain. Solve for the case T = .5. For the case T = .5, compute the gain of a proportional feedback controller so that the M_p is 1.05.

B.8 Derive the equations for the sensitivity and complementary sensitivity functions for a process with transfer function $p(s) = 1/(s+1)$ and controller $c(s) = 1+1/s$. Sketch the amplitude ratio of both functions as a function of frequency. What is the M_p for this system? At what frequency is the closed-loop system most sensitive to disturbances? Is the system stable? Show using a Nyquist plot.

References

Brosilow, C., L. Popiel, and T. Matsko. 1986. "Coordinated Control." Proceedings of the Third International Conference on Chemical Process Control. Edited by M. Morari and T. J. McAvoy. Elsevier, NY.

Luyben, W. L. 1989. *Process Modeling, Simulation, and Control for Chemical Engineers*, 2nd Ed. McGraw-Hill, NY.

Marlin, T. E. 1999. *Process Control: Designing Processes and Control Systems for Dynamic Performance.* McGraw-Hill, NY.

Morari, M., and E. Zafiriou. 1989. *Robust Process Control.* Prentice Hall, NJ.

Seborg, D. E., T. F. Edgar, and D.A. Mellichamp. 1989. *Process Dynamics and Control.* John Wiley & Sons, NY.

Stephanopoulos, G. 1984. *Chemical Process Control: An Introduction to Theory and Practice.* Prentice Hall, NJ.

Vegte, J. V. 1986. *Feedback Control Systems.* Prentice Hall, NJ.

Review of Linear Least-Squares Regression

Objectives

The objective of this appendix is to briefly review the theory of linear least-squares. Topics discussed include

- The concept of linear least-squares solution.
- Properties of the least-squares estimates.
- Overcoming collinearity problems with principal component and partial least-squares regression.

C.1 DERIVATION OF THE LINEAR LEAST-SQUARES ESTIMATE

The problem of finding a least-squares solution to a set of linear equations arises at a number of places in this text.
- In process identification where we want to determine parameters in a linear model from empirical data.
- In model-predictive control where we want to determine the control moves that will minimize the future deviations of the output variable from its setpoint.
- In inferential estimation where we want to develop a linear relationship between the secondary measurements and the primary output variable.

In this section we give a brief introduction to linear least-squares theory. For the sake of brevity we have sacrificed mathematical and statistical rigor for simplicity of the presentation. The interested reader is referred to one of the numerous books on linear regression.

Let $x_1, x_2, ..., x_n$ represent a set of independent variables, and y represent a variable that can be modeled as a linear combination of these independent variables:

$$y = \alpha_1 x_1 + \alpha_2 x_2 ..... + \alpha_n x_n \tag{C.1}$$

Suppose we are given a set of N experimental data points,

$$\{x_{1i}, x_{2i}, ..., x_{ni}, y_i\} i = 1, ..., N \tag{C.2}$$

Let us say that due to noise in the measurements and other experimental errors, these points do not exactly fit the model of Eq. (C.2) perfectly. Instead,

$$y_i = a_1 x_{i1} + a_2 x_{2i} + a_n x_{ni} + e_i \quad i = 1, ..., N, \tag{C.3}$$

where e_i represent the error in a linear fit, and $a_1 ... , a_n$ are estimates of the unknown model parameters. A least-squares solution seeks to find values of $a_i, i = 1, ..., N$ that minimizes the sum of the square of the errors,

$$SS_E = \sum_{i=1}^{N} e_i^2 = \|e\|^2 = e'e \tag{C.4}$$

We can write the equations in expanded form:

$$
\begin{aligned}
y_1 &= x_{11}a_1 + x_{12}a_2 + x_{1n}a_n + e_1, \\
y_2 &= x_{21}a_1 + x_{22}a_2 + x_{2n}a_n + e_2, \\
&..., \\
y_N &= x_{N1}a_1 + x_{N2}a_2 + x_{Nn}a_n + e_N.
\end{aligned}
\tag{C.5}
$$

If $N \leq n$, then we can make $\|e\|^2 = 0$. However, we prefer to have an excess number of measurements, $N \gg n$, so that we can overcome any error in the model defined as $\|a - \alpha\|$ by having a redundant set of equations.

In matrix notation,

$$Y = Xa + e \tag{C.6}$$

where

$$Y = \begin{bmatrix} y_1 \\ y_2 \\ . \\ . \\ y_N \end{bmatrix}; \quad X = \begin{bmatrix} x_{11} & x_{12} & \cdots & x_{1n} \\ x_{21} & x_{22} & \cdots & x_{2n} \\ \cdots & & & \\ \cdots & & & \\ x_{N1} & x_{N2} & \cdots & x_{Nn} \end{bmatrix}; \quad a = \begin{bmatrix} a_1 \\ a_2 \\ . \\ . \\ a_n \end{bmatrix}; \quad e = \begin{bmatrix} e_1 \\ e_2 \\ . \\ . \\ e_n \end{bmatrix}. \quad \text{(C.7)}$$

Now

$$e = Y - Xa, \tag{C.8}$$

so that

$$
\begin{aligned}
\|e\|^2 &= \|y - Xa\|^2 \\
&= (y - Xa)^T (y - Xa) \\
&= y^T y + a^T X^T Xa - y^T Xa + a^T X^T y \\
&= \left[a^T - y^T X \left(X^T X \right)^{-1} \right] X^T X \left[a - \left(X^T X \right)^{-1} X^T y \right] \\
&\quad + y^T y - y^T X \left(X^T X \right)^{-1} X^T y,
\end{aligned}
\tag{C.9}
$$

which is minimized by making the first term zero, by choosing

$$a = (X^T X)^{-1} X^T y, \tag{C.10}$$

This is called the least-squares solution or estimate of the model parameters. If X is square and invertible, then

$$a = X^{-1} y, \tag{C.11}$$

which fits the data perfectly $(e = 0)$. This does not mean we have a perfect model ($\mathbf{a} \neq \mathbf{\alpha}$ necessarily), but that we have a perfect fit.

C.2 PROPERTIES OF THE LINEAR LEAST-SQUARES ESTI-MATE

Assume that the error in observations e_i are independent and randomly distributed with zero mean, that is,

$$E\{e\} = 0,$$
$$E\{ee^T\} = I\sigma^2. \tag{C.12}$$

Then we can derive some properties of the parameter estimates (we drop the underscore for convenience):

$$
\begin{aligned}
E\{a\} &= E\{(X^TX)^{-1}X^Ty\}, \\
&= E\{(X^TX)^{-1}X^T(X\alpha+e)\}, \\
&= E\{\alpha+(X^TX)^{-1}X^Te\}, \\
&= \alpha+(X^TX)^{-1}X^TE\{e\}, \\
&= \alpha.
\end{aligned}
\tag{C.13}
$$

Hence we say that a is an unbiased estimate of α. The variance of a (an estimate of the accuracy of the parameters) can be computed as

$$
\begin{aligned}
\mathrm{var}\{a\} &= E\left[(a-\alpha)(a-\alpha)'\right], \\
&= \sigma^2(X^TX)^{-1}.
\end{aligned}
\tag{C.14}
$$

σ^2 can be estimated as follows:

$$
\begin{aligned}
SS_E &= \sum e_i^2 = e^Te, \\
&= y^Ty - a^TX^Ty.
\end{aligned}
\tag{C.15}
$$

This residual has $N-n$ degrees of freedom associated with it, since n parameters have been calculated from the N data. So the residual mean square value is

$$MS_E = \frac{SS_E}{N-n} = \hat{\sigma}^2. \tag{C.16}$$

This is an unbiased estimated of σ^2 (Draper & Smith, 1981). Using these estimates we can also compute confidence intervals for the estimates. As a very rough approximation, we can take 95% confidence interval to be

$$a_i \pm 2\hat{\sigma}_i^2, \tag{C.17}$$

where

$$\hat{\sigma}_i = \sqrt{\hat{\sigma}^2 C_{ii}}. \tag{C.18}$$

C_{ii} is the ith diagonal element of $\left(\mathbf{X^TX}\right)^{-1}$.

C.3 Measures of Model Fit

The basic measures of model fit are the residuals e_i. Assuming these residuals are random variables, they have variance given by

$$MS_E = \frac{SS_E}{N-n} = \frac{\sum e_i^2}{N-n}. \tag{C.19}$$

A plot of e_i versus y_i should be used to determine if these are indeed randomly distributed. Any pattern here will indicate that the model is inadequate to describe the data.

A common measure of the goodness of fit is the correlation coefficient or coefficient of multiple determination, R^2, defined as

$$R^2 = 1 - \frac{SS_E}{S_{yy}}, \tag{C.20}$$

where

$$S_{yy} = \sum_{i=1}^N (y_i - \bar{y})^2 = \sum_{i=1}^N (\hat{y}_i - \bar{y})^2 + \sum_{i=1}^N (y_i - \hat{y}_i)^2$$
$$\tag{C.21}$$
$$= SS_R + SS_E,$$

where $\hat{\mathbf{y}} = \mathbf{Xa}$. Small values of R^2 (close to 0) indicate poor fit, whereas values closer to 1 indicate better fit. The fit must be weighed against n, the number of parameters used to fit the model.

C.4 Weighted Least-Squares

If some of the observations are more accurate than others, then we can apply increased weight to these observations during the minimization procedure. Let W represent a weighting matrix. We seek a minimum of the weighted residual.

$$\underset{a}{Min} \ \ e' \ We. \tag{C.22}$$

This leads to the least-squares solution

$$a = \left(X^T W^{-1} X \right)^{-1} X^T W^{-1} y. \tag{C.23}$$

C.5 ROBUSTNESS

By robustness, we mean the sensitivity of the parameters a to measurement errors in x and y. Consider the equation for a:

$$Xa = y \tag{C.24}$$

If there are some measurement errors associated with x and y, then the equation becomes

$$(X + X_e)(a + a_e) = y + y_e. \tag{C.25}$$

It can be shown that

$$\frac{\|a_e\|}{\|a\|} \leq \rho(X)\left[\frac{\|y_e\|}{\|y\|} + \frac{\|X_e\|}{\|X\|}\right], \tag{C.26}$$

where the Euclidean norm of the vector and matrix are defined as

$$\|z\| = \sqrt{z_1^2 + z_2^2 + ... + z_n^2} = \sqrt{z^T z}, \tag{C.27}$$

$$\|X\| = \max_z \frac{\|Xz\|}{\|z\|} = \text{Max singular value of } X = \sqrt{\text{Max eigen value of } X^T X}. \tag{C.28}$$

For a discussion on singular value decomposition, see the next section.

$$
\begin{aligned}
\rho(X) \quad &= \quad \text{Condition number of } X \\
&= \quad \frac{\text{Max singular value of } X}{\text{Min singular value of } X}
\end{aligned}
\tag{C.29}
$$

This result indicates that if the matrix $\mathbf{X}$ is ill conditioned (has a large condition number), then the small errors in $\mathbf{X}$ or $\mathbf{y}$ will result in large changes in the estimated parameters $\boldsymbol{a}$ of the model. A large condition numbers indicate that the matrix is close to being singular. The smallest possible value of $\rho(X)$ is 1, and $\rho(X)$ is infinite for a singular matrix. Large condition number indicates that the rows (or columns) of $\mathbf{X}$ are correlated with each other. Collinearity can arise if the variables in the model $x_1, x_2 ...., x_n$ are not truly independent.

Example C.1 An Estimation Problem

Consider the problem of estimating a distillate composition from two tray temperature measurements. If the trays are close to each other, then these two temperatures (we will call them

x_1 and x_2) will be closely related to each other. If we propose a predictive model of the form

$$y = \alpha_1 x_1 + \alpha_1 x_2, \; y = \text{composition},$$
$$x_1 = \text{temp on tray 1}, x_2 = \text{temp on tray 2}.$$

we can expect the resulting least-squares estimation problem to be ill-conditioned. Let us say that the two measurements are located on adjacent trays. Consider two data points,

Data Set No	x_1	x_2	y
1	1.0	1.0	2.0
2	2.0	2.0	4.0

$$X = \begin{bmatrix} 1.0 & 1.0 \\ 2.0 & 2.0 \end{bmatrix}$$

is singular, and hence a least-squares solution is not possible. Now suppose there was a small measurement error in the second set of measurements, resulting in

$$X = \begin{bmatrix} 1 & 1 \\ 2 & 2.001 \end{bmatrix}.$$

This matrix has a condition number = 10,000. Solving for a, we get an estimate of α,

$$a = X^{-1} y = \begin{bmatrix} 1.5 \\ 0.0 \end{bmatrix}.$$

This estimate of parameters in the model will change drastically if x changes even a little bit. If, for example, x is measured as

$$X = \begin{bmatrix} 1.001 & 1.0 \\ 2.0 & 2.002 \end{bmatrix},$$

the solution changes to

$$a = \begin{bmatrix} .5 \\ 1.0 \end{bmatrix}.$$

If the matrix to be inverted is ill conditioned, then some of the columns are correlated with each other. If such collinearity exists, then some steps can be taken to alleviate this problem. One such tool is the Principal Component Regression (PCR) technique, described next.

♦

C.6 PRINCIPAL COMPONENT REGRESSION

Consider the solution of a set of equations of the form

$$Xa = y. \tag{C.30}$$

where X is an $n \times m$ matrix. The least-squares solution is given by

$$a = (X^T X)^{-1} X^T y. \tag{C.31}$$

If some of the columns in X are nearly dependent, then the matrix to be inverted will be nearly singular. This may be determined by a singular value decomposition of X, given by (the command *svd* (X) does this in MATLAB):

$$X = U \Sigma V^T,$$

where

$$U = \begin{bmatrix} u_1 & u_2 & ... & u_m \end{bmatrix}$$

and

$$V = \begin{bmatrix} v_1 & v_2 & ... & v_m \end{bmatrix}. \tag{C.32}$$

Here Σ is a diagonal matrix of the singular values of X. U and V are orthogonal matrices. X can also be written as a sum of rank 1 matrices in decreasing order of the singular values:

$$X = u_1 \sigma_1 v_1^T + u_2 \sigma_2 v_2^T + ..., + u_m \sigma_m v_m^T, \tag{C.33}$$

where m is the rank of X and $\sigma_1 \geq \sigma_2 \geq ..., \geq \sigma_m$. If some of the singular values are zero, the matrix is rank deficient. Similarly, if some of the singular values are very small, then the matrix is ill conditioned and some of the columns of X are nearly parallel or collinear to each other. We can rewrite the expansion as

$$X = t_1 p_1^T + t_2 p_2^T + ..., + t_m p_m^T, \tag{C.34}$$

where $t_i = u_i \sigma_i$ and $p_i = v_i$. In PCR terminology, t_i are called the **scores vector** or **latent variable**, and p_i are called as the **loadings vector**. If some of the singular values are close to zero, so that it is within the noise level of the coefficients in X (recall that we cannot determine X precisely, due to measurement errors), then it makes sense to neglect the terms containing these singular values in the expansion shown in Eq. (C.34). Dropping these terms should lead to a better conditioned matrix. Hence we take ($k < m$)

$$X \approx t_1 p_1^T + t_2 p_2^T + \dots + t_k p_k^T$$
$$= TP^T .$$

(C.35)

The parameters a are then computed by solving the better conditioned system of equations,

$$TP^T a = y,$$

(C.36)

with the least-squares solution

$$a = P(T^T T)^{-1} T^T y .$$

(C.37)

Note that $P^T P = I$ (P consists of orthonormal vectors).

Example C.2 Collinearity

Consider the case

$$X = \begin{bmatrix} 1 & 1 \\ 2 & 2 \\ 3 & 3 \end{bmatrix}; \ y = \begin{bmatrix} 1 \\ 2 \\ 3 \end{bmatrix}.$$

A least-squares solution does not exist in this case because $\mathbf{X}^T \mathbf{X}$ is singular. However, we can get an approximate solution using PCR analysis.

A singular value decomposition yields

$$X = \begin{bmatrix} .2673 & -.5345 & -.8018 \\ .5345 & .7745 & -.3382 \\ .8018 & -.3382 & .4927 \end{bmatrix} \begin{bmatrix} 5.2915 & 0 \\ 0 & 0 \\ 0 & 0 \end{bmatrix} \begin{bmatrix} .707 & -.707 \\ .707 & .707 \end{bmatrix} .$$

There is only one nonzero singular value, indicating that A is of rank 1. Neglecting the second zero singular value, we can write

$$X \approx \begin{bmatrix} .2673 \\ .5345 \\ .8018 \end{bmatrix} [5.2915][.7071 \quad .7071] .$$

Hence

$$T = \begin{bmatrix} 1.4142 \\ 2.8284 \\ 4.2426 \end{bmatrix} \quad P = \begin{bmatrix} .7071 \\ .7071 \end{bmatrix},$$

and the solution

$$a = P(T^T T)^{-1} T^T y,$$

$$= \begin{bmatrix} 0.5 \\ 0.5 \end{bmatrix}.$$

PCR recognizes the linear dependency of the two variables and suggests a solution that takes this dependency into account. Ill conditioning of the matrix X is usually a sign that some of the variables under consideration are related to each other.

Program *ex_C_2.m* on the website implements this MATLAB. In addition, a program called *pcr_bj.m* implements the PCR algorithm for a general case.

♦

Problems

C.1   The composition analyzers used in distillation columns are very unreliable. As a backup, it is proposed to estimate the composition y using the temperature measurements on three trays x_1, x_2, and x_3. These measurements respond to disturbances in a very similar way and so are correlated with each other. Some data was collected as follows to correlate y with x.

Data for this problem is available on the website.

Data Set No	x_1	x_2	x_3	y
1	1.2750	0.5696	1.6277	3.4713
2	1.5517	-0.0744	2.3655	3.8484
3	0.1865	0.1668	0.1963	0.5530
4	0.5451	0.2675	0.6837	1.5042
5	1.1390	0.7333	1.3419	3.2166
6	1.1159	0.7190	1.3151	3.1523

a. Determine the rank of the matrix (use MATLAB) $X = \begin{bmatrix} x_1 & x_2 & x_3 \end{bmatrix}$.

b. Compute the singular value decomposition of $X = U \Lambda V^T$, using MATLAB.

c. How many singular values are close to zero? What is the condition number of X?

d. How many columns of X are independent? Explain.

e. Compute an estimator for y using x_1, x_2, and x_3 by PCR. Express your answer in the form $y_{est} = a_1 x_1 + a_2 x_2 + a_3 x_3$, where a_1, a_2, a_3 are regression coefficients.

f. Repeat by using direct least-squares inversion of x to calculate the estimator. Compare your answers with part c. Check the accuracy of your estimators using the following new data:

x_1	x_2	x_3	y
0.6812	0.0178	0.8131	1.2192
0.8845	−0.1037	1.3531	1.8987
0.8109	0.6832	0.9431	2.1348

Discuss your results.

g. Compute MS_E, SS_E, and R^2 for both estimators.

h. For the direct least-squares estimator, estimate the variance in the parameters and compute the 95% confidence interval.

References

Draper, H. R., and H. Smith. 1981. *Applied Regression Analysis,* 2nd Ed. John Wiley, NY.

Graybill, F. A., and H. K. Iyer. 1994. *Regression Analysis: Concepts and Applications.* Duxbury Press, CA.

Jolliffe, I. T. 1986. *Principal Component Analysis.* Springer-Verlag, NY.

Mendenhall, W., and T. Sincich. 1996. *A Second Course in Statistics: Regression Analysis.* Prentice Hall, NJ.

Plackett, R. L. 1981. *Principles of Regression Analysis.* Clarendon Press, Oxford, Eng.

Sen, K. K., and M. S. Srivastava. 1994. *Regression Analysis: Theory, Methods, and Applications.* Springer-Verlag, NY.

Wise, B., and N. B. Gallagher. 1998. *PLS_Toolbox Version 2.0.* Eigenvector Research, Inc., WA.

Review of Random Variables and Random Processes

Objectives

- Introduce some basic notions of probability.
- Introduce random variables and random processes.
- Introduce white noise and colored noise.
- Introduce frequency (spectral) decomposition of random processes.

D.1 INTRODUCTION TO RANDOM VARIABLES

Most measurements contain a random component that fluctuates from one measurement to the next. We refer to this as *measurement noise*. In order to properly account for the effect of this noise on identification, estimation, and control, we need to introduce a few concepts and terms associated with the mathematical description of such noise processes. The *disturbances* that enter a process also exhibit random behavior. We need tools to model and describe such random disturbances as well. This appendix is a brief introduction to probability theory and modeling of random variables and random processes. We will omit details and mathematical rigor and include only results and terms that are pertinent to the developments in the main body of this text.

Consider a random variable x. Associated with it is a population space $\mathbf{P}$ that contains all the values that x can take. For example, if x = result of a coin toss, then

$$x \in \mathbf{P} \; ; \; \mathbf{P} = \{H, T\} \, .$$

We can associate with each element of the population space the probability (fraction of times) of its occurrence. For the case of a coin toss, the associated probabilities for a fair coin are

$$\text{Prob}\{x = H\} = 1/2,$$
$$\text{Prob}\{x = T\} = 1/2.$$

Note that

$$\text{Prob}\{x \in \text{P}\} = 1.$$

This is an example of a *discrete random variable*. An example of a *continuous random variable* is the angle of a bottle that is spun and comes to rest. Let x be the angle. If the probability of getting any angle is the same as any other, then by definition,

$$\text{P} = \{x \mid 0 \le x \le 360\},$$
$$\text{Prob}\{0 \le x \le 360\} = 1,$$
$$\text{Prob}\{x \le \theta\} = \frac{\theta}{360},$$

where θ is any angle measured from a reference position.

Let u be a random variable, $-\infty < u < \infty$. Then let

$$\text{Pr } ob\{x < u < x + dx\} = p(x)dx. \tag{D.1}$$

$p(x)$ is called the *probability density function*. Let

$$\text{Pr } ob\{-\infty < u < x\} = P(x). \tag{D.2}$$

$P(x)$ is called the *cumulative probability density function*.   Note that

$$P(x) = \int_{-\infty}^{x} p(u)du \tag{D.3}$$

$$\int_{-\infty}^{+\infty} p(u)du = 1. \tag{D.4}$$

The mean and variance associated with a random variable is defined as (E is called the *expectation* operator):

Mean or Expected value: $$E\{x\} \equiv \int_{-\infty}^{+\infty} xp(x)dx = \mu \tag{D.5}$$

Variance:
$$E\{(x-\mu)^2\} \equiv \int_{-\infty}^{+\infty}(x-\mu)^2\, p(x)dx \qquad \text{(D.6)}$$

D.1.1 The Normal Distribution

A common probability density function that is used to describe noise in measurements is the *Gaussian Normal probability distribution*:

$$p(x) = \frac{1}{\sigma\sqrt{2\pi}}e^{-\frac{1}{2}(x-\mu)^2/\sigma^2} \qquad \text{(D.7)}$$

This is denoted using $N(\mu,\sigma)$. This can be put in dimensionless form:

$$z = \frac{x-\mu}{\sigma} \qquad \text{(D.8)}$$

$$p(z) = \frac{1}{\sqrt{2\mu}}e^{-z^2/2}. \qquad \text{(D.9)}$$

Note that for Normal distribution $x \in N(\mu,\sigma)$,

$$E\{x\} = \mu, \qquad \text{(D.10)}$$

$$E\{(x-\mu)^2\} = \sigma^2. \qquad \text{(D.11)}$$

D.1.2 Sampling from a Population

The mean and variance can be estimated from a sample $x_1, x_2..., x_N$, taken from the population. Based on the sample we can define two quantities,

$$\text{Sample Average } = \overline{x} = \frac{(x_1 + x_2 + ... + x_N)}{N}, \qquad \text{(D.12)}$$

$$\text{Sample Standard Deviation} = s^2 = \frac{(x_1 - \overline{x})^2 + (x_2 - \overline{x})^2 + ...(x_N - \overline{x})^2}{N-1} \qquad \text{(D.13)}$$

As the sample size gets large (approaches the population size), then $\bar{x}$ approaches μ and s^2 approaches σ^2.

D.2 RANDOM PROCESSES AND WHITE NOISE

Suppose we want to model a disturbance that is entering a process over a period of time. We can take measurements of the disturbance x at fixed intervals of time (sample times). This constitutes a random sequence or a random process.

$$x_1, x_2, \ldots, x_n.$$

Stationary random processes are those processes whose mean does not vary with time (e.g., the sample mean estimated from the first n values is roughly the same as the mean estimated from the next n values). If these successive values of x are uncorrelated with each other, that is

$$E\{x_n x_{n-k}\} = 0, k \neq n, \tag{D.14}$$

$$E\{x_n x_n\} = \sigma_x^2.$$

Such a random sequence is called a *white noise*. Thus each sampled value of the noise is independent of all others and have zero mean with variance σ_x^2.

If we perturb a linear system with white noise, the output is said to be *colored noise*. For example, consider a discrete system with transfer function

$$G(z) = \frac{b_1 z^{-1} + b_2 z^{-2}}{1 - a_1 z^{-1} - a_2 z^{-2}}. \tag{D.15}$$

When this system is perturbed by a white noise sequence, we get the output

$$y_n - a_1 y_{n-1} - a_2 y_{n-2} = b_1 x_{n-1} + b_2 x_{n-2},$$

where x_i is the input sequence. This is called an ARMA (auto-regressive moving average) sequence. Many *disturbances* can be modeled in this way. If

$$y_n = y_{n-1} + x_{n-1}, \tag{D.16}$$

we have an *integrated* white noise. An ARIMA model is a series resulting from a combination of integrated process with an ARMA process.

Next we define a few parameters that characterize a random process with mean μ:

$$C_k = E\{(x_n - \mu)(x_{n-k} - \mu)\} \tag{D.17}$$

is called the *autocovariance* function.

$$\rho_k = C_k / \sigma_x^2 \tag{D.18}$$

is called the *autocorrelation* function. For a white noise sequence,

$$\rho_k = 1, \quad k = 0.$$
$$\rho = 0, \quad k \neq 0.$$

D.3 SPECTRAL DECOMPOSITION OF RANDOM PROCESSES

Given a random signal (say, sampled values of a random process), we can examine its frequency content using Fourier transform analysis. A Fourier transform expands a signal in terms of sine and cosine harmonic functions. This type of analysis is useful in determining the dominant frequencies contained in the signal. In this book, we focus on discrete signals and discrete Fourier transforms. The discrete Fourier transform is defined by the pair of relations

$$X(k+1) = \frac{1}{N} \sum_{n=0}^{N-1} x(n+1)e^{\frac{-2\pi ikn}{N}}, \tag{D.19a}$$

$$x(n+1) = \frac{1}{N} \sum_{k=0}^{N-1} X(k+1)e^{\frac{-2\pi ikn}{N}}, \tag{D.19b}$$

where $x(n)$ is the signal in time domain and N is the number of sample points. The discrete frequencies used in the expansion are

$$\omega_k = \frac{2\pi k}{NT_s}; k = 1, \ldots, N/2, \tag{D.20}$$

where T_s is the sample time. The set of complex numbers $X(1), \ldots, X(N)$ represents the frequency content of the original time series $x(1), \ldots, x(N)$ at frequencies starting at $2\pi/NT_s$ and ending at π/T_s. The highest frequency recovered, π/T_s, is called the Nyquist frequency. This forms the basis of the **Nyquist sampling theorem**, which states that to recover a signal from its sampled values, it must be sampled at $2h$ samples/sec where h cycles/sec is the highest frequency present in the signal.

Spectral decomposition is a useful diagnostic tool to use for analyzing noise present in sampled values of a process signal. We make extensive use of this tool in process identification (see Chapter 16). The following example illustrates how to use MATLAB to probe a signal for its frequency content.

Example D.1 Matlab Script for Spectral Decomposition

Consider the random signal shown in Figure D.1(a). This signal was generated by combining two sine waves with white noise. This Fourier transform yields the complex sequence $X(k)$. If we graph the amplitude of $X(k)$ [magnitude], we can identify the dominant two frequencies. This is shown in Figure D.1(b). The corresponding MATLAB commands are

```
X=fft(x);
% matlab computes frequency range from 2pi/(t_samp)/N to
2Pi/t_samp
% The transform is symmetric about the nyquist frequency
pi/t_samp
% only the first half is necessary.
freq=0:2*pi/(t_samp*N):pi/t_samp;
amp=abs(X(1:N/2+1))*2/N;
% Takes the first half and normalizes it
% so that unit amplitude in time corresponds to unit amplitude in
% frequency
loglog(freq,amp);% use loglog since power varies a great deal
```

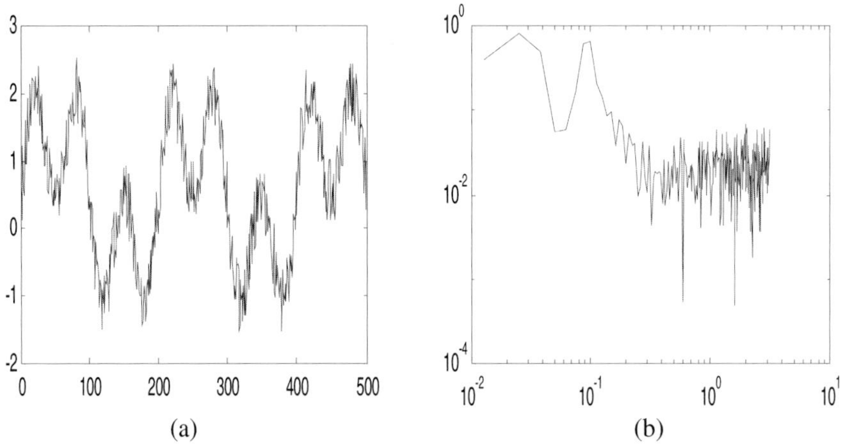

(a)   (b)

Figure D.1 (a) Random signal with two dominant sinusoidal variations. (b) Amplitude of its Fourier transform.

In practice, this method of extracting the frequency content is difficult to use because of the large random noise present in a signal. Instead the frequency content is extracted using averaging techniques. The Signal Processing toolbox in MATLAB contains a number of such programs (see the command `psd.m` for example).

A plot of the magnitude of the Fourier transform x at each frequency is called the power spectrum of the signal (since it shows contribution of each frequency to the total energy contained in the signal).

$\blacklozenge$

D.4 MULTIDIMENSIONAL RANDOM VARIABLES

Given a vector of random variables

$$x = [x_1 x_2 ..... x_n],$$

where each x_i is a random variable, we can define properties of the random vector x. These include the mean of x and covariance of x.

$$E\{x\} = \overline{x} \tag{D.21}$$

$$\varphi_{xx} = E\{(x - \mu)(x - \mu)^T\}. \tag{D.22}$$

Given two random vectors x and y, we can also define a cross covariance matrix,

$$\varphi_{xy} = E\{(x - \overline{x})(y - \overline{y})^T\}. \tag{D.23}$$

These quantities are used in Chapter 13 where we need to estimate one random variable in terms of another.

MATLAB and Control Toolbox Tutorial

Objectives

- Provide a brief introduction to MATLAB commands.
- Introduce the Control System toolbox.

MATLAB is an interactive program for scientific and engineering numeric calculations. It includes graphics (2D and 3D), programmable macros, interface to C and Fortran objects, and a graphical user interface (GUI) for block diagram simulation of linear and nonlinear systems (called SIMULINK). It is used for numeric computation, algorithm prototyping, control system analysis, signal processing, statistics, and optimization. MATLAB is available on a wide variety of platforms including Windows and Unix.

To get full benefit from this tutorial, the reader should enter the commands suggested and execute them while studying this Tutorial. It may take a few repetitions before you are completely familiar with the various commands. The problems at the end of the appendix should further reinforce the commands and their uses. This tutorial will take about 2 hours to complete (excluding problems).

E.1 MATLAB RESOURCES

There is a considerable amount of online help with MATLAB. There are a number of excellent websites with help and tutorials on MATLAB. See *www.mathworks.com* for additional help, if needed. The following are some tutorial web sites:

www.engin.umich.edu/group/ctm
www.rpi.edu/dept/chem-eng/WWW/faculty/bequette/edu.html (for SIMULINK)

   Student versions of MATLAB and SIMULINK are available directly from Mathworks, Inc. There are many good books that discuss MATLAB and SIMULINK in conjunction with control system design:

Using MATLAB, SIMULINK and Control System Toolbox, A. Cavallo et al., Prentice Hall, New Jersey, 1996.
Control System Design Using MATLAB, B. Shahian, and M. Hassul, Prentice Hall, New Jersey, 1993.
Mastering SIMULINK 2, J. B. Dabney, and T. L. Harman, Prentice Hall, New Jersey, 1998.
Mastering MATLAB 5, Hanselman and Littlefield, Prentice Hall, New Jersey, 1998. Highly recommended if you plan to use MATLAB heavily in your work.

For a complete listing of all MATLAB related books, visit the *mathworks.com* site.

E.2 BASIC COMMANDS

MATLAB is a command-line oriented program. When you start MATLAB, it responds with the MATLAB prompt » on the Command window. The commands you type are executed immediately. You can also store commands in a file to be executed later. To terminate MATLAB, type

```
»exit
```

Alternatively, you may type quit. They are synonymous. *Important: MATLAB commands are case sensitive.*
To get help, type

```
»help    % (This gives a list of help topics)
»help topic %  to obtain help on a specific topic (e.g.,
help plot)
»lookfor word % to obtain commands pertinent to a word
```

Also see the hypertext help available through the Help menu.

E.2.1 Creating Matrices

```
»    a=[1 2 3; 4 5 6; 8.5 10 12.5]
```

enters a 3 x 3 matrix. To see the matrix created, simply type

```
»a
```

Typing a variable name displays that variable. You can also enter matrices as follows:

```
»a = [                1    2    3
                      4    5    6
                      8.5 10  12.5]
```

You can create larger matrices already defined. Try these:

```
»b = [a a]
»b = [a; a]
```

Smaller matrices may be extracted from a larger matrix.
Try these:

```
»b = a(1,1)  ; % extract one element
»b = a(1,:)  ; % extract first row
»b = a(:,1)  ; % extract first column
»b = a(1:2,1) ; % extract first two rows of first column
»b = a(1:2,:) ; % extract first two rows of all columns
```

Note the significance of the : operator. Using this operator, you can extract selected rows or columns from a matrix. How would you extract an individual element of a matrix, say a_{23}? How would you extract the cofactor of a_{11}? Experiment and see if you can accomplish these tasks.
Matrices may be made up of complex numbers:

```
»Z1 = 3 + 4 * i

»Z2 = 1 + 2 * i
»Z  = [Z1 Z2]
```

E.2.2 Manipulating Matrices

To transpose a matrix	»b = a'
To get complex conjugate	»zc = conj (z)
To add the matrices	»b = a + a
To multiply matrices	»b = a * b
To divide (invert a matrix)	» b = inv(a)
To divide b by a (multiply by the inverse)	»c = a\b
To get determinant	»c = det(a)
To get eigen values	»[v,c]=eig(a)

Sine function	»c = sin(a)
Magnitude of a vector	»c=abs(a)

Try each one of these commands. What did you get? Try also the commands length(z),
size(a),norm(z). You can also use the logical operators (<, >, = etc.) with elements
of a matrix.

E.2.3 Creating Matrices

```
Identity matrix              »a  = eye(3)
Matrix of zeros              »a  = zeros (4,1)
Matrix of ones               »a  = ones(2,3)
Matrix of random elements  »rand(2,3)
Array,uniformly spaced       »a  =12:1:25
```

What do you get if you type rand(4)? How would you create a vector of random numbers
distributed uniformly between 0 and 100? Note that help is available on each command. Try
help rand. What is the difference between rand and randn?

E.2.4 Operations on Matrix Elements

```
»b = a .* 2
```

will multiply each element of the matrix *a* by 2. Note the period (.) before the *. This convention holds for the / operator as well.

```
b = a .* b
```

will multiply each element of *a* by each element of *b*; *a* and *b* must be the same size.
»b = max(a) is a column operator that will find the maximum value in each column of *a*.
Other column operators include sum and mean
Use the help command to determine the meaning of the following commands:

```
sort
median
std
any
all
```

E.2.5 Programming in MATLAB

The commands

```
» for i = 1:4
   x(i) = i ^ 2
   end
```

create a loop in MATLAB. Note that the variable x is created and augmented as needed by MATLAB. Display it by typing x at the command line. Similarly, to find the smallest integer n such that $2^n > = 20$,

```
»n = o
»while   2^ n < 20
    n = n + 1
    end
»n
```

Now try the same statements with a semicolon inserted after each statement. What is the difference? This feature is very useful in debugging MATLAB programs.

The following program creates an identity matrix of size n.

```
»n = 4
»for i = 1:n
    for j = 1:n
        if i==j
                x(i,j) = 1
        else
                x(i,j) = 0
        end
    end
end
»x
```

Note the = = sign. This implies a comparison, as opposed to the = sign, which implies an assignment of a value to a variable. Note the three end statements, one for the `if` statement and two for the two `for` statements. You could also have used the `eye` command. The latter is far more efficient (computationally and otherwise). Type "help eye" to learn how to use this command.

Note the use of the arrow keys on your keyboard. What do you get when you press these?

E.2.6 File Management, Search Paths

```
»who
```

will display all the variables you currently have in the workspace (memory). You can also use the Workspace Browser located on the top menu toolbar. Try it as well.

```
»save temp
```

will save all variables in memory (workspace) in a file called *temp.mat*. Check that it is saved by typing

```
»dir
```

This will list all files in your current directory. You can change the current directory using the Path Browser icon located on the top menu toolbar. This also allows you to add new directories to the paths, which MATLAB uses while searching for words that you type. When you type a word, MATLAB first searches current memory for the variable. If the file is not found, then the current directory is searched for a file with that name. If not found, MATLAB will search other directories listed in your path. If the file is not found, MATLAB will try to see if it is a built-in MATLAB command. If the file is still not found, MATLAB will issue an error message saying that the command is not recognized. For example, say you typed the command

```
»b=mean(a)
```

MATLAB first looks for a file named *mean.m* or *mean.p* or *mean.mex* in the current directory. If no such files exist, it will look in other directories listed in the path. Eventually, it finds that it is a built-in MATLAB command and executes it. It passes along the matrix *a* from the workspace. The result is stored in a matrix named *b*. MATLAB will also print *b* on the screen, since you did not put a semicolon after the command. If *a* is not in the workspace, MATLAB will issue an error message.

You can clear the current workspace with the command

```
»clear
```

Now check your workspace with the command

```
who
```

You can reload your variables from disk using the command

```
»load  temp.mat
```

To see what is defined in the workspace, type

```
»who
```

To store specified variables in a file, type

```
»save temp    a b c
```

To create a diary of all activities on the screen, type

```
»diary temp2.dia
```

Now try some of the commands you learned and then stop logging activity by typing

```
»diary off
```

Check the directory by typing

```
dir
```

Also use an editor to check the contents of your file *temp2.dia*. MATLAB has a built-in editor. Use File|Open from the menu from the top menu bar.

E.2.7 Storing MATLAB Statements in a File: m-files

Often it is convenient to store your program and data in a file. MATLAB can take instructions from a file instead of from the keyboard. These are called m-files, since they use that extension. Using the MATLAB editor, create a file called *temp.m* containing the following lines:

```
% file temp.m
% This is a comment line
% This is an example m file
% This file creates an nxn identity matrix
n=input('enter a number'); % user prompt
for i=1:n
   for j=1:n;
        x(i,j) = o;
   end;
        x(i,i) = 1;
end;
x   % display x.
```

Now, if you save this file in your current directory and type

```
»temp
```

this file will be read and executed.

E.2.8 Creating New Functions

You can define new function files. Unlike m-files, all variables in a function file are temporary and values are passed via arguments. Create the following file (called *mean1.m* in your directory) using the MATLAB editor:

```
function y = mean1(x)
% function to find mean of a vector
m = length (x);
y = sum(x)/m;
```

Now, if you type

```
»x = [1 2 3]
»Y = mean1(x)
```

it will display the mean of a vector, using your definition. However, the variable m is not in the workspace (check it). Note that any function defined in your directory gets precedence over built-in MATLAB function definitions.

E.2.9 Graphics

MATLAB can be used to create 2D and 3D plots. First choose the type of graph paper (e.g., plot for a linear *x-y* lot). Other choices include loglog, semilogx, semilogy, polar, mesh, and bar. Use the Help command to see what they are.
The following example plots a sine wave:

```
»x = 0:0.1:2*pi; % Note pi is already defined in MATLAB.
»y = sin(x);
»plot (x,y)
»% you can created multiple plots on the same window.
»z = cos(x)
»plot (x, y, x, z)
»plot (x, y, '*', x, z, '+')
```

Use the following commands to enhance the plot:

```
»title ('sinewave plot')
»xlabel ('time')
»ylabel ('y and z')
»grid
```

You can also use the menu bar available on the plot window to enhance the plot. Other useful commands are shown in the following table (type `help plot` for more info):

»shg	show graphic screen
»clc	clear command screen
»clg	clear graph screen
»axis	manual scaling of axes, this needs some arguments
»hold	hold the plot on screen
»subplot	for breaking graphic screen into multiple windows

E.2.10 Demonstrations

To see a demo of other capabilities of MATLAB, type

```
»demo
```

This will lead you to a menu of choices and some tutorials that explain how to use MATLAB.

E.2.11 Numerical Analysis

Create a file in your directory called *f1.m*, containing the following statements:

```
function y = f1(x)
y = 1 + x + x.*x
```

To numerically integrate this function from $x = -2$ to $x = 2$, use the command

```
»q = quad ('f1', -2,2)
```

This uses an adaptive Simpson's rule.

```
»xmin = fmin ('f1', -2,2)
```

finds the value of x at which $f1$ is a minimum in the interval given.

E.2.12 Solution of Ordinary Differential Equations

To solve the system of odes,

$$\frac{dx_1}{dt} = - 2x_1 - x_2,$$

$$\frac{dx_2}{dt} = - 2x_1 - 3x_2,$$

$x_1(0) = 1$, $x_2(0) = 1$, $t = 0$ to 10,

first create a function file *deriv.m*:

```
function xdot = deriv (t,x)
xdot(1) = -2.* x(1) - x(2);  % first equation
xdot(2) = -2.*x(1) - 3.*x(2); % second equation
xdot=xdot'; % Transpose the vector
```

Next define the initial values and time intervals:

```
»ti = 0;   % initial time
»tf = 10; % final time
»xi(1) = 1: initial value of x(1)
»xi(2) = 1
```

The solution is obtained using the fourth-order Runge-Kutta solver command ode45 .

```
» [t,x] = ode45 ('deriv', [ti tf] , xi); plot (t,x)% solve
```

Other solvers are available for stiff equations.

E.3 Control System Toolbox Tutorial

The Control System toolbox extends the commands in MATLAB to incorporate functions for the design and analysis control systems. This is sold as an add-on option to the basic MATLAB system.

E.3.1 Entering a System Transfer Function Model

Given a process with transfer function $G(s) = (s + 1)/(2s^2 + s + 1)$, we can enter this into MATLAB using

```
» num = [0 1 1]
» den = [2 1 1]
» sys = tf(sys)
```

The pair (num, den) is used to represent the transfer function. *num* contains the coefficients of the numerator polynomial and *den* contains the same for the denominator. *num* must be padded with a zero to make it equal in length to *den*.

Time delay can be entered as an optional property of the Control System object:

```
»g = tf(1,[1 1],'InputDelay',1)
```

results in the creation of a transfer function

$$g(s) = \frac{1.0e^{-s}}{s+1}.$$

E.3.2 Conversion to Other Representations

The transfer function may be converted to state-space form using the command

```
»sys_ss = ss(sys)
```

This converts the transfer function to a model of the form

$$\frac{dX}{dt} = AX + BU$$

$$y = CX + DU$$

tf does the reverse:

```
»sys2 = tf(sys_ss)
```

How does sys compare with sys2?

E.3.3 Analysis of Control Systems

```
»step (sys)
```

will generate and plot a step response. This is equivalent to

```
» step (sys_ss)
```

If you want to store the results, use

```
» [y,t] = step (sys)
```

Vector *y* will contain the output vector and *t* will contain the time vector. You can plot the results using

```
» plot (t,y)
```

where *y* contains the response of the state variables. If you want *y* at specific times given in a vector *t*, use the command

```
[y,x,t] = step (sys,t)
```

To obtain impulse responses, use the command

```
» impulse (sys)
```

Impulse can be used in various formats, just as step.

If you want the system response to arbitrary inputs given by a vector *u* which contains input at various times *t*, use the command lsim(num,den,u,t). For example, to get response to a sine function,

```
» t = 0:0.1:2*pi    %  generate a t vector
» u = sin(t)
» lsim(sys,u,t)
```

Alternatively, use

```
» [y,x] = lsim(sys,u,t)
» plot (t,y,t,u)
```

To obtain a frequency response of a system, use any one of the following commands:

```
» bode(sys)
» bode(sys_ss)
» [mag,phase,w] = bode (sys)
```

To compute the steady-state gain of a system, use

```
» k = dcgain (sys)
```

The program LTIVIEW provides a convenient GUI to examine the response characteristics of systems. First create all the system transfer functions or state-space models in the MAT-LAB workspace. Then type

```
» ltiview
```

which brings up the GUI. The control systems to be examined can then be imported using the File|Import command from the menu.

E.3.4 Block Diagram Analysis

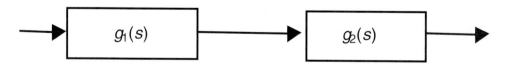

To combine two systems in series, use the command

```
[sys3] = series (sys2,sys2)
```

For example, let

$$g_1(s) = 1/(s + 1),$$
$$g_2(s) = 1/(s + 1),$$
$$g_3(s) = 1/(s + 1).$$

Then the following commands will generate a model of the combined system $g(s) = g_1(s)g_2(s)g_3(s).$

```
» num1 = [0 1]; den1 = [1 1]; sys1=tf(num1,den1)
» num2 = [0 1]; den2 = [1 1] ; sys2=tf(num2,den2)
» num3 = [0 1]; den3 = [1 1]; sys3=tf(num3,den3)
» sys4 = series(sys1,sys2)
» sys5 = series(sys4,sys3)
```

`series` can also be used with state-space models.
An alternate way to do this is to use the * operator:

```
» sys5 = sys1*sys2*sys3
```

To obtain the closed-loop representation of the system shown in the following figure,

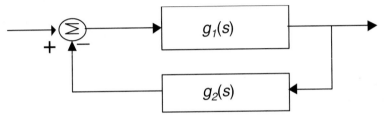

use the commands

```
» [sysfb] = feedback(sys1,sys2)
```

To obtain unity feedback, set $g_2(s) = 1$ (i.e., *num2* = 1, *den2* = 1). Let

$$g_1(s) = 1/(s^3+3s^2+3s+1)$$
$$g_2(s) = 1.$$

Then the commands are

```
» num1 = [0 0 0 1]
» den1 = [1 3 3 1]; sys=tf(num1,den1);sys1=tf([1];[1]);
» sysfb= feedback(sys,sys1)
```

E.3.5 Root Locus Analysis

To find the root locus of the system $g(s)$ (i.e., to find the roots of the equation $1 + kg(s) = 0$) for various values of k use the command

```
» r = rlocus(sys)
```

or

```
» r = rlocus(a,b,c,d)
```

You can also specify a gain vector k at which roots are to be computed:

```
» r = rlocus(num,den,k)
```

Note that r is a vector of complex numbers. As an exercise, plot the root locus of the system
$$g(s) = 1/(s^3 + 3s^2 + 3s + 1).$$
To get a root locus plot, use the command

```
» rlocus(sys)
```

If you want to know the gain k corresponding to a point, use the command

```
» rlocfind(sys)
```

immediately after using `rlocus`. This command will put up a crosshair on the graph. Use the mouse to position the cursor on the graph and press a key. MATLAB will print k. As an exercise, find the ultimate gain of the system just described. (Answer: $k = 8$.)

E.3.6 Phase and Gain Margins

Let us say that we used a controller gain of $k = 4$ to control the system described in the previous section:

$$g(s) = 1/(s^3 + 3s^2 + 3s + 1)$$

Find the gain and phase margin. Since the controller is in series as shown in the following figure,

```
» num = [0 0 0 4]
» den = [1 3 3 1]; sys=tf(num,den)
» % get frequency response
» [mag,phase,w] = bode [sys]
» [gm,pm,wcg,wcp] = margin (mag,phase,w)
```

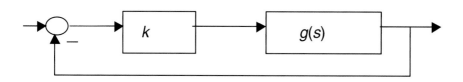

To see the margins on the screen, use

```
» margin (mag,phase,w)
» shg
```

E.3.7 Approximating Time Delays

Time delays can be approximated using the Padé approximation. The first-order Padé approximation is given by

$$e^{-Ds} \approx \frac{(D/2)\,s - 1}{(D/2)\,s + 1}.$$

Higher order approximations are possible. To get the equivalent transfer function, use the command pade:

```
» delay = 5.0
» [sys1] = Pade (delay,1)
» [sys2] = Pade (delay,2)
```

As an exercise, compare the frequency responses of the two approximations with the frequency response of a pure delay.

E.3.8 Discrete System Modeling

Discrete systems are represented using z-transfer functions (see Chapter 14). For example,

$$g(z) = \frac{(-3\,z + 1)}{(z^2 - 2\,z + 1)}$$

can be modeled in MATLAB using

```
»num = [0 -3 1]
»den = [1 -2 1]
ts=1; specify sample time
sys=tf(num,den,ts)
```

To convert a continuous system to discrete system, use the command c2dm. To do the reverse, use the command d2cm. A zero-order hold is assumed.
For example, consider the transfer function
$$g(s) = 1/(s + 1)$$
The following commands first convert this to discrete domain, using a sample time of 0.1 unit, and then reconvert it back into the time domain.

```
» numc = [0 1]
» denc = [1 1]; sysc=tf(numc,denc)
» sysd = c2d (sysc,0.1)
» sysc1 = d2c(sysd)
»
```

E.3.9 Analysis of Discrete Systems

The commands for analysis of discrete systems are as follows:

tf2ss	Convert to state-space form
ss2tf	Convert to transfer function form
dstep	Get step response
dimpulse	Get impulse response
dlsim	Do discrete system simulation
dbode	Get frequency response
ddcgain	Get steady-state gain
series	Connect systems in series
parallel	Connect systems in parallel
feedback	Get closed-loop transfer function
rlocus	Get root-locus

The use of these commands is quite similar to the corresponding continuous-time equivalents.

E.3.10 Multivariable System Modeling

The transfer function representation is suitable primarily for SISO systems. For MIMO systems, it is better to use the state-space representation in MATLAB. The general representation is

$$d\mathbf{x}/dt = a\mathbf{x} + b\mathbf{u}$$

$$\mathbf{y} = \mathbf{c}\mathbf{x} + \mathbf{d}\mathbf{u}$$

Example:

```
» a = [1   0
       0   1]
» b = [1
       0]
» c = [1   0
       0   1]
» d = [0
       0]
» sys=ss(a,b,c,d);printsys(sys)
```

This defines a single input, two-output, two-state variable system.

E.3.11 Parallel Connections

Two systems can be connected in parallel, using the command `parallel`. The syntax is (note the use of ellipses " ..." to indicate continuation to a new line)

```
[a,b,c,d] = parallel …
(a1,b1,c1,d1,a,2,b2,c2,d2,inp1,inp2,out1,out2)
```

Here `inp1` and `inp2` contain two vectors which specify which inputs are common among the two systems. For example, if

```
inp1 = [1 2] ; inp2 = [3 2]
```

then the combined system will have input 1 of system 1 connected to input 3 of system 2 and input 2 of system 1 connected with input 2 of system 2. The resulting system has three independent inputs. See the following figure:

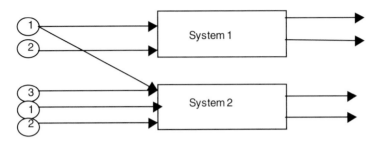

As an example, create a model of the system shown in the following figure:

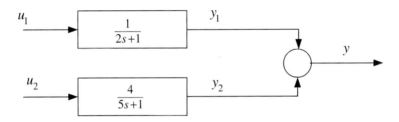

```
» % example of parallel connection
» num1 = [0 1] ; den1 = [2 1]
» num2 = [0 4] ; den2 = [5 1]
» [a1,b1,c1,d1] = tf2ss(num1,den1)
» [a2,b2,c2,d2] = tf 2ss(num2,den2)
» inp1 = []
» inp2 = []
» out1 = [1]
» out2 = [1]
» [a,b,c,d] = parallel (a1,b1,c1,d1,a2,b2,c2,d2,inp1,inp2,
out1,out2)
» printsys (a,b,c,d)
» k = dcgain (a,b,c,d) %  ss gain
» pzmap (a,b,c,d) % poles + zeros
```

E.3.12 Closed-Loop Systems

While closed-loop systems may be generated using append and connect, sometimes it is easier to use the cloop command. The syntax is

```
» [ac,bc,cc,dc] = cloop(a,b,c,d,out,in)
```

where out contains a list of outputs that are to be connected to the inputs named in the vector in. Negative values in the input vector imply negative feedback.

Examples

Model the system shown in the following figure:

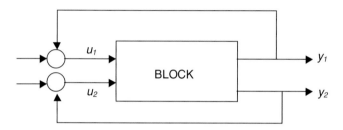

The block is modeled by

$$\dot{x}_1 = -x_1 - x_2 + u_1$$
$$\dot{x}_2 = -x_1 + x_2 + u_2$$
$$y_1 = x_1$$
$$y_2 = x_2$$

The MATLAB commands are

```
» a = [-1 -1;  -1  1]
» b = [ 1  0;  0 1]
» c = [ 1  0;  0 1]
» d = [ 0  0;   0 0]
» printsys (a,b,c,d)
» out = [1 2]
» in = [-1 -2]
» [ac,bc,cc,dc] = cloop(a,b,c,d,out,in)
» printsys (ac,bc,cc,dc)
» pzmap(ac,bc,cc,dc)
```

Problems

E.1 Use MATLAB to compute the following properties of the matrix

$$A = \begin{bmatrix} 4 & 3 \\ 2 & 5 \end{bmatrix}.$$

a. Inverse of A.

b. Determinant of A.

c. Eigen values and eigen vectors of A.

    d.   Sum of each column of A.

    e.   Mean of each column of A.

    f.   Rank of A.

    g.   Suppose $b = \begin{bmatrix} 10 \\ 14 \end{bmatrix}$ and $x = \begin{bmatrix} x_1 \\ x_2 \end{bmatrix}$; write the set of linear algebraic equations described by $Ax = b$.

E.2   Using MATLAB, solve the system of equations using MATLAB.

$$x_1 + x_2 + x_3 = 3$$
$$x_1 + 2x_2 + 3x_3 = 6$$
$$2x_1 + 2x_2 + x_3 = 5$$

Hint: First write the equations in matrix form: $Ax = b$; The solution is given by $x = A^{-1}b$.

E.3   Create matrix called A of random elements of size 4 x 4. The random numbers should be normally distributed.

    a.   What MATLAB command is used to set b = first row of A? first column of A? first two columns of A?

    b.   Using the Help facility in MATLAB, learn how to save this matrix in a file in the ASCII format. Print this file and attach it to your solution.

E.4   Create a function file to compute the sum of all elements in a given matrix. Hint: Use the MATLAB command `size` to get the size of a matrix.

E.5   Use MATLAB to solve the system of equations

$$\frac{dx_1}{dt} = -x_1,$$

$$\frac{dx_2}{dt} = -x_2 + x_1,$$

$$\frac{dx_3}{dt} = -x_3 + x_2,$$

$$x_1(0) = x_2(0) = x_3(0) = 1$$
$$0 < t < 10.$$

Submit a plot of x_1, x_2, and x_3 versus time on one graph.

E.6   Consider the process with the transfer function $g_p(s) = \dfrac{1}{s+1}$. The transfer function of the measurement/transmitter is given by $g_m(s) = \dfrac{1}{s+2}$. The transfer function of the control valve is given by $g_v(s) = \dfrac{1}{s+3}$. The system is represented by the following block diagram:

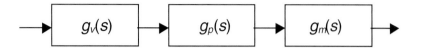

a. Use MATLAB to define g_p, g_m, and g_v, then connect them in series to obtain the overall transfer function

$$g(s) = g_v(s)g_p(s)g_m(s).$$

What is $g(s)$?

b. Using MATLAB, obtain and print a step response of the system $g(s)$.

c. Now add a proportional controller to the system $g_C = k_C$.

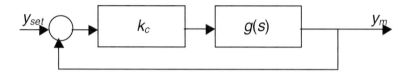

Using MATLAB obtain the closed-loop transfer function $g_{C1} = y_m/y_{set}$. Verify this by a hand calculation. Use $k_c = 30$.

d. Find the closed-loop poles of the system under proportional control using the `rlocus` command. For what values of k_c do you get poles in the RHP (i.e., the real-valued poles)

e. Obtain a step response of the closed-loop system for $k_c = 30, 60, 120$. Submit the plots.

SIMULINK Tutorial

Objectives

- Learn how to model nonlinear systems using SIMULINK.
- Learn how to model discrete time systems.
- Learn how to set up masks for systems.

F.1 BASICS

Consider the process shown in Figure F.1. We want to study how tank level $h(t)$ varies with inlet flow $f_{in}(t)$. A mass balance around the tank yields (assuming area of tank $A = 1m^2$, and that the density ρ is constant)

Accumulation of fluid mass in tank = Input of fluid − Output of fluid:

$$\frac{d(\rho Ah)}{dt} = \rho f_{in}(t) - \rho f_{out}(t),$$

$$\frac{dh}{dt} = f_{in}(t) - f_{out}(t).$$

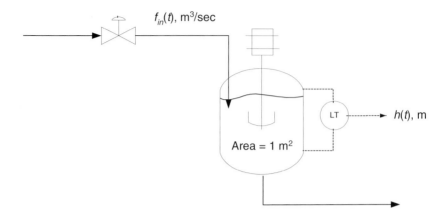

Figure F.1 Model of level variations in a tank.

Let us assume that outlet flow is proportional to the height, given by $f_{out}(t) = Kh(t)$, where K depends on the outlet valve opening. We will use $K = 2 \ m^3/s/m$ for this simulation. Increasing the opening leads to an increased value of K. The differential equation now becomes

$$\frac{dh}{dt} = f_{in}(t) - Kh(t)$$

or

$$h(t) = \int \left[f_{in}(t) - Kh(t) \right] dt.$$

We can represent this system using the block diagram shown in Figure F.1. This block diagram has the following elements:

Sum	A block to add or subtract signals.
$1/s$	A block that integrates.
Step	A block that supplies a step input (a signal source) used to model $f_{in}(t)$.
Scope	A block that is used to display a variable (a signal). The name derives from the use of oscilloscopes used by electrical engineers to probe signals.

We can create a simulation of this system by first generating the block diagram on the screen, using the building blocks supplied with SIMULINK, as shown in Figure F.2.

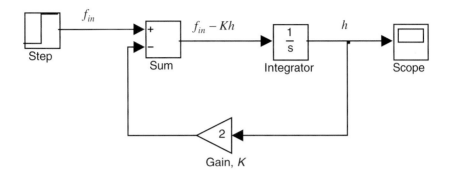

Figure F.2 Block diagram representation of the model.

Note that we have used a step block to simulate a step change in the inlet flow rate. Using this simulation, we will study the response characteristics (change in level as a function of the changes in inlet flow). The scope allows us to graph the time-dependent behavior of the level. The lines connecting blocks represent signals.

F.2 CREATING A BLOCK DIAGRAM IN SIMULINK

1. Start up MATLAB. Click on the SIMULINK icon (shown on the top menu bar) to start SIMULINK. Create a new model by clicking on the blank page icon at the top of the SIMU-LINK Library Browser Window. Save this as *tank* using the File/Save option in the menu bar of the new window called *Untitled* that opens up. It will be saved as *tank.mdl* in your current MATLAB directory. **.mdl** is the extension used by all SIMULINK **model** files.

2. Double-click the SIMULINK Library in the Library Browser Window (in the newer versions of SIMULINK, this library may already be open). You should see a library of models displayed. These include

> **Sources** - to generate input signals
> **Sinks** - to display or store outputs
> **Continuous** - linear blocks used in simulation, such as gain, integration, sum, etc.

3. Open up Continuous Block Library by double-clicking on it. Click and drag the following elements to your tank screen. Place them as shown in Figure F.1.

> 1 Integrator (from Continuous Block Library)
> 1 Sum (from Math Block Library)
> 1 Gain (from Math Block Library)

4. Open Sources. Click and drag a "step" input to the tank window. Close library.

5. Open Sinks Library. Click and drag a "scope" element to the tank window. Close library.

6. Rearrange and connect the blocks to achieve the model shown in Figure F.1. Blocks are linked by clicking and dragging arrows from source to destination. You can flip a block by clicking on it and using the Format command from the top menu bar.

You can *branch* out from a link by holding down the *Ctrl* key and clicking and drawing from the link.

This completes the block diagram construction.

7. Now we need to define *parameters* for each block. Start with the *Step* block. Double-click on it. A menu opens up. Define the *Step Time* (time at which the step increase in signal occurs) as 1. Define *Final Value* (the step size) as 2.0. This varies the inlet flow f_{in} as a step function that is zero until time = 1 sec, and then jumps to a value 2.0 m^3/s from then on. Close the menu.

Next, open up the *Sum* block. Define list of signs as + –, indicating that positive (+) value of the first input is combined with negative (–) value of second input. This gives us the output of the block as output = input 1 – input 2. Close the menu.

Next open up the menu for *Integrator*. Note that initial condition is taken as zero by default. There is no need to change any parameters here. Next open up the *gain* block. Set gain = 2. This is equivalent to setting $K = 2$ in our model.

Next open up the *Scope* block. This screen will display the output of the integrator as the simulation proceeds. Leave it open. Move it to a side, if necessary, by clicking and dragging the top menu bar.

Next open the *Simulation|Parameters* menu. Note that the start time is zero and the stop time is 10.0. There is no need to change any of these. Close the menu.

8. Start the simulation by clicking on the menu *Simulation|Start*.

You should see the result displayed on your scope screen. If the scale is not correct, click on the Binoculars button. You can also adjust the scale using the right-most button on the menu of the scope.

That is it! You now have completed the simulation of the tank. The plot represents the change in level for a step change in the input.

Answer the following questions:

1. What is the steady-state value reached by the level? Note that the level is measured in meters.
2. What is the steady-state reached if the input step is doubled? Go back to the step block, double the input size, and repeat the simulation.
3. What is the steady-state *gain of the process* (defined as change in output/change in input)? Does it depend on the step size of the input? What are the units of this process gain?
4. Approximately how long does it take to reach 63% of the steady-state change (note the time taken *after the step change was made*). Does it depend on the step size? This is called the *time constant* of the process.

In order to answer the last question, you should produce a hard copy of the plot. The best way to do this is to use the MATLAB plotting capability. First you must save the result of the simulation in the workspace. To do this, add a block called *To workspace* (taken from the *Sinks* library) to your diagram. Double-click on it and change its name to *level*. Change the save format to *Matrix*. Close it and link it to the output of the integrator, as shown in Figure F.3.

Now repeat the simulation. Go to the MATLAB window and examine the variables present by typing *who*. (You can also use the Workspace Browser icon on the top menu bar of the MATLAB Command window.) Plot *level* using the plot command. You may need to see *Help* on plotting. You can print it using *File|Print* from the menu.

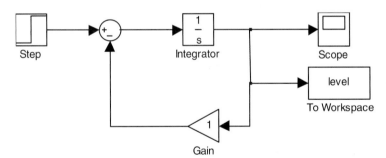

Figure F.3 SIMULINK model of the tank level system.

F.3 LAPLACE TRANSFORM MODELS

Another way to simulate this tank is to take the Laplace transform of the differential equation model

$$\frac{dh}{dt} = f_{in}(t) - Kh(t), \quad h(0) = 0$$

to get the transfer function

$$h(s) = \frac{f_{in}(s)}{(s+K)},$$

which can be modeled in SIMULINK using a transfer function block:

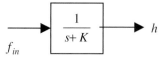

f_{in}

This block can model a general transfer function of the term

$$G(s) = \frac{a_n s^n + a_{n-1} s^{n-1} + ... + a_0}{b_m s^m + b_{m-1} s^{m-1} + ... + b_0}.$$

The parameters of this block are the coefficients of the numerator and denominator polynomials entered as a vector array of elements separated by spaces as $[a_n \ a_{n-1} \ ... \ a_0]$.

As an exercise, set up the tank model using this method. You will need to connect a step input function and a scope to the transfer function. Verify that it gives the same result as the earlier model you created.

F.4 SIMULATION OF DISCRETE SYSTEMS USING SIMULINK

Consider the tank level system we modeled earlier:

$$\frac{dh}{dt} = f_{in}(t) - Kh(t)$$

Discretizing this system using a forward finite differencing yields

$$\frac{h_{n+1} - h_n}{T_s} = f_{in,n} - Kh_n,$$

where T_s = sampling time (or discretization time step) and $h_n = h(n \cdot T_s)$; that is, $h(t)$ evaluated after n time steps. Let z^{-1} denote the discrete time delay operator:

$$z^{-1} h_{n+1} = h_n.$$

SIMULINK has a built-in discrete time delay operator z^{-1}. It has one parameter, namely, the sample time T_s. The default value is 1 unit. We can rewrite the model as

$$h_{n+1} = T_s(f_{in,n} - Kh_n) + h_n = T_s f_{in,n} + (1 - T_s k) h_n.$$

This is translated into a SIMULINK model as follows:

$$h_n = z^{-1}[h_{n+1}] = [z^{-1}][T_s f_{in,n} + (1 - KT_s)h_n].$$

Implement this model in SIMULINK, as shown in Figure F.4. *Note that T_s and K are defined in the workspace before the simulation is started.*

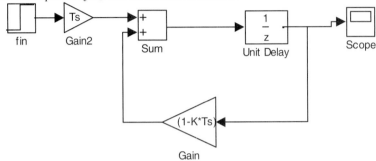

Figure F.4 Discrete model of the tank.

Compare the step response obtained by this model for T_s varying from 0.1 to .4 (see Figure F.5). When $K = 2$ was used for this simulation, a step change of 1 was introduced in f_{in} at time = 1.

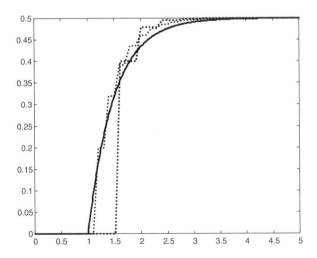

Figure F.5 Response of level to a step increase in f_{in} as computed using increasing values of sample time T_s.

At 0.1 the response is close to that of the continuous system. As T_s increases, the finite difference approximation gets worse. At high values of T_s, the system may become unstable due to numerical instability. Verify this by simulation.

F.4.1 Use of Discrete Transfer Functions

Another way to simulate the system is to define a discrete transfer function. We have

$$h_{n+1} = zh_n.$$

Substituting this into our model equation,

$$h_{n+1} = T_s f_{in,n} + (1 - KT_s)h_n.$$

Using the z-domain notation, we can write this as

$$zh(z) = T_s\, f_{in}(z) + (1 - kT_s)\, h(z).$$

Solving for h,

$$h(z) = \frac{T_s}{z + (kT_s - 1)}\, f_{in}(z),$$

$$= G(z) f_{in}(z),$$

where

$$G(z) = \frac{T_s}{z + (kT_s - 1)}.$$

$G(z)$ is called the discrete transfer function relating $f_{in}(z)$ and $h(z)$. In general,

$$G(z) = \frac{a_n z^n + a_{n-1} z^{n-1} + \cdots + a_0}{b_m z^m + b_{m-1} z^{m-1} + \cdots + b_0}.$$

SIMULINK has a discrete transfer function block capable of simulating this type of block. The block has two vector parameters: The numerator coefficients are specified as vector

$$[a_n \quad a_{n-1} \quad \cdots \quad a_0],$$

and the denominator as a second vector,

$$[b_m \quad b_{m-1} \quad \cdots \quad b_0].$$

Figure F.6 shows the tank level simulation using a transfer function block. The sample time is also a parameter associated with this block.

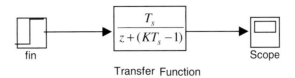

fin

Transfer Function

Scope

Figure F.6 Discrete transfer function model of the tank level.

Verify that this model gives the same results as the previous model. Make sure the parameters, such as T_s and K, are defined in your workspace. Also, the transfer function block must have the sample time defined as one of the block parameters.

F.4.2 Subsystems and Masks

You can condense a large SIMULINK step into a single block by drawing a rectangle around the blocks you want to condense, using the mouse and then choosing *Create Subsystem* from the Edit menu. This will hide all the details and show only the inputs and outputs from the subsystem.

Open up the tank simulation you created earlier. Create a subsystem that includes the integrater, sum, and gain blocks.

Masking is one way to hide the details of the subsystem from the user. See the Help file on masking. Masking allows you to enter parameters used by the subsystem. For example, the gain K can be made one of the parameters that can be entered through the mask (see SIMULINK Helpdesk to get a tutorial on how to create masks).

As an exercise, create a mask for the system just described that allows the user to change K.

Problems

F.1 Replace the step input in the previous example with a sinusoidal input function. Observe the output response of the system to increasing frequencies of the input sine wave. Qualitatively summarize your observations (note the amplitude of the output wave and how it changes with the frequency). Try frequencies of 1, 2, 4, 10, and 20.

F.2 Create a SIMULINK diagram to generate a pulse input of the form shown in the following figure. You can do this by first generating two step signals of opposite magnitudes and then summing them together using a *Sum* block. The Pulse function in the SIMULINK Sources Library is not suitable, as it produces a series of pulses, not just one pulse, as shown in the figure.

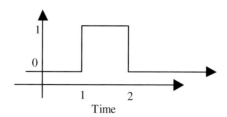

Time

F.3   Set up a SIMULINK model of a second-order system modeled by

$$\tau^2 \frac{d^2x}{dt^2} + 2\tau\zeta \frac{dx}{dt} + x = u.$$

One way to do this is by using two integrator blocks; the first block will integrate

$$d^2x/dt^2 = [+u - x - 2\tau\zeta\dot{x}]/\tau^2$$

to get dx/dt and then integrate this signal to get x:

$$dx/dt = \int (d^2x/dt^2)\, dt.$$

u can be generated using a step input source block. Another way to solve this problem is to use a transfer function block. Obtain the response for $\tau = 1$, $\zeta = 0.5$, 1, 2 and $u =$ step input of size 1.

F.4   Set up the SIMULINK scheme to simulate the following control system:

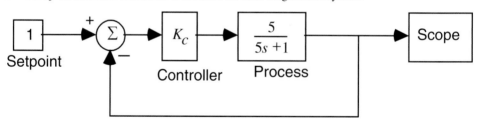

Hint: Use *Constant* from the *Sources* blockset for the setpoint . Use *Sum* from linear blockset for the summer. Use *Transfer Function* for the process. Use *Gain Block* for the controller.

 a. Simulate the system for 40 seconds for various values of K_c. Observe responses and comment on the results.

 b. Create a subsystem for a proportional integral (PI) controller with the elements as shown in the following figure:

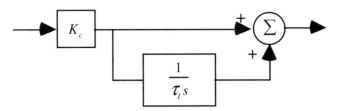

c. Replace the proportional controller in part a with PI controller. Vary K_c, τ_I and study its effect on the system response.

d. What values of K_c and τ_I do you recommend for this system? Why? Submit plots of the process output response and controller output response. What is the largest value reached by the controller output? In practice, controller outputs are limited by physical limitations.

Tutorial on IMCTUNE Software

Objectives

- Provide an introduction to IMCTUNE software.
- Describe the tfn and tcf commands for MATLAB that are provided in IMCTUNE to assist in IMC controller selection and to facilitate transfer function analysis.

G.1   INTRODUCTION

IMCTUNE facilitates the design and tuning of the following types of controllers with or
without model uncertainty.
- 1DF IMC controllers
- 2DF IMC controllers
- MSF IMC controllers
- 1DF and 2DF PID controllers
- IMC and PID Feedforward and Cascade controllers

The IMCTUNE software package is a collection of MATLAB m-files that can be
downloaded from http://www.phptr.com/brosilow/. It requires MATLAB 5.3 or higher, and
the Control System and Optimization Toolboxes. The software can also compute and
display
- single-loop and cascade IMC, MSF, and PID closed-loop responses to step setpoint
  and disturbance changes,
- the SIMULINK diagrams used to simulate such structures,
- various closed-loop upper and lower-bound frequency responses, and
- individual process closed-loop frequency responses.

We recommend installing IMCTUNE in its own directory and creating a subdirectory
called *data* under the IMCTUNE directory to store the data files for IMCTUNE.

G.2   GETTING STARTED ON 1DF SYSTEMS

Either put the IMCTUNE directory in the MATLAB path, or change directories inside
MATLAB to the IMCTUNE directory. Start IMCTUNE by typing *imctune* in the MATLAB
Command Window. This should open up the IMCTUNE's primary interface shown in
Figure G.1.
The primary IMCTUNE interface (c.f. Figure G.1) has been structured to eventually
accommodate multi-input, multi-output processes. However, the version of IMCTUNE
available on the Prentice Hall website is only for single-input, single-output processes, and
that is all that is needed for this text. Thus, the number of inputs and outputs should remain
at the default of one. The drop down menu labeled A in Figure G.1 offers the following
structure options: one-degree of freedom, two-degree of freedom, and cascade system. We
shall focus on the default one-degree of freedom system, and comment only on the major
differences in the other two structures. The drop down menu labeled B in Figure G.1 offers
a single-term transfer function for the process and model, or a two-term transfer function.
The latter allows entry of process descriptions like $(1+.5e^{-s})/(s+1)(2s+1)$. Other inputs
are similar to those for the single-term transfer function.

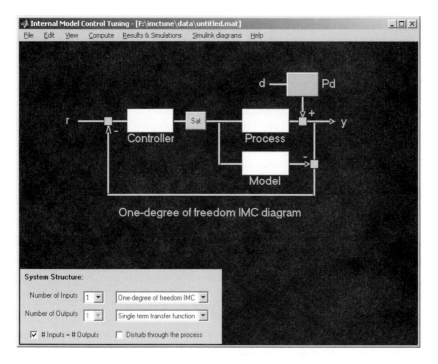

Figure G.1 IMCTUNE 1DF primary interface.

G.2.1 Data Input

Clicking on any block allows the user to enter the transfer function and parameters for that block. For example, clicking on the process block brings up the window shown in Figure G.2.

The coefficients of the numerator polynomial, and denominator polynomial are entered as vectors, and the time delay as a scalar. Consider the process, model and controller given by

$$p(s) = \frac{1}{s+1} e^{-Ts} \quad .5 \le T \le 1.5, \tag{G.1}$$

$$\tilde{p}(s) = \frac{1}{s+1} e^{-s}, \tag{G.2}$$

$$q(s) = \frac{(s+1)}{(\varepsilon s+1)}. \tag{G.3}$$

The process window for Eq. (G.1) is shown in Figure G.3. Notice that the uncertain dead time is entered as the letter u. This shows up in the *current process* portion of the window as exp(−*x*(1)*s*). The upper and lower-bounds for *x*(1) are entered as shown in Figure G.3.

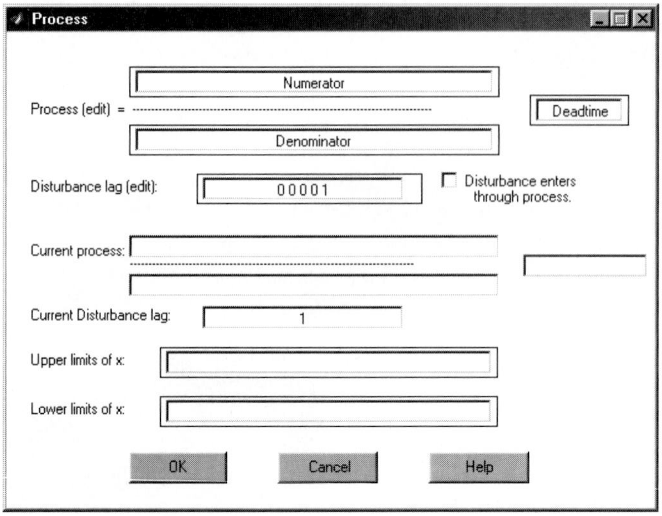

Figure G.2 Blank process screen.

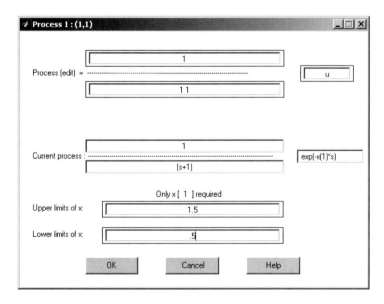

Figure G.3 Screen after entering the process of Eq. (G.1).

G.2.2 Entering General Numerator and Denominator Polynomials for a Process or Model

A polynomial may be entered in the following two ways.

G.2.2.1 Expanded Form

The polynomial $\alpha_n s^n + \alpha_{n-1} s^{n-1} + \ldots \alpha_1 s + \alpha_0$ is entered using only its coefficients as $\alpha_n \; \alpha_{n-1} \ldots \alpha_1 \; \alpha_0$. That is, the coefficients are entered separated by spaces. Any uncertain coefficient is entered as u and its limits are given in the rows labeled upper and lower limits. IMCTUNE will assign a variable $x(i)$ to each uncertain coefficient you enter. For example, if α_n and α_1 are uncertain coefficients, the polynomial would be entered as $u \; \alpha_{n-1} \ldots \; u \; \alpha_0$. If any uncertain coefficients are the same they are entered using $u1$, $u2$, etc. For example, if the coefficients $\alpha_n, \alpha_{n-1}, \alpha_1,$ and α_0 of the polynomial $\alpha_n s^n + \alpha_{n-1} s^{n-1} + \ldots \alpha_1 s + \alpha_0$ are uncertain, and α_{n-1} and α_0 are the same, but different from α_n and α_1, which are also the same, then the polynomial is entered as $u1 \; u2 \; \alpha_{n-2} \ldots \alpha_2 \; u1 \; u2$.

G.2.2.2 Factored Form

The polynomial $(s + \alpha_1)(s + \alpha_2)(s^2 + \alpha_3 s + \alpha_4)$ can be entered as $1\,\alpha_1 ; 1\,\alpha_2 ; 1\,\alpha_3 \; \alpha_4$. Again, each independent uncertain coefficient is entered as u.

    If the disturbance passes through a lag, this lag can be entered in either of two ways. First, it can be entered just like the process by clicking on the Pd block in Figure G.1, which brings up Figure G.4.

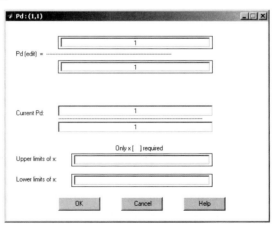

Figure G.4 Disturbance lag screen.

Numerator and denominator polynomials are entered as described above. There is no deadtime element since an unmeasured delayed disturbance cannot be distinguished from one that simply enters later without passing through a deadtime.

Second, if the disturbance passes through the process, check the box labeled *disturbance through the process* in the primary screen (Figure G.1).

G.2.1.3 Parallel Processes

If the process is modeled as a sum of two transfer functions, it is entered by opening the pull down menu labeled B in Figure G.1, and selecting *two-term transfer function*. Clicking on the process block, and entering the process given by Eq. (G.4), results in the screen of Figure G.5.

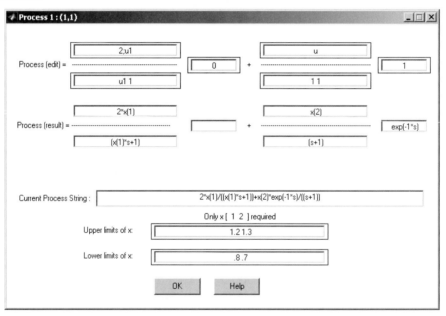

Figure G.5 Screen after entering the process of Eq. (G.4).

In this process we have entered the process model

$$p(s) = \frac{2x(1)}{x(1)*s+1} + \frac{x(2)}{s+1}e^{-s}, \, 0.8 \le x(1) \le 1.2, 0.7 \le x(2) \le 1.3 \qquad (G.4)$$

Note that the uncertainty in gain and time constant of the first term are correlated. This correlation is entered using $u1$ rather than u for the uncertain variable. If you have more correlated uncertain variables, use $u1$, $u2$, and so on. IMCTUNE automatically assigns variable names $x(1)$, $x(2)$, and so on, for each independent uncertain variable, as shown in Figure G.5.

G.2.1.4 Uncertainty Bounds

Upper and lower-bounds must be provided for each uncertain parameter (entered as u or u#) in the transfer function. See Figures G.3 and G.5. Such bounds are entered as arrays of numbers separated by spaces just like the coefficient vectors.

G.2.1.5 Entering the Model

By clicking once in the block labeled *Model*, you can enter the model. The model transfer function is entered using the same format as described previously, either as the coefficients of a single polynomial or as the coefficients of factored polynomials separated by semicolons. Since a future implementation of IMCTUNE may find optimal values for model parameters, the current IMCTUNE interface accepts uncertain parameters for the model, which become variables $y(1)$, $y(2)$, …, $y(n)$. However, we recommend that in using the current version, only constant values should be entered for model parameters. Figure G.6 shows the Model window for the model of Eq. (G.2) using a mid-range deadtime.

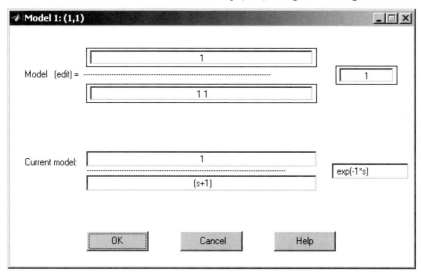

Figure G.6 Model window for Eq. (G.2).

G.2.1.6 Controller

The IMC controller is entered in the controller window by specifying the part of the process model to be inverted by the controller. The filter order is generally chosen to be the relative degree of the transfer function to be inverted. Any filter time constant may be entered here, but it is usually calculated by IMCTUNE either to satisfy an Mp criterion for an uncertain process, or to satisfy a high frequency maximum noise amplification specification for a perfect model. Figure G.7 shows the IMC controller window for the controller given by Eq. (G.3) using a filter time constant of 0.05 for a maximum noise amplification of 20.

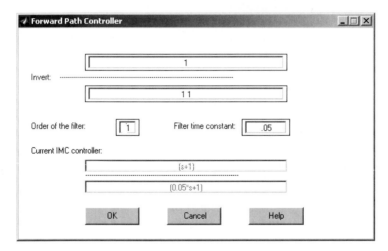

Figure G.7 IMC controller window for the process model of Figure G.6.

G.3   MENU BAR FOR 1DF SYSTEMS

The menu bar has the following options:

File

- ➢ New: Start a new design.
- ➢ Load: Load a file containing data describing a previous problem and controller design.
- ➢ Save: Save current state of IMCTUNE as a *mat* file in the data subdirectory of the current directory.
- ➢ Save as: Save current state of IMCTUNE with a file name, and in a directory as specified by the user.
- ➢ Exit: Quit IMCTUNE.

Edit

> ➤ Saturation bounds: For entering upper and lower limits of the manipulated variable (same as clicking on the *sat* block).
> ➤ Default values: You can enter values for maximum allowed noise amplification, frequency range for closed loop frequency response calculations, the number of points per decade used in the calculation, and the desired accuracy of the calculation. Any number larger 10^5 is treated as infinity during calculations. The default values are shown in Figure G.8. Any of these may be changed. The frequency range is in powers of 10 (e.g., 0.1 to 10) for Figure G.8.

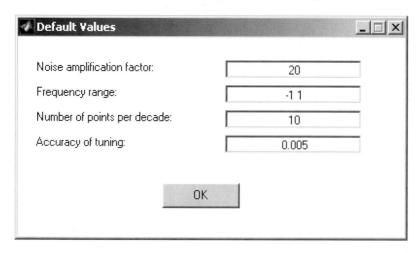

Figure G.8 Default values.

View

View brings up the summary of all parameters entered so far as in Figure G.9 for the process of Eq. (G.1), and the model and controller of Equations (G.2) and (G.3).

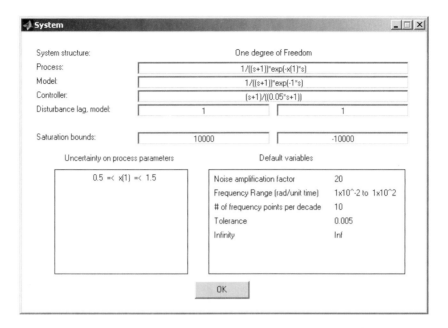

Figure G.9 View window for Equations (G.1), (G.2) and (G.3) with $\varepsilon = .05$.

Compute

The choices under the Compute menu option are

> Tuning: Computes filter time constant via Mp Tuning (see Chapter 7). Figure G.10 shows the tuning results for an Mp specification of 1.05 for Eq. (G.1).
> Noise amplification filter: Computes filter time constant that satisfies the noise amplification set in the default values window (see Chapter 3).
> MSF K, K_{sp}: Computes the MSF coefficients (see Chapter 5).
> Find uncertainty bounds: Given a model, a filter time constant, and a set of nominal values of the uncertain parameters, IMCTUNE computes the fractional variation of the uncertain parameters around the nominal values for which the specified Mp will be met (see Chapter 3).
> Tuning for lower-bound saturation: Computes a safe lower-bound for the IMC filter time constant for MSF (see Chapter 5).
> PID controller: Computes the parameters of an ideal PID controller, and PID controllers cascaded with 1st and 2nd order lags (see Chapter 6).
> Frequency response: Computes upper- and/or lower-bounds for the sensitivity or the complementary sensitivity functions. Individual process frequency responses can be added to the upper and/or lower-bounds results via the add button (see Chapter 7 and Figure G.10).

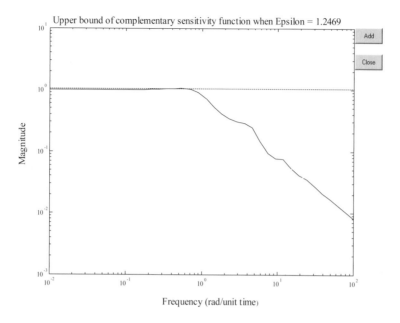

Figure G.10 Results from Mp Tuning of Eq. (G.1) for Mp = 1.05.

Results & Simulations

➤ PID controller: Shows the results of the most recent calculations of PID parameters.

➤ MSF gain, K, K_{sp}: Computes the MSF feedback parameters (see Chapter 5).

➤ IMC & MSF step responses: Provides, and allows comparison of, IMC and MSF step setpoint and disturbance changes as shown in Figures G.11a and G.11b.

➤ IMC & PID step responses: Provides, and allows comparison of, IMC and MSF step setpoint and disturbance changes as shown in Figure G.12.

The responses in Figures G.11a, G.11b, and G.12 are for the following process, model, and IMC controller

$$\widetilde{p}(s) = p(s) = \frac{e^{-s}}{(s+1)}, \quad q(s) = \frac{(s+1)}{(\varepsilon s+1)} \quad 0 \le u(t) \le 1.1. \tag{G.5}$$

The model and IMC controller in Eq. (G.5) are the same as for Equations (G.2) and (G.3), but the process has no uncertainty so as to show up the difference between IMC and MSF responses. Also, the control effort is constrained as shown. The button windows in Figure G.12 are the same as for Figure G.11b, except for the *add and remove* window, which is shown on the figure.

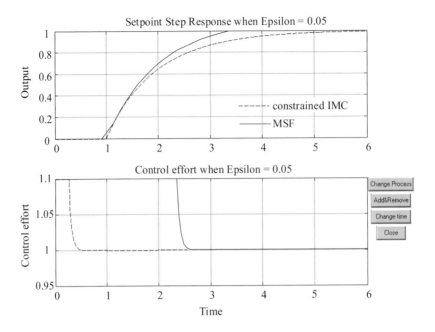

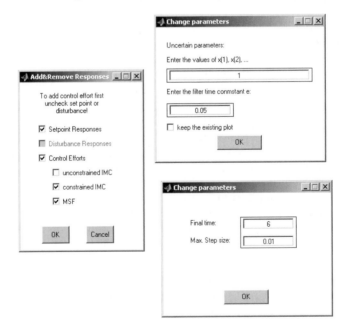

Figure G.11a IMC & MSF step responses.

Figure G.11b Screens associated with the buttons of Figure G.11a.

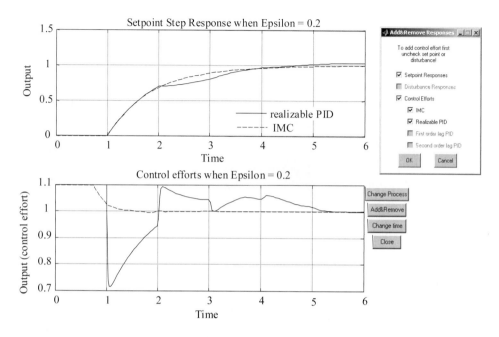

Figure G.12 IMC and PID step responses.

SIMULINK diagrams

The user has access to the SIMULINK programs that carry out the IMC, MSF, and PID simulations, and can modify these diagrams as desired. However, to maintain the integrity of the IMCTUNE software, all the diagrams can be restored to their original form.

G.4   GETTING STARTED ON 2DF SYSTEMS

Figure G.13 shows the primary IMCTUNE interface for 2DF control systems for the case where the disturbance passes through the process. As the reader can see from Figure G.13, the controller is now split into two parts: a forward path part and a feedback path part. While the menu bar for 1DF and 2DF systems are the same, the items under View and Compute are different, as are the model and controller windows. This section describes the differences in the controllers and the model. The next section describes the differences in View and Compute menus.

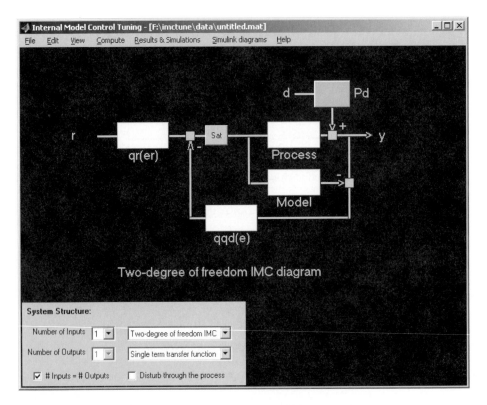

Figure G.13 IMCTUNE 2DF primary interface.

G.4.1 The Model

Figure G.14 shows the model window for a 2DF system using the same model as in Eq. (G.2). There is one very important difference with the model window for 1DF systems. There is now a model *disturbance lag*. This lag actually has nothing to do with the model, but rather establishes the form of the $q_d(s)$ part of the feedback path controller. Future versions of IMCTUNE will therefore have the *disturbance lag* moved to the feedback path controller window. Its location in the model is due to the history of the development of IMCTUNE. The $q_d(s)$ part of the feedback path controller is selected so as to have the zeros of $(1 - \tilde{p}qq_d(s))$ the same as the roots of the disturbance lag.

Figure G.15 shows the feedback path controller window for the model of Figure G.14 for a filter time constant of .7. Figure G.16 shows the forward path controller. Figure G.17 shows the process, model, and controllers for all windows in Figures G.14 through G.16.

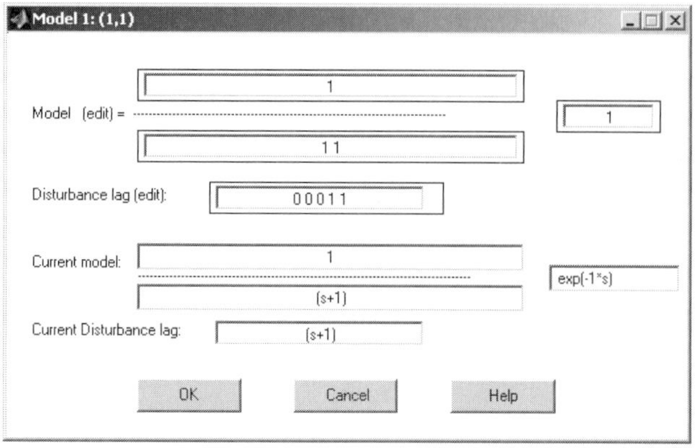

Figure G.14 2DF model window.

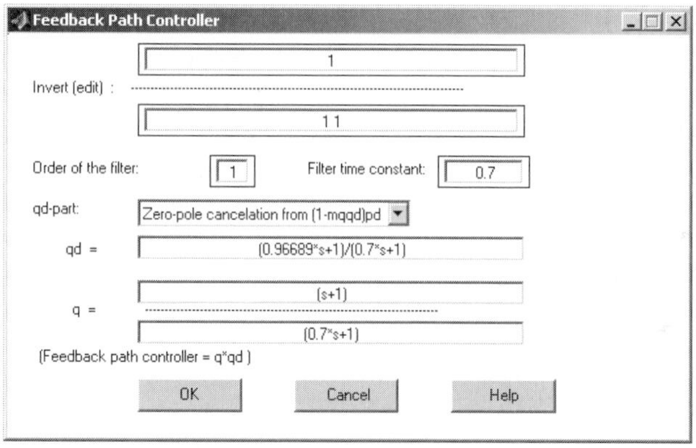

Figure G.15 2DF feedback path controller window.

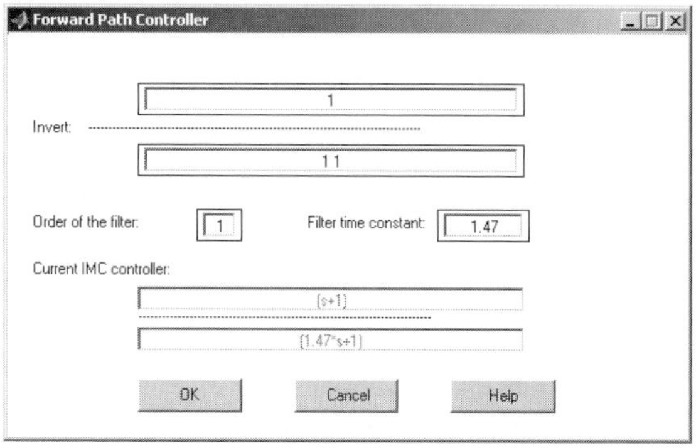

Figure G.16 2DF forward path controller window.

G.5   MENU BAR FOR 2DF SYSTEMS

View

Figure G.17 shows the current system. Notice that the feedback controller has two parts, q, and q_d. Also, the filter time constants for the forward path and feedback path controllers are not the same.

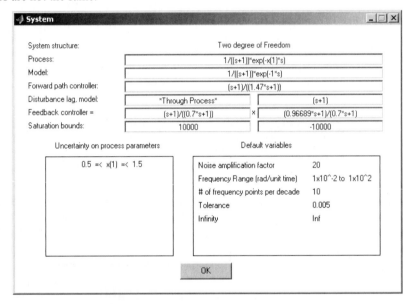

Figure G.17 View screen for 2DF system.

Compute

The choices under the Compute menu option are similar to those for 1DF systems. Therefore, below we emphasize the differences

> ➢ 2DF tuning: Has both feedback path controller tuning (inner loop tuning) via the partial sensitivity function and forward path controller tuning via the complementary sensitivity function (see Chapter 8).
> ➢ 2DF PID controller: Provides the setpoint filter as well as the PID controller approximation for $c(s)$ in Figure 6.7 of Chapter 6.
> ➢ Frequency response: For the response of the output to the disturbance, computes the sensitivity function, the integrated sensitivity function, the normalized integrated sensitivity function, and the partial sensitivity function.

G.6   GETTING STARTED ON CASCADE CONTROL SYSTEMS

Figure G.18 shows the primary window for cascade control systems.

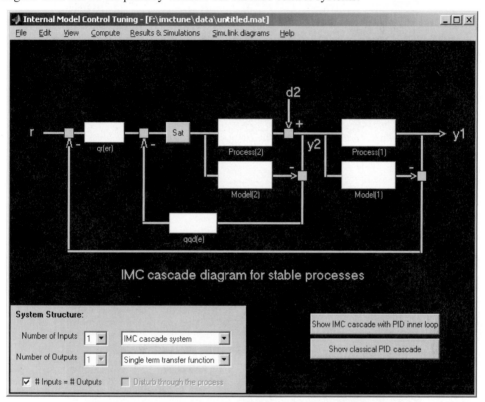

Figure G.18 Cascade control system primary interface.

Cascade control system design and tuning starts with the IMC cascade structure shown in Figure G.18. Once the controllers $qq_d(s,\varepsilon)$ and $q_r(s,\varepsilon_r)$ have been designed and tuned, the user can select two other cascade configurations: (1) an IMC cascade with a PID inner loop as in Figure G.19 or (2) a classical PID cascade control system as in Figure G.20. Clicking on either of the *Show* buttons on the lower right of the primary interface brings up the diagrams in Figures G.19 and G.20. However, neither of these diagrams is active until the computations under the compute menu are activated as described in the next section.

After computation of the various PID controllers, clicking on a PID controller icon brings up a screen containing descriptions of the various possible PID controllers. Clicking inside the screen allows scrolling up and down within the screen.

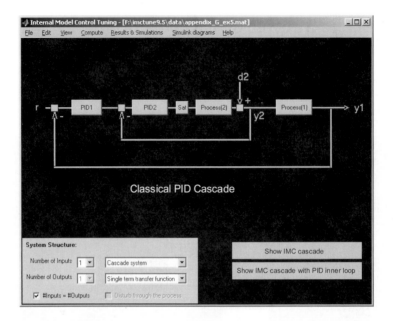

Figure G.19 IMC cascade with PID inner loop.

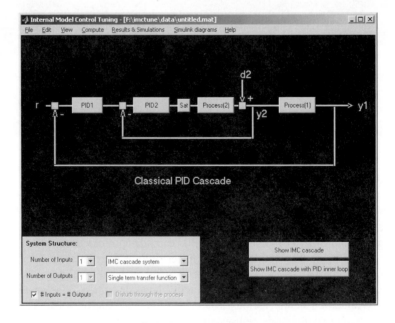

Figure G.20 Traditional PID cascade.

G.7   MENU BAR FOR CASCADE SYSTEMS

Here again, we review only those items in the menu bar that differ significantly from the menu bar of 1DF systems.

View

Figure G.21 shows a typical view window for uncertain inner loop and outer loop processes. The deadtime in the outer loop process is uncertain, while the gain is the uncertain element in the inner loop.

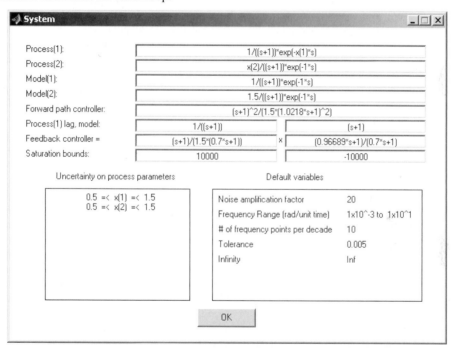

Figure G.21 Typical view window.

Compute

> 2DF tuning: Allows design and tuning of the inner loop controller as a 2DF controller based on the lag of the outer loop process, as well as the design and tuning of the outer loop controller as a 1DF controller.
> Noise amplification: Computes the minimum filter time constant for the inner loop to achieve the desired maximum noise amplification.

➢ IMC controller with PID inner loop: Converts the IMC cascade of Figure G.18 to the diagram in Figure G.19. Clicking on the controller icons shows their transfer functions.

➢ Classical PID cascade: Converts the IMC cascade of Figure G.18 to the classical cascade structure of Figure G.20. Clicking on the controller icons shows their transfer functions.

➢ Frequency response: Computes the frequency response of the transfer functions (1) between the setpoint r and the primary output y_1, and (2) from the disturbance d_2 and the output of the inner loop y_2, with the outer loop open.

Results and Simulations

➢ Step responses for cascade with outer loop IMC controller: Provides, and allows comparison of, IMC cascade and IMC cascade with PID inner loop. The *Add & Remove* menu permits selection of just IMC cascade or just IMC cascade with a PID inner loop or both. It also permits selection of different controllers for the inner loop (e.g., a PID controller cascaded with a first or second order lag). The uncertain parameters in both inner and outer loops can also be changed via the change process button.

➢ Step responses for cascade with outer loop PID controller: Provides, and allows comparison of, IMC cascade and classical PID cascade. The *Add & Remove* menu permits selection of just IMC cascade or just PID cascade or both. It also permits selection of different controllers for the inner loop (e.g., a PID controller cascaded with a first or second order lag). The uncertain parameters in both inner and outer loops can also be changed via the change process button.

SIMULINK Diagrams

Allows access to all the SIMULINK diagrams used to generate the time responses obtained from the Results and Simulations menu.

G.8 OTHER USEFUL .m FILES INCLUDED WITH IMCTUNE

The files *tfn.m* and *tcf.m* were created in order to facilitate entering into, and manipulating transfer functions in MATLAB. The file, *tcf.m* puts a transfer function into time constant form so that right half plane zeros can be conveniently reflected around the imaginary axis. It also cancels common factors in the numerator and denominator and allows linear and quadratic factors to be easily modified. We recommend copying *tfn.m* and *tcf.m* into the Control System toolbox so that they are always available. As is usual in MATLAB, the commands *help tfn* and *help tcf* bring up instructions on how to use the commands.

G.8.1   TFN.m: Create Transfer Functions as Products of Polynomials Cascaded with a Deadtime

The MATLAB m-file *tfn* takes a scalar gain k, matrices n and d, and dead time td to form a SISO transfer function, using the rows of n and d to form the numerator and denominator polynomials, respectively. If the dead time is omitted, it is taken as zero. For example,

   num=[1 3]

   den=[1 2 5; 0 1 2]

   td=6

   g= tfn(5,num,den,td)

forms:

$$g(s) = \frac{5(s+3)e^{-6s}}{(s^2 + 2s + 5)(s+2)}$$

MATLAB returns the transfer function in three forms as:

Transfer Function as Entered

```
        5*(s + 3)
   ----------------------------
   (s^2 + 2 s + 5)*(s + 2)
```

input delay: 6

Time Constant Form:

```
        1.5*(0.33333 s + 1)
   ---------------------------------------
   (0.2 s^2 + 0.4 s + 1)*(0.5 s + 1)
```

input delay: 6

Transfer function:

```
                      5 s + 15
   exp(-6*s) * ---------------------------
                  s^3 + 4 s^2 + 9 s + 10
```

Notice that each row of *num* or *den* must have the same number of elements. Therefore, lower order polynomials must have explicit zero coefficients for the higher order terms.

G.8.2 TCF.m Time Constant Form

tcf(g) is used to display a transfer function g in time constant form. For example, if

$$g = \frac{5s + 15}{s^3 + 4s^2 + 9s + 10} ,$$

then, *tcf*(g) returns:

$$\frac{1.5*(\ 0.33333\ s + 1)}{(0.2\ s^2 + 0.4\ s + 1)*(\ 0.5\ s + 1)}$$

The command [k, n, d] = *tcf*(g) displays the time constant form of g, prints the dead time associated with g, if any, and returns the transfer function gain as k and its numerator and denominator as the polynomial matrices n and d.

The time constant form will also cancel common factors, if any. Each row of num and den contains the coefficients of a polynomial in the numerator and denominator of g, respectively, with the common factors removed. For example, [k,n,d] = *tcf*(g) for the above transfer function, g, returns:

$$\frac{1.5*(0.33333\ s + 1)}{(0.2\ s^2 + 0.4\ s + 1)*(0.5\ s + 1)}$$

input delay: 0

k = 1.5000

n = 0   0.3333   1.0000

d = 0.2000   0.4000   1.0000
        0   0.5000   1.0000

Identification Software

Objectives

The objective of this appendix is to provide an introduction to the polynomial discrete-finite impulse response (DFIR) model and the first-order plus dead time (FOPDT) model identification software that is on the book website.

Prerequisite Reading

Chapter 15, "Identification: Basic Concepts"
Chapter 16, "Identification: Advanced Concepts"

H.1 INTRODUCTION

This appendix describes three programs included in the website for this book:

1. POLYID is a MATLAB program to identify a DFIR model for an SISO process using the polynomial identification procedure outlined in Section 16.2.
2. MODELBUILDER is a menu-driven program to develop an FOPDT model, given input/output data. It uses POLYID software.
3. PIDTUNER uses the FOPDT model to tune a process, using various tuning rules.

H.2 POLYID: DESCRIPTION

H.2.1 System Requirements

POLYID works under the MATLAB package. You need MATLAB 5.2 or a higher version with the Control System toolbox. The Identification toolbox is needed if you want to POLYID to estimate initial values of dead time and time to steady state. The Optimization toolbox is also needed is you want to get optimal estimates of dead time and time to steady state. Two versions are provided. One for MATLAB 5.2 and another one for MATLAB 6.0.

H.2.2 Input Requirements

    Usage: `[hm, hh, hl, K, theta, tau]= polyid (yd, ud, time)`

Inputs		Outputs	
`yd`	Process output vector	`hm`	Identified impulse response
`ud`	Process input vector	`hh`	95% confidence limit (high)
`time`	Time vector	`hl`	95% confidence limit (low)
		`k, theta, tau`	Gain, delay, and time constant for an FOPDT model

You will be asked to provide the optimization option file named *opt.dat*. You may modify the *opt.dat* file that determines the options used in the identification process. The options that you can specify include the following:

Option Number	Default	Choices
1	0	%remove mean. 0 = no, 1 = yes
2	0	%remove trend. 0 = no, 1 = yes
3	1	%remove steady state. 0 = no, 1 = yes
4	1	%use ARX to give initial estimate of dead time and time constant. 0 = no, 1 = yes
5	1	%automatic estimate the polynomial order. 0 = no, 1 = yes
6	0	%first order process. 0 = no, 1 = yes
7	1	%get FOPDT model. 0= no, 1 = yes % the following is parameter used for optimization
8	0	%Display parameters. 0 = no, 1 = yes

Option 1 should be set to 1 if you want to remove the mean from the input/output data before doing the identification. This is used with *prbs* inputs. If you want to remove any linear trends in the data, use option 2. If you started from a steady state and made changes, such as a step change, then set option 3 to 1. If you do not want to provide any initial guesses for the dead time or time constant, then set option 4 to 0. POLYID will then use an ARX program from the Identification toolbox to estimate these quantities.

Use option 5 if you want to specify the polynomial order used. Higher order polynomials will lead to oscillatory DFIR models. Use option 7 to specify whether or not you want POLYID to estimate the FOPDT constants.

If you type

```
[hm, hh, hl, K, theta, tau]=polyid ([], [], [])
```

then you will be asked to provide the input/output data file. The file must list output data, input data, and time vector in columns 1,2 and 3 respectively. An example file is provided: *iodata.dat*.

H.2.3 Subroutines Used

Function `est1`

Usage: `[tmin, tmax]= est1(yd, ud, N, ts,tmin, tmax, options)`
Purpose: This function uses the Identification toolbox.

Inputs		Outputs	
`yd`	Process output vector	`tmin`	Optimal estimate of dead time
`ud`	Process input vector	`tmax`	Optimal estimate of time to reach steady state
`ts`	Sampling time		
`N`	Polynomial order		
`tmin`	Initial estimate of dead time		
`tmax`	Initial estimate of time to reach steady state		

Function `estn`

Usage: `N = estn(yd, ud, tmin, tmax)`
Purpose: This function is used to estimate the polynomial order based on input/output data.

Inputs		Outputs	
yd	Process output vector	N	Polynomial order
ud	Process input vector		
tmin	Dead time		
tmax	Time to steady state		

Function `id1`

Usage: `[K, theta, tau, yp]= id1(y, u, t, dmin)`

Purpose: This function is used to calculate the FOPDT model.

Inputs		Outputs	
y	Noise-free data generated by polynomial model	K	Process gain
u	PRBS input	theta	Process dead time
t	Time vector	tau	Process time constant
dmin	Dead time	yp	Predicted output vector

Function `polyid`

Usage: `[hm, hh, hl, K, theta,tau]= polyid(yd, ud, time)`

Purpose: This is the main function used to estimate the FIR polynomial model.

Inputs		Outputs	
yd	Process output vector	hm	Identified impulse response
ud	Process input vector	hh	95% confidence limit (high)
time	Time vector	hl	95% confidence limit (low)
		K	Process gain
		theta	Process dead time
		tau	Process time constant

Function `polyid1`

Usage: `[hm, hh, hl, ee]= polyid1(yd, ud, N, tmin, tmax)`

Purpose: This is the core engine of polynomial model identification.

Inputs		Outputs	
yd	Process output vector	hm	Identified impulse response
ud	Process input vector	hh	95% confidence limit (high)
N	Polynomial order	hl	95% confidence limit (low)
tmin	Dead time	ee	Prediction error
tmax	Time to steady state		

Function `polyid11`

Usage: [hm1, hh1, hl1, ee1]= polyid11(yd, ud, N, tmin, tmax)

Purpose: This serves the same purpose as `polyid1`. The only difference is that `polyid11` works for a first-order system while `polyid1` works for a higher order system.

Function `timest`

Usage: [tmin, tmax]= timest(yd, ud, ts)

Purpose: This function is used to get an initial estimate of dead time and time to steady state.

Figure H.1 shows the schematic of the calculation sequence. See Chapter 16 for details of the algorithm used. See also Ying and Joseph (1999).

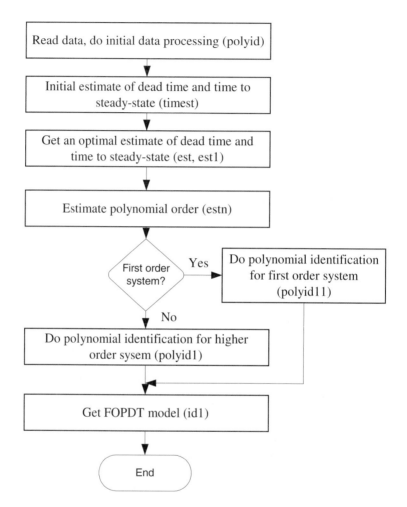

Figure H.1 Schematic of the procedure used by the POLYID software package.

H.3 MODELBUILDER

This program puts a graphical, menu-driven front end to the POLYID program described in Section H.2. However, its main output is in the form of an FOPDT model. If you desire DFIR models, you should use POLYID.

Figure H.2 shows the GUI that pops up when the user types *modelbuilder* at the MAT-LAB prompt.

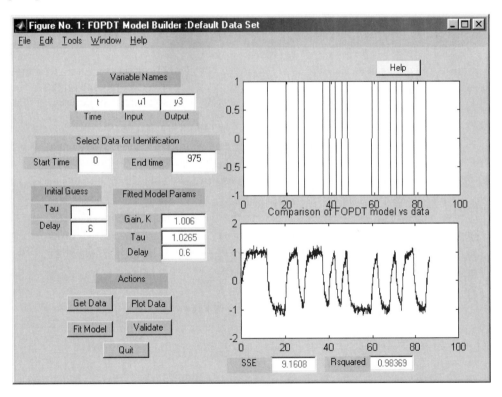

Figure H.2 MODELBUILDER menu page.

You begin by pressing the Get Data button. This brings up a menu of choices. The data may be obtained from the MATLAB workspace, a MATLAB *.mat* file, or an ASCII file. The data must be stored as three-column vectors (*y, u, t*) in the ASCII file. Once the data is loaded, it is plotted and displayed. The program also checks the data for errors and issues warnings if needed.

Next you must choose the Start time and End time for the data segment to be used in the identification. Thus unwanted parts of the data series can be eliminated. Hitting the Fit Model button will start the identification algorithm. If there are errors or warnings, these will appear on your MATLAB window.

The resulting FOPDT model parameters are displayed in the appropriate boxes, as shown in Figure H. In addition, the program plots the DFIR model-based output versus the actual output so you can compare the goodness of fit. Two parameters that measure the goodness of fit are computed. One is the R-squared value for the fit, which should be close to 1 for a good fit. The other is the sum of the squares of the error prediction.

H.4 PIDTUNER

This program calculates PID tuning constants for FOPDT processes and integrating processes, using a variety of classical tuning rules, including one that is based on IMC. The latter takes into account the model uncertainty as well. This software may be used as a shortcut method for tuning of PID controllers. Figure H.3 shows the GUI for this program.

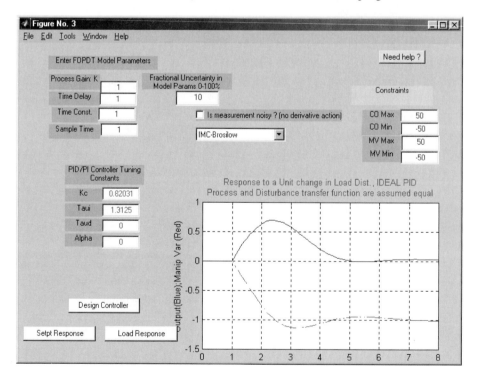

Figure H.3 Opening menu page for PIDTUNER software.

The quantities that appear on your screen in blue are user-entered and those in red are computed by the program. Note that the controller output (CO) limits and measured variable (MV) limits are used in the calculation of the step responses but not for tuning purposes. The Alpha is the derivative filter time constant. The sample time is used in the calculation of the step responses.

Fractional uncertainty is assumed to be the same in all three parameters of the model. This information is used only by the IMC-Brosilow Tuning Rule which is a crude approximation of the tuning constants recommended by IMCTUNE software package for FOPDT systems with equal fractional uncertainty in all three model parameters. If you want a more precise tuning you should use IMCTUNE.

Reference

Ying, C. Y., and B. Joseph. 1999. "Identification of Stable Linear Systems Using Polynomial Kernels." *I&EC. Res. 38*, 4712–4728.

SIMULINK Models for Projects

Objectives

This appendix discusses various SIMULINK process simulation models that can be used for course projects and homework assignments. You will need MATLAB version 5.3 or higher to run these simulations. The models can be downloaded from the book website *http://www.phptr.com/brosilow/*. These simulations provide an excellent way to learn concepts by implementation and testing. Concepts that can be explored using these case studies include

- Internal model control
- Feedforward control
- Cascade control
- Inferential control
- Override and constraint control
- Model-predictive control

I.1 NAPHTHA CRACKER

The objective of this simulation is to provide a realistic platform for implementing and testing various multiloop control strategies. This case study uses the model of a typical chemical process to introduce various control concepts and their importance in industry.

The SIMULINK model simulates real-time operation of a naphtha cracking furnace. You are provided control measurements from various sensors located throughout the process. You can manipulate two inputs to the process, namely, the fuel and air flow valve positions. Naturally occurring disturbances such as ambient temperature variations are included in the simulation. Such disturbances cause the process to deviate from its desired optimum operating conditions.

The control objective is to minimize variations in the operating temperature of the reactor. The SIMULINK model also incorporates a PID controller and a real-time clock to slow the simulation so that you can make modifications to the operation of the process while the simulation is running.

Two "money meters" are included to visualize the economic impact of good control vs. poor control. Both are *virtual* sensors for display purposes only. They are the ultimate indicators of successful control performance. The money meters indicate the amount of money saved or lost as a result of the process variations.

I.1.1 Process Description

Naphtha is a lightoil refined from crude oil. It is composed of a mixture of straight-chained, branched, and aromatic hydrocarbons anywhere from the range of 5–10 carbons long. (Leffler, 1990). Cracking is the process by which extreme heat breaks down these long-chained, higher-boiling hydrocarbons into shorter-chained, lower-boiling hydrocarbons. In the case of naphtha cracking, naphtha is mixed with high-pressure steam and fed through a furnace to begin the cracking process. The C5-C10 hydrocarbons of naphtha break down to C4, C3, and C2 hydrocarbons of which ethylene and propylene are the most valuable. Figure I.1 illustrates the process.

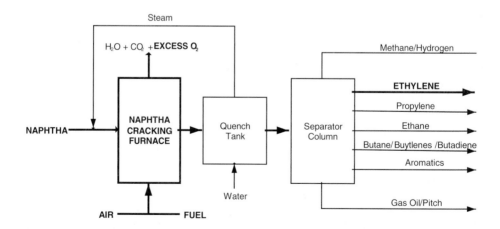

Figure I.1 Block diagram of the naphtha cracking process.

In 2001 approximately 48 billion pounds of ethylene were produced (Hoffman, 2001). Nearly half of this is produced from cracking naphtha (Leffler, 1990). Typical ethylene production in a plant may reach anywhere from 400,000 to 800,000 tons per year. Considering that ethylene currently sells for an average of $0.27/lb, one 400,000 ton/year-plant can make up to $216 million/year in revenue. When designing a control strategy, you may find that an expensive control scheme "only" increases product yield by 1% over a cheaper one. You may abandon the expensive control strategy thinking that 1% improvement is not worth the extra cost. However, in the case of a 400,000 ton/year-naphtha cracker, 1% improvement means an extra $2.16 million per year! Figure I.2 illustrates the process flow diagram of the naphtha cracking furnace. Shown below in Table I.1 are typical operating conditions for a naphtha cracker (Hatch, 1981).

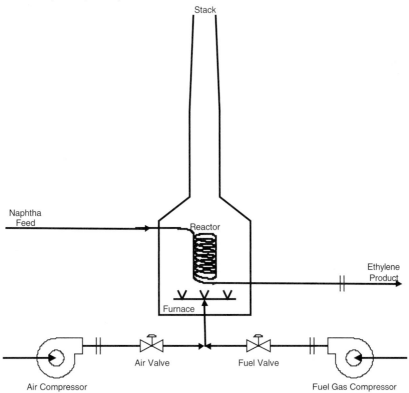

Figure I.2 Process flow diagram of the naphtha cracker.

Naphtha feed enters the reactor where it is heated inside a furnace to reaction temperature and cracked into ethylene product. The heat within the furnace is produced by combusting a mixture of fuel and air. Stack gases produced by the combustion exits the furnace through the top. Figure I.3 illustrates the variables and their relationships to each other in the naphtha cracking process.

Table I.1 Naphtha Cracker – Typical Operating Conditions.

Outlet Temperature		840°C
Catalyst		none
Pressure		atmospheric+
Steam Dilution		0.6 kg steam/kg naphtha
Residence Time		0.35 sec
Yields	Ethylene	31.0%
	Propylene	14.7%
	Butadiene	4.4%
	Other	49.9%

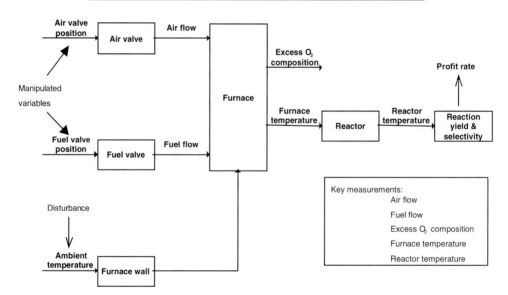

Figure I.3 Block diagram of the naphtha cracker.

The two variables manipulated by the operator, air valve position and fuel valve position, determine the air flow and fuel flow respectively. Air flow and fuel flow affect the combustion reaction, which in turn affects furnace temperature and the composition of excess O_2 in the exiting stack gases. Ambient temperature is a disturbance variable that affects furnace temperature. The furnace temperature determines the reactor temperature that controls the conversion and yield which in turn affects the profitability of the process. Figure I.4 shows the physical location of each variable listed in Figure I.3.

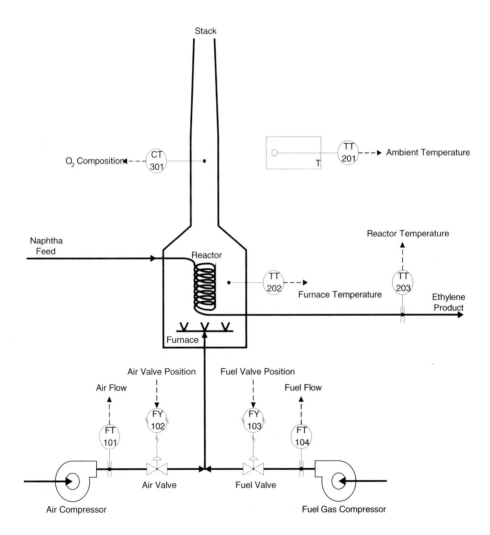

Figure I.4 Process instrumentation diagram of the naphtha cracker.

I.1.3 Control Objectives

A. Safety Issues
1. Maintain 10% excess O_2 within the furnace.
2. Maintain constant air-to-fuel ratio in the furnace.
3. Maintain safe operating conditions in the furnace.

B. Regulatory Issues
1.   Maintain reactor temperature constant.
2.   Reject disturbances due to variations in ambient temperature.
3.   Reject changes in fuel and air flow due to fluctuations in feed air supply pressure and fuel supply pressure.
4.   Reject disturbances due to fuel composition variations.

C. Economic Issue
1.   Operate the reactor to maximize profitability of the process. The profitability depends on the yield and selectivity in the reactor and the percent excess oxygen in the stack gases.

I.1.4 Control Constraints

Table I.2 lists each variable in the process instrumentation diagram, their steady-state values and limits.

Table I.2 Naphtha Cracker— Measured Variables.

Process Variable	Units	Type	Steady State Value	Range
Fuel valve position	%	Manipulated input	50	0–100
Air valve position	%	Manipulated input	50	0–100
Fuel flow	lbmol/hr	Controlled output	50	0–100
Furnace temperature	°F	Controlled output	1200	980–1410
Reactor temperature	°F	Controlled output	800	580–1010
Air flow	lbmol/hr	Controlled output	500	0–1000
% excess O_2	%	Controlled output	10	−90–110
Ambient temperature	°F	Disturbance	70	60–80
Profit rate	$/hr	N/A	N/A	N/A
Cumulative change in profit	$	N/A	N/A	N/A

If reaction temperature is too far below the optimum, the naphtha cracker will not crack enough of the large hydrocarbons into ethylene leading to lower profits. At higher operating temperatures the ethylene and propylene will break down into smaller hydrocarbons again resulting in a loss of profits. The challenge is to keep reactor temperature at its optimum point. Figure I.5 illustrates this concept.

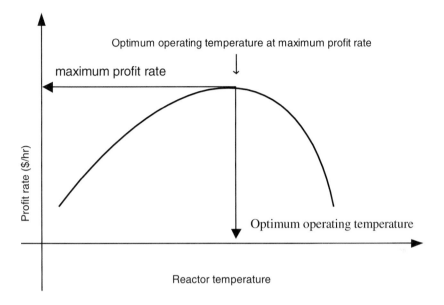

Figure I.5 Optimum operating temperature for the naphtha cracker.

The energy source that raises the reaction temperature is the furnace. The combustion reaction is given below. Here a, b, and c are the molar distribution of hydrocarbons in natural gas:

$$(natural\ \ gas) + O_2 \xrightarrow{\ heat\ } CO_2 + H_2O + excess\ \ O_2$$

$$(aCH_4 + bC_2H_6 + cC_3H_8) + nO_2 \xrightarrow{\ \Delta\ } (a + 2b + 3c)CO_2 + (2a + 3b + 4c)H_2O + (n - 4a - 7b - 10c)O_2$$

Percent excess O_2 is a measurement important to the safe and efficient burning of fuel gas:

$$\%\ excess\ \ O_2 = \frac{O_2\ \ left\ \ over\ \ after\ \ combustion}{O_2\ \ fed\ \ to\ \ fuel\ \ stream} * 100\% = \frac{(n - 4a - 7b - 10c)}{n} * 100\%$$

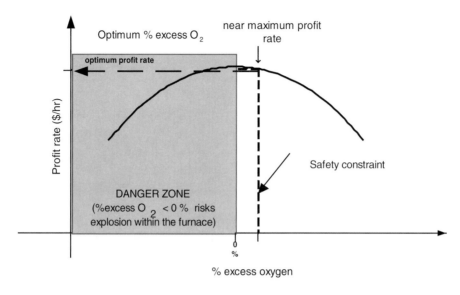

Figure I.6 Optimum percent excess O_2 for the naphtha cracker.

In the equation above, n is the molar amount of O_2 fed and *a, b,* and *c* are the molar amounts of O_2 consumed by the hydrocarbons in the fuel fed to reaction. Having excess O_2 in the presence of fuel prevents the possibility of fuel gas accumulating in the furnace, which is a waste of fuel and an explosion hazard. To maximize profit, fuel cannot be wasted (unconsumed). However excess O_2 reduces furnace efficiency, which results in a loss of maximum profit. A small% excess O_2 is an optimum value from an economic standpoint *and* a safety standpoint, as illustrated in Figure I.6.

I.1.5 SIMLINK Model

Figure I.7 shows the SIMLINK model of the naphtha cracking process. It contains the following components:

1. **Naphtha cracking furnace model.** Included are two inputs (valve blocks) and eight outputs, six of which are measured. The profit rate and cumulative profit are calculated quantities and are used for the evaluation of the control system. All outputs are color coded and attached to scopes and numeric displays.
2. **Real-time clock**. The clock is intended to make one minute of process time (the time represented on the scopes and any graphs) equal to one second computer time (the time you experience as the simulation is running) so that changes can be made while the simulation is running. The .mfile "Wait3.m" must be in the same MATLAB path as the model *naphthacracker.mdl* in order

for the clock to be functional. By default the simulation runs for 300 seconds of simulation time.

3. **PID controller**. This is a discrete industrial proportional-integral-derivative feedback controller with a noise filter.

4. **Feedforward controller**. This is a lead/lag transfer function model that may be used as a simple feedforward controller.

5. **plotall.m** (not shown in Figure I.7) This is a file that came with the Zip file *NapthaCracker.zip* downloaded from the website. It can be executed by typing *plotall* in the MATLAB command window after a simulation is run. This m-file generates a figure of the response curves of all current simulation variables in the MATLAB workspace. The command *plotwho* plots the same graphs in black and white. This is useful for printing in black and white.

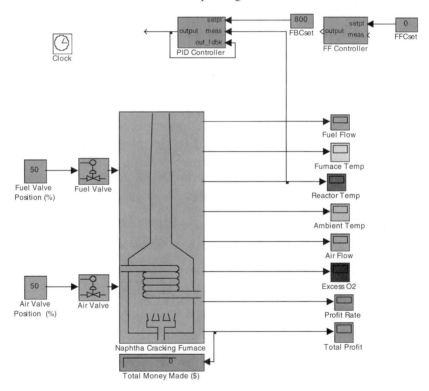

Figure I.7 SIMULINK diagram of the naphtha cracker model.

Every output measurement has its own characteristic level of noise. In addition to the measurement noise, the ambient temperature fluctuates sinusoidally over a 24-hour period (1440 seconds simulation time), with a step drop in temperature occurring to simulate a rainstorm. The initial values of the fuel and air valve position are 50 (%), as can be seen on the left of the SIMULINK diagram (Figure I.7).

I.1.6 Suggested Project Work

A number of activities are suggested using this simulation. It can be used to introduce the basic concepts of feedback control, cascade control, feedforward control, and multivariable control. The goal is to meet the control objectives discussed in Sections I.1.1 through I.1.5.

1. Start by running the simulation to understand the process behavior characteristics. This can be done by keeping the valves on manual control and seeing how the disturbance affects the process variables and profitability. The effect of the manipulated variables can be investigated next. One initial task is to identify a process model that relates the fuel valve position to the reactor temperature. This task is compounded by the presence of external disturbances in the system. This would require a careful planning and execution of the identification test. An estimate of the uncertainty in the process model can also be obtained by conducting the test at a number of different operating points.

2. Having obtained a model one can investigate the tuning of a feedback controller to regulate the furnace temperature using the fuel flow valve. One can immediately see the payback due to improved regulation of the reactor temperature. The payback depends on how well the control system performs.

3. Another point to investigate is the safety issue. It is important to maintain the excess oxygen around the target 10% to avoid dangerous operating conditions within the furnace. This will require keeping the ratio of fuel to air flow constant.

4. Control can be improved by designing and implementing various feedforward and cascade control strategies. SIMULINK allows easy configuration and testing of such strategies. Design of such strategies requires a good understanding of the control concepts and the process at hand.

5. Finally, the simultaneous control of the excess O_2 percent and the reactor temperature should be investigated.

I.2 SHELL HEAVY OIL FRACTIONATOR

I.2.1 Introduction

This problem was first published in Prett and Morari (1987). It provides a transfer function model of a heavy oil fractionator with multiple inputs, disturbances, and measurements. The uncertainty in the model is also specified. The problem formed the basis of a conference and the proceedings have been published in Prett and Garcia (1990). The description below is taken from Prett and Garcia (1988) with permission from Butterworths.

I.2.2 Process Description

Figure I.8 shows the schematic of the heavy oil fractionator and associated instrumentation. Feed vapor enters at the bottom of the column and three products leave the column. Energy is recovered from the column through three reflux recirculation exchangers. The top two act as disturbances, whereas the bottom one (used to make steam) acts as a manipulated input variable. The economic objective is to maximize the steam made in this bottom reflux boiler while maintaining product quality specifications. A SIMULINK model of the column is provided on the website (see Figure I.9). The objective is to synthesize a control system that will meet all of these objectives:

1. Maintain top product endpoint y_1 at 0.0 ± 0.005 at steady-state.
2. Maintain side draw endpoint y_2 at 0.0 ± 0.005 at steady-state.
3. Reject variations in reflux duties in l_1 and l_2, which can vary between -0.5 and 0.5.
4. Reject disturbances when both composition analyzers fail.

I.2.3 Control Constraints

1. The draws have hard upper and lower bounds:
   $$-0.5 \le u_1 \le 0.5$$
   $$-0.5 \le u_2 \le 0.5$$
2. The side draws have a maximum move size of 0.05/min.
3. The bottoms reflux duty has bounds:
   $$-0.5 \le u_3 \le 0.5$$
4. The fastest sampling time is 1 min.
5. Top endpoint must be controlled between -0.5 and 0.5 during disturbances.
6. The bottom reflux draw temperature has a minimum value of -0.5.

I.2.4 Economic Objective

Maximize steam make in the bottom reflux exchanger. Steam make is maximized by making u_3 as small as possible.

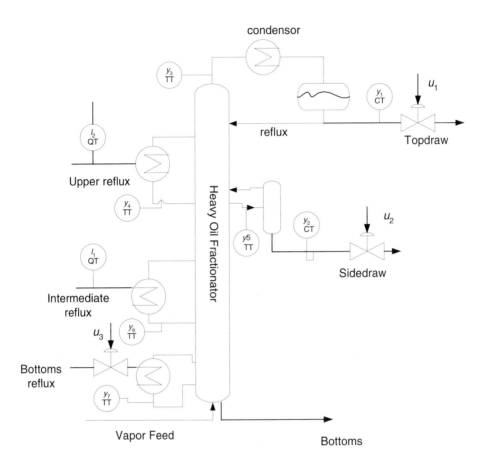

Figure I.8 Shell heavy oil fractionation column instrumentation diagram. From Prett and Garcia (1988). Adapted with permission.

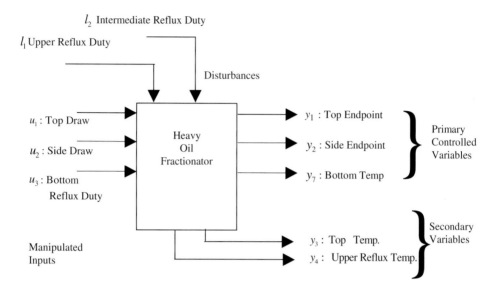

Figure I.9 Heavy oil fractionator schematic diagram.

I.2.5 SIMULINK Model

Figure I.9 shows the schematic of the fractionator and Figure I.10 shows the SIMULINK diagram. This diagram contains the following blocks:

1. Heavy oil fractionator model.
2. Real-time clock. This block slows the simulation so that you can make changes while the simulation is running.
3. A PID block, which implements a PID feedback controller. This is a discrete controller and takes control action once every second (1 second of computer time = 1 minute of process simulation time).

Double-clicking the block brings up the mask where you can enter the model errors if any are needed to run the simulation. These model errors are expressed as fractional changes in the process gains. ε_i applies to the change in gains corresponding to transfer functions relating u_i to the outputs. A value of zero indicates no model error.

The initial values of the inputs are shown on the left. With these values, the column is operating at steady-state at the start of the simulation.

I.2.6 Prototype Test Cases

Prett and Garcia (1988) propose five prototype test cases. In each case the control system
must meet the control and economic objectives subject to the constraints.

1. $\varepsilon_i = 0, l_1 = l_2 = .5$
2. $\varepsilon_1 = \varepsilon_2 = \varepsilon_3 = -1; \varepsilon_4 = \varepsilon_5 = 1; l_1 = l_2 = -0.5$
3. $\varepsilon_1 = \varepsilon_3 = \varepsilon_4 = \varepsilon_5 = 1; \varepsilon_{2=-1}; l_1 = l_2 = -0.5$
4. $\varepsilon_i = 1; l_1 = 0.5; l_2 = -0.5$
5. $\varepsilon_1 = -1; \varepsilon_2 = 1; \varepsilon_3 = \varepsilon_4 = \varepsilon_5 = 0; l_1 = l_2 = -0.5$

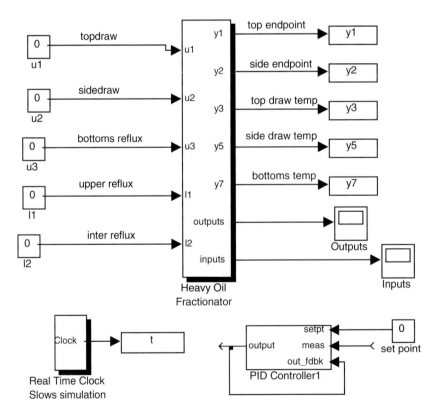

Figure I.10 SIMULINK model of the Shell fractionator.

I.2.7 Suggested Project Assignments

The use of IMCTUNE is recommended. The project can be divided into the following parts.

1. Consider the problem of controlling y_1 using u_1. Suggest an IMC controller. What nominal model and what filter time constant would you recommend, considering the given uncertainties? What is the bandwidth of the resulting controller? What is the peak value of the sensitivity function and at what frequency does it occur?

2. Suppose the disturbance l_1 has a frequency range of 1 to 3 cycles/hr (note units) with an amplitude of 1.0. What will be the maximum deviations expected in y_1 as a result? At what frequency does this maximum occur? Assume the model is perfect for this analysis.

3. Suggest a feedforward controller to suppress the disturbance in l_1. Assume l_1 can be measured for this feedforward control system. Give its transfer function and show the block diagram indicating all the transfer functions in the control system. If a feedforward plus feedback controller designed as in Part 1 above is used, what will be the maximum deviations expected in y_1 under the conditions described in Part 2 above? Again, assume a perfect model.

4. Consider a cascade control system to control y_1. What are the choices for the secondary variable? Which variable is recommended? Why? Design a cascade controller using y_3 as the secondary variable. What are the parameters for an inner-loop PID controller? You should take into account the model uncertainty in this design. Would it be reasonable to replace the IMC controllers with equivalent PID controllers? Why or why not? Compare the disturbance response of an IMC control system with that of a PID control system for the nominal plant.

5. For conditions described in Part 2, compute the maximum deviations in y_1 when using an IMC cascade control. Assume perfect models.

6. Consider an inferential control system using a single secondary variable. What secondary variable would you recommend? Why? Design an inferential control system to control y_1 using y_3 as the secondary variable. Show the block diagram of the inferential control system, including in it all the transfer functions.

7. For conditions described in Part 2, compute the maximum deviations in y_2 when using the inferential controller. Assume perfect models.

8. Based on your results in Parts 1 through 7, make recommendations regarding the type of control system to use in this case.

9. Using the SIMULINK model provided, develop and implement two PID controllers to control y_1 and y_2, using u_1 and u_2 respectively. Tune these controllers using IMC tuning methods. Compare their performance to the prototype test cases 1 and 2. Are any constraints violated?

10. Add a constraint (override) controller that maintains y_7 above -0.5, using u_3 as the manipulated variable. Tune the constraint controller; u_3 should be kept at its minimum of -0.5 if the constraint on y_7 is not violated. Check this system using prototype test cases 1 and 2. Are any constraints stated in the problem statement violated? (A select switch is provided in a file named *sel_switch.mdl*.)

11. Develop an MPC solution to the Shell problem using MPC toolbox. Test your implementation using prototype test cases 1 and 2. Compare your results for the decentralized solution in Part 10.

I.3 TEMPERATURE AND LEVEL CONTROL IN A MIXING TANK

I.3.1 Process Description

The objective in this simulation is to control the level and temperature in a hot and cold water mixing tank by simultaneously adjusting the hot and cold water flow rate. Figure I.11 shows the schematic of the process. Hot and cold water are mixed in a stirred tank. The control objective is to regulate the temperature and level in the tank when the feed temperature and outlet flow vary. The inlet flow of hot and cold water can be manipulated. Figure I.11 also shows the available measurements and manipulated inputs. See Example 2.3 in Chapter 2 for a mathematical model of this process. Figure I.12 shows the SIMULINK model of the process.

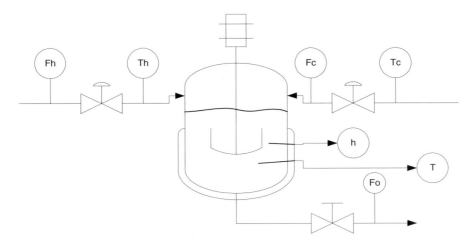

Figure I.11 Signal diagram for mixing tank experiment.

I.3.2 Suggestions for Projects

The control concepts that can be illustrated using this process include the following:

1.   Control loop pairing.
2.   Interaction in multivariable control.
3.   Reducing interaction by advanced control techniques.
4.   Feedforward control.
5.   Decoupling control.

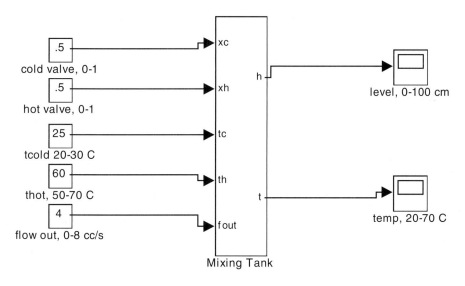

Figure I.12 SIMULINK model of hot and cold water mixing tank.

Your objective should be to obtain the best possible control of temperature and level in the tank given the constraints on the system.

1.   Try independent control of temperature and level. How should the variables be paired; that is, which valve should be used for level control? You may want to design and tune a level controller first. Level is an integrating process and hence should be tuned accordingly. (See Appendix A for suggestions for tuning integrating processes.) Then implement temperature control. Design tests to evaluate the control system performance.
2.   Design, implement, and test advanced control strategies that attempt to minimize the interaction between the variables. One strategy that has been suggested in the literature is to control the level by using the total flow into the tank and to control

the temperature by adjusting the ratio of hot to cold water flow. This essentially re-defines the input variables.

3. Design, implement, and test a feedforward control that utilizes the inlet temperature of the water.

4. How does the actual performance of the level controller compare with a theoretical prediction of the controller performance obtained, say, from a MATLAB simulation?

5. Is the sample time important for the controller? If so, how?

When testing controllers, do not make very large changes in the setpoint or disturbance, since this will cause controller output saturation. Choose your tests wisely.

I.4 PRESSURE AND LEVEL CONTROL EXPERIMENT

I.4.1 Process Description

Air and water stream enter a closed tank. The control objective is to simultaneously control the liquid level and gas pressure in a tank at given set points. The air flow out and the water flow in can be manipulated using control valves. This apparatus may be used to explore such concepts as multivariable control, decoupling control, tuning an unstable process, and over-ride control.

Figure I.13 shows the schematic of the control system and Figure I.14 shows the SIMULINK diagram. There are two measurement sensors and two manipulated variables. The objective is to maintain pressure in the tank at a setpoint and the level at a setpoint.

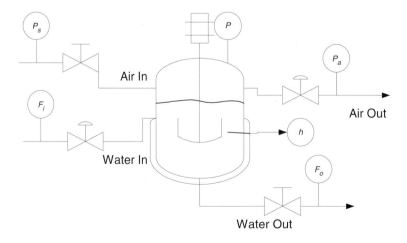

Figure I.13 Schematic diagram of pressurized tank.

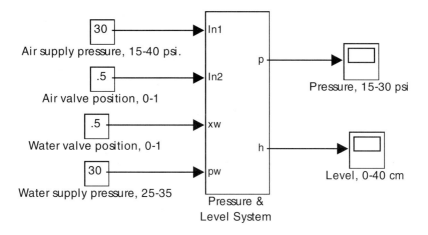

Figure I.14 SIMULINK model of pressurized tank.

I.4.2 Suggested Projects

This is a multivariable system that lends itself to studying a number of challenging projects. A few possibilities are listed below.

1. Implement and test a pressure control system. Document your results with suitable graphs.
2. Implement and test a level control system. Level usually responds like an integrating process. Hence it is inherently unstable in that the tank either fills up or drains empty without a level controller in place. What kind of tests would you conduct to model this system and tune the level controller?
3. Implement and test an override control scheme that will maintain the level above a low limit if the outlet valve is opened more fully. If the water level gets below a preset minimum, the override controller should reduce the air pressure setpoint in the tank so that the tank does not run empty.
4. Other possible projects include designing and testing a decoupling control scheme to minimize the interaction between the pressure control and level control. This should keep changes in level to a minimum when pressure setpoint is changed, and vice versa.

I.5. Temperature and Level Control

I.5.1 Process Description

The objective of this simulation is to simultaneously control level and temperature in a tank with minimal interaction. The presence of multiple disturbances, interaction among the variables, and nonlinearities makes it difficult to regulate the temperature and level in the tank.

Figure I.15 shows the schematic of the process. The process consists of a stream of water entering a tank where it is heated using an electric heater. The temperature of the inlet stream can vary. The temperature and level of the water in the tank are the controlled output variables. The manipulated inputs are the inlet flow and the heater duty.

You are asked to control the temperature and level in the tank as tightly as possible in the presence of changes in inlet water pressure, temperature, and outlet valve position. You can use any or all of the measurements provided. It is also desirable to minimize the interaction between the temperature control and the level control; that is, we would like to minimize the impact of changing level set point on the temperature in the tank. A further complicating factor is that the system is inherently unstable in that without control, the tank will either fill up or run empty, depending on the outlet valve position.

Figure I.16 shows the SIMULINK diagram of the process.

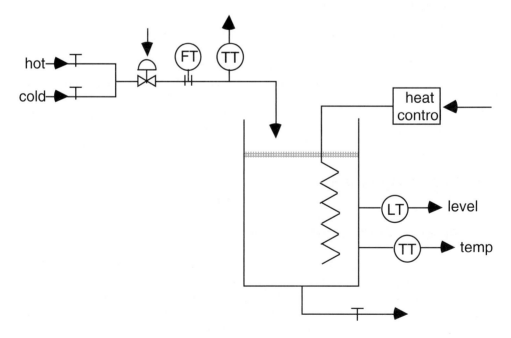

Figure I.15 Schematic diagram of heated tank system.

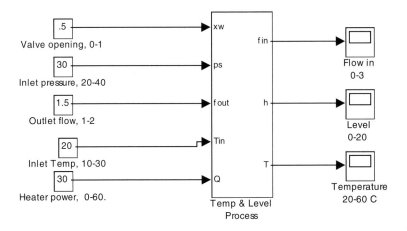

Figure I.16. SIMULINK model of the heated tank.

I.5.2 Suggested Projects

Possible projects that can be studied using this simulation include:

1. Implement and test feedback controllers to keep the level and temperature in the tank constant despite variations in inlet water temperature, inlet water pressure variations, and outlet flow changes.
2. Implement a feedforward controller to counteract the changes in inlet water temperature.

I.6 HEAT EXCHANGER

I.6.1 Process Description

A process flow diagram of the heat exchanger system is shown in Figure I.17. The liquid flowing through the inner pipe of the heat exchanger will be called the hot-side liquid, while the liquid in the outer pipe will be called the cold-side liquid. The hot-side liquid enters the system from its supply line with temperature Tin and flow rate $Fhin$, both of which are disturbance variables. The hot-side liquid then flows past an electric heater. The heater is controlled by varying the input power *Qin*, a manipulated variable. The hot-side liquid then flows through the double-pipe exchanger.

The cold-side liquid entering the heat exchanger has an inlet temperature of *Tin* (the same *Tin* as the temperature of the entering hot-side liquid). A control valve is used to manipulate the flow rate of the entering cold-side liquid to *Fcin*.

All three of the controlled variables are temperatures of the hot-side liquid. They are the inlet temperature of the liquid to the exchanger (*Thin*), the temperature in the middle of the exchanger (*Thm*), and the temperature exiting the exchanger (*Thout*).

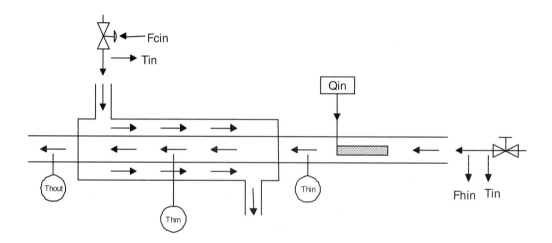

Figure I.17 Process flow diagram of the double pipe heat exchanger.

I.6.2 Implementation

The simulation of the heat exchanger system is performed in MATLAB using SIMULINK as shown in Figure I.18. An S-function block is used in SIMULINK, which utilizes MATLAB .mfiles. The four inputs to the SIMULINK model consist of the two disturbance variables (*Tin* and *Fhin*) and the two manipulated variables (*Qin* and *Fcin*). The five outputs include the three controlled variables. Each of the other two outputs is a vector that is sent to a scope to be displayed: one for the inputs to the process, the other for the measurements. The system parameters, such as the length of the double-pipe exchanger, the diameters of the pipes, and the overall heat transfer coefficients, can be changed by double-clicking on the mask.

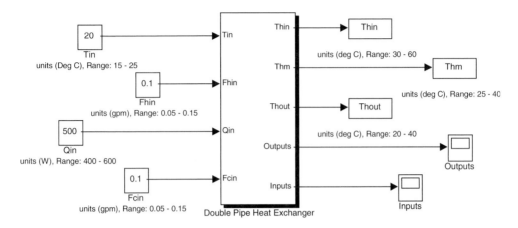

Figure I.18 SIMULINK model of the double-pipe heat exchanger.

I.6.3 Suggested Projects

This simulation can be used to implement and test a variety of control schemes. The specifications for the control system are deliberately left open so that you may have the freedom to choose alternative objectives and check their feasibility. The process is amenable to implement and test a variety of feedback, feedforward, cascade, and multivariable control strategies.

I.7 THE TENNESSEE EASTMAN PROJECT

This problem was published by Down and Vogel (1993). It represents a truly challenging control problem. It has all the nuances that one might encounter in the real-world application of advanced control strategies. The problem has spawned a number of research papers, some of which are listed at the end of this appendix.

I.7.1 SIMULINK Model Description

In this section we provide an overview of the SIMULINK model provided at the book website. Refer to the original paper by Downs and Vogel for a description of the process. The authors originally provided a set of Fortran subroutines to model the process. For convenience, we have compiled these routines into SIMULINK and MATLAB codes that are easier to use. These SIMULINK models are included on the website with permission from the authors. With these files, one can develop control structures using the MATLAB code or using SIMULINK control blocks. A detailed study of the problem was done by Palavajjhala as part of his doctoral thesis (Palavajjhala, 1994). *An excerpt from this thesis containing a discussion of this control problem is included on the book website with permission from the author.* The interested reader should refer to this excerpt for a better understanding of the problem. The SIMULINK models were developed by Nick Graham and Andrew Tillinghast (2000). *Their report is also included on the website with permission from the authors.* Figure I.19 shows the block diagram schematic of the process. Table I.3 lists all the input variables and their steady-state values.

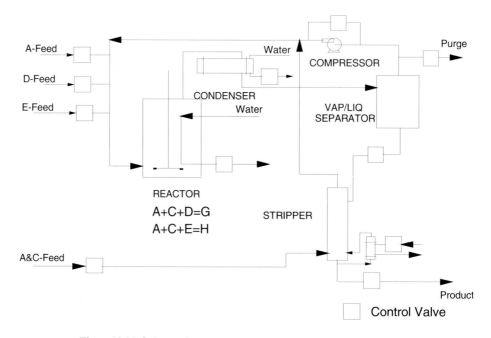

Figure I.19 Schematic representation of the T-E problem.

Table I.3 Steady-State Valve Positions.

Input Number	Manipulated Variable	Base Case Values (Percent open)
1	D Feed Flow Valve	63.053
2	E Feed Flow Valve	53.980
3	A Feed Flow Valve	24.644
4	A/C Feed Flow Valve	61.302
5	Compressor Recycle Valve	22.210
6	Purge Valve	40.064
7	Separator Liquid Underflow Valve	38.100
8	Stripper Liquid Underflow Valve	46.534

9	Stripper Steam Valve	47.446
10	Reactor Cooling Water Valve	41.106
11	Condenser Cooling Water Valve	18.144
12	Agitator Speed	50.000 rpm

The last 20 input variables correspond to the disturbances entering the process. Under normal conditions, these variables are set to zero. These disturbances must be specified in a vector named *disturb* in the MATLAB workspace. For a complete description of the input disturbances, please refer to the original article by Downs and Vogel (1993). For example

```
disturb = zeros(1,21)
```
will create a no-disturbance scenario.

The descriptions corresponding to these variables are listed Table I.4.

Table I.4 Possible Set Of Disturbances For The T-E Process.

Distur-bance Number	Type	Description
1	Dummy variable	Dummy variable
2	Step	A/C feed ratio
3	Step	B composition
4	Step	D feed temperature
5	Step	Reactor cooling water inlet temperature
6	Step	Condenser cooling water inlet temperature
7	Step	A feed loss
8	Step	C header Pressure loss
9	Random variation	A, B, C feed composition
10	Random variation	D feed temperature
11	Random variation	C feed temperature
12	Random variation	Reactor cooling water inlet temperature
13	Random variation	Condenser cooling water inlet temperature
14	Slow Drift	Reaction kinetics
15	Sticking	Reactor cooling water valve
16	Sticking	Condenser cooling water valve
17	Unknown	Unknown
18	Unknown	Unknown
19	Unknown	Unknown
20	Unknown	Unknown
21	Unknown	Unknown

We recommend that a *T_samp* of 1 second be used for all of these SIMULINK models. This may be specified by opening up the S-function block that calls the model.

We have provided three SIMULINK files:

1. **Te_nocontrol.mdl:** This simulates the system with no internal controllers. This system is unstable, even with no disturbances, due to slow accumulation of level in the reactor. The inputs to this model are the valve positions as shown in Figure I.20.
2. **Te_int_control.mdl:** This simulates the system with local flow control loops present. This stabilizes the system under normal conditions. The inputs to this model are the set points of the flow controllers, as shown in Figure I.21.
3. **Te_ext_control.mdl:** This adds two external controllers to stabilize the system even in the presence of disturbances.

I.7.2 Suggested Project Work

Please refer to the original article by Downs and Vogel (1993) for a number of case studies for project work using this model.

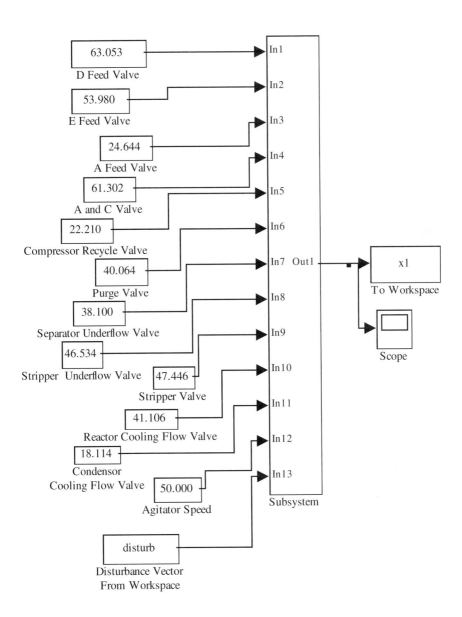

Figure I.20 Schematic of SIMULINK model with no internal controllers.

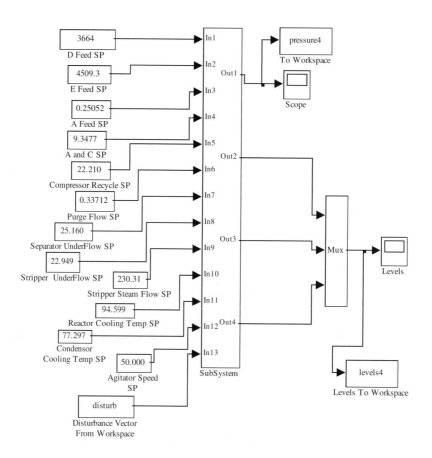

Figure I.21 Schematic of T-E process with internal, local flow control loops installed.

References

Banerjee, A., and Y. Arkun. 1995. "Control Configuration Design Applied to the Tennessee Eastman Plant-Wide Control Problem." *Comput. Chem. Eng. 19*(4), 453–80.

Downs, J. J., and E. F. Vogel. 1993. "A Plant-Wide Industrial Process Control Problem." *Comp. and Chem. Eng., 17*(3), 245–355.

Duvall, P. M., and J. B. Riggs. 1990. "Online Optimization of the Tennessee Eastman Challenge Problem." *J. Process Control, 10*(1), 19–33.

Ellis, C. 1934. *The Chemistry of Petroleum Derivatives.* The Chemical Catalog Company, Inc. New York, NY.

Graham, N., and A. Tillinghast. 2000. *Simulink Model of the Tennessee Eastman Process.* ChE 462 Course Project Report. Washington University, St. Louis, MO.

Hatch, L. F. 1981. *From Hydrocarbons to Petrochemicals.* Gulf Publishing Company. Houston, Texas.

Hoffman, J. 2001. "Olefins Business Hits Deep Depression." *Chemical Market Reporter.* Vol. 260, No. 6.

Kookos, I. K., and J. D. Perkins. 2001. "Heuristic-Based Mathematical Programming Framework for Control Structure Selection." *Ind. Eng. Chem. Res., 40*(9), 2079–2088.

Leffler, W. L. 1990. *Petrochemicals in Non-Technical Language.* Penn Well Publishing Company. Tulsa, OK.

McAvoy, T. J. 1998. "A Methodology for Screening Level Control Structures in Plantwide Control Systems." *Comput. Chem. Eng., 22*(11), 1543–1552.

McAvoy, T. J. 1999. "Synthesis of Plantwide Control Systems Using Optimization." *Ind. Eng. Chem. Res., 38*(8), 2984–2994.

McAvoy, T. J., and N. Ye. 1994. "Base Control for the Tennessee Eastman Problem." *Comput. Chem. Eng., 18*(5), 383–413.

Palavajjhala, S. 1994. *Studies in Model-Predictive Control Using Wavelet Transforms.* D.Sc. Thesis. Washington University, St. Louis, MO.

Prett, D. M., and M. Morari. 1987. *The Shell Process Control Workshop*, Butterworths, MA.

Prett, D.M., and C. E. Garcia. 1988. *Fundamental Process Control.* Butterworths, Boston, MA.

Prett, D. M., and C. E. García. 1990. *The Second Shell Process Control Workshop: Solutions to the Shell Standard Control Problem.* Butterworths, MA.

Ricker, N. L. 1995. "Optimal Steady-State Operation of the Tennessee Eastman Challenge Process." *Comput. Chem. Eng., 19*(9), 949–59.

Ricker, N. L. 1996. "Decentralized Control of the Tennessee Eastman Challenge Process." *J. Process Control, 6*(4), 205–21.

Ricker, N. L., and J. H. Lee. 1995. "Nonlinear Modeling and State Estimation for the Tennessee Eastman Challenge Process." *Comput. Chem. Eng., 19*(9), 983–1005.

Srinivas, G. R., and Y. Arkun. "Control of the Tennessee Eastman Process Using Input-Output Models." *J. Process Control, 7*(5), 387–400.

Tyreus, B. D. 1999. "Dominant Variables for Partial Control, 1: A Thermodynamic Method for Their Identification." *Ind. Eng. Chem. Res., 38*(4), 1432–1443.

Index

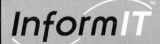